Advances in 3D Printing

Edited by Ashutosh Sharma

Published in London, United Kingdom

Advances in 3D Printing
http://dx.doi.org/10.5772/intechopen.104046
Edited by Ashutosh Sharma

Contributors
Hector Garcia Gonzalez, Mª Teresa Lopez Pola, Chung-Dann Kan, Wei-Ling Chen, Jieh-Neng Wang, Tsung-Lung Yang, Maria Victória Rocha, Jorge Mauricio Fuentes, Ilona Pavlovska, Žanna Martinsone, Lāsma Akūlova, Linda Paegle, Aneka Kļaviņa, Inese Mārtiņsone, Anna Lece, Klinta Luīze Sprudža, Silvana Gallinaro, Kirthanashri S V, Prachi Agarwal, Vidhi Mathur, Meghana Kasturi, Varadharajan Srinivasan, Tatsuhiko Aizawa, Tomomi Shiratori, Yohei Suzuki, Xiangjia Li, Tengteng Tang, Dylan Joralmon, Liselore N.M. Tissen, Przemysław Kustra, Andrey Koptyug, Mikael Bäckström, Reza Shamsabadi, Adolfo Nadal Serrano, José María Espejo Bares, Valentina Pusateri, Constantinos Goulas, Stig Irving Olsen, Jun Xu, Jing Tian, Yanyan Zheng, Baohua Guo, Qing Ouyang, Ping Xue, Joan Josep Roa Rovira, Guiomar Riu Perdrix, Carlos Aravena, Thomas R. Gildea, Seyed Hamid Zoljalali Moghaddam

First published in London, United Kingdom, 2023 by IntechOpen
IntechOpen is the global imprint of INTECHOPEN LIMITED, registered in England and Wales, registration number: 11086078, 5 Princes Gate Court, London, SW7 2QJ, United Kingdom

British Library Cataloguing-in-Publication Data
A catalogue record for this book is available from the British Library

Additional hard and PDF copies can be obtained from orders@intechopen.com

Advances in 3D Printing
Edited by Ashutosh Sharma
p. cm.
Print ISBN 978-1-80355-843-1
Online ISBN 978-1-80355-844-8
eBook (PDF) ISBN 978-1-80355-845-5

For EU product safety concerns:
IN TECH d.o.o., Prolaz Marije Krucifikse Kozulić 3, 51000 Rijeka, Croatia, info@intechopen.com or visit our website at intechopen.com.

Meet the editor

Ashutosh Sharma is an assistant professor in the Department of Materials Science and Engineering, at Ajou University, Suwon, South Korea. He earned his Ph.D. in Metallurgical and Materials Engineering from the Indian Institute of Technology (IIT) Kharagpur. His research interests include electrochemical deposition, lead-free soldering and brazing, additive manufacturing, high-entropy alloys, gas sensors, and biomaterials. Dr. Sharma is a life member of various scientific and professional bodies. He has more than 160 international journal articles, 17 patents, 8 book chapters, and 1 book to his credit. Dr. Sharma received an Extraction and Processing Division Award from The Minerals, Metals and Materials Society (TMS) for his outstanding contribution to nonferrous materials processing.

Contents

Preface

Three-dimensional (3D) printing has witnessed tremendous growth in the past few decades. This technology offers unprecedented freedom to design and fabricate complex geometric parts in a variety of applications in aerospace, automotive, energy, building and construction, microelectronics, and biomedicine industries.

Recent developments and achievements demonstrate that this emerging technology can potentially transform traditional subtractive manufacturing processes. Numerous materials and products consisting of metals, ceramics, polymers, and composites have already been applied in diverse engineering applications.

This book presents and discusses the recent developments and applications of 3D printing in different fields such as biomedicine, health care, electronics, energy storage, and environmental remediation. It highlights advancements in 3D printing technology and offers a platform for materials researchers and scientists to share outstanding research advances in this discipline.

I take this opportunity to thank all the contributing authors, co-authors, and publication staff from IntechOpen for their efforts in preparing this book. I hope that the new research and knowledge presented herein will be the foundation for the innovation of materials and technologies for humankind. I hope that readers enjoy this book and that it will serve as an aid in creating and printing exciting new materials with unique properties.

Dr. Ashutosh Sharma
Department of Materials Science and Engineering,
Ajou University,
Suwon, South Korea

Section 1

Introduction to 3D Printing

Chapter 1

3D Printing and the Art World: Current Developments and Future Perspectives

Liselore N.M. Tissen

Abstract

3D printing's rapid technological development is starting to impact the art field because, for the first time, it has become possible to exactly reproduce and reconstruct artworks without any loss of their physical features. Yet, a coherent overview of how 3D printing is used within the art field while paying attention to ethical considerations does not exist. This study will provide an overview of the current developments of 3D printing in the art world, its use, and the direction it is moving toward. Within this study, the technologies that enable, influence, and will continue to affect the 3D reproduction of artworks, namely technologies necessary to capture an artwork's materials on a chemical and physical level, artificial intelligence (AI), 3D printing technology itself, and the rise of the non-fungible token (NFT) are analyzed to be able to understand what 3D printing implies for our changing perception of art in the future.

Keywords: 3D printing, art, cultural heritage, authenticity, 3D scanning

1. Introduction

In 2011 CE, the British paper *The Economist* announced that 3D printing would change the world:

> *"Three-dimensional printing makes it as cheap to create single items as it is to produce thousands and thus undermines economies of scale. [...] Just as nobody could have predicted the impact of the steam engine in 1750 CE — or the printing press in 1450 CE, or the transistor in 1950 CE — it is impossible to foresee the long-term impact of 3D printing. But the technology is coming, and it is likely to disrupt every field it touches" [1].*

Nowadays, it is possible to mass produce a large variety of objects in different materials, complexities, shapes, and sizes: from boat propellers to nanochips as small as a 1μm; from pizzas to pills, prosthetics and organs; and from firearms to entire glass houses. As *The Economist* rightly predicted, it would be a matter of time before 3D printing would enter the art world. In contrast to other reproduction techniques (e.g., photography, augmented reality (AR), virtual reality (VR), handmade), 3D printing not only replicates the whole three-dimensional object including its color but also offers the ability to vary the texture of surfaces, translucency, and glossiness. Moreover, due to the digital nature of this technology, it becomes easy to manipulate and alter these material elements, making it possible to endlessly reproduce a diversity of versions of one single artwork [2]. These factors also immediately appeal to contemporary artists to use this fascinating technique as a new medium for artistic expression and cultural reinterpretation, for example, American artist Mark Dion's *Waiting for the Extraordinary* (2013 CE), *The Natural Sciences* (2015 CE), and *Leiden University Phantom Cabinet* (2017 CE). In this work, he used 3D printing to replicate specimens from early scientific collections of Museum de Lakenhal, Natural History Museum Naturalis, and Leiden University (e.g., a saw of a sawfish, a microscope, and a sword) and covered them with glow-in-the-dark paint (**Figure 1**). Furthermore, by combining 3D printing with digital imaging techniques almost indistinguishable copies of oil-painted paintings can be created. Moreover, 3D prints can be made of materials that are more resistant to environmental conditions (like variations in humidity, exposure to (day)light and temperature fluctuations) than traditional materials found in artworks, making them possibly more durable and far less fragile than the original, thus adding to the longevity of artifacts. Moreover, artificial intelligence (AI) and the way it can use digital art historical and material information to be able to alter, predict, and shed new light on the material state of an artwork and an artist's creative process is rapidly developing. Combining AI with 3D printing will undeniably have its effects on the art field and our perception of art history.

This chapter aims to shed light on the way 3D printing currently impacts the art field and the role it will have in the future. Firstly, it will be discussed which ethical considerations to keep in mind when reproducing cultural heritage (CH) objects and artworks. Based on the latter, an overview will be provided of how the technology is

Figure 1.
Mark Dion, Leiden University Phantom Cabinet, 2017 CE, detail, wood, polyamide, fluorescent paint, 2 × 3 × 3 m. Image by author.

currently being used within the art field. Furthermore, the current status and developments within the technologies that enable, influence, and will continue to affect the 3D reproduction of art will be discussed. These enablers are the technologies necessary to capture an artwork's materials, AI, and 3D printing technology itself. This chapter ends by showing what is on the horizon and what the continuous development of these technologies potentially signifies for our perception of art in the future.

2. Art and authenticity

When considering the reproduction of art, one aspect that automatically comes to mind is the authenticity granted to artworks. Authenticity is a complex and heavily debated concept and continues to be of importance, especially with the introduction of high-quality reproduction methods such as AR, VR, and 3D printing. Describing the term authenticity is challenging as it is diverse, culturally determined, and personal [3]. In the case of Western culture (the context in which this chapter is written), Walter Benjamin's essay (1936 CE) on "mechanical reproduction"—in his time photography—has laid the foundation for the current way of thinking about the connection between reproductions and original works of art [4]. He described that this type of reproduction makes it possible to capture reality instantly at a high quality, facilitating the infinite display of artworks at any place and at any time. Consequently, the exclusivity of art and heritage disappears, because the mechanical reproductions—3D printing in our case—extracts the artwork from its tradition and meaning in history because it can be anywhere at any time. This way, art loses its authentic experience or "aura," which is encapsulated in its unique materials. This way, "authenticity," can still be linked to the quality of being authentic, unique, or genuine. This aspect is automatically linked to the material of the original artwork as the only true element that forges a connection between the past and the now, with the beholder and the artist.

Although we have become more used to seeing artworks everywhere through facsimiles, this has not led to a decreasing interest in the original. The popularity of museums, where the original is highlighted, and the importance of conservation and restoration to keep an artwork's materials intact show that this has been increasing. Seeing the massive crowds in front of Leonardo da Vinci's *Mona Lisa* (1503 CE), it could be argued that art's popularity has resulted in the objectification of authenticity and a greater emphasis on the artwork's uniqueness, rejecting almost any form of (physical) reproduction. With this definition of authenticity, 3D printing art and CH objects cause a complex discussion. Namely, a 3D print can never comply with originality in materials and could thus be labeled as "anti-authentic."

However, there are an increasing number of studies that show a shift in the way authenticity can be granted [5–7]. In an increasingly connected world through digitalization and technological advancements, it would be short-sighted to consider authenticity as something singular, static, and based on tangible and physical aspects alone. Furthermore, the ephemeral character of contemporary artworks indicates that art and CH's value are also dependent on intangible and more conceptual qualities. For instance, emotional and/or religious connections to an artwork, its significance in a cultural context, the idea of the artist, or the functional qualities of the object. Thus, only a small part of the notion of authenticity depends on the artwork as a material expression of a fixed moment in time and space. Rather, "aura" should be seen as a network, an interplay between the physical historical artifact and its intangible and emotional and social values which is ever-changing.

In this way, 3D printing artworks could be considered useful for the CH field. Indeed, there are a growing number of studies on the perception of 3D-printed facsimiles confirming the significance of this technology in creating sustainable and durable bonds between the beholder and the object [8]. Although the interest in the authenticity of art and the unique object has not diminished, yet, because only one true material version of the artwork exists and should be conserved for as long as possible, a 3D print could be acceptable as long as it enhances the authentic value of the artwork and does not pretend to be the original. What this means is that when considering the use of an ethical 3D print, it should display or do something the original artwork cannot do due to its fragile materials. For instance, the 3D print can make it possible to display an artwork which is otherwise too fragile to travel; show an artwork without discolorations and closer to what it must have looked like when it left the artist's studio; present an object in its original setting instead of in a carefully conditioned museum; or provide the possibility to touch the artwork or artifact. With this in mind, through a 3D print, new experiences and narratives can take place both inside and outside the museum, with the indication of course that it is a copy and why it is there. Important to always keep in mind, however, is the fact that reproductions both extend an artwork's history and connection with the public but, at the same time, freeze the artwork's image and might imprint a specific significance for generations to come. Therefore, consider what the 3D print communicates and why as it might impact the perception of the artwork in the future. Of course, a 3D print will and can never replace the original, but it allows us to think about the value of the artwork and the range of new connections between visitors and artwork, making the original even more important. In this way, a 3D print could provide a best-of-both-worlds scenario. By eliminating the need to make compromises as every material aspect of the artwork can be reproduced, 3D printing is important to the desire to keep in touch with the physical. At the same time, it can fulfill the need for a more multifaceted story of one artwork sustainably and durably [5–9].

3. Current applications

Currently, 3D printing products are highly accurate and precise and can be made with a wide variety of durable and inexpensive materials. Furthermore, in comparison with traditional processes of creating physical reproductions, 3D printing is highly flexible: The relative ease with which digital data can be modeled, scaled, modified, and used has made the technology irresistible for the cultural heritage field [10]. Moreover, as Moritz Neumüller (Ph.D. in New Media) says: "3D Printing will not only become vital in the field of reconstruction of objects, but also for research, documentation, preservation and educational purposes, and it has the potential to serve these purposes in an accessible and all-inclusive way" [11]. This rapid diffusion of 3D printing has already had its effects in the field of cultural heritage and fine art museums as there are various reasons for creating reproductions. The overview will be divided as follows:

- Restoration and reconstruction
- Documentation and accessibility
- Research and education
- Museum presentation and interaction

3.1 Restoration and reconstruction

As various studies on the application of 3D technologies for conservation show, 3D printing is considered useful for the conservation, restoration, and preservation of artworks and CH [11, 12]. Yet, in the case of more 2D art forms such as painting, the application of 3D printing is more challenging due to the complexity and large range of the different materials used to create one single artwork [13]. Therefore, the exploration of 3D printing has mainly been applied in archaeology, architecture, and sculpture for the conservation, restoration, and preservation of damaged and fragmented or objects that do not exist anymore. An exemplary case is the restoration of French impressionist Auguste Rodin's (1840–1917 CE) sculpture *The Thinker* (1881 CE). In 2007 CE, the statue was heavily vandalized in the museum, and Singer Laren (The Netherlands) was burglarized [14]. In this case, scanning the still existing original mold of the sculpture that was located at the Rodin Museum in Paris made it possible to create positives of the missing elements. This was used to create a cast (a negative) to fabricate corresponding facsimiles of the sections of the statue that were missing, such as parts of its leg and face. Similarly and one of the rare instances, when 3D printing is used to restore a painting, is Vincent van Gogh's *Zeezicht bij Scheveningen* (1881–1883 CE) in 2019 CE [15]. Here, 3D printing was combined with photographs of the artwork before it was damaged to reconstruct a mold for the piece of paint that was missing. In both instances, 3D printing was used to recreate missing parts using similar materials to the original artwork. Additionally, there are multiple examples where 3D-printed copies were directly attached to the original artwork's materials. A 3D print's compatibility and long-term stability make it an interesting method as there is no intention to replace the parts frequently. An often referred to example is the restoration of the fingers of one of the figurines on the tomb in the chapel of Villa Borromeo d'Adda in Arcore (Italy) [16]. Based on a 3D scan of the broken hand which was combined with a drawing of the complete statue when its fingers were not missing, the restorers were able to reconstruct and eventually 3D print the missing parts. The 3D-printed parts were painted in the colors of the marble and attached to the statue using magnets, thus resulting in a non-invasive (and reversible) restoration treatment.[1]

However, 3D printing has not only proven to be a useful technique for saving artworks that are partially incomplete, fragmented, or damaged. *The 3D reconstruction of Nineveh* is an interesting example to show that 3D printing is a welcome method to reconstruct artifacts or archaeological sites that have been lost entirely. Here, black-and-white photographs made by archaeologists were used to reconstruct the texture of the Bas-reliefs of the Southwest Palace of Sennacherib (700 BCE) in Nineveh (Iraq) [18]. This photographic information was translated into 3D printable data to recreate the reliefs that were destroyed by the Islamic State. Furthermore, the reproduction of existing artworks may help restore the original context they were displayed in. For example, renaissance painter Paolo Veronese's (1528–1588 CE) *Wedding at Cana* (1562–1563 CE) in Fondazione Giorgio Cini (Venice)—of which the original hangs on opposite Leonardo da Vinci's (1452–1519 CE) *Mona Lisa* (1503 CE) in the Louvre (Paris)—restored the original context the artwork was made for [10, 19].

[1] To read more about similar projects executed by Mattia Mercante in collaboration with FormLabs see: [17].

3.2 Documentation and accessibility

Not only did the previous examples show that 3D printing can be used to physically reconstruct parts or entire artworks for restoration purposes, but it also shows that 3D printing can document lost artifacts and make them accessible once again. This is exemplified by the Spanish company Factum Foundation which used 3D printing to reproduce the Egyptian pharaoh Tutankhamun (2009) and pharaoh Seti's (2016) tombs located in Egypt's Valley of the Kings in their entirety, including all the details of the tombs' hieroglyph walls (**Figure 2**).[2] Not only did 3D scanning and 3D printing facilitate in making these tombs visible and tangible again without the (further) destruction of the original site, but it also documented the current state of the artifact's materials both digitally and physically [12].

Yet, large-scale documentation of artifacts utilizing 3D printing is not solely limited to accurately measuring, documenting, and sharing the current state of precious artifacts [12]. For example, as the Mauritshuis' (The Hague) exhibition *Rembrandt? The Case of Saul and David* (2015 CE) showed, 3D printing attributed to the reconstruction of the original composition of Rembrandt's *Saul & David* (1660 CE), a painting that was sawn into pieces in the nineteenth century and later reassembled [21]. This example shows 3D printing's possibility of creating tangible representations of the current and past features of an artwork or artifact without damaging the current physical state of the original object [5].

Figure 2.
Facsimile and physical reconstruction of the "Hall of Beauties" for the exhibition Scanning Seti I: the Regeneration of a Pharaonic Tomb (Antikenmuseum, Basel, 2017–2018 CE). Photo credits: Oak Taylor-Smith for Factum Foundation [20].

---

[2] In this case, Factum Foundation used 3D printing to create a negative with an acrylic gesso. Subsequently, they adhere a highly detailed 2D colored skin to this negative in order to create a high-definition facsimile.

In addition, 3D printing's digitization and the possibility to reproduce these CH objects in various ways are automatically of great benefit for engagement with a larger audience and for facilitating public programming. For example, the Smithsonian Institution (Washington D.C.), The National Museum of Antiquities (Leiden), and the Uffizi Gallery (Florence) digitized parts of their collections in 3D and present these models on open-access forums (e.g., Sketchfab) or their own websites [22–24]. This way, their collections become accessible to people at home, in museums, or in schools who can use these data to 3D print their artworks.

3.3 Research and education

The digitization and the physical documentation and preservation of historical objects and artworks and their consequent accessibility are undeniably beneficial for research on and about these objects as well as an aid for cultural education. In the case of conservation and restoration, 3D printing can be considered useful in assisting the monitoring of an artifact's past, current, and future state, thus making it possible to research the artwork's material behavior [14]. The 3D prints at the Mauritshuis' *Facelifts & Makeovers exhibition* (2022 CE) effectively demonstrated a work of art's metamorphosis due to restoration treatments [25]. The fact that one-to-one reproductions of fragile artworks and artifacts can be made in large quantities makes it possible to study these objects more extensively and, more importantly, in a different way [12]. For example, for the *Scanning for Syria* project (2015–2018 CE), the fragile Syrian clay tablets were 3D printed to aid researchers all over the world in researching what was written on them. Instead of having to physically travel to an object, 3D printing facilitates the accessibility of these objects for a larger group of researchers, making it easier for various researchers to research the same object at the same time. Furthermore, 3D printing's materials and the use of a digital model as a starting point allow the researcher to interact with and handle the cultural heritage objects freely and differently, thus stimulating different kinds of approaches and ways of analyzing them (e.g., through magnification or rearrangement). For example, 3D printing was used as a practical method to better understand and scrutinize the physical qualities of a few Palaeolithic limestone slabs [13]. 3D printing the heads of the figurines of *Laocoon and his sons* (1540 CE) separately and at a larger scale, for the first time, it became possible to study their expressions, which are otherwise inaccessible due to the height of the statue (**Figure 3**). This way, 3D printing facilitates getting a better sense of the object's dynamics, shape, materials, and composition in space [8, 13].

Consequently, this can be beneficial in communicating these findings by using these objects as a means of education. In the case of education, a project initiated by Cornell University in 2008 CE is exemplary, in which 3D printing was used to enable students to access and manipulate the archive of their archaeological collection [11, 13]. It was shown that the use of three-dimensional and physical models was more helpful to the education of children and students than traditional methods [26]. Some museums, such as the Metropolitan Museum of Art (New York) and The Victoria & Albert Museum (London), have been experimenting with this technology as a method to communicate with and educate their users (ranging from visitors to researchers) within and outside of the museum's walls [27]. As a few studies show, 3D printing's use effectively stimulated the creative process of users and this could potentially increase the ease with which cultural heritage objects around us are understood [28]. For instance, the *La Riscoperta di un Capolavoro* (2020–2021 CE) exhibition at Palazzo Fava, Bologna, showed that only printing the surface makes

Figure 3.
3D print of Laocoon and his sons and 3D-printed heads.

it possible to understand the techniques used by artists to create relief and depth without the distraction of color [29]. Thus, 3D prints can potentially stimulate new ways of interacting with and studying art objects, therefore facilitating the possibility to learn from them in a non-invasive way.

3.4 Museum presentation and interaction

In the case of museum presentations, 3D printing has been implemented in various ways already [9]. One way is presenting a 3D-printed model next to the original artwork. In this sense, a 3D print can fulfill something the original material does not allow, such as the possibility to touch the object or show additional information [11]. Furthermore, they can enhance interaction within the museum. An example of this is the Museum Boijmans van Beuningen's (Rotterdam) *Sgrafitto in 3D exhibition* (2009 CE) [30]. Here, 3D-printed reproductions of ceramic plates were placed in front of the original artifacts, allowing the visitors to experience the weight, dimension, and topography of the object in front of them and thus interact with the artwork indirectly. Furthermore, as 3D printing provides a way of handling an artwork without having to use (thus damage) its materials while being easily adjustable as it is based on a digital model—it might be an interesting technology to make art more accessible for people who cannot rely on sight only [10, 12, 31]. *Hoy toca el Prado* (Touch the Prado today) exhibition was hosted in Madrid in 2015 and

DOI: http://dx.doi.org/10.5772/intechopen.109107

2021 CE permanently installed *Feeling van Gogh* exhibition at the van Gogh museum (Amsterdam) used 3D prints of paintings to let the visitor explore their collection through touch.[3] By enhancing a portrait's contours, 3D printing can be used to make the paintings more accessible to visually impaired visitors [32]. Simultaneously, the fact that artworks can be touched could make the experience of art in the museum not only more interesting for children and students, but also possibly more effective [8–10]. Furthermore, 3D-printed facsimiles can replace an original when it is absent because it cannot travel or because it is not available. For instance, a 3D print of *The Anatomy Lesson of Dr. Deijman* (1723 CE) painted by Rembrandt van Rijn (1606–1669 CE) (collection Amsterdam Museum, Amsterdam) was displayed during Fondazione Prada's *Human Brains* exhibition at the 59th Venice Biennale in 2022. Since the gallery does not own a permanent collection, they also displayed a 3D-printed copy of Hieronymus Bosch' The Extraction of the Stone of Madness (1501–1505 CE) (collection El Museo del Prado, Madrid) and many other copies

Figure 4.
3D print of Rembrandt van Rijn, The Anatomy Lesson of Dr. Deijman, 1656, oil on canvas, 100 × 134 cm, Amsterdam Museum, Amsterdam—Photo by author, 3D print by Factum Foundation.

[3] Touching the Prado (Hoy toca el Prado) in El Museo del Prado in Madrid is an ongoing exhibition that exhibits 3D facsimiles of various paintings in the collection and invited the visitor to touch the works of art. The exhibition was first held on January 20th, 2015. Feeling van Gogh in the Van Gogh Museum is "an interactive programme developed especially for blind and partially sighted visitors and their sighted friends, families and carers." The exhibition has been part of the museum since 2021.

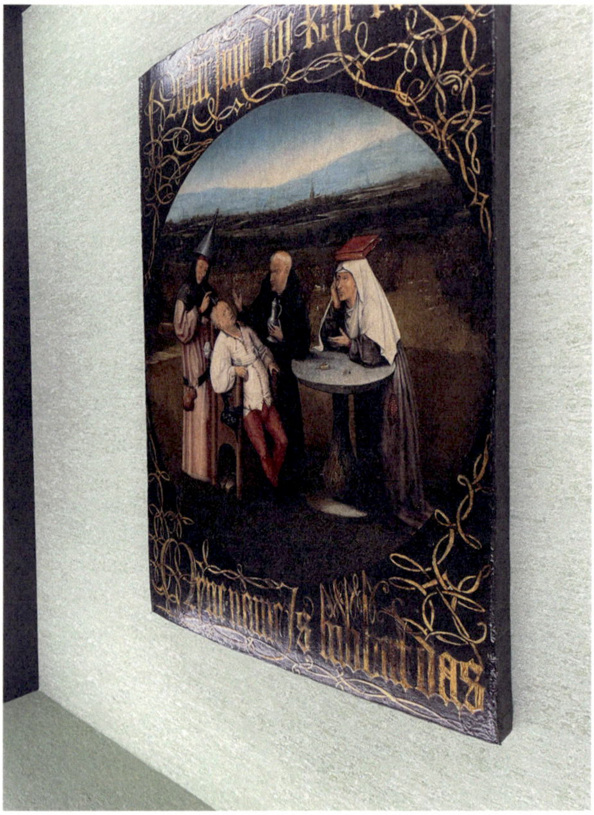

Figure 5.
3D print of Hieronymus Bosch, The Extraction of the Stone of Madness, 1501–1505, oil on oak panel, 48.5 × 34.5 cm. Museo del Prado, Madrid—Photo by author, 3D print by Factum Foundation.

as a way of durable and sustainable exhibition building (**Figures 4** and 5). In this regard, the curators argued that to them, the story of the exhibition is more important than showing original works of art. This, subsequently, can open up a way for museums to rethink the repatriation or decolonization of artifacts to their communities of origin for museums can become less attached to the originals. The *Tlingit* prints in the Smithsonian, (Washington D.C.) are a great example of how creating high-quality 3D prints helped rethink the role of objects and the museum in different societies [33].

As this brief exploration of the use of 3D printing within the cultural heritage field shows, 3D printing is starting to be implemented and its applications are being explored. Furthermore, whereas the technology's potential for architecture, three-dimensional objects, and artworks such as sculptures is developing rapidly and is being researched widely, the 3D reproduction of paintings has not received the same amount of attention. A reason for this could be the fact that the technologies needed to scan and 3D print paintings were not available or developed enough until 2015 CE. Furthermore, the developments of 3D printing and other types of (digital) technologies will undeniably increase the use of 3D printing within the field. The following section will analyze the technologies that enable the 3D reproduction of CH objects by elaborating on the current status and developments of 3D scanning, and the involvement of AI and 3D printing technologies.

4. Capturing and digitizing artwork

To reproduce artworks, they must be digitally captured. Artworks are appreciated because of many aspects, yet their aesthetic qualities are considered to be one of the (if not the) most important features. To be able to scan artworks that all have their unique dimensions, materials, and complexities, the following methods can be considered the most relevant within the cultural heritage field [34]:

- Photogrammetry
- 3D scanning using structured light
- 3D laser scanning
- Computed tomography (CT)

4.1 Photogrammetry

Photogrammetry is a photographic method that measures data about an object by analyzing the changes in position between several pictures taken from various places and angles (with a minimum of two) [35]. After the images are taken, the pictures are aligned and data points plotted, the distance and location of each point in the 3D space can be calculated. Nowadays, this technology is often used within the cultural heritage field because it is rather easy to use and set up as it only requires taking high-quality (high pixels) digital photographs of an object to digitize it into a 3D model.

Although this technology is very useful for capturing (large) three-dimensional objects, it is not as useful in the case of surfaces that are flatter, and uniform without a distinct appearance (e.g., paintings) as it is hard to create multiple angles. For this reason, it is difficult to triangulate the small topographical irregularities, such as craquelure patterns, which can be found on painted surfaces. Furthermore, capturing highly reflective surfaces can be difficult. Another downside of this method is the fact that the optimization of the camera's position relative to the object is not calibrated, making this method oftentimes less accurate than other 3D methods. Moreover, the quality of the 3D model greatly depends on the underlying algorithm and assumptions or optimization used.

4.2 3D scanning using structured light

3D scanning through structured light uses the photography of a single point of a (flat) surface from different angles. This way of 3D scanning uses the projection of a geometric pattern onto the artwork—usually a painting—of choice [35]. By photographing the irregularities and deformations in the projected pattern from multiple angles, it is possible to generate the 3D data of the object. This way, this scanning method is capable of capturing smaller topographic features such as crack patterns. By using a fringe pattern, the data gathered are more precise than the 3D information provided by photogrammetry. With the addition of a polarization camera that takes images both with and without the reflection of the surface, a painting's glossiness can be captured simultaneously [36, 37]. This topography, color and gloss information can be translated into 3D printable data.

Yet, like photogrammetry, this method is less suitable for capturing highly reflective surfaces. Furthermore, because it is a photographic method, it can only record where light can be reflected, making it hard if not impossible to obtain accurate information on deeper sections.

4.3 3D laser scanning

Moving away from photographic methods, laser scanning uses either a static or moving platform that controls the deflection of visible and invisible laser beams. The laser source casts sequences of laser lines onto the surface [35]. These light waves bounce off the surface of the 3D object and are reflected to the sensor. Subsequently, the scanner can determine the three-dimensionality in two ways. The first method is called the time-of-flight (ToF) method, in which the amount of time it took the light beam to go back and forth is calculated. The second method is called triangulation. Here, the known distance between the laser and the camera together with the known angle at which the camera was placed is used to calculate the exact distance between the laser and the object millions of times [38]. Subsequently, when all of these points are processed and mapped together, it becomes possible to form a digital representation of the scanned area. An advantage of this method as opposed to the beforementioned photographic methods is that the light beams can scan through the surface. Therefore, even objects, structures, and areas invisible to the naked eye can be captured and measured with high accuracy (**Figure 6**). However, the laser beams do not detect color or glossiness simultaneously and thus will have to be captured and processed individually. This can lead to difficulties and inaccuracies when integrating the color data with the 3D data, making this technology's use limited and rather exclusive [39].

Figure 6.
Recording the surface relief of Ercole De'Roberti's Stories of Saint Vincent Ferrer, predella of the Politico Griffoni, at the Pinacoteca Vaticana. Photo credits: Grégoire Dupond for Factum Foundation.

4.4 Computed tomography (CT)

Computed tomography, which is mostly known for its purposes in medicine, uses X-rays to take a series of radiographs from different angles. By combining these individual 2D images into a 3D image using reconstruction algorithms, it becomes possible to get high-resolution information on the three-dimensionality of an object both internally and externally [40]. This makes the technology costly and not suitable for larger objects. Furthermore, the data are usually presented as a stack of 2D slices that are layered on top of each other. Therefore, because it takes an extra step to create a digital 3D model which is also suitable for 3D printing is more time-consuming than the other technologies mentioned. Lastly, this technology uses X-rays and outputs greyscale values linked to the attenuation of different materials in the object. Metal is highly attenuating and therefore introduces image artifacts in the reconstruction, which makes 3D printing the model problematic. This means that other features that contribute to an artwork's material appearance are not included in the scan, such as its colors. Yet, projects such as the collaboration between the Centrum Wiskunde & Informatica (CWI, Center for Mathematics & Computer Science, Amsterdam)'s FleX-ray Lab and the Rijksmuseum (Amsterdam) reveal that these types of scans can be interesting for getting high-quality 3D information about both the inside and the outside of objects. The scan of the *Holy woman with lantern* of the Rijksmuseum, Amsterdam, shows that this method makes it possible to record and digitize complex shapes and show the internal structure (**Figure 7**). Furthermore, they show that surface scans can be added to the CT data, thus overcoming the issue that CT scanning does not record the object's material appearance [41, 42].

Figure 7.
(a) Holy woman with lantern, Rijksmuseum collection BK-NM-9253, (b) 3D rendering and orthogonal slices through the 3D reconstruction showing the internal structure (tree rings). Through the CT scan, this object could be dated using dendrochronological methods. Photo credits: Francien Bossema.

5. 3D printing artworks

After having captured an artwork digitally, 3D printing can be used to materialize it. Right now, although 3D printing is still developing and has certainly not reached its peak capacity, there are various methods with which one can create three-dimensional shapes, according to ISO/ASTM [43, 44]:

- Vat photopolymerization (VP)
- Material extrusion (ME)
- Material jetting (MJ)
- Binder jetting (BJ)
- Powder bed fusion (PBF)
- Directed energy deposition (DED)
- Sheet lamination (SL)

To convincingly reproduce an artwork using 3D print and match its esthetical qualities, there are a few requirements to be met by the method of choice:

- Full color and appearance control should be supported;
- High resolution should be guaranteed;
- The printing material and technique should imitate the artwork's material appearance as best as possible.

Based on these requirements, VAT photopolymerization, directed energy deposition, and sheet lamination can be excluded right away. Although material extrusion does not meet these requirements either, it is, however, frequently used in the CH field because it is easy to use and low cost. The method has been used when polychromy or high-quality resemblance is not a requirement, for example, when creating replicas for education purposes. In terms of high-resolution, full color, and appearance control, material jetting, binder jetting and powder bed fusion could meet all three requirements to a certain degree.

5.1 Material jetting

Like material extrusion, material jetting creates an object layer by layer until the right height is reached. However, instead of using a spool of one material, it uses multiple inkjet nozzles—similar to the one used by regular paper printers—with which individual droplets of different colors and materials are selectively deposited [44]. Furthermore, material jetting uses photosensitive polymer instead of thermosensitive polymer [45]. This means that the liquid ink does not harden as it cools off, but instead, a UV light is used to solidify the individual droplets [46]. This results in a method that allows a high level of detailing and in which a large range of colors, materials, glossiness, gradients, and transparencies can be used, making it a more suitable technology in the case of reproducing near flat artworks such as paintings and low-bas reliefs. However, a downside to using various photosensitive materials is the fact that their products are more sensitive and brittle, which means that these prints are considerably less durable and functional.

5.2 Binder jetting

Similar to material jetting, this type of jetting uses a printer head with multiple nozzles depositing droplets of liquid ink(s) onto a printer platform, making it possible to print high-quality and polychrome three-dimensional objects [47]. However, this process differentiates from material jetting as it does not use the solidification of photosensitive materials. Instead, a liquid bonding agent, which functions as a sort of glue, is deposited to bind the powdered particles together [44, 45]. Because the powder is used as a basis, the materials suitable for this type of printing are limited: Only finely granulated ceramics, sand, and metals are suitable for this technology. Furthermore, because these small particles are glued together instead of heated together, there are no new chemical bonds forged between the individual components. For this reason, these prints are porous and their surface finish is rough and grainy, which also results in lesser color quality [46]. These elements make this technology less useful for 3D reproducing paintings as the right material appearance and "feel" can hardly be achieved. Yet, for the reproduction of statues or artworks made of stone, this method can be considered suitable.

5.3 Powder bed fusion

Like binder jetting, powder bed fusion uses granulated materials as its base material. However, the difference lies in the solidification of the material [44, 45]. Whereas binder jetting uses a bonding agent in liquid form, with powder bed fusion, the heat of a laser is used to melt and/or cure the layer of powder. The advantage of using a heat source to bind materials instead of a "glue" is that a larger variety of materials can be used to create more durable high-quality reproductions (e.g., titanium, aluminum, nylon) as they have been chemically bonded together [47]. However, like binder jetting, the result is grainy and rough. Furthermore, because the melting of materials is used, it is hard, if not impossible, to integrate color within the AM process [48]. Because of this reason—the limited use of materials and the graininess of the end product—powder bed fusion is currently not as suitable for 3D reproducing all types of artworks (e.g., paintings). Yet, for the reproduction of statues or artworks made of stone, this method can be considered suitable.

As this overview shows, there are various ways of 3D printing artwork. However, the categories mentioned here are arbitrary and can change rapidly over time as new technologies enter the market and open up new opportunities. 3D printing has just started to move beyond its infancy stage and as such is not completely mature, while still developing. Right now, each technology still experiences some major limitations. For instance, most 3D printing technologies are still commonly limited to one color, which is the same as its printing material [46]. Furthermore, the limitation of using just one material makes it hard to vary in transparency and opacity between individual layers. Although some techniques offer polychrome printing or the use of multiple inks, the production of realistic colors or surface effects is still hard as it requires extra finishing processes to give synthetic material the desired appearance. Particularly, in the case of CH and the reproduction of art, color printing and resolution are hurdles to overcome as accuracy, good quality, high resolution, and the right material feel are often a top priority for correctly curating narratives with these delicate artifacts [8]. However, it is worth noticing that larger companies such as Hewlett-Packard (HP) have taken on the challenge of developing new technologies to overcome these issues [49]. Since 2016 CE, HP has been working on Multi Jet Fusion printing, a method with which full-color 3D printing with a larger variety of materials is starting to become possible [50]. This way, powder bed fusion might be able to compete with material jetting in terms of material properties and appearance as it would meet the most basic requirements for creating a 3D reproduction of art, namely the ability to print color and varying appearance.

Aside from the lack of aesthetic quality in trying to print in full color, currently, there is still a lot of manual work and technical knowledge needed to create and print a 3D printable model [47]. This makes the method expensive, exclusive, and time-consuming. Furthermore, the standardization of the process and, consequently, guaranteeing continuous product performance remains challenging and one to be overcome [48]. Nevertheless, right now, the main focus of the 3D printing realm will largely remain on the integration of 3D printing in the use and production of parts and products. Moreover, it is discussed that in the future, the exploration of the materials which can be used in 3D printing to convincingly express the object's materials, colors, "feel" and, additionally, to understand 3D printing's materials behavior in the longer term might gain more interest [47, 48].

6. Enabling technologies

Although 3D scanning and 3D printing are useful in understanding and mimicking an object's current appearance, they cannot fabricate its past and/or future material state. Understanding the way an object looks and what it has looked like over time cannot solely be attributed to what is currently visible on the surface. Artwork consists of layers of many different elements (e.g., its canvas, underdrawings, the various types of paint, and varnish) that mostly remain invisible to the naked eye. Moreover, due to external factors such as exposure to light, humidity and careless handling resulting in mechanical losses, an artwork's appearance can change more rapidly and in different ways. These internal and external effects that take place on a molecular and surface level attribute to the optical changes in the artwork's materials, such as its colors, relief, glossiness, and the transparency of individual layers. Because 3D printing uses digital modeling to materialize artworks, this information together with other digital technologies may be used to manipulate, alter, and change this information. This section will discuss the technologies that currently attribute to the possibility of reconstructing and predicting the past and future states of the same artwork.

6.1 Material analyses methods

Recently, there is a growing involvement of scientific methods in the art world which make it possible to visualize elements that have remained invisible for centuries. The first signs date back to the 1930s CE when X-radiography (IR) was undertaken for the analysis of paintings for the first time. Later, thanks to Dutch physicist van Asperen de Boer's research in the 1960s CE, infrared reflectography (IRR) made it possible to see through a painted surface or a painting's upper layers and

DOI: http://dx.doi.org/10.5772/intechopen.109107

analyze its underdrawings [51]. This non-invasive process involves analysis with an infrared camera, which uses light in the near-infrared region of the electromagnetic spectrum. Simply put, as an object consists of layers where each one is covering the other, infrared light decreases the opacity of most pigments [52]. One element that is highly infrared absorbent is carbon black, which is often used by artists to sketch the initial design [53]. Using IR, the artist's creative process or sometimes even an entirely different painting can be revealed. This was the case with Vincent van Gogh's *Still life with meadow flowers and roses* (1886–1887 CE), where X-ray scanning revealed a scene of two wrestling men hidden underneath the visible paint layers [54].

In the following decades, many other, more scientific analysis methods providing more information on an artwork's materials would follow, such as dendrochronology, paint sampling, neutron activation audio radiography (1970 CE), and cross-sectional fluorescent staining (1980 CE) [55]. However, these technologies were still not very visually appealing and understandable for the majority of people working with CH objects. This changed with macro-X-ray fluorescence scanning (macro-XRF). Already existing since the beginning of the 1930s CE, this technology only very recently started to attract the attention of art historians, scholars, and conservators as computers and visualization technologies have become increasingly better [56]. By focusing an X-ray beam on a designated area, the emitted fluorescence radiation of individual materials can be measured and analyzed, offering a new non-invasive way of studying paintings based on their material components [57]. The added value of macro-XRF in comparison with the beforementioned and more traditional imaging techniques (e.g., the beforementioned X-radiography and IR) relies on the fact that with this method information on the artwork's composition and the identification of multiple material components can be collected and visualized simultaneously instead of individually [53]. Consequently, the visual relationship of each layer of material becomes clearer.

Nowadays, thanks to the development of digital technology we have succeeded in creating understandable and easy-to-use tools with which we can navigate this information. This way, these imaging technologies have been transformed into mobile instruments that can be deployed in museums, galleries, or conservation studios, which increases the availability of these techniques. For example, Robert Erdmann (professor at the University of Amsterdam and senior scientist at the Rijksmuseum) has created an app with a "curtain viewer," with which you can look at and interact with the visualized results of various material scans (e.g., IR, IRR), research and treatments of Jheronimus Bosch' *Saint Wilgefortis Triptych* (1500–1504 CE), and many other paintings [58].

The large-scale application of these scientific methods and the unveiling of many artworks' materials still have to take place.[4] Yet, the fact that we can look through an object's layers, dissect each material component present in the artwork, and calculate how its materials change or have changed, undeniably changes our perception of artworks and art history. With the increasing power of computers, improvement of visualization technologies, and the growing efficiency with which neural and deep learning networks can be applied to this data, it will be a matter of time before it will become possible to predict and visualize what an artwork's material may have looked like in the past or what it will look like in the future. A successful example is a

[4] Although this overview mentions some material examination technologies used for the analysis of paintings, it most certainly does not cover all available methods. For more information, see: Stoner and Rushfield, Conservation of Easel Paintings.

reproduction of Rembrandt van Rijn the *Night Watch's* missing panels (2021 CE) [59]. Artificial neural networks were used to train a computer to recreate the missing pieces based on a still-existing sketch of the entire artwork. The following part sheds light on the current developments within the field of artificial intelligence.

6.2 Artificial intelligence

Ever since artist Harold Cohen created *AARON* in 1960 in which he tried to code the act of drawing, AI's influence on the art field has been undeniable [60]. Examples like Obvious' series of portraits of the fictive Belamy family (2018 CE) show that with AI and machine learning, one can construct entirely new works of art (**Figure 8**) [61]. The application of neural networks and AI are numerous, yet the focus will be limited to generative methods for they can be considered the most relevant and influential for 3D printing artworks.

Being the most popular and most developed method in the art field hitherto, generative visualization models are capable of generating new examples that plausibly could have been drawn from the original dataset, simply put. Such models are trained to recognize the regularities or patterns in the data it is fed in an unsupervised manner [62]. Previously, generative adversarial network (GAN), variational autoencoder (VAE), and flow-based models have been used (e.g., GAN was used for the construction of the Belamy portraits), yet their influence in the art world has remained limited for they were unstable, complex, or too hard to use [63]. However, this has changed due to the recent development of diffusion models which use a neural network that denoises images blurred with Gaussian noise and reverses the diffusion process. They steer the diffusion process with information learned from image-caption pairings gathered from the internet. This allows them to generate images conditioned on a text prompt. The influence of AI within the field of art has taken flight. Imagen,

Figure 8.
Edmond de Belamy, Obvious, 2018 CE, digital artwork.

DOI: http://dx.doi.org/10.5772/intechopen.109107

DALL-E2, Midjourney, and other open-source diffusion systems show that it has become possible to easily and rapidly generate new images and artworks based on simple concepts and keywords [64–66]. Evidently, these generative visualizations are interesting for creating new works of art and for experimentation. Yet, by supplementing the right data (e.g., based on material research or by uploading a large set of artworks by one artist) they can also become a helpful tool to get an understanding of an artwork, the way an artist has worked and to predict what an artwork must have looked like at some point, as is exemplified by the previously mentioned reconstruction of the *Night Watch*.

However, artworks are not two-dimensional. Hitherto, many models, including diffusion models, use operations that are well-defined on a flat domain with a regular grid, such as images or voxel grids. However, many disciplines work with data that lies on a curved or stretched surface (e.g., the canvas of a painting or the surface of a sculpture) or data that has no spatial embedding, such as citation graphs or friendship networks. Geometric deep learning attempts to transfer deep learning techniques from Euclidean space to non-Euclidean domains [67], thereby allowing for the models to work more closely with real-world data, including 3D data such as meshes and point clouds. The developments have not been as fast as is the case with generative visualizations. One reason is the fact that a three-dimensional space is rather complex and a large quantity of data are necessary to generate one single 3D object, which is not as readily available as is the case with Euclidean data. The neural network in a NeRF is optimized to best represent existing views of a 3D scene and can be sampled continuously from new viewing directions [68]. NeRF can generate new views of 3D scenes, by using partial sets of 2D images from angles along with the corresponding positions from which the images are taken [69]. By using a rendering loss, the input views of a scene can be reproduced. Currently, NeRF is still computationally-intensive.[5] Yet, new algorithms are consistently improving their performance, making it more accessible. This way, it will become easier and faster to scan, store, and share 3D information. Thus, more 3D CH objects can become available online, which is beneficial for training neural networks in the future. Furthermore, the quick and easy rendering of 3D models could make the computer-aided design (CAD) process used for 3D printing easier and more accurate. Subsequently, as *The Next Rembrandt* (2016 CE) shows, AI combined with 3D printing could lead to the creation of entirely new works of art in the style of a well-known artist [70].

It must be emphasized, though, that the term "AI" in these examples often means that an algorithm is *data-driven*. There are examples of "smart" algorithms, yet they do not fall under the term AI. The techniques that are now known as AI are often algorithms that are optimized on a dataset. This is not necessarily the case with traditional algorithms: Each step is then programmed and determined based on theory or analysis (and sometimes arbitrariness). Keeping this in mind, it is important to realize that the application of AI in the art world is often complimented by existing and more traditional tools, such as methods in computer graphics and geometry. An example of this is a tool that helps to naturally pose human bodies [71]. Previously, each arm and/or leg had to be placed in exactly the right place. Adding a data-driven tool that has been trained on many human poses makes it easier to generate a pose based on an "expected" natural pose. In this way, combining data-driven AI with

[5] NeRF uses images representing a scene and interpolates these images in order to render one complete scene. This way, a NeRF network directly maps from viewing direction and spatial location (5D input) to opacity and color (4D output).

these classical tools will make it easier to show examples and automate tasks that otherwise required a lot of expertise and mathematical knowledge. This will make the 3D printing process easier and more accessible to the art field. Furthermore, as Case Western Reserve University shows, a neural network can be trained to distinguish and generate different painters' brush strokes in both 2D and 3D [72]. This way, combining data-driven AI with methods in geometry helps predict lost information about an artwork or generate past or future versions of existing artworks, which can be physicalized with 3D printing. For instance, the color and geometry of the pounces in the background of the medieval *Crucifixion* of the Lindau Master (c. 1425 CE) can be reconstructed and 3D printed the way it must have looked before the blue azurite layer was applied (**Figures 9** and **10**) [73]. A painting like Johannes Vermeer's *Girl with a Pearl Earring* (1665 CE) can be 3D printed without a craquelure pattern. Furthermore, using a neural network trained in identifying Rembrandt's brushstrokes could help reconstruct the flattened surfaces in *Study of an Old Man* (1650 CE) and the artwork can be 3D printed the way it must have looked like.

Yet, there are still some obstacles to overcome. Currently, to be able to generate a 3D object with AI, one needs a very large dataset to compensate for information that has not been recorded. Unlike the case with generative visualization models based on Euclidean data, 3D data is not as easily accessible nor available in large quantities. The digitalization of artworks and open-access sharing of these data are not yet a common practice in museums and galleries. This way, generating entire 3D objects is still hardly possible. What seems more likely soon, however, is automatically adding a 2D texture map to a 3D mesh or a topography map. Furthermore, it must be emphasized that these generated visualizations and 3D objects will always be partly based on

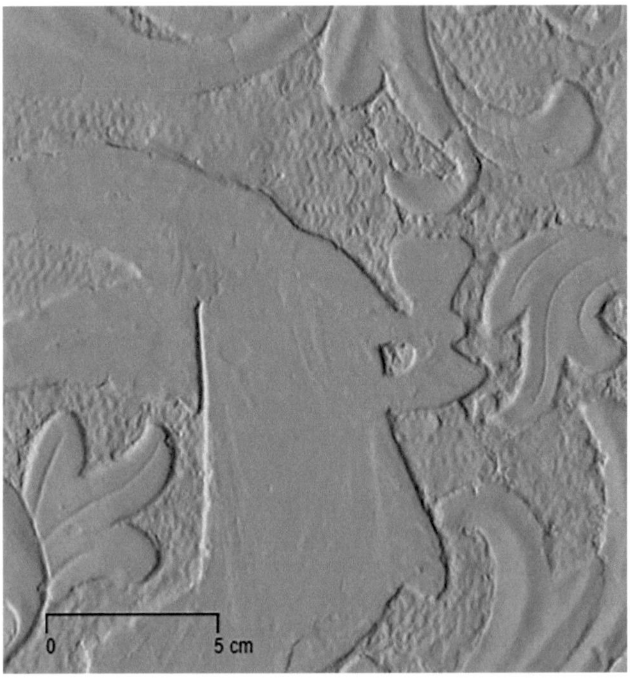

Figure 9.
Master of the Lamentation of Christ in Lindau, The Crucifixion, ca. 1425 CE. Tempera on panel, 125 × 89 cm, Museum Catharijneconvent, 3D detail without color the way the painting looks now.

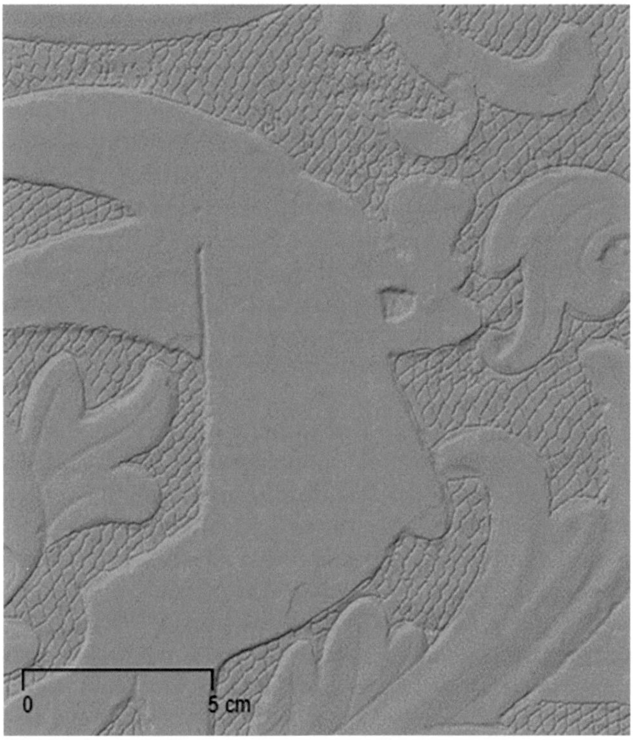

Figure 10.
Reconstruction of the pattern in the background. Photo credits: Ruben Wiersma.

estimates, predictions, the data supplemented, and the algorithm itself. Therefore, it might thus not be suitable for scientific research purposes in all cases.

Nevertheless, as 3D scanning and printing technology also improves which aids the digitalization of more objects, one could imagine how these developments combined with AI would yield far more accurate and precise results in the future. Furthermore, with the growing influence of the Metaverse, there is a growing necessity to easily and quickly digitalize objects, pushing the developments within the field. Additionally, with more open-source 3D modeling websites such as Blender and Adobe's Substance 3D and the possibility to experiment with diffusion models, the future impact of AI on the art world is undeniable.

7. Future perspectives

As this chapter has shown, various technologies enable the creation of 3D-printed reproductions of paintings, each attributing to what the final product might look like. It is undeniable that the continuous development of scientific methods within material science will uncover more of the hidden layers of artworks invisible to the naked eye. Furthermore, as computational power will continue to increase, it will become easier to visualize the discoveries made due to material and scientific research and examination. With the help of better neural networks, this could potentially result in better "curtain view"-like applications in which more scans can be integrated and better stitched together in a more user-friendly way. Furthermore, the growth in deep learning systems together with these material examination technologies could potentially attribute to better understanding and predicting the painting's current, past, and future material state. Moreover, as mentioned before, large printing companies will have taken on the challenge of developing new technologies to come up with full-color 3D printing technologies with a larger variety of materials. When more materials become available as a base to print textures on (e.g., canvas) or to print with (e.g., gold and other metals) in a more automated way, 3D reproductions of paintings will become more convincing. This will make the technology more interesting for the art field, as the lack of the right "feel" of the technology is often considered a hurdle to overcome. Furthermore, advancements in better printing materials will make it an interesting method for visualizing the data gathered and potentially manipulated using neural networks and user-friendlier methods, making it a more applicable method for conservation and restoration purposes.

However, the beforementioned increase in computational power is not solely limited to the increasing resolution, better visualization methods, and the data which can be gathered and processed in a short amount of time, but it will undeniably have its effects on other technologies in the digital realm as well. Right now, extended reality (XR) technologies, which is an umbrella term that covers all of the various technologies such as AR, VR, and mixed reality (MR) that enhance our senses by either providing additional information about the material world we live in or by creating entirely simulated worlds, are only at the brim of their existence.[6] Yet, their rapid development together with the advancements within social media and the rise of virtual worlds (e.g., the Metaverse) will make it possible to immerse oneself entirely into online realms that look similar to the "real" world but are computer generated. Although these technologies are not directly involved in the fabrication process of CH objects, they will have their influence on the perception of and engagement with artworks and 3D prints. This growing digitization and digitalization of the world, objects, and artworks around us using 3D technologies make it possible for one to find a variety of options for looking at the same artwork and find easier ways of either digitally or physically engaging with them.[7]

After the increasing digitalization of the world around us and the involvement within the virtual realm, the *non-fungible token* (NFT)—a digital seal of authenticity—is on the rise and becoming increasingly important. In short, an NFT can be seen as a token or a sign which can be used as a validator of ownership and one unique (digital) item. The way an official owner can secure its ownership and the unicity of the digital object is by securing the object with the Ethereum blockchain [74]. This way, it becomes impossible to modify the record or copy-paste its ownership. Right now, digital artworks are already being sold at renowned auction houses such as Christie's, for example, *Beeple's Everydays: The First 5000 Days* which sold in 2020 CE for over 69 million dollars (**Figure 11**) [75]. Not only can digital models and data become valuable by themselves, but it also becomes possible to transform material artworks into valuable digital formats. This happened for example, with Banksy's print called Morons, where a collective called BurntBanksy filmed the destruction of the original material artwork and sold the

[6] Augmented reality (AR): adds a layer of digital elements and/or information to a live situation, usually by using a smartphone. Virtual reality (VR): creates a complete immersive experience shutting out the physical world completely, usually by wearing VR glasses. Mixed reality (MR): includes the interaction of elements of both AR and/or VR, real-world and digital objects.

[7] Digitization: the conversion from an analogue format to a digital version. Digitalization: making digitized workable. This way, the digitized information is converted in a way to make processes more effective or productive.

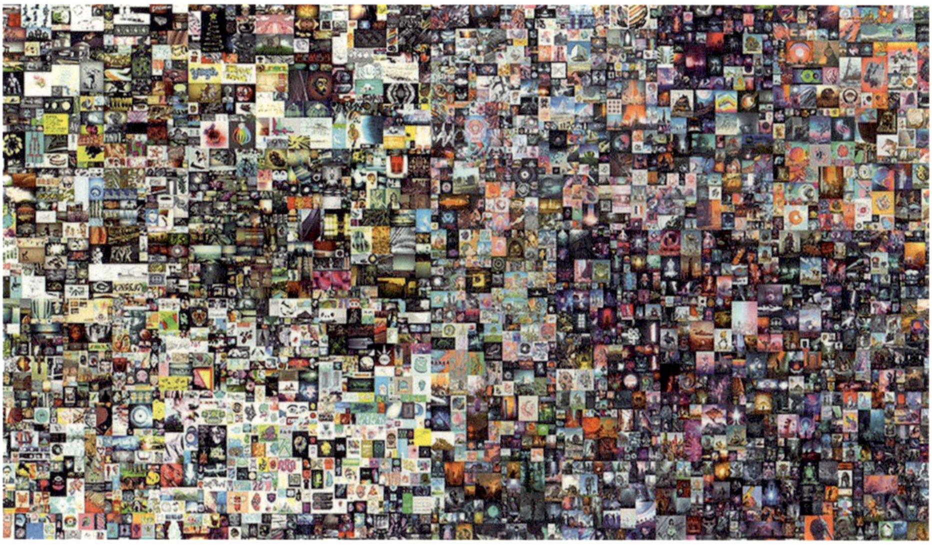

Figure 11.
Everydays: The First 5000 Days, Beeple, 2007–2020 CE, digital artwork.

video as an NFT [76]. A similar case has happened with a drawing by Frida Kahlo [77]. Furthermore, artists like Damien Hirst have also started to sell their art similarly [78]. In this way, the artwork's value or authenticity has moved from the physical manifestation to a digital form as an NFT. What this development will mean in the case of the 3D models, digitized paintings and cultural heritage objects are that they may become valuable and highly prized assets existing next to the material original artwork and outside of cultural institutions and museum walls and authority. Moreover, it could mean that they could potentially take over the "aura" of the existing artworks.

With the potential rise of the value of these 3D digital art models caused by phenomena such as NFTs, questions can arise about the actual ownership of the 3D scanned objects under study. Currently, there is no clear idea to which these data belong: would this be the museum that owns the original artwork, the person(s) behind the scanning, or the computer processing the data? This could lead to potential issues regarding the intellectual property rights of the scanned data and future 3D prints. Although this is an interesting and important facet of 3D printing's development, the legal and copyright-related questions fall out of the scope of this particular research.

Right now, it is hard to predict how these technologies will change or have changed the perception of artworks exactly. However, discoveries thanks to material research, alterations made with the help of artificial intelligence, and the translation of this information into high-quality 3D printed reproductions could drastically change art history at large. A parallel can be drawn with the perception of dinosaurs. Until recently, the prevalent idea was that dinosaurs must have looked like reptiles with scales and that they were rather tough looking. Yet, as a result of research and by discovering feathers in fossils, the whole idea of their physical appearance changed drastically: They might have looked more like birds, feathered, and possibly brightly colored. In the case of 3D printing artworks, the same thing could be said: We believe that what we know now is "authentic" and "real," yet discoveries and revelations due to these technologies combined with 3D printing can and will drastically change what we consider to be the truth in the future.

Author details

Liselore N.M. Tissen
Leiden University and Delft University of Technology, Leiden & Delft, The Netherlands

*Address all correspondence to: l.n.m.tissen@tudelft.nl

References

[1] The Economist. Print Me a Stradivarius—How a New Manufacturing Technology will Change the World. 2011. Available from: https://www.economist.com/node/18114327/.%21%21?page=2 [Accessed: September 1, 2022]

[2] Elkhuizen W et al. Digital Manufacturing of Fine Art Reproductions for Appearance. Delft: TU Delft; 2018

[3] Tissen LNM. Authenticity and Meaningful Futures for Museums: The Role of 3D Printing. In: Journal of the LUCAS Graduate Conference. Leiden: Leiden University; 2021. pp. 94-122

[4] Benjamin W. The Work of Art in the Age of Mechanical Reproduction. London: Penguin Books; 2008

[5] Malik US, Tissen LNM, Vermeeren APOS. 3D reproductions of cultural heritage artefacts: Evaluation of significance and experience. In: Studies in Digital Heritage. 4th ed. Bloomington: Indiana University; 2021. pp. 24-28. ISSN: 2574-1748

[6] Gao Q, Jones S. Authenticity and Heritage Conservation: Seeking Common Complexities beyond the "Eastern" and "Western" Dichotomy. International Journal of Heritage Studies. 2021;**2021**:90-106

[7] Lowe A. The Aura in the Age of Digital Materiality—Rethinking Preservation in the Shadow of an Uncertain Future. Milan: Silvana Editoriale; 2020

[8] Tissen LNM, van Veldhuizen MJ. Picture-Perfect—The Perception and Applicability of Facsimiles in Museums: Art & Perception. Leiden: Brill; 2022

[9] Tissen LNM. 3D Printing Works of Art: An Opportunity or Nightmare for Curators. CODART. 2022. Available from: https://www.codart.nl/feature/museum-affairs/3d-printing-works-of-art-an-opportunity-or-nightmare-for-curators/ [Accessed: October 22, 2022]

[10] Scopigno R et al. Digital fabrication techniques for cultural heritage: A survey: Fabrication techniques for cultural heritage. Computer Graphics Forum. 2017;**2017**:1-6

[11] Neumüller M et al. 3D Printing for Cultural Heritage: Preservation, Accessibility, Research and Education. Heidelberg: Springer; 2014

[12] Ballarin M, Balletti C, Vernier P. Replicas in cultural heritage: 3D printing and the museum experience. In: ISPRS—International Archives of the Photogrammetry, Remote Sensing and Spatial Information Sciences. Göttingen: Copernicus GmbH; 2018. pp. 55-62

[13] Acke L et al. Survey and literature study to provide insights on the application of 3D technologies in objects conservation and restoration. Journal of Cultural Heritage. 2021;**49**:1-15

[14] Beentjes TPC. Casting Rodin's Thinker. Sand Mould Casting, the Case of the Laren Thinker and Conservation Treatment Innovation. Amsterdam: University of Amsterdam; 2019. pp. 281-288

[15] Van Gogh Museum. Restauratie "Zeegezicht Bij Scheveningen". 2019. Available from: https://www.vangoghmuseum.nl/nl/over/kennis-en-onderzoek/restauraties/restauratie-zeegezicht-bij-scheveningen [Accessed: October 22, 2022]

[16] How 3D Printing Brings Antiquities Back to Life. Formlabs. 2018. Available

at: https://formlabs.com/blog/how-3d-printing-brings-antiquities-back to-life/ [Accessed: October 22, 2022]

[17] Speranza L. The Gap between Ethics and Aesthetics in Italian Restoration: Experiences in Laboratories of the Opificio Delle Pietre Dure in Florence—Della Robbia Sculpture Case Studies. Washington D.C.: American Institute for Conservation of Historic and Artistic Works. 2017

[18] Van Apeldoorn N. 3D Reconstruction of Niniveh. Delft: TU Delft; 2017

[19] Scanning for Syria Project. Global Heritage and Development. 2015. Available from: https://www.globalheritage.nl/news/scanning-for-syria-project. [Accessed: October 22, 2022]

[20] Factum Foundation. The Thomb of Seti: Recording and Facsimile. Available from: https://www.factumfoundation.org/pag/1502/the-tomb-of-seti-recording-and-facsimile. [Accessed: October 22, 2022]

[21] Elkhuizen W et al. A 3D Printed Reconstructing of a Painting's Original Size: Showing the Original Size of Saul and David by Rembrandt. Barcelona, Spain; 2016

[22] Smithsonian Institute 3D Digitization. Available from: https://3d.si.edu/ [Accessed: October 22, 2022]

[23] Uffizi in 3D. Available from: https://sketchfab.com/tags/uffizi. [Accessed: October 22, 2022]

[24] The National Museum of Antiquities. Available at: https://sketchfab.com/rmo_leiden. [Accessed: October 22, 2022]

[25] Mauritshuis. Facelifts En Makeovers. 2022. Available from: https://www.mauritshuis.nl/nu-te-doen/tentoonstellingen/tentoonstellingen-uit-het-verleden/facelifts-makeovers/ [Accessed: October 22, 2022]

[26] Knapp M, Wolff R, Lipson H. Developing printable content: A repository for printable teaching models. In: Proceedings of the 19th Annual Solid Freeform Fabrication Symposium. Austin, TX; 2008

[27] Victoria & Albert Museum. Taking Sculptures for a Walk: 3D Printing and Museum Outreach. Available from: https://www.vam.ac.uk/blog/digital/taking-sculptures-for-a-walk-3d-printing-and-museum-outreach [Accessed: October 22, 2022]

[28] Menano L et al. Integration of 3D printing in art education: A multidisciplinary approach. Computers in the Schools. 2019;**3**:222-236

[29] Factum Foundation. The Materiality of the Aura. New Technologies for Preservation. 2019. Available from: https://www.factumfoundation.org/pag/1598/. [Accessed: October 22, 2022]

[30] Museum Boijmans van Beuningen. Sgraffito in 3D. 2009. Available from : https://www.boijmans.nl/tentoonstellingen/sgraffito-in-3d [Accessed: October 22, 2022]

[31] Balletti C, Ballarin M, Guerra F. 3D printing: State of the art and future perspectives. Journal of Cultural Heritage. 2017;**26**:176-178

[32] Cantoni V et al. Art Masterpieces Accessibility for Blind and Visually Impaired People. In: Computers Helping People with Special Needs. Cham: Springer International Publishing; 2018. pp. 267-274

[33] Solly M. This Replica of a Tlingit Killer Whale Hat Is Spurring Dialogue

About Digitization. Smithsonian Magazine. 2017. Available from: https://www.smithsonianmag.com/smithsonian-institution/replica-tlingit-killer-whale-hat-spurring-dialogue-about-digitization-180964483/. [Accessed: October 22, 2022].

[34] Daneshmand M et al. 3D Scanning: A Comprehensive Survey. New York: Cornell University. 2018. DOI: 10.48550/arXiv.1801.08863

[35] Akca D et al. High Definition 3D-Scanning of Arts Objects and Paintings, Zürich; 2007. DOI: 10.3929/ethz-a-005748653

[36] Elkhuizen WS et al. Reproduction of gloss, color and relief of paintings using 3D scanning and 3D printing. In: Eurographics Workshop on Graphics and Cultural Heritage, Barcelona; 2017. pp. 183-187

[37] Tissen LNM et al. Using 3D scanning to support conservation treatments for paintings. In: Materials Science and Engineering. Bristol: IOP Publishing; 2020

[38] Rakitina E et al. An overview of 3D laser scanning technology. In: Proceedings of the International Scientific Conference. Chiteseer; 2008

[39] Means BK. 3D recording, documentation and management of cultural heritage. Historical Archieves. 2017;**51**:582-583

[40] Elkhuizen WS et al. Comparison of three 3D scanning techniques for paintings, as applied to Vermeer's "Girl with a Pearl Earring". Heritage Science. 2019;7:5-9

[41] Bossema FG et al. Integrating expert feedback on the spot in a time-efficient explorative CT scanning workflow for cultural heritage objects. Journal of Cultural Heritage. 2021;**49**:38-47

[42] Domínguez-Delmás M et al. Dating and provenancing the Woman with lantern sculpture – A contribution towards attribution of Netherlandish art. Journal of Cultural Heritage. 2021;**50**:179-187

[43] ISO/ASTM. 52900:2015 Additive Manufacturing—General Principles—Terminology. 2015. Available from: https://www.iso.org/obp/ui/#iso:std:iso-astm:52900:ed-1:v1:en. [Accessed: October 22, 2022]

[44] ISO/ASTM. Additive Manufacturing e General Principles e Part 2: Overview of Process Categories and Feedstock. 2022. Available from: https://www.iso.org/standard/61626.html [Accessed: October 22, 2022]

[45] Yuan J et al. Accurate and computational: A review of color reproduction in full-color 3D printing. Materials & Design. 2021;**209**:109943

[46] Doubrovski EL. Design Methodology for Additive Manufacturing: Supporting Designers in the Exploitation of Additive Manufacturing Affordances. Delft: TU Delft; 2016

[47] Deshmukh K et al. Chapter 1—Introduction to 3D and 4D printing technology: State of the art and recent trends'. In: Sadasivuni KK, Deshmukh K, Almaadeed MA, editors. 3D and 4D Printing of Polymer Nanocomposite Materials. Amsterdam: Elsevier; 2020. pp. 1-24. DOI: 10.1016/B978-0-12-816805-9.00001-6

[48] Wang X et al. Color reproduction accuracy promotion of 3D-printed surfaces based on microscopic image analysis. International Journal of Pattern Recognition and Artificial Intelligence. 2020;**34**(01):2054004

[49] Molitch-Hou M. Overview of additive manufacturing process. In:

Zhang J, Jung Y, editors. Additive Manufacturing. Oxford: Butterworth-Heinemann; 2018. pp. 1-38. DOI: 10.1016/B978-0-12-812155-9.00001-3

[50] HP. HP Multi Jet Fusion Technology. Available from: https://www.hp.com/us-en/printers/3d-printers/products/multi-jet-technology.html [Accessed: October 22, 2022]

[51] Legrand S et al. Examination of historical paintings by state-of-the-art hyperspectral imaging methods: From scanning infra-red spectroscopy to computed X-ray laminography. Heritage Science. 2014;**2**(1):1-8

[52] Groen K. Paintings in the Laboratory: Scientific Examination for Art History and Conservation. Amsterdam: University of Amsterdam; 2011. pp. 26-28

[53] The National Gallery. Infrared. Glossary. Available from: https://www.nationalgallery.org.uk/paintings/glossary/infrared. [Accessed: October 22, 2022].

[54] Kröller Müller. Stilleven Met Akkerbloemen En Rozen. Available from: https://krollermuller.nl/en/vincent-van-gogh-still-life-with-meadow-flowers-and-roses. [Accessed: October 22, 2022]

[55] Hermens E. Technical art history: A synergy of art, conservation and science. In: Art History and Visual Studies in Europe: Transnational Discourses and National Frameworks. Leiden: BRILL; 2012

[56] Stoner JH. Turning points in technical art history. In: American Art. Chicago: University of Chicago; 2012

[57] Stoner JH, Rushfield R. Conservation of Easel Paintings. In: Routledge Series in Conservation and Museology. London: Routledge; 2020. pp. 317-318

[58] Saint Wilgefortis Triptych. Curtain viewer showing visible light, infrared photography, and infrared reflectography from before restoration. Available from: http://boschproject.org/view.html?layout=top-major&mode=curtain&pointer=0.5,0.5&i=23_24_25_MCPVIS%5Bl=visible%20light%20photography%5D,23_24_25_MCPIRP%5Bl=infrared%20photography%5D,23_24_25_IRREFL%5Bl=infrared%20reflectography%5D [Accessed: October 22, 2022]

[59] Rijksmuseum. Nachtwacht de Missende Stukken. 2020. Available from: https://www.rijksmuseum.nl/nl/stories/operatie-nachtwacht/story/nachtwacht-de-missende-stukken [Accessed: October 22, 2022]

[60] Cohen H. The further exploits of Aaron, Painter. Stanford Humanities Review. 1995;**4**(2):1-18

[61] Obvious. La Famille de Belamy. 2019. Available from: https://obvious-art.com/la-famille-belamy/. [Accessed: October 22, 2022]

[62] Geng X, Kang BH. PRICAI 2018: Trends in Artificial Intelligence: 15th Pacific Rim International Conference on Artificial Intelligence, Nanjing, China, August 28-31, 2018, Proceedings, Part I. In: Lecture Notes in Computer Science. Berlin, Heidelberg: Springer; 2018

[63] Bond-Taylor S et al. Deep Generative Modelling: A Comparative Review of VAEs, GANs, Normalizing Flows, Energy-Based and Autoregressive Models. New York: IEEE Transactions on Pattern Analysis and Machine Intelligence. 2021

[64] Imagen. Available from: https://imagen.research.google/ [Accessed: October 22, 2022]

[65] Dall-E 2. Available from: https://openai.com/dall-e-2/ [Accessed: October 22, 2022]

[66] Midjourney. Available from: https://www.midjourney.com/home/ [Accessed: October 22, 2022]

[67] Bronstein MM et al. Geometric deep learning: Going beyond Euclidean data. IEEE Signal Processing Magazine. 2017;**34**(4):18-42

[68] Barron JT et al. Mip-Nerf: A Multiscale Representation for Anti-Aliasing Neural Radiance Fields. New York: IEEE Transactions on Pattern Analysis and Machine Intelligence. 2021. pp. 5855-5864

[69] Mildenhall B et al. Nerf: Representing scenes as neural radiance fields for view synthesis. Communications of the ACM. 2021;**65**(1):99-106. DOI: 10.1145/374308.374363

[70] The Next Rembrandt. 2015. Available from: www.thenextrembrandt.com. [Accessed: October 22, 2022]

[71] Bocquelet F et al. AI and Physics Assisted Character Pose Authoring. New York, NY, USA: Association for Computing Machinery; 2022

[72] Ji F et al. Discerning the painter's hand: Machine learning on surface topography. Heritage Science. 2021;**9**(1):2-22

[73] Tissen LNM, Frequin S, Wiersma RT. The case of the golden background, a virtual restoration and a physical reconstruction of the medieval crucifixion of the Lindau Master (c. 1425). Digital Humanities Quarterly. 2022;**16**(2):1-16

[74] Nadini M et al. Mapping the NFT revolution: Market trends, trade networks, and visual features. Scientific Reports. 2021;**11**(1):1-11

[75] Christies. Beeple's Opus Created over 5,000 Days by the Groundbreaking Artist, This Monumental Collage Was the First Purely Digital Artwork (NFT) Ever Offered at Christie's. 2021. Available from: https://www.christies.com/features/Monumental-collage-by-Beeple-is-first-purely-digital-artwork-NFT-to-come-to-auction-11510-7.aspx. [Accessed: October 22, 2022]

[76] Tissen LNM, Frequin S. Digitalisering van Kunstwerken Is van Waarde, Maar Waak Voor Cowboys En Destructie van de Kunstmarkt. Amsterdam; 2021. Available from: https://www.volkskrant.nl/cultuur-media/digitalisering-van-kunstwerken-is-van-waarde-maar-waak-voor-cowboys-en-destructie-van-de-kunstmarkt~bb96615b/?referrer=https%3A%2F%2Fwww.google.com%2F [Accessed: October 22, 2022]

[77] Renders W. Miljonair Verbrandt Tekening van Frida Kahlo, Om Die Als NFT Te Kunnen Verkopen. Amsterdam; 2022. Available from: https://www.volkskrant.nl/nieuws-achtergrond/miljonair-verbrandt-tekening-van-frida-kahlo-om-die-als-nft-te-kunnen-verkopen~bc8d827b/ [Accessed: October 22, 2022]

[78] Holland O, Dolan L. "That Is Art!": Damien Hirst Has Begun Burning Thousands of His Own Paintings. London: CNN; 2022

Chapter 2

Assessment of Occupational Exposures in the 3D Printing: Current Status and Future Prospects

Ilona Pavlovska, Lāsma Akūlova, Anna Lece, Žanna Martinsone, Linda Paegle, Aneka Kļaviņa, Klinta Luīze Sprūdža and Inese Mārtiņsone

Abstract

3D (three-dimensional) printing technologies are widespread and rapidly evolving, creating new specific working conditions, and their importance has been highlighted by increasing publications in recent years. The report provides a compilation of current information on 3D technologies, materials, and measurements, considering the determination of the potential actual exposure dose of chemicals through airborne inhalation and dermal exposure, including workers' exhaled breath condensate and urine data. Noninvasive assessment methods are becoming increasingly popular, as they are painless, easy to perform, and inexpensive. Investigation of biomarkers reflecting pulmonary inflammation and local and systemic oxidative stress in exhaled breath, exhaled breath condensate, and urine are among them. It is also important to consider the occupational health and safety risks associated with the use of various new materials in 3D printing, which are associated with skin irritation and sensitivity risks. Therefore, EDI (estimated daily intake) calculations for assessment of the potential occupational health risk purposes *via* inhalation and dermal exposure are critical in future. The assessment of occupational exposure and health risks of 3D printing processes is essential for the proper identification, control, and prevention of working conditions, also for the diagnosis and monitoring of occupational diseases among workers to improve public health and well-being in general.

Keywords: additive manufacturing, 3D printing, occupational exposure, exhaled breath condensate, dermal exposure, biomarkers, exposure assessment, health risk evaluation

1. Introduction

3D (three-dimensional) printing has become an integral part of today's market and service industry since the 1980s, producing versatile and widespread products not only in industry but also in the home. Nowadays, highly specialized industries, such as medicine, automotive, military, and aerospace engineering, use 3D printing for prototyping and facilitating the manufacturing process. Originally, the first material used was synthetic plastic, but as printing techniques have evolved, materials such as metals (steel), conductive materials, glass, ceramics, wood as well as organic tissue are also used. Besides, if 3D printing is combined with nanotechnologies, the future is under development, making it possible to create objects at nano and molecular level. One of the main type of emissions from desktop 3D printers is ultrafine particles (UFP), whose uniform distribution is related to the end of the printing process. In addition, about 50 different volatile organic compounds (VOC) have been identified from printer cameras of which the three most commonly emitted are styrene, caprolactam, and lactide [1]. Emissions occur especially when polymers are thermally treated. In general, exposure to contamination occurs *via* inhalation, food and beverage, water consumption, and dermal contacts. 3D emission main exposure routes are inhalation and dermal exposure, and usually, employees could be exposed at highest levels due to operations and maintenance of 3D printers, for example, when using acrylonitrile butadiene styrene (ABS) and polylactic acid (PLA), the worker is already exposed to emissions at an exposure time of 1 h [2]. Exposure *via* inhalation is breathing in a substance as a gas or vapor or as airborne particles. This includes small amounts of soil and dust that can be inhaled into the lungs. The lungs often absorb gases and vapors quickly and efficiently. Hereabouts, the estimated daily intake (EDI) for a chemical contaminant can represent the total exposure from all known or suspected exposure pathways for an average person to consider also their occupational exposures.

One of the possible methods to improve occupational health and safety in the 3D printing process to limit the spread of emissions is to seal heating, ventilation, and air condition systems with high-performance filters impregnated with activated carbon; to use a portable air purifier in the immediate operating areas; to install ventilation systems and enclosures in the emerging operating area for the storage of finished products [3]. Thus, ensuring local shock ventilation is an essential aspect of occupational safety. Furthermore, metal powder materials are at risk of ignition and it is recommended to use EX-devices (detection devices for explosive atmospheres) to identify the risk of explosion at workplaces [4].

Organizations (e.g., National Institute of Occupational Safety and Health (NIOSH), Occupational Safety and Health Agency (OSHA), etc.) with an objection to maximizing health protection from exposure to contaminants play an important role in identifying health risks and controlling emissions. Thus, NIOSH is a part of the Occupational Safety and Health Research of the United States of America. NIOSH's operational goals include understanding the health risks of advanced materials and manufacturing processes; studying the initial risks of engineered nanomaterials, advanced materials, and additive materials; supporting the development of guidelines and epidemiological studies to inform stakeholder identification of health risks and assessing progress to date; and evaluating risk management [5]. Further, OSHA's main focus is safety, health, and productivity of European workplaces, while promoting risk prevention and improving working conditions in Europe [6]. Working with 3D printers, existing occupational chemical risks compromising immune system function (cytokines-interferon and lymphokines), which play a key role in providing inflammatory responses and monitoring other abnormal processes. Considerable awareness of complex health assessment—cardiovascular, nervous, reproductive, etc. plays a role in the emergence of health risks and efficiency of occupational hazard controlling and risk prevention [7]. In addition, well-being and productivity could be affected by poor working conditions caused by 3D printing activities at work, schools [8], and homes as well [9].

This chapter takes a complex look at the safety aspects of work-associated risks of 3D printing technologies.

2. 3D printing technologies, materials, and methods

Basic computer skills are needed for semiprofessional or home office printer users, while technicians also need knowledge to use 3D printers (in professional use). These printers come in different shapes, sizes, and prices—home office printers can fit on an office table but office or professional printers require a lot more space and it is suggested to keep them in a separate room (to limit noise and emission exposure) [10]. In order to attain the proper design goals, the materials in this special process must be functional. Material properties are highly important as the product undergoes conversion from design and functional prototype development to production [11].

Materials and technologies are a subject of constant improvement and modification, but the main and most commonly used materials and methods in 3D printing technologies will be discussed in the following chapter.

There are multiple industrial 3D printing processes: stereolithography (SL), powder-bed fusion (PBF), fused-deposition modeling (FMD), laminated object manufacturing (LOM), and direct energy deposition (DED) [11]. Each of these processes uses a different technology to produce the desired outcome. And each of them has many modifications, improvements, and varieties.

2.1 Stereolithography (SL)

Materials used in SL are usually photopolymers, and thermoset resins improved by ultraviolet (UV) light. There are numerous plastics that can be printed in 3D and have a variety of mechanical properties. SL is a fine-resolution and high-quality process commonly used in biomedical prototyping, yet the materials are quite scarce because of the relatively sluggish and expensive printing [11].

2.2 Powder-bed fusion (PBF)

PBF processes utilize tightly packed layers of ultra-fine powders on a platform. The polymer powder layers are fused together with a laser beam or a binder [12, 13]. Despite high cost, slow printing, and layer-by-layer finish, due to its high resolution and high-quality outcomes, PBF is a method that is widely used in a variety of industries, for example, biomedicine, electronics, aerospace, and to produce lightweight structures (lattices) [13].

2.3 Fused-deposition modeling (FMD)

In FMD, a continuous filament of thermoplastics and fiber-reinforced polymers are used. The thermoplasticity of the filament is crucial for FDM, allowing the filaments to fuse together during the printing process, and afterward, usually, a copolymer is used to solidify at layers. Due to its qualities and the materials used, FDM is a low-cost and high-speed prototyping method [14].

2.4 Laminated object manufacturing (LOM)

LOM can be used for a variety of materials, such as polymers, ceramics, paper, and metals. The printing method is based on layer-by-layer cutting and lamination of sheets or rolls, or the other way around. Reduced tooling and manufacturing time, a vast range of materials, and low cost allows to produce large structures and use for foundry industries, paper manufacturing, as well as in electronics and printing smart structures [11].

2.5 Direct energy deposition (DED)

DED method is also known as laser-engineered net shaping (LENS) and it is used to melt a feedstock material simultaneously. The melted material is then deposited and fused into the melted substrate and solidified after the movement of the laser beam [15]. This method is commonly used for stainless steel, Inconel, or with titanium, aluminum, and related alloys [11] and can be used for producing materials for repair, aerospace, biomedicine, etc.

2.6 "Hobby-class" 3D printing

3D printers are becoming more popular to be used at home as well, especially some portable low-cost desktop 3D printers (called "hobby-class"). In particular, the emergence of inexpensive portable 3D pens in the private and school sectors, using materials, such as PLA and ABS, is of concern due to the lack of data confirming their safe use [16].

Very few reports are available, which conclude a biosafety assessment of the 3D printer emissions, particularly the ones popular among young age users, such as 3D printing pens [17, 18].

However, a recent study has highlighted the potential risk posed by high particle emissions, including thermoplastic material nanoparticles as well as metal emissions from 3D pen usage [18].

3. Emissions from materials in 3D printing

All the above-mentioned methods use a variety of materials to produce the desired outcome products. The printing process uses high-temperature and the treated material emits VOCs, such as styrene, aldehydes, ethylbenzene, xylenes, caprolactam, lactide, and others, as well as UPFs [19–21]. It is known that ABS filaments are more hazardous to human health than PLA filaments. These emissions of particles and gases are higher when using ABS. ABS filaments produce VOCs, for example, styrene is a possible human carcinogen, but nylon filaments emit a compound called caprolactam (**Table 1**), which can lead to irritation of eyes and respiratory system and may cause effects on central nervous system. The total emission level and composition of VOCs depend on the filament material. PLA filaments have lower particle and total VOC emission rates than ABS and nylon filaments. There are some carbon emissions from the fumes when printing with polyethylene terephthalate glycol (PETG), and the main emitted VOC is acetaldehyde [29].

Some 3D printers use photopolymers (photosensitive liquid resins that become solid after exposure to laser or UV light) that are known to be toxic. During 3D metal

Filament	Type of VOCs	Reference
ABS	Aldehydes, acetone, diethyl phthalate, ethylbenzene, styrene, xylenes, etc.	[14, 22–27]
PLA	Acethyladehyde, butadiene, propionic acid, toluene, etc.	[28, 29]
NYLON	Caprolactam	[14, 26]
PETG	Acetone, ethylbenzene, formaldehyde, toluene, xylene, etc.	[15, 30, 31]
nGEN	Ethylbenzene, heptane, toluene, xylene, etc.	[31]

Table 1.
Some examples of emitted types of VOCs drives from some filament material used in 3D printing.

printing fine (including ultrafine) metal powder is used that is capable to become a respiratory hazard. Not only chemical risks but also well-known occupational health risks: high temperature, low humidity, high CO_2 levels, forced posture, eyestrain, and injuries can pose health risks among 3D printers' workers [32].

Depending on their content, other filaments may content additional chemical and metal exposures. The decision must be emphasized before using a particular type of filament, including 3D printers in the working environment. Considering the growing adoption of 3D printing and the growing mix of print materials with complex chemical properties, it is important for manufacturers to better educate themselves on the health risks to protect workers in their facilities [28].

UFPs may be of particular importance for the toxicity of fumes emitted from the melting of some thermoplastics [31, 33], although the latest literature reports show the release of UFPs and VOCs during printing with printers using nGEN (made by ColorFabb from Eastman Amphora™ AM3300 polymer) filament as well [31].

4. Occupational exposure

4.1 Risk identification

Occupational exposure data characterizing and description (e.g., profile of workplace, location, materials, equipment, methods, exposure time, ventilation, personal protective equipment, etc.) is an important issue to evaluate potential exposure, control, and prevention.

Furthermore, working environment description is essential to evaluate occupational exposure. The first step is to determine the exposure routes. During 3D printing process, workers can be exposed through inhalation or skin contact to various pollutants that can affect health and work abilities, oral exposure is less common in this industry. Also, to evaluate occupational exposure risk, 3D printing methods have to be categorized to determine the right pollutants in air in the previous paragraph. ISO standard defines 3D printing methods into seven categories: material extrusion, material jetting, binder jetting, sheet lamination, vat photopolymerization, bed fusion, and directed energy deposition [34].

During risk assessment and exposure determination, printing materials must be evaluated as well. In addition, exposure to printing materials, particulate matter (PM), and various chemicals can happen during cleaning and maintaining 3D printers and workspace. The common approach, for clarifying potential chemical substances' emissions of used materials and other agents during cleaning, maintenance, etc., is to check a chemical products safety data sheet [35].

4.2 Toxicity and health effects

As it is a new industry, only few publications indicate adverse health effects associated with 3D printing processes. One article mentioned that exposure to material extrusion fumes causes asthma [36]; however, other case reports mentioned chronic hypersensitivity pneumonitis with exposure to PBF nylon powder [37]. Various articles that used health questionnaires found a high incidence of 3D operator self-reported respiratory symptoms [38]. Animal studies show pulmonary and systemic toxicity in rats following exposure to 3D printing process [17, 39]; however, human studies on volunteers exposed to 3D printing fumes showed no acute changes in inflammatory markers [40]. Nevertheless, all this information draws attention to the risks of 3D printers.

Studies involving real occupational exposure measurements in 3D industrial settings are limited; however, the emission of material extrusion 3D printing process has been studied extensively, but the impact from other 3D printing technologies has not been documented as comprehensively. Emission and exposure in a work environment depend on differences in printing methods and related activities, including print materials. The thermal 3D printing process is a known source of VOCs and semi-volatile organic compounds (SVOCs). The typical VOC emitted during 3D printing can attach to airborne particles [41]. In addition, 3D printing is a source of UFP emissions [42–45].

VOCs and UFPs both have the potential adverse effect on human health through inhalation and skin contact, causing irritation, inflammation, lung disease, and various skin problems. It can also affect the central nervous system [36, 38, 46]. In addition, UFPs can pose health risks in a delayed manner even 1–5 days because of UFPs airborne exposure lag time. And UFPs toxicity depends on smaller size, larger surface area, absorbed substances, and physical properties [47]. Nowadays, 3D printing includes also nanomaterial usage in filaments because of their physicochemical and microbiological properties [48]. Further exposure measurements and the development of an appropriate measurement strategy for specific printing methods are needed to determine full exposure during the 3D printing process.

4.3 Prevention

The main principles of control/prevention measures at workplaces are elimination, replacement, technical, organizational measures, and personal protective equipment.

- Elimination of emissions or substitution with other materials.

- Technical measures that prevent emission sources. The most effective measure is a protective barrier for the emission source, closed-production systems are used and employees are physically isolated from contact with emissions. However, the potential for leakage with these activities should be considered. A mandatory requirement is the existence of local and general ventilation with special air filtration systems (e.g., HEPA) that retain dust, as well as forced exhaust systems.

- Organizational measures are the main measures to reduce risk. Employees' workplaces can be isolated, separated from the place where the process of 3D printing takes place. Isolation can take place, for example, by installing and creating walls.

- Personal protective measures: This control measure would be the last resort. Respiratory protection requires respirators and masks with filters or full-face or partial-covering masks with carbon filters, etc. [49, 50].

5. Exposure measurements and assessment in 3D printing processes

5.1 General data

In order to evaluate the exposure of the working environment of 3D printing, air sample analysis of the working environment is performed to determine the chemical composition (PM, VOCs, metals, and inorganic elements) and thermal comfort parameters (air temperature, humidity and air velocity) as well. Samples are taken in the breathing zone of the worker (i.e., within a radius of 30 cm around the face) to assess worker exposure or as close as possible to the workplace under normal working conditions. As it is listed and described into several technical standards and methods [51–54]. Main principles of measurement/sampling include purpose of measurement/sampling; selection of measurement/sampling location; selection of measurement/sampling technique; duration of measurements/sampling; selection of parameters to characterize worker exposure and potential health effects. It is important to consider the type of technologies, materials used during production and work processes, which allow us to guess what kind of chemical emissions could be released in the work environment. Measurements/samples should be taken in all work areas, where the work is performed [55].

Measurements done for risk assessment are a combination of traditional industrial hygiene methods and newly developed sampling techniques. Traditional methods include personal and area air sampling for the total PM concentration, inhalable dust [42, 45], respirable PM [43, 44, 56], VOCs or other potential exposures of interest [57]. Other exposure limit detection is done using direct reading instruments for PM count, sizing, and classification. Techniques used for this are condensation particle counters (CPC), optical particle counter (OPS), dust trackers, nano-scans, fast mobility particle sizer (FMPS™), electrical low-pressure impactor (Dekati® ELPI®+), etc. [56].

In addition, air temperature, relative air humidity and air velocity play a role regarding occupational exposure and employees' thermal comfort. Increasing air temperature also increases the emission of VOCs, as well as increases breathing rate and occupational exposure risk *via* inhalation as the main exposure route. And these parameters are influenced by outdoor conditions, premises' occupants, activities, etc. [58, 59].

5.2 Dust measurements

To determine the total PM concentration or inhalable dust in the breeding area of workers, personal samplers (plastic cassettes with filters inside of them) are used. During the measurement process, samplers are connected to air pumps with an airflow of 1.0 to 2.5 L/min. For the respirable dust level measurements, use the same type of air pump, but with a cyclone and membrane filter. Air flow for respirable dust measurements is from 2.0 to 3.0 L/min. Analysis of respirable or total PM concentration is done gravimetrically [42, 43, 60–62]. Individual work processes to calculate an 8-hour (shift) exposure level. The duration of measurement/sampling should reflect the real situation at the workplace, which includes all work processes that can cause the emission of chemical substances into the air of the working environment. But "not overloading" the samples with dust, especially if it is necessary to further use these samples for the analysis of the chemical and physical characterization of aerosols. Work intensity and visual assessment of the work process allow one to choose the duration of sampling in non-emitting processes. The duration of the measurements depends on the homogeneity (uniform dispersion) of the pollution in the working environment. The inhomogeneous occupational environment precedes the longer measurements providing procedure (preferably the entire shift) and includes all processes.

Particulates in 3D industry are measured using direct reading instruments to determine particle size, count, and distribution, for example, Dekati® ELPI®+. And Dekati® ELPI® + measures during 3D printing show occupational exposure from $4x10^3$ to $26x10^3$ particles/cm^3 during 8-h shift. Furthermore, in all tested premises, the median diameter for particle number were detected: 0.014, 0.015, and 0.019 μm, but for mass concentrations: 4.394, 4.433, and 4.677 μm [46].

A condensation particle counter is used to measure the total number of particles with diameter from 10 to 1000 nm depending on the CPC model. An optical particle counter is used to measure the distribution of larger-size particles, usually from 1 to 10 μm, depending on the OCP model, as well as scanning mobility particle sizer is used to measure the distribution of particles. Using these techniques, particle surface area and volume concentration can be calculated from the measured number distribution assuming that particles are spherical; however, research showed trough 3D printer particle imaging it is not the case, as particles can be in different shapes [63, 64], this could lead to uncertainty in analyses [14, 42, 43].

5.3 Chemical measurements

The VOCs in 3D printing processes are done using traditional techniques with slight modifications. Measurements are done using adsorbent tubes Tenax TA type to capture VOCs or SVOCs; tubes are connected with sampling devices with airflow from 50 to 200 ml/min. After sampling, tubes are analyzed using gas chromatography (GC) with flame ionizing detector (FID)/mass spectrometry (MS) detectors [44, 65, 66]. Thus, in 3D printing offices, total VOC concentration was found at high concentration in comparison with indoor air quality recommendations, in general. In addition, there were estimated specific substances, such as toluene and formaldehyde, with concentrations 0.56 ± 0.1 and 0.23 ± 0.034 mg/m^3, respectively [46].

Passive sampling method to determine VOCs also can be used for monitoring purposes and in restricted areas, for example, drug development facilities, where 3D printing is used or in public buildings, for example, schools [67, 68].

Studies involving occupational exposure measurements in 3D industrial settings are limited. As emission and exposure in work environments depend on differences in printing methods and related activities, including print materials (also nanomaterials), further exposure measurements and the development of an appropriate measurement strategy for specific printing methods are needed.

6. Potential health effects and biomarker detection

6.1 Health surveys

In the scientific literature, questionnaires for the assessment of health disorders of employees working in the 3D industry have not been widely used. In 2016, Chan et al. conducted a survey of 17 companies in Toronto, Canada, surveying 46 employees, who work daily in 3D printing companies [38]. The questionnaire included demographic information, work/ exposure, and symptom/ health history. The work/exposure section included questions about working hours per week, exact job descriptions and duration ratio, materials, and personal protective equipment used when working in 3D printing. The symptom/health section asked about previous diagnoses and whether they were aggravated by working at the 3D printing company, as well as their smoking status. If the employees noted that they tend to feel unhealthy during work, they were also asked about various symptoms and their frequency. Sixty-five percent (65%) of the surveyed employees experienced some symptoms more than once a week in the past year. In 59% of cases, employees noted various respiratory symptoms, 17% headaches, and 20% skin problems. Thirty percent (30%) of the respondents had already had some respiratory or skin diseases in the past prior to their work with 3D printers. Fifty-two percent (52%) of the surveyed employees admitted that they do not use any personal protective equipment during their daily duties [38]. Guemperlein et al. conducted an experimental study with 26 healthy volunteers. Each participant spent 1 hour in a special exposure chamber near the working 3D printer, approximately 40 cm away from the face. Before exposure, samples of urine, exhaled CO, nasal secretion, FeNO, and spirometry were taken. Immediately after exposure, volunteers completed a survey, repeated FeNO, urine sample, and spirometry. After 2–3 hours, another urine sample, FeNO and nasal secretion were taken. Participants were assessed for chemical sensitivity before entering the study by the chemical and general environmental sensitivity (CGES) questionnaire. After the exposure, the participants filled out a questionnaire noting their subjective feelings during the exposure. The survey had two sections: general feelings and a specific offer of symptoms, to determine the specifics of the experience. Proposed symptoms included various respiratory system disorders, skin and eye irritation, and headache. A visual analog scale (VAS) of 1–10 cm was offered for each of the variants. Most often, participants noted irritation in the nose and throat, eye fatigue, and headaches [40]. Two more studies are available where individual complaints after working with 3D printers are studied. House et al. have described a relapse of asthma after a 20-year hiatus in a 28-year-old businessman who had previously experienced a last asthmatic episode at age 8 [36]. Creytens et al. described an episode of allergic contact dermatitis of two 3D printing company workers [69]. The complaints of these employees are collected individually.

6.2 Noninvasive methods for health effect assessment

Noninvasive healthcare technologies are an important part of research and development today because they are cost-effective and offer benefits to both healthcare recipients and providers.

Various strategies are available to study the association between air pollutant inhalation and human health effects, the most promising of them is using noninvasive painless tests relatively simple in sample collection. Investigation of biomarkers reflecting pulmonary inflammation and local and systemic oxidative stress in exhaled breath condensate (EBC) and urine are among them.

6.2.1 Exhaled breath condensate (EBC)

The collection of EBC is a completely noninvasive method for obtaining samples from the lungs. EBC is a matrix containing numerous biomolecules, including nitric oxide-derived products, hydrogen ions, hydrogen peroxide, prostaglandins, and leukotrienes. The sample collection is easy to perform, does not require special training of the participant, and has no influence on airway function or inflammation. EBC is obtained when the breath is exhaled from the lungs into a cooled collecting device; thus, condensing the vapor. In terms of biomarkers, determination in EBC focus must be put on the temperature in the cooling device and freezing of the samples due to the presence of chemically unstable substances (e.g., leukotrienes).

Information relating to some detectable biomarkers in EBC is provided below in **Table 2**.

6.2.1.1 Acidity and nitrogen/oxygen species

Assessment of EBC pH for detection of airways acidity purposes is widely described in research studies and it is related with airways inflammation [24, 84, 85]. Low EBC pH is found in patients with a variety of inflammatory lung disorders, including cystic fibrosis, chronic obstructive pulmonary disease (COPD), and asthma [84]. Acid stress can also be used to understand the effect of environmental

Biomarker (s)	Detection method/ Measurement device	Benefits/ Drawbacks	Subjects, range	Year, Ref.
Acidity and nitrogen/oxygen species				
Acidity (pH)	pH was recorded with a calibrated and validated glass microelectrode on stabilization/glass microelectrode.	Simple, quick, easy to perform, and highly reproducible.	404 healthy subjects, all ages, mean value 7.83, median value 8.0 (7.7–8.1), no differences based on age, sex, or race.	2007, [24]
NO_2/NO_3	Commercially available colorimetric assay kit, LOD 1 mcM/spectrophotometer, microplate reader.	Low values in healthy subjects; samples need concentration.	15 healthy controls; age 39 ± 10 (M6/F9); NO_2 < 1 mcM NO_2/NO_3 11.4 mcM	2005, [70]
NO_2/NO_3	Nitrate is measured as nitrite after enzymatic conversion by nitrate reductase using Griess reagent; LOD till 1.0 mkM/ spectrophotometer.	Complicated, long preparation, but informative.	10 normal, nonatopic subjects; age 23 ± 4 (males); NO_2/NO_3 0.63 (0.20-0.41) mcM;	2001, [71]
H_2O_2	Enzymatic assay; LOD 0.1 mkM/ double beam spectrophotometer.	Low values in healthy subjects.	10 normal, nonatopic subjects; age 23 ± 4 (males); H_2O_2 0.3 (0.2-0.4) mkM	2001, [71]
H_2O_2	The fluorescence of the reaction product (dimer 2,2'- dihydroxybiphenyl-5,5'-diacetate) measured with an automated sampler, low injection, and scanning fluorescence detector (295/405 nm); LOD 0.02 mcmol/L; fluorimeter with automated sampler.	Reproducible; H_2O_2 concentration in most healthy persons is below the LOD.	20 stable chronic obstructive pulmonary diseases (COPD) patients; no healthy subjects.	2002, [72]
NO_2/NO_3	Colorimetric assay based on the Griess reaction; LOD 2.5 mcM/ microplate reader at 540 nm.	Relatively simple; good variability.	20 healthy nonsmokers, age median 60 (52–73); NO_2/NO_3 15.06 (10.73–23.30) mcM.	2010, [73]

Biomarker (s)	Detection method/ Measurement device	Benefits/ Drawbacks	Subjects, range	Year, Ref.
3-Nitrotyrosine	A specific enzyme immunoassay (EIA) (Cayman Chemical, Ann Arbor, MI, USA); LOD 3.9 ng/mL/ freeze dryer for concentration.	Low values in healthy subjects; sample concentration is necessary.	14 nonsmoking healthy volunteers (6 male), age 34 ± 2 yrs.; 6.3 ± 0.8 ng/mL.	2001, [74, 75]
Nitrosothiols (RS-NOs)	Commercially available colorimetric assay kit (Oxonon, Emeryville, CA) based on classic reaction of Saville and Griess; LOD kit 0.025 mcM/ spectrophotometer, microplate reader, at 540 nm.	Relatively easy to perform. Low values in healthy subjects.	Nonsmoking control subjects (n = 10), age 30.2 ± 1.5 yrs.; smoking control subjects (n = 7), age 36.1 ± 3.2; 0.11 ± 0.02 mcM;	2001, [75]
Eicosanoids (prostaglandins and leukotrienes)				
Cysteinyl-leukotrienes (cys-LTs: $LTC_4/D_4/E_4$); leukotriene B_4	Specific enzyme immunoassay (EIA) kits; LOD: cys-LTs 15 pg./mL; LTB_4 4.4 pg./mL.	Relatively easy to perform. Expensive, high variability.	15 normal controls; age 32.3 ± 4.5 (7 male); cys-LTs 15.5 ± 0.2 pg./mL; LTB_4 63.1 ± 17.3 pg./mL	2000, [76]
cysteinyl-leukotrienes/ prostaglandin E_2/ leukotriene B_4	Specific enzyme immunoassay (EIA) kits (Cayman Chemical, Ann Arbor, MI); LOD: cys-LTs 13 pg./mL; PGE_2 15 pg./mL; LTB_4 4.4 pg./mL	Relatively easy to perform. Expensive	16 healthy subjects; age 45 ± 17 yrs.; cys-LTs 19.4 ± 2.8 pg./mL; PGE_2 and LTB_4—no information	2002, [77]
8-iso-prostaglandin $F_{2\alpha}$	Competitive enzyme immunoassay kit; sensitivity 4 pg./mL/Specific reader	Variable (CV 11–15%)	Healthy nonsmokers, n = 20; age median 60 (52–73); 9.09 (6.63–11.43) pg./mL	2010, [73]
8-iso-prostaglandin $F_{2\alpha}$	Solid-phase extraction using liquid chromatography-electrospray ionization-tandem spectrometry (LC-ESI-MS/MS); LOD 1.2 ± 0.2 pg./mL	Precise. Long; expensive; requires expensive instruments.	21 healthy controls (15 male); age 38.7 ± 9.1 yrs.; values not presented, only comparison	2020, [78]
Inflammatory mediators				
Cytokines interleukin (IL)-1β, IL-4, IL-6, IL-8, IL-10, tumor necrosis factor-alpha (TNF-α), C reactive protein (CRP)	ELISA or cytometric bead array (CBA) assays, according to the manufacturer's guidelines; need normalization to protein concentration/ specific device (e.g., Randox evidence).	Cytokine concentration is around the assay LOD; high variability; expensive; depends on the assay kit manufacturer.	CRP 0.075 ± 0.03 mg/L; IL-1β 3.31/3.74 pg./mL; IL-4 31.6–40.8 pg./mL; IL-6 2.6–5.2 pg./mL; IL-8 3.15–16.3 pg./mL; IL-10 1.0–24.3 pg./mL; TNF-α 0.4–4.84 pg./mL.	2022, review article [79]
Lipid peroxidation end products and damage to nucleic acids				
Malondialdehyde condensed with thiobarbituric acid (MDA-TBA)	HPLC with fluorescent detection (ex 532 nm/em 553 nm); LOD 4.1 nM	Relatively easy to perform; linear, selective. Samples stability: at −20°C 1 month; at −80°C 3 month;	Only patients; range 0.15– 0.23 pmol/s	2002, [80]

Biomarker (s)	Detection method/ Measurement device	Benefits/ Drawbacks	Subjects, range	Year, Ref.
Malondialdehyde condensed with thiobarbituric acid (MDA-TBA)	HPLC with fluorescent detection (ex 532 nm/em 553 nm); LOD 1.8 nM	Relatively easy to perform, linear, selective, precision 2.2%.	125 healthy adults, age median 24 (19–33); male/female = 64/64; 16.0–22.3 nM	2013, [81]
8-iso-prostaglandin $F_{2\alpha}$, malondialdehyde (MDA), 4-HNE - complex detection	Liquid chromatography-electrospray ionization-mass spectrometry/mass spectrometry (LC-ESI-MS/MS); LOD: 8-iso $PGF_{2\alpha}$ - 2 pg./mL; MDA – 21 pg./mL 4-HNE – 26 pg./mL	Pretreatment part: solid-phase extraction for biomarkers isolation from matrix; samples are spiked with biomarkers. Requires expensive instruments.	10 control subjects, age average 63 ± 5 yrs.; male/ female similar; Median: 8-iso $PGF_{2\alpha}$ - 47 pg./mL; MDA – 43 ng/mL 4-HNE – 162 ng/mL	2009, [82]
8-hydroxy-2-deoxyguanosi-ne (8-OHdG)	Liquid chromatography-electrospray ionization-tandem spectrometry (LC-ESI-MS/MS); LOD 7 pg./ mL.	High sensitivity and specificity, reproducible. Requires expensive instruments	Healthy individuals, 9.0–21.0 pg./mL	2022, review article [83]
	ELISA immunoassay; LOD 41 pg./mL / spectrophotometer or fluorimeter, microplate reader.	Simple, low cost; variable, less specific; not validated for EBC.	Healthy individuals, 360 ± 90 pg./mL	
4-hydroxynonenal (4-HNE)	4-Hydroxynonenal ELISA immunoassay kit; Sensitivity: 18.75 pg./ml Range: 31.25–2000 pg./ mL. Spectrophotometer/ microplate reader.	Variable; not standardized for detection in EBC.	—	2022, [83]

Table 2.
Biomarkers in exhaled breath condensate (EBC). LOD—Limit of detection.

conditions on the airways. However, the presence of CO_2 in the samples interferes with the results and must be removed using deaeration or degasification as recommended by the European Respiratory Society technical standard [85] to avoid high variability of the results. pH values less than 7.4 are considered to be abnormal [24].

Production of nitric oxide (NO) is generally increased in inflammatory conditions, but it is difficult to measure because it is a free radical, which reacts rapidly with oxygen, superoxide, water, thiols, amines, and lipids to form products with biochemical activities, but nitrites and nitrates are products of nitric oxide metabolism, and these products can be detected in EBC. Investigators use different methods to assess nitrite/ nitrate levels in EBC, examples are presented in **Table 2**.

Nitrotyrosine is formed when nitric oxide and superoxide anions create peroxynitrite, which can then react with tyrosine residues on proteins. Nitrotyrosine, which is formed in the airways, can be collected in the EBC and serves as a marker of oxidative stress. Increased production of nitrotyrosine may reflect the increased formation of reactive nitrogen species, such as peroxynitrite in airways and also in the whole organism. Some investigators consider that nitrotyrosine formation in EBC may be a more sensitive marker to evaluate the contribution of oxidative stress to airway inflammation than exhaled NO [76, 86].

Nitrosothiols (RS-NOs) are formed by the interaction of nitric oxide with thiol-containing macromolecules (glutathione and cysteine) and may limit the detrimental effect of NO. RS-NOs are detectable in EBC of healthy subjects and of individuals with various inflammatory airway diseases. Increased levels of RS-NOs may be related to enhanced nitrosative stress. A positive correlation between smoking history (packs/year) and RS-NOs levels suggest that nitrosative stress induced by smoke stimulates RS-NOs production, and also indicates that RS-NOs can be used as a biomarker of ambient air pollution and pulmonary nitrosative stress [76].

Exposure to exogenous oxidants, such as inhaled cigarette smoke and/or air pollution, increase oxidative stress in exposed individuals. The degree of oxidative stress can be determined by measuring hydrogen peroxide (H_2O_2) concentration in EBC [71, 72]. Also, H_2O_2 is released by neutrophils and eosinophils and by macrophages and epithelial airway cells; it provides one line of defense against infection and is therefore an important marker of airway infection. Thus, measurements of H_2O_2 in EBC can be used as a maker of oxidative stress and/or inflammatory processes in occupational medicine (**Table 2**).

6.2.1.2 Eicosanoids (prostaglandins and leukotrienes)

Prostaglandins and thromboxanes are synthesized by the cyclooxygenase (COX) pathway in arachidonic acid cascade. Studies of the eicosanoid pathway and COX inhibition suggest that this biomarker may be used to understand processes of airway inflammation. The 8-iso-prostaglandin F2α (F2-isoprostane or 8-iso-PGF2α) is a lipid peroxidation product of arachidonic acid and the representative marker of oxidative damage. However, there is an alternate enzymatic pathway to generate 8-iso-PGF2α catalyzed by prostaglandin endoperoxide synthase, which is independent of free radical-mediated peroxidation. This fact can impact the conclusions based on the 8-iso-PGF2α increase concentration under exposed conditions reflect in literature analysis by Hemmendinger et al. [83]. So, measurement of only esterified 8-iso-PGF2α or total 8-iso-PGF2α, may be a more indicative marker of oxidative stress [83, 87]. Van 't Erve and coworkers provide a detailed literature meta-analysis of the levels of 8-iso-PGF2α in systemic oxidative damage across human disease and in response to environmental exposures and concluded that there is a general increase in the levels of both free and total 8-iso-PGF2α associated with mentioned conditions [88].

Cysteinyl-leukotrienes (cys-LTs) and prostaglandin E2 (PGE2) are eicosanoids implicated in the development of airway disease mechanisms [76, 77]. PGE2 relaxes airway smooth muscles and exerts potent anti-inflammatory activity. Activated eosinophils and mast cells are capable of producing several inflammatory mediators, including cysteinyl-leukotrienes (cys-LTs) (LTC4, LTD4, LTE4), which, in turn, are potent bronchoconstrictors.

Cytokines measurements in EBC are widely used as inflammatory biomarkers. The main cytokines with a pro-inflammatory role are interleukin (IL)-1β, IL-6, tumor necrosis factor α (TNF-α), and C reactive protein (CRP). Anti-inflammatory cytokines, instead, such as IL-4 and IL-10, play a crucial role in controlling the regulation of pro-inflammatory cytokines. Positive associations of these biomarkers with workplace aerosols, including nanocomposite materials, were reported [89].

Recently, a thorough literature systematic review was done by Ghelli and coauthors comparing baseline values of pro/anti-inflammatory cytokines measured in healthy, nonsmoking adults in order to elucidate issues that interfere with the obtained results evaluation: standardization of sampling and test preparation are the most actual [79].

6.2.1.3 Lipid peroxidation end products

Lipid peroxidation is a well-defined mechanism of cellular damage, lipid peroxides are formed under oxidative stress conditions and form more complex and reactive compounds, such as malondialdehyde (MDA) and 4-hydroxynonenal (4-HNE), which have been shown to be capable of binding to proteins and forming stable adducts—advanced lipid peroxidation end products. These products are generally accepted biomarkers of oxidative stress. Since oxidative stress has been recognized as an important mechanism by which air pollution exposure leads to adverse health effects, the positive association of lipid peroxidation product levels in EBC and air pollutants concentration was observed. Increased MDA-TBA was associated with fine particles of PM2.5, elementary carbon, and sulfur dioxide concentration [81]. The author's finding suggests that gaseous pollutants may lead to lipid peroxidation in the respiratory tract more rapidly than particulate pollutants. MDA is a marker of oxidative stress also in different pulmonary diseases [90]. The following analytical methods are used for MDA detection in EBC: high-performance liquid chromatography/thiobarbituric acid (HPLC/TBA), spectrophotometry/TBA, HPLC-mass spectroscopy (HPLC-MS), HPLC-MS/initrophenylhydrazine (DNPH) [91], but HPLC/TBA is considered the reference method (**Table 2**). Due to methodological discrepancies in sampling and detection technics, a reference interval in healthy adult populations for MDA in EBC is not established.

4-hydroxynonenal (4-HNE) measured in EBC was not significantly affected by systemic disorders, but was increased in patients with occupational respiratory disorders due to asbestos and silica expose; therefore, can serve as a good biomarker for oxidative stress [82, 92].

8-hydroxy-2-deoxyguanosine (8-OHdG) is commonly used for identifications of oxidative damage to nucleic acids. Two main analytical approaches are available for 8-OHdG: immunoassay-based methods (ELISA) and chemical analytical methods, such as gas or liquid chromatography with MS detection. 8-OHdG was found to increase the EBC of engineered nanomaterials exposed workers, smokers, COPD patients, and patients with occupational respiratory disorders due to asbestos and silica expose [82, 83, 93].

All above mentioned allow us to consider that in order to perform qualitative reproducible investigation/study of biomarkers in exhaled breath condensate it is necessary to work out elaborated sampling collection protocol according to the European Respiratory Society/American Thoracic Society recommendations, including participant training and total volume of air exhaled, select samples storage conditions, to choose more informative biomarkers with appropriate detectable methods and use standardized analytical protocols for sample preparing and analysis.

6.2.2 Biomarkers in urine

Biomarkers measurement in urine has many advantages: urine collection is noninvasive, can be obtained in large quantities, contains relatively low concentrations of metals and organic content, and can be stored for a long time. On the other hand, the content of biosubstances in urine can be influenced by diet, there are some cells or cell debris, and bacterial growth can occur during urine storage.

The most commonly measured urine biomarkers in air pollution or occupational health studies are biomarkers of oxidative stress products: F2-isoprostanes, aldehydes in forms of MDA, 4-HNE adducts, 4-oxo-2-nonenal (4-ONE), DNA oxidation modification product 8-OHdG (**Table 3**).

Biomarker (s)	Detection method/ Measurement device	Benefits/ Drawbacks	Subjects, range	Year, Ref.
8-iso-prostaglandin $F_{2\alpha}$ (8-iso-PGF2α)	Commercial ELISA kit; Inter-assay precision: 0.53% to 6.08% Intra-assay precision 8%. LOD 16.3 pg./ mL. Specific reader	Variable; cross-reactivity occurs	None-exposed group, Chinese N = 210; range 1.1–2.9 mcmol/ mol creatinine	2021, [94]
	Gas chromatography-tandem mass spectrometry (GC-tandem MS); LOD 1.2 ± 0.2 pg./mL. Requires expensive instruments	Precise. Long; expensive; spiked samples	93 healthy controls (M/F 160/26); age 63.5 ± 7.9 yrs.; 80–130 pmol/ mmol creatinine	2004, [95]
	Liquid chromatography-tandem mass spectrometry (LC-tandem MS). Requires expensive instruments.	Precise. Long; expensive; sample pretreatment; spiked samples	12 healthy men; 69.25 ± 3.81 yrs.; 103.38 ± 32.18 ng/ mg creatinine.	2022, [96]
Malondialdehyde condensed with thiobarbituric acid (MDA-TBA adducts)	MDA-TBA adducts commercial kit Inter-assay precision: 0% to 2.02% Intra-assay precision: 0.12% LOD 0.5 nmol/mL. Spectrophotometer or fluorimeter, microplate reader	Relatively easy to perform; linear, selective	None-exposed group, Chinese N = 210; range 345.6–794 mcmol/mol creatinine	2021, [94]
	High-performance liquid chromatography HPLC with mass spectrometry (MS), fluorescence detection, or UV photometry;	Depend on the selected method; high heterogeneity	Healthy populations, 0.07–0.12 mg/g creatinine	2022, review [97]
8-iso-prostaglandin $F_{2\alpha}$, malondialdehyde (MDA), and 4-hydroxynonenal (4-HNE) - complex detection	Liquid chromatography-electrospray ionization-mass spectrometry/mass spectrometry (LC-ESI-MS/ MS); LOD: 8-iso $PGF_{2\alpha}$ - 17 pg./mL; MDA – 87 pg./mL 4-HNE – 91 pg./mL	Samples pretreated and spiked with biomarkers. Long. Requires expensive specific instruments	10 control subjects, age average 63 ± 5 yrs.; male/ female similar; Median: 8-iso $PGF_{2\alpha}$ – 24 ng/ mmol creatinine; MDA – 6.8 mcg/ mmol creatinine 4-HNE – 182 mcg/mmol creatinine	2009, [82]

Biomarker (s)	Detection method/ Measurement device	Benefits/ Drawbacks	Subjects, range	Year, Ref.
8-hydroxy-2-deoxyguanosine (8-OHdG)	Ultra-high-performance liquid chromatography-mass spectrometry (UPLC-MS/MS); LOD 0.5 ng/mL.	High sensitivity and specificity, reproducible. Requires expensive instruments	Healthy individuals, Chinese144; age < 45 yrs.; about 0.4 mcmol/mol creatinine.	2019, [98]
	Liquid chromatography-tandem mass spectrometry (LC-tandem MS). Requires expensive instruments.	Precise. Long; expensive; sample pretreatment; spiked samples	12 healthy men; 69.25 ± 3.81 yrs.; 1.87 ± 0.77 ng/mg creatinine	2022, [96]
	ELISA commercial kit; Sensitivity 0.59 ng/mL; LOD 41 pg./mL / spectrophotometer or fluorimeter, microplate reader.	Simple. Variable, less specific.	Healthy individuals, Body mass index >25; 5.9 to 19.8 ng/mg creatinine.	2020, review [99]

Table 3.
Biomarkers in urine. LOD—Limit of detection, F—Females, and M—Males.

F2-isoprostanes are the products of non-enzymatic oxidation of arachidonic acid by different free radicals, also the additional enzymatic pathway from esterified phospholipids not related to oxidative stress occurs [100]. There are 4 F2-isoprostane isomers presented and detectable in urine [101].

Three main techniques used to assay F2-isoprotanes in urine are gas-chromatography with mass spectrometry detection (GC-MS), liquid chromatography with tandem mass spectrometry detection (LC-MS/MS), and enzyme-linked immunosorbent assay (ELISA). Mostly, the measurements are assigned to 8-iso-prostaglandin F2α.

MDA is a frequently used biomarker that is measured in urine using different methods. Detailed analysis of urinary MDA levels for healthy adult populations is presented by Toto et al. [97]. MDA levels are evaluated after air pollution exposure [81]. However, some investigators pay attention to non-specificity of MDA formation in oxidative conditions and possible dietary, body mass index (BMI), and age as confounding factors, concluding that MDA cannot be recommended as a systemic biomarker of oxidative stress [101].

4-HNE and 4-oxo-2(E)-nonenal (4-ONE) are products of polyunsaturated fatty acids oxidation, highly reactive aldehydes, easily form covalent bonds with protein thiol and amino groups and with other biological molecules, so their levels are unstable [101], but measuring of their stable adducts is used as oxidative stress biomarkers. 1,4-Dihydroxynonane mercapturic acid (DHN-MA), the major urinary metabolite of 4-HNE, is an additional biomarker that may be assayed [102], and commercially available kits may be used. Another possibility is to use high-tech techniques, such as liquid chromatography/gas chromatography, in tandem with mass-spectrometry, depending on the study purpose [103, 104]. Measurements of urinary 8-OHdG include chromatography-based methods (HPLC MS/MS, HPLC ECD, GC/MS, etc.) and ELISA, with chromatography-based techniques showing low inter-assay variability, but cross-reactivity and inter-individual variability for all [101, 105, 106]. Diet and cell death have minimal, if any, influence on urinary levels of 8-oxodG, but results may be age-dependent [107]. The pooled median value for urinary 8-OHdG concentrations in healthy adults with a mean BMI ≤ 25 measured using chemical methods was 3.9 ng/mg creatinine (range from 3 to 5.5 ng/mg) [99]. Increased urinary 8-OHdG was associated with air and work-place pollutants; however, data are contradictory [108].

In conclusion, it is necessary to pay attention to the standardization of urine sample collection (spot or 24-h urine), storage conditions (time and temperature), pretreatment of samples (urine is a very complex matrix), appropriate choice of biomarkers, results in creatinine normalization—the rigorous study design/protocol is essential to obtain repeatable reliable results.

7. Estimated daily intake (EDI) and health risk modeling

Pollutants pose a potential hazard to public healthcare, particularly human exposure. It is important to understand to what extent the population is exposed to environmental pollutants at work and home. It is important to track the level of environmental pollutants in the population, which allows an assessment of how exposure to pollutants in this population changes over time [109, 110].

Human contact occurs *via* inhalation, ingestion, or dermal exposure [111].

Different approaches are used to measure possible human exposure, for example, exposure modeling, personal monitoring, and ambient concentration measurements. Exposure models estimate exposure by combining information about environmental contaminant concentrations, including surveys about people's activities and locations (working shifts, exercising indoors/outdoors, sleeping, food consumption, etc.) to account for possible contact with contaminants [68, 112, 113]. As a possibility, exposure indices can be developed to evaluate relative changes in environmental contaminant exposure over time [113].

Biomonitoring measures how much a pollutant or its metabolites or reaction product—biomarkers—are present in the human body. Several environmental contaminants, for example, heavy metals, pesticides, and other organic pollutants, can accumulate in the body. Measurements are commonly made in blood and tissues obtained by biopsy or autopsy (invasive method) or using urine, feces, breast milk, hair, nails, skin contaminants, and exhaled air condensate (noninvasive methods) [114, 115].

Occupational hygiene practices have generally focused on the inhalation of exposure pathway. It was assumed that inhalation was the main route of exposure, with some exceptions (pesticides, certain solvents, etc.). Many methods have been developed to measure inhalation exposure and there is a clear idea of how to interpret these values in the context of a risk reduction strategy. Attempts have been made to reduce inhalation exposure over the years, and some authors have suggested that dermal exposure may be more important compared [116, 117].

7.1 Estimated daily intake (EDI)

Evaluation of the level of health and safety risk during 3D printing based on the data of the EDI calculation data [118].

Exposure by inhalation is the inhalation of a substance as a gas or vapor or as particles in the air. This includes small amounts of soil and dust that can be inhaled into the lungs. The lungs often absorb gases and vapors quickly and efficiently. We are all exposed to low levels of contamination in the air we breathe, food we eat, and water we drink. The total exposure across all suspected exposure pathways for a chemical pollutant for an average person describes the EDI [111, 119–121]. The calculations of the EDI method are directly adjusted to the occupational exposure data for recommendations [38, 122–129]. The EDI value of chemicals can be calculated by adding up all. In summary, each EDI is the amount of pollutant absorbed through a different combination of exposure routes and exposure pathways (**Figure 1**).

All equations for calculating the estimated amount of the pollutant taken up *via* the combinations of exposure pathway and exposure route are similar, but require a different equation for each combination [130, 131].

Health risks from occupational dermal exposure to hazardous substances may also occur at many workplaces. In addition to the local effects that chemicals can directly have on the skin, the skin also acts as a pathway for hazardous chemicals to be absorbed into the body. The dermal dose absorbed is the amount of a chemical absorbed into the body through the skin [132]. In recent decades, several methods have been developed to assess dermal exposure. These methods can be broadly divided into direct (interception and sampling methods and visualization techniques) and indirect methods (surface sampling methods (nonhuman), dermal exposure modeling, and biomonitoring) [133, 134].

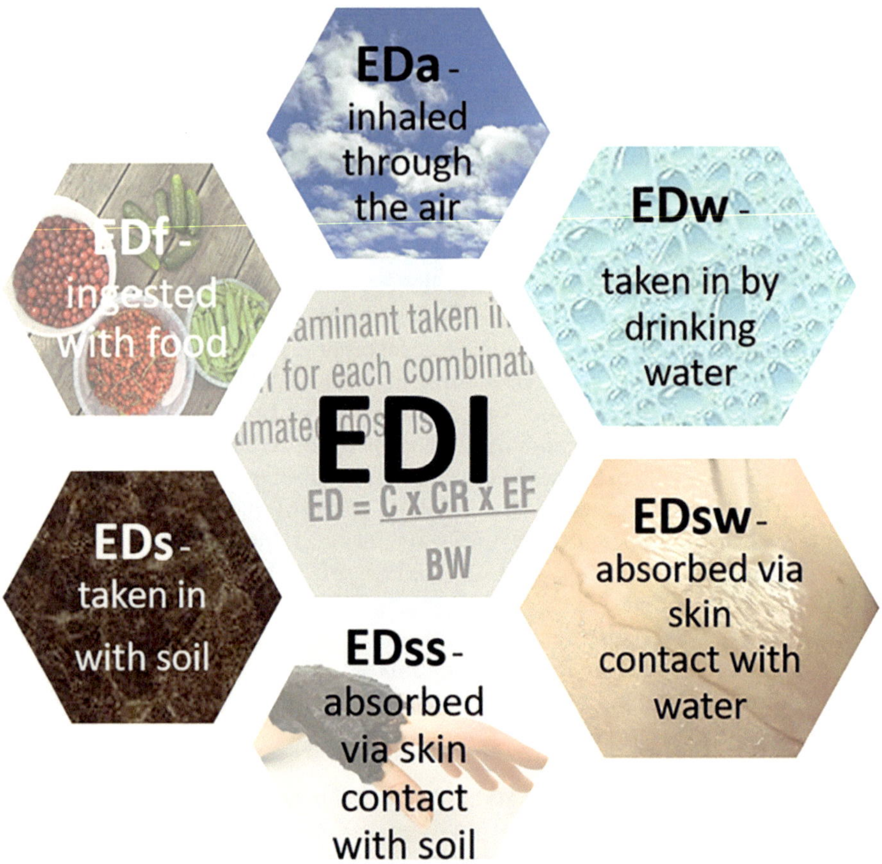

Figure 1.
Combination of exposure routes forming EDI.

The dose-response relationship is the relationship between the amount of a contaminant that is given and the health effect. It is possible to measure short-term exposure or exposure as a function of time. Exposure means contact of chemicals with the outer perimeter of an individual, for example, skin, nose, mouth, and doses. There are four ways to analyze dermal exposure that has occurred: the first is the potential dose, which is the amount of contaminant applied to the skin. The second, the applied dose indicates the amount of the pollutant at the absorption barrier (e.g., the skin) that can be absorbed by the body. The third and fourth doses are the internal dose (the amount of the pollutant that is absorbed and available to interact with biological receptors, e.g., organs and tissues) and the biologically effective dose (the amount of the pollutant that interacts with the internal target tissue or organ), respectively [135]. Depending on the exposure assessment, it may be necessary to assess the effects and dose in different ways [136]. Calculations of dermal exposure can be made as lifetime average daily potential dose (LADD), average daily potential dose (ADD), and acute potential dose rate (ADR). The potential skin dose rate can be determined by dividing the potential dose rate (PDR) by the body weight to obtain the normalized potential dose, with the body weight chosen to fit the specific population of individuals [132, 134, 137, 138].

Another important meaning is applied by governments to set guidelines that protect human health from the potential health effects of exposure to environmental pollutants: the tolerable daily intake (TDI). Acceptable daily intake (ADI) is also used interchangeably with TDI. However, the two terms TDI and ADI describe the same value. Guidelines for TDI are regularly reviewed and revised, especially when new scientific information on pollutants becomes available.

The TDI calculations depend on how comprehensive the studies (usually on animals) were and how confident the researchers were in extrapolating the data to humans.

For contaminants known to be carcinogenic, the risk-specific dose (RSD) is required. The RSD is the amount of contaminant to which people can be exposed daily throughout their lifetime without exceeding the accepted risk of cancer.

8. Future prospective for 3D printing risk assessment

The technology of 3D printing will alter how we create, produce, and distribute our consumer items in this digital age. In recent years, expectations for 3D printing have been so high that nothing less than a new industrial revolution has emerged. [139].

Food industry started using 3D printing opportunities because of food's flexible production and freedom in creation. For example, it is used for liquid food (chocolate, pancake batter, etc.) production [140]. The literature highlighted that the latest directions in the field of the 3D food industry are to introduce new bioactive ingredients in addition to basic nutrients. In nearest future, raw food also will be produced by 3D technologies next to heating, fermentation, or germination [141]. Therefore, it will arise new challenges regarding hygiene, safety, and working conditions (air quality, ergonomics, etc.), in general. This will present new difficulties for hygiene, workplace safety, and general conditions (clean air, ergonomics, etc.). The controlling of smart materials processing by temperature, power, light, moisture, pH, and electric or magnetic fields can change smart materials' properties, for example, changing shape, tactility, or hardness [142]. Thus, this transformation process leads to 4D printing because of material availability changes again over time. And these changes will be initiated by sensitivity to light, pressure, or temperature. Furthermore, some of smart materials will have "a memory," when these materials can change back to their initial shape under the changing of circumstances again. Considering these types of materials' experimental stage at the moment, their impact on health and hygiene is uncertain [143]. There have been calls for regulation [144].

One of the potential challenges is to detect errors in 3D printing with the help of artificial intelligence. Brion et al. [145] proposed a multi-head neural network employing labeled pictures that allows for the detection of deviation from ideal printing settings and permits reliable and generalizable, real-time extrusion additive manufacturing mistake detection, and quick rectification.

Developing 3D printing of organic and/or living tissues is essential and very promising because of inserting cells into requested places. And this technology is based on bioprinters' head movements and cell insertion. Bioprinting will offer to create many very thin layers of tissues. Thus, algae and fungi already are successfully used in experiments to manufacture "living" materials, but same as regarding smart materials also "living" materials potentially could cause risks to health and hygiene. Furthermore, it raises ethical issues [146]. Also, nanotechnologies are combined with 3D printing to shape objects at nano or molecular size receiving opportunities through additive manufacturing to create any form, material, shape, or volume objects. However, this technology is still theoretical without a well-known nano-printing effect on to work environment, in general [4].

Exposure assessment future perspectives based on accurate personal sampling, specific pollutants' assessment, and sensitive biomarkers' evaluation with a purpose of highlighting 3D printing emissions impact on employees' health. There are important existing sensitive methods (e.g., exhaled breathing condensate collection and analysis) and evaluation of health risk according to calculations approaches (e.g., EDI) implementation for occupational exposure assessment in 3D printing industry.

9. Conclusions

The growing popularity and widespread use of 3D printers raise concerns about their health effects. It is extremely important to identify the chemical substances released by the 3D printing process to more accurately determine the extent of exposure and assess potential health risks. The placement of the workplace where 3D printing is performed is also important, particularly a sufficiently efficient ventilation system, the technical condition of the 3D equipment itself, the materials used, and the amount of time the worker spends on the equipment. To assess the potential health effects of 3D printer emissions, information from safety data sheets and exposure analyses, including exposure assessment and dose calculations (e.g., EDI), can be used (adapted for specific occupational exposures, including 3D printing). Future prospects include sensitive and noninvasive biomaterial analyses to assess health risks, such as exhaled breath and urine biomarkers, for example, oxidative stress, etc. More research is needed to select the most appropriate biomarkers for risk assessment. As 3D printing technologies evolve, their emissions and the materials used also change, so it is important to continue their research and emission determination. Sensitive and appropriate methods, including equipment, for sampling and analytical analysis will be promoted and developed in the future. Hobby-class 3D printers (also 3D pen printing), which are most commonly used at home, are a separate issue because in most cases there is not adequate ventilation and controlled environmental conditions at home, especially if the users are children.

Acknowledgements

The research for this article was partly carried out within a project of National Program Grants (RSU Internal Grants) "Occupational health and safety risks during 3D printing" 6-ZD-22/22/2022.

Author details

Ilona Pavlovska[1*], Lāsma Akūlova[1], Anna Lece[2], Žanna Martinsone[1], Linda Paegle[1], Aneka Kļaviņa[3], Klinta Luīze Sprūdža[1] and Inese Mārtiņsone[1]

1 Institute of Occupational Safety and Environmental Health, Laboratory of Hygiene and Occupational Diseases, Rīga Stradiņš University, Latvia

2 Institute of Occupational Safety and Environmental Health, Scientific Laboratory of Biochemistry, Rīga Stradiņš University, Latvia

3 Department of Occupational and Environmental Medicine, Rīga Stradiņš University, Latvia

*Address all correspondence to: ilona.pavlovska@rsu.lv

References

[1] The Built Environment Research Group. VOC & Particle Emissions from 3D Printers [Internet]. 2022. Available from: http://built-envi.com/portfolio/ultrafine-particle-emissions-from-3d-printers/ [Accessed: November 11, 2022]

[2] Wojtyła S, Klama P, Śpiewak K, Baran T. 3D printer as a potential source of indoor air pollution. International Journal of Environmental Science and Technology. 2020;**17**:207-218. DOI: 10.1007/s13762-019-02444-x

[3] The National Institute for Occupational Safety & Health (NIOSH). Hydrocarbons, Aromatic [Internet]. 2003. Available from: https://www.cdc.gov/niosh/docs/2003-154/pdfs/1501.pdf [Accessed: November 10, 2022]

[4] Shaw M. 3D printing technology: Its application and scope in fashion industry. Man-Made Textiles in India. 2016;**XLIV**(1):7-10

[5] The National Institute for Occupational Safety and Health (NIOSH). Nanotechnology [Internet]. 2020. Available from: https://www.cdc.gov/niosh/topics/nanotech/ [Accessed October 12, 2022]

[6] Berman B. 3D printing: The new industrial revolution. IEEE Engineering Management Review. 2004;**41**(4):72-80. DOI: 10.1109/EMR.2013.6693869

[7] Carll AP, Salatini R, Pirela S v., et al. Inhalation of printer-emitted particles impairs cardiac conduction, hemodynamics, and autonomic regulation and induces arrhythmia and electrical remodeling in rats. Particle and Fibre Toxicology. 2020;**17**:1-21. DOI: 10.1186/s12989-019-0335-z

[8] Byrley P, Boyes WK, Rogers K, Jarabek AM. 3D printer particle emissions: Translation to internal dose in adults and children. Journal of Aerosol Science. 2021;**154**:154. DOI: 10.1016/j.jaerosci.2021.105765

[9] United States Environmental Protection Agency. 3D Printing Research at EPA [Internet]. 2021. Available form: https://www.epa.gov/chemical-research/3d-printing-research-epa [Accessed: November 12, 2022]

[10] Leso V, Ercolano ML, Mazzotta I, Romano M, Cannavacciuolo F, Iavicoli I. Three-dimensional (3D) printing: Implications for risk assessment and Management in Occupational Settings. Annals of Work Exposures and Health. 2021;**65**(6):617-634. DOI: 10.1093/annweh/wxaa146

[11] Ustundag A, Cevikcan E. Conceptual Framework for Industry 4.0. Springer Series in Advanced Manufacturing. Additive Manufacturing. Industry 4.0: Managing The Digital Transformation. Book Chapter. 2018, pp. 3-23. DOI: 10.1007/978-3-319-57870-51

[12] Utela B, Storti D, Anderson R, Ganter M. A review of process development steps for new material systems in three dimensional printing (3DP). Journal of Manufacturing Processes. 2008;**10**(2):96-104. DOI: 10.1016/j.jmapro.2009.03.002

[13] Ngo TD, Kashani A, Imbalzano G, Nguyen KTQ, Hui D. Additive manufacturing (3D printing): A review of materials, methods, applications and challenges. Composites Part B: Engineering. 2018;**143**:172-196. DOI: 10.1016/j.compositesb.2018.02.012

[14] Azimi P, Zhao D, Pouzet C, Crain NE, Stephens B. Emissions of

ultrafine particles and volatile organic compounds from commercially available desktop three-dimensional printers with multiple filaments. Environmental Science&Technology. 2016;**50**:1260-1268. DOI: 10.1021/acs.est.5b04983

[15] Pinheiro ND, Freire RT, Conrado JAM, Batista AD, da Silveira Petruci JF. Paper-based optoelectronic nose for identification of indoor air pollution caused by 3D printing thermoplastic filaments. Analytica Chimica Acta. 2021;**1143**:1-8. DOI: 10.1016/j.aca.2020.11.012

[16] Singh AV, Maharjan RS, Jungnickel H, et al. Evaluating particle emissions and toxicity of 3D pen printed filaments with metal nanoparticles As additives: In vitro and in silico discriminant function analysis. ACS Sustainable chemistry and Engineering. 2021;**9**:11724-11737. DOI: 10.1021/acssuschemeng.1c02589

[17] Farcas MT, McKinney W, Qi C, et al. Pulmonary and systemic toxicity in rats following inhalation exposure of 3-D printer emissions from acrylonitrile butadiene styrene (ABS) filament. Inhalation Toxicology. 2020;**32**:403-418. DOI: 10.1080/08958378.2020.1834034

[18] Sigloch H, Bierkandt FS, Singh AV, Gadicherla AK, Laux P, Luch A. 3d printing - evaluating particle emissions of a 3d printing pen. Journal of Visualized Experiments. 2020:164. Article number: e61829. DOI: 10.3791/61829

[19] Graff P, Ståhlbom B, Nordenberg E, Graichen A, Johansson P, Karlsson H. Evaluating measuring techniques for occupational exposure during additive manufacturing of metals: A pilot study. Journal of Industrial Ecology. 2017;**21**(S1):S120-S129. DOI: 10.1111/jiec.12498

[20] Wan GH, Wu CL, Chen YF, Huang SH, Wang YL, Chen CW. Particle size concentration distribution and influences on exhaled breath particles in mechanically ventilated patients. PLoS One. 2014;**9**(1):1-9. Article number: e8708 DOI: 10.1371/journal.pone.0087088

[21] Das S, Pal M. Review—Non-invasive monitoring of human health by exhaled breath analysis: A comprehensive review. Journal of The Electrochemical Society. 2020;**167**(3):167. DOI: 10.1149/1945-7111/ab67a6

[22] Stefaniak AB, Lebouf RF, Yi J, et al. Characterization of chemical contaminants generated by a desktop fused deposition modeling 3-dimensional printer. Journal of Occupational and Environmental Hygiene. 2017;**14**:540-550. DOI: 10.1080/15459624.2017.1302589

[23] Potter PM, Al-Abed SR, Lay D, Lomnicki SM. VOC emissions and formation mechanisms from carbon nanotube composites during 3D printing. Environmental Science & Technology. 2019;**53**:4364-4370. DOI: 10.1021/acs.est.9b00765

[24] Davis MD, Hunt J. Exhaled breath condensate pH assays. Immunology and Allergy Clinics of North America. 2012;**32**(3):377-386. DOI: 10.1016/j.iac.2012.06.003

[25] Kim Y, Yoon C, Ham S, Park J, Kim S, Kwon O, et al. Emissions of nanoparticles and gaseous material from 3D printer operation. Environmental Science & Technology. 2015;**49**:12044-12053. DOI: 10.1021/acs.est.5b02805

[26] Floyd EL, Wang J, Regens JL. Fume emissions from a low-cost 3-D printer with various filaments. Journal of Occupational and Environmental Hygiene. 2017;**14**:523-533. DOI: 10.1080/15459624.2017.1302587

[27] Byrley P, Geer Wallace MA, Boyes WK, Rogers K. Particle and volatile

organic compound emissions from a 3D printer filament extruder. Science of the Total Environment. 2020;**736**:1-10. DOI: 10.1016/j.scitotenv.2020.139604

[28] Davis AY, Zhang Q, Wong JPS, Weber RJ, Black MS. Characterization of volatile organic compound emissions from consumer level material extrusion 3D printers. Building and Environment. 2019;**160**:160. DOI: 10.1016/j.buildenv.2019.106209

[29] Wojnowski W, Kalinowska K, Majchrzak T, Zabiegała B. Real-time monitoring of the emission of volatile organic compounds from polylactide 3D printing filaments. Science of the Total Environment. 2022;**805**:150181. DOI: 10.1016/j.scitotenv.2021.150181

[30] Gu J, Wensing M, Uhde E, Salthammer T. Characterization of particulate and gaseous pollutants emitted during operation of a desktop 3D printer. Environment International. 2019;**123**:476-485. DOI: 10.1016/j.envint.2018.12.014

[31] Bernatikova S, Dudacek A, Prichystalova R, Klecka V, Kocurkova L. Characterization of ultrafine particles and VOCs emitted from a 3D printer. International Journal of Environmental Research and Public Health. 2021;**18**:929-944. DOI: 10.3390/ijerph18030929

[32] Industry Week. 3D Printing Presents Health Risks: Tips on Protecting Your Workers [Internet]. Available from: https://www.industryweek.com/operations/safety/article/21138777/3d-printing-presents-health-risks-tips-on-protecting-your-workers [Accessed: November 29, 2022]

[33] Afshar-Mohajer N, Wu CY, Ladun T, Rajon DA, Huang Y. Characterization of particulate matters and total VOC emissions from a binder jetting 3D printer. Building and Environment. 2015;**93**:293-301. DOI: 10.1016/j.buildenv.2015.07.013

[34] The International Organization for Standardization. ISO/ASTM 52900:2015 Additive Manufacturing — General Principles — Terminology [Internet]. 2015. Available from: https://www.iso.org/obp/ui/#iso:std:iso-astm:52900:ed-1:v1:en [Accessed: November 28, 2022]

[35] Chemical Inspection & Regulation Service. REACH SDS and eSDS [Internet]. Available from: https://www.cirs-reach.com/REACH/REACH_SDS_Safety_Data_Sheet.html [Accessed: November 22, 2022]

[36] House R, Rajaram N, Tarlo SM. Case report of asthma associated with 3D printing. Occupational Medicine (Oxford, England). 2017;**67**(8):652-654. DOI: 10.1093/occmed/kqx129

[37] Rezayat T, Wallace WD, Lynch JP, Johannes J. Chronic hypersensitivity pneumonitis associated with inhaled exposure to nylon powder for 3-D printing: A variant of nylon flock Worker's lung disease? American Journal of Respiratory and Critical Care Medicine. 2016;**193**:A7071

[38] Chan FL, House R, Kudla I, Lipszyc JC, Rajaram N, Tarlo SM. Health survey of employees regularly using 3D printers. Occupational Medicine (Oxford, England). 2018;**68**(3):211-214. DOI: 10.1093/occmed/kqy042

[39] Farcas MT, Stefaniak AB, Knepp AK, et al. Acrylonitrile butadiene styrene (ABS) and polycarbonate (PC) filaments three-dimensional (3-D) printer emissions-induced cell toxicity. Toxicology Letters. 2019;**317**:1-12. DOI: 10.1016/j.toxlet.2019.09.013

[40] Gümperlein I, Fischer E, Dietrich-Gümperlein G, Karrasch S,

Nowak D, Jörres R, et al. Acute health effects of desktop 3D printing (FDM) using ABS and PLA materials: An experimental exposure study in human volunteers. Indoor Air. 2018;**28**(4):611-623. DOI: 10.1111/ina.12458

[41] Andersson T, Stålbom B, Wesslén B. Degradation of polyethylene during extrusion. II. Degradation of low-density polyethylene, linear low-density polyethylene, and high-density polyethylene in film extrusion. Journal of Applied Polymer Science. 2004;**91**:1525-1537. DOI: 10.1002/app.13024

[42] Jensen AC, Harboe H, Brostrøm A, Jensen KA, Fonseca AS. Nanoparticle exposure and workplace measurements during processes related to 3D printing of a metal object. Frontiers in Public Health. 2020;**8**:608718. DOI: 10.3389/fpubh.2020.608718

[43] Zhang Q, Wong JPS, Davis AY, Black MS, Weber RJ. Characterization of particle emissions from consumer fused deposition modeling 3D printers. Aerosol Science and Technology. 2017;**51**(11):1275-1286. DOI: 10.1080/02786826.2017.1342029

[44] Runström Eden G, Tinnerberg H, Rosell L, Möller R, Almstrand AC, Bredberg A. Exploring methods for surveillance of occupational exposure from additive manufacturing in four different industrial facilities. Annals of Work Exposures and Health. 2022;**66**(2):163-177. DOI: 10.1093/annweh/wxab070

[45] Mendes L, Kangas A, Kukko K, et al. Characterization of emissions from a desktop 3D printer. Journal of Industrial Technology. 2017;**21**:S94-S106. DOI: 10.1111/jiec.12569

[46] Pavlovska I, Martinsone Ž, Kļaviņa A, Akūlova L, Paegle L. Emissions from 3D printers as occupational environmental pollutants. Environmental and Climate Technologies. 2021;**25**(1):1018-1031. DOI: 10.2478/rtuect-2021-0077

[47] Schraufnagel DE. The health effects of ultrafine particles. Experimental & Molecular Medicine. 2020;**52**:311-317. DOI: 10.1038/s12276-020-0403-3

[48] Taylor AA, Freeman EL, van der Ploeg MJC. Regulatory developments and their impacts to the nano-industry: A case study for nano-additives in 3D printing. Ecotoxicology and Environmental Safety. 2021;**111458**:1-11. DOI: 10.1016/j.ecoenv.2020.111458

[49] The National Institute for Occupational Safety and Health (NIOSH). Hierarchy of Controls [Internet]. 2022. Available from: https://www.cdc.gov/niosh/topics/hierarchy/default.html [Accessed: November 20, 2022]

[50] United States Department of Labor. Occupational Safety and Health Administration. Chemical Hazards and Toxic Substances [Internet]. Available from: https://www.osha.gov/chemical-hazards [Accessed: November 22, 2022]

[51] The International Organization for Standardization. ISO 16000-1:2004 Indoor air — Part 1: General aspects of sampling strategy [Internet]. 2004. Available from: https://www.iso.org/standard/39844.html [Accessed: November 28, 2022]

[52] The International Organization for Standardization. ISO 16000-5:2007 Indoor air — Part 5: Sampling strategy for volatile organic compounds (VOCs) [Internet]. 2007. Available from: https://www.osha.gov/chemical-hazards/controlling-exposure [Accessed: November 29, 2022]

[53] The International Organization for Standardization. ISO 22065:2019 Workplace Air — Gases and

Vapours — Requirements for Evaluation of Measuring Procedures using Pumped Samplers [Internet]. 2019. Available from: https://www.iso.org/standard/72487.html [Accessed: December 1, 2022]

[54] The International Organization for Standardization. ISO 16017-1:2000 Indoor, Ambient and Workplace Air — Sampling and Analysis of Volatile Organic Compounds by Sorbent Tube/ Thermal Desorption/Capillary Gas Chromatography — Part 1: Pumped Sampling [Internet]. 2000. Available from: https://www.iso.org/standard/29194.html [Accessed: December 1, 2022]

[55] United States Department of Labor. Occupational Safety and Health Administration. OSHA Technical Manual (OTM) Section II: Chapter 1 [Internet]. Available from: https://www.osha.gov/otm/section-2-health-hazards/chapter-1 [Accessed: December 2, 2022]

[56] Koehler KA, Peters TM. New methods for personal exposure monitoring for airborne particles. Current Environmental Health Reports. 2015;**2**:399-411. DOI: 10.1007/s40572-015-0070-z

[57] Mohammadian Y, Nasirzadeh N. Toxicity risks of occupational exposure in 3D printing and bioprinting industries: A systematic review. Toxicology and Industrial Health. 2021;**37**(9):573, 584. DOI: 10.1177/07482337211031691

[58] U.S. Environmental Protection Agency. Factors Affecting Indoor Air Quality [Internet]. Available from: https://www.epa.gov/sites/default/files/2014-08/documents/sec_2.pdf [Accessed: November 28, 2022]

[59] The International Organization for Standardization. ISO 7730:2005 Ergonomics of the Thermal Environment — Analytical Determination and Interpretation of Thermal Comfort using Calculation of the PMV and PPD Indices and Local Thermal Comfort Criteria [Internet]. Available from: https://www.iso.org/standard/39155.html [Accessed: December 3, 2022]

[60] ASTM International. ASTM D 4532:2010. Standard Test Method for Respirable Dust in Workplace Atmospheres Using Cyclone Samplers [Internet]. Available from: https://www.astm.org/d4532-15.html [Accessed: November 20, 2022]

[61] The National Institute for Occupational Safety and Health (NIOSH). NIOSH METHOD 0500 [Internet]. 1994. Available from: https://www.cdc.gov/niosh/docs/2003-154/pdfs/0500.pdf [Accessed: November 1, 20221]

[62] United Stated Department of Labor. Occupational Safety and Health Administration. Sampling and Analytical Methods [Internet]. Available from: https://www.osha.gov/chemicaldata/sampling-analytical-methods [Accessed: November 20, 2022]

[63] Zontek TL, Ogle BR, Jankovic JT, Hollenbeck SM. An exposure assessment of desktop 3D printing. Journal of Chemical Health and Safety. 2017;**24**(2):15-25. DOI: 10.1016/j.jchas.2016.05.008

[64] Steinle P. Characterization of emissions from a desktop 3D printer and indoor air measurements in office settings. Journal of Occupational and Environmental Hygiene. 2016;**13**(2):121-132. DOI: 10.1080/15459624.2015.1091957

[65] Väisänen AJK, Alonen L, Ylönen S, Lyijynen I, Hyttinen M. The impact of thermal reprocessing of 3D printable polymers on their mechanical performance and airborne pollutant profiles. Journal of Polymer Research. 2021;**28**(436):1-13. DOI: 10.1007/s10965-021-02723-7

[66] Chaiklieng S, Suggaravetsiri P, Autrup H. Risk assessment on benzene exposure among gasoline station workers. International Journal of Environmental Research and Public Health. 2019;**16**(14):2545. DOI: 10.3390/ijerph16142545

[67] El-Hashemy MA, Ali HM. Characterization of BTEX group of VOCs and inhalation risks in indoor microenvironments at small enterprises. Science of the Total Environment. 2018;**645**:974-983. DOI: 10.1016/j.scitotenv.2018.07.157

[68] Chan WR, Parthasarathy S, Fisk WJ, Mckone TE. Estimated effect of ventilation and filtration on chronic health risks in U.S. offices, schools, and retail stores. Indoor Air. 2016;**26**(2):331-343. DOI: 10.1111/ina.12189

[69] Creytens K, Gilissen L, Huygens S, Goossens A. A new application for epoxy resins resulting in occupational allergic contact dermatitis: The three-dimensional printing industry. Contact Dermatitis. 2017;**77**(5):349-351. DOI: 10.1111/cod.12840

[70] Ojoo JC, Mulrennan SA, Kastelik JA, Morice AH, Redington AE. Exhaled breath condensate pH and exhaled nitric oxide in allergic asthma and in cystic fibrosis. Thorax. 2005;**60**(1):22-26. DOI: 10.1136/thx.2003.017327

[71] Ganas K, Loukides S, Papatheodorou G, Panagou P, Kalogeropoulos N. Total nitrite/nitrate in expired breath condensate of patients with asthma. Respiratory Medicine. 2001;**95**(8):649-654. DOI: 10.1053/rmed.2001.1117

[72] van Beurden WJC, Harff GA, Dekhuijzen PNR, van den Bosch MJA, Creemers JPHM, Smeenk FWJM. An efficient and reproducible method for measuring hydrogen peroxide in exhaled breath condensate. Respiratory Medicine. 2002;**66**(3):197-203. DOI: 10.1053/rmed.2001.1240

[73] Roca O, Gómez-Ollés S, Cruz MJ, Muñoz X, Griffiths MJD, Masclans JR. Mechanical ventilation induces changes in exhaled breath condensate of patients without lung injury. Respiratory Medicine. 2010;**104**(6):822-628. DOI: 10.1016/j.rmed.2010.01.013

[74] Balint B, Kharitonov SA, Hanazawa T, Donnelly LE, Shah PL, Hodson ME, et al. Increased nitrotyrosine in exhaled breath condensate in cystic fibrosis. European Respiratory Journal. 2001;**17**(6):1201-1207. DOI: 10.1183/09031936.01.00072501

[75] Corradi M, Montuschi P, Donnelly LE, Pesci A, Kharitonov SA, Barnes PJ. Increased nitrosothiols in exhaled breath condensate in inflammatory airway diseases. American Journal of Respiratory and Critical Care Medicine. 2001;**163**(4):854-858. DOI: 10.1164/ajrccm.163.4.2001108

[76] Hanazawa T, Kharitonov SA, Barnes. Increased nitrotyrosine in exhaled breath condensate of patients with asthma. American Journal of Respiratory and Critical Care Medicine. PJ. 2000;**162**:1273-1276. DOI: 10.1164/ajrccm.162.4.9912064

[77] Antczak A, Montuschi P, Kharitonov S, Gorski P, Barnes PJ. Increased exhaled cysteinyl-leukotrienes and 8-isoprostane in aspirin-induced asthma. American Journal of Respiratory and Critical Care Medicine. 2002;**166**(3):301-306. DOI: 10.1164/rccm.2101021

[78] Pelclova D, Zdimal V, Komarc M, et al. Three-year study of markers of oxidative stress in exhaled breath condensate in workers producing nanocomposites, extended by plasma and urine analysis in last two years.

Nanomaterials. 2020;**10**(12):2440. DOI: 10.3390/nano10122440

[79] Ghelli F, Panizzolo M, Garzaro G, Squillacioti G, Bellisario V, Colombi N, et al. Inflammatory biomarkers in exhaled breath condensate: A systematic review. International Journal of Molecular Sciences. 2022;**23**(17):9820. DOI: 10.3390/ijms23179820

[80] Lärstad M, Ljungkvist G, Olin AC, Torén K. Determination of malondialdehyde in breath condensate by high-performance liquid chromatography with fluorescence detection. Journal of Chromatography B. 2002;**766**(1):107-114. DOI: 10.1016/S0378-4347(01)00437-6

[81] Gong J, Zhu T, Kipen H, et al. Malondialdehyde in exhaled breath condensate and urine as a biomarker of air pollution induced oxidative stress. Journal of Exposure Science & Environmental Epidemiology. 2013;**23**(3):322-327. DOI: 10.1038/jes.2012.127

[82] Syslová K, Kačer P, Kuzma M, Najmanová V, Fenclová Z, Vlčková Š, et al. Rapid and easy method for monitoring oxidative stress markers in body fluids of patients with asbestos or silica-induced lung diseases. Journal of Chromatography B. 2009;**877**(24):2477-2486. DOI: 10.1016/j.jchromb.2009.06.008

[83] Hemmendinger M, Sauvain JJ, Hopf NB, Suárez G, Canu G. Challenges in quantifying 8-OHdG and 8-Isoprostane in exhaled breath condensate. Antioxidants. 2022;**11**(5):830. DOI: 10.3390/antiox11050830

[84] Koczulla R, Dragonieri S, Schot R, Bals R, Gauw SA, Vogelmeier C, et al. Comparison of exhaled breath condensate pH using two commercially available devices in healthy controls, asthma and COPD patients. Respiratory Research. 2009;**10**(1):78. DOI: 10.1186/1465-9921-10-78

[85] Horváth I, Barnes PJ, Loukides S, et al. A european respiratory society technical standard: Exhaled biomarkers in lung disease. European Respiratory Journal. 2017;**49**(4):1600965. DOI: 10.1183/13993003.00965-2016

[86] Kostikas K, Koutsokera A, Gourgoulianis K, Loukides S. Biomarkers in the exhaled breath condensate of healthy adults: Mapping the path towards reference values. Current Medicinal Chemistry. 2008;**15**(6):620-630. DOI: 10.2174/092986708783769768

[87] Shoman Y, Wild P, Hemmendinger M, Graille M, Sauvain JJ, Hopf NB, et al. Reference ranges of 8-isoprostane concentrations in exhaled breath condensate (Ebc): A systematic review and meta-analysis. International Journal of Molecular Sciences. 2020;**21**(11):3822. DOI: 10.3390/ijms21113822

[88] van 't Erve TJ, Kadiiska MB, London SJ, Mason RP. Classifying oxidative stress by F2-isoprostane levels across human diseases: A meta-analysis. Redox Biology. 2017;**12**:582-599. DOI: 10.1016/j.redox.2017.03.024

[89] Pelclova D, Zdimal V, Schwarz J, et al. Markers of oxidative stress in the exhaled breath condensate of workers handling nanocomposites. Nanomaterials. 2018;**8**(8):611. DOI: 10.3390/nano8080611

[90] Paggiaro PL, Bartoli ML, Novelli F, et al. Malondialdehyde in exhaled breath condensate as a marker of oxidative stress in different pulmonary diseases. Mediators of Inflammation. 2011;**2011**:891752. DOI: 10.1155/2011/891752

[91] Turcu V, Wild P, Hemmendinger M, Sauvain JJ, Bergamaschi E, Hopf NB, et al. Towards reference values for malondialdehyde on exhaled breath condensate: A systematic literature review and meta-analysis. Toxics. 2022;**10**(5):258. DOI: 10.3390/toxics10050258

[92] Pelcova D. Oxidative stress markers in exhaled breath condensate in lung fibroses are not significantly affected by systemic diseases. Industrial Health. 2011;**49**(6):746-754. DOI: 10.2486/indhealth.ms1237

[93] Hemmendinger M, Wild P, Shoman Y, Graille M, Bergamaschi E, Hopf N, et al. Reference ranges of oxidative stress biomarkers selected for non-invasive biological surveillance of nanotechnology workers: Study protocol and meta-analysis results for 8-OHdG in exhaled breath condensate. Toxicology Letters. 2020;**327**:41-47. DOI: 10.1016/j.toxlet.2020.03.021

[94] Hu W, Wang Y, Wang T, et al. Ambient particulate matter compositions and increased oxidative stress: Exposure-response analysis among high-level exposed population. Environment International. 2021;**147**:106341. DOI: 10.1016/j.envint.2020.106341

[95] Schwedhelm E, Bartling A, Lenzen H, Tsikas D, Maas R, Brümmer J, et al. Urinary 8-iso-prostaglandin F2α as a risk marker in patients with coronary heart disease: A matched case-control study. Circulation. 2004;**109**(7):843-848. DOI: 10.1161/01.CIR.0000116761.93647.30

[96] di Minno A, Aveta A, Gelzo M, Tripodi L, Pandolfo SD, Crocetto F, et al. 8-Hydroxy-2-Deoxyguanosine and 8-Iso-prostaglandin F2α: Putative biomarkers to assess oxidative stress damage following robot-assisted radical prostatectomy (RARP). Journal of Clinical Medicine. 2022;**11**(20):6102. DOI: 10.3390/jcm11206102

[97] Toto A, Wild P, Graille M, Turcu V, Crézé C, Hemmendinger M, et al. Urinary malondialdehyde (MDA) concentrations in the general population—A systematic literature review and meta-analysis. Toxics. 2022;**10**(4):160. DOI: 10.3390/toxics10040160

[98] Wang Y, Zhao H, Wang T, et al. Polycyclic aromatic hydrocarbons exposure and hematotoxicity in occupational population: A two-year follow-up study. Toxicology and Applied Pharmacology. 2019;**378**:114622. DOI: 10.1016/j.taap.2019.114622

[99] Graille M, Wild P, Sauvain JJ, Hemmendinger M, Canu IG, Hopf NB. Urinary 8-OHDG as a biomarker for oxidative stress: A systematic literature review and meta-analysis. International Journal of Molecular Sciences. 2020;**21**(11):3743. DOI: 10.3390/ijms21113743

[100] Halliwell B, Lee CYJ. Using isoprostanes as biomarkers of oxidative stress: Some rarely considered issues. Antioxidants & Redox Signaling. 2010;**13**(2):145-156. DOI: 10.1089/ars.2009.2934

[101] Il'yasova D, Scarbrough P, Spasojevic I. Urinary biomarkers of oxidative status. Clinica Chimica Acta. 2012;**413**(19-20):1446-1453. DOI: 10.1016/j.cca.2012.06.012

[102] Caymal Chemical. Stressed about Picking an Oxidative Damage Assay? Find the Right Biomarker to Detect in your Application [Internet]. 2017. Available from: https://www.biomol.com/dateien/Cayman--Guide-to-Oxidative-Stress-Assays.pdf [Accessed: November 10, 2022]

[103] Dator RP, Solivio MJ, Villalta PW, Balbo S. Bioanalytical and mass spectrometric methods for aldehyde profiling in biological fluids. Toxics. 2019;7(2):32. DOI: 10.3390/TOXICS7020032

[104] Bin P, Shen M, Li H, et al. Increased levels of urinary biomarkers of lipid peroxidation products among workers occupationally exposed to diesel engine exhaust. Free Radical Research. 2016;**50**(8):820-830. DOI: 10.1080/10715762.2016.1178738

[105] Song MF, Li YS, Ootsuyama Y, et al. Urea, the most abundant component in urine, cross-reacts with a commercial 8-OH-dG ELISA kit and contributes to overestimation of urinary 8-OH-dG. Free Radical Biology and Medicine. 2009;**47**:41-46. DOI: 10.1016/j.freeradbiomed.2009.02.017

[106] Cooke MS, Olinski R, Loft S. Measurement and meaning of oxidatively modified DNA lesions in urine. Cancer Epidemiology Biomarkers and Prevention. 2008;**17**(1):3-14. DOI: 10.1158/1055-9965.EPI-07-0751

[107] Gan W, le Liu X, Yu T, Zou YG, Li TT, Wang S, et al. Urinary 8-oxo-7,8-dihydroguanosine as a potential biomarker of aging. Frontiers in Aging Neuroscience. 2018;**10**:34. DOI: 10.3389/fnagi.2018.00034

[108] Mirowsky J, Gordon T. Noninvasive effects measurements for air pollution human studies: Methods, analysis, and implications. Journal of Exposure Science & Environmental Epidemiology. 2015;**25**(4):354-380. DOI: 10.1038/jes.2014.93

[109] Arata C, Misztal PK, Tian Y, Lunderberg DM, Kristensen K, Novoselac A, et al. In: Goldstein AH, editor. Volatile organic compound emissions during HOMEChem. Indoor Air. 2021;**31**(6):2099-2117. DOI: 10.1111/ina.12906

[110] American Lung Assocciatiion. Toxic Air Pollutants [Internet]. 2022. Available from: https://www.lung.org/clean-air/outdoors/what-makes-air-unhealthy/toxic-air-pollutants [Accessed: November 11, 2022]

[111] Government of Canada. Health Canada. Investigating human exposure to contaminants: a handbook for exposure calculations [Internet]. 1995. Available from: https://publications.gc.ca/collections/Collection/H49-96-1-1995E-1.pdf [Accessed: November 10, 2022]

[112] United States Environmental Protection Agency. Human Exposure and Health [Internet]. Available from: https://www.epa.gov/report-environment/human-exposure-and-health [Accessed: December 1, 2022]

[113] United States Environmental Protection Agency. Exposure to Environmental Contaminants [Internet]. 2022. Available from: https://www.epa.gov/report-environment/exposure-environmental-contaminants [Accessed: November 8, 2022]

[114] Landrigan PJ, Kimmel CA, Correa A, Eskenazi B. Children's health and the environment: Public health issues and challenges for risk assessment. Environmental Health Perspectives. 2004;**112**(2):257-265. DOI: 10.1289/ehp.6115

[115] World Health Organization. Principles for Evaluating Health Risks in Children Associated with Exposure to Chemicals. 2006. Available from: https://apps.who.int/iris/handle/10665/43604 [Accessed: December 1, 2022]

[116] Vermeulen R, Heideman J, Bos RP, Kromhout H. Identification of dermal exposure pathways in the rubber manufacturing industry. Annals of Occupational Hygiene. 2000;**44**(7):533-541. DOI: 10.1016/S0003-4878(00)00039-9

[117] Vermeulen R, de Hartog J, Swuste P, Kromhout H. Trends in exposure to inhalable particulate and dermal contamination in the rubber manufacturing industry: Effectiveness of control measures implemented over a nine-year period. Annuals of Occupatioanl Hygiene. 2000;**44**(5):343-354. DOI: 10.1093/annhyg/44.5.343

[118] Health Canada. Investigating Human Exposure to Contaminants in the Environment: A Handbook for Exposure Calculations. 1995. Available from:

https://publications.gc.ca/collections/Collection/H49-96-1-1995E-1.pd [Accessed: November 10, 2022]

[119] Konstantinidi EM, Lappas AS, Tzortzi AS, Behrakis PK. Exhaled breath condensate: Technical and diagnostic aspects. The Scientific World Journal. 2015;**2015**(435160):1-25. DOI: 10.1155/2015/435160

[120] Horváth I, Hunt J, Barnes PJ, et al. Exhaled breath condensate: Methodological recommendations and unresolved questions. European Respiratory Journal. 2005;**26**(3):523-548. DOI: 10.1183/09031936.05.00029705

[121] Ministry of Environment and Food Protection of Denmark. Survey and Risk Assessment of 3D Pens [Internet]. 2018. Available from: https://www2.mst.dk/Udgiv/publications/2018/12/978-87-7038-021-8.pdf [Accessed: November 13, 2022]

[122] Su WC, Chen Y, Xi J. Estimation of the deposition of ultrafine 3D printing particles in human tracheobronchial airways. Journal of Aerosol Science. 2020;**149**:105605. DOI: 10.1016/j.jaerosci.2020.105605

[123] Wojtyła S, Klama P, Baran T. Is 3D printing safe? Analysis of the thermal treatment of thermoplastics: ABS, PLA, PET, and nylon. Journal of Occupational and Environmental Hygiene. 2017;**14**(6):D80-D85. DOI: 10.1080/15459624.2017.1285489

[124] Ljunggren SA, Karlsson H, Ståhlbom B, Krapi B, Fornander L, Karlsson LE, et al. Biomonitoring of metal exposure during additive manufacturing (3D printing). Safety and Health at Work. 2019;**10**:518-526

[125] Zhou Y, Kong X, Chen A, Cao S. Investigation of ultrafine particle emissions of desktop 3D printers in the clean room. Procedia Engineering. 2015;**121**:506-512

[126] Kakar A, Hayat MT, Abbasi AM, et al. Risk assessment of heavy metals in selected marine fish species of gadani shipbreaking area and Pakistan. Animals. 2020;**10**(10):1738. DOI: 10.3390/ani10101738

[127] Gardner DE, Kirkpatrick DT. Respiratory Tract. Encyclopedia of Toxicology [Internet]. 2005. Available from: https://www.sciencedirect.com/referencework/9780123864550/encyclopedia-of-toxicology [Accessed: November 10, 2022]

[128] Montuschi P. Measurement of biomarkers of oxidative stress and airway inflammation in exhaled breath condensate: Methodology and potential applications in patients with COPD and healthy smokers. Volatile Biomarkers. 2013;**19**:360-381. DOI: 10.1183/09031936.00116107

[129] Chang JW, Lee CC, Pan WH, Chou WC, Huang HB, Chiang HC, et al. Estimated daily intake and cumulative risk assessment of phthalates in the general Taiwanese after the 2011 DEHP food scandal. Scientific Reports. 2017;7:45009. DOI: 10.1038/srep45009

[130] Chilakapati J, Mehendale HM. Acceptable Daily Intake (ADI). Encyclopedia of Toxicology: Third Edition 8-9 [Internet]. 2014. Available from: https://www.sciencedirect.com/referencework/9780123864550/encyclopedia-of-toxicology [Accessed: November 11, 2022]

[131] Chen S, Gao S, Jing J, Lu Q. Designing 3D biological surfaces via the breath-figure method. Advanced Healthcare Materials. 2018;**7**(6):e1701043. DOI: 10.1002/adhm.201701043

[132] United States Environmental Protection Agency. Dermal Exposure Assessment: A Summary of EPA Approaches [Internet]. Available from: https://cfpub.epa.gov/ncea/risk/recordisplay.cfm?deid=183584 [Accessed: November 11, 2022]

[133] Behroozy A. On dermal exposure assessment. International Journal of Occupational and Environmental Medicine 4 2013;**4**(3):113-127. [Internet]. Available from: https://pubmed.ncbi.nlm.nih.gov/23860542/ [Assessed: November 20, 2022]

[134] Van-Wendel-De-Joode B, Brouwer DH, Vermeulen R, van Hemmen JJ, Heederik D, Kromhout H. DREAM: A method for semi-quantitative dermal exposure assessment. Annals of Occupational Hygiene. 2003;**47**(1):71-87. DOI: 10.1093/annhyg/meg012

[135] United States Environmental Protection Agency. Exposure Assessment Tools by Routes – Dermal [Internet]. Available from: https://www.epa.gov/expobox/exposure-assessment-tools-routes-dermal [Accessed: December 1, 2022]

[136] Simon T. Environmental risk assessment: A toxicological approach. Environmental Risk Assessment: A Toxicological Approach. 2nd Edition. 2014:520. DOI: 10.1201/b16454

[137] Environmental Protection Agency. Dermal Exposure Assessment: A Summary of EPA Approaches [Internet]. 2007. Available from: https://cfpub.epa.gov/ncea/risk/recordisplay.cfm?deid=183584 [Accessed: December 1, 2022]

[138] Lawal O, Ahmed WM, Nijsen TME, Goodacre R, Fowler SJ. Exhaled breath analysis: A review of 'breath-taking' methods for off-line analysis. Metabolomics. 2017;**13**:110. DOI: 10.1007/s11306-017-1241-8

[139] 3Dprint.com. Can 3D Printing Save the World? A Look at 3D Printing and the Environment [Internet]. 2022. Available from: https://3dprint.com/144928/3d-printing-environmental/ [Accessed: November 25, 2022]

[140] Karyappa R, Hashimoto M. Chocolate-based ink three-dimensional printing (Ci3DP). Scientific Reports. 2019;**9**:14178. DOI: 10.1038/s41598-019-50583-5

[141] Kewuyemi YO, Kesa H, Meijboom R, Alimi OA, Adebo OA. 3D food printing improves color profile and structural properties of the derived novel whole-grain sourdough and malt biscuits. Scientific Reports. 2022;**12**(1):12347. DOI: 10.1038/s41598-022-16659-5

[142] Liu WC, Chou VHY, Behera RP, le Ferrand H. Magnetically assisted drop-on-demand 3D printing of microstructured multimaterial composites. Nature Communications. 2022;**13**:5015. DOI: 10.1038/s41467-022-32792-1

[143] Liu K, Ding H, Li S, Niu Y, Zeng Y, Zhang J, et al. Gu Z 3D printing colloidal crystal microstructures via sacrificial-scaffold-mediated two-photon lithography. Nature Communications. 2022;**13**:4563. DOI: 10.1038/s41467-022-32317-w

[144] Post MJ, Levenberg S, Kaplan DL, Genovese N, Fu J, Bryant CJ, et al. Scientific, sustainability and regulatory challenges of cultured meat. Nature Food. 2020;**1**:413-415. DOI: 10.1038/s43016-020-0112-z

[145] Brion DAJ, Pattinson SW. Generalisable 3D printing error detection and correction via multi-head neural

networks. Nature Communications. 2022;**13**:4654. DOI: 10.1038/s41467-022-31985-y

[146] Computerworld. Bio-Printing Human Parts will Spark Ethical, Regulatory Debate [Internet]. 2022. Available from: https://www.computerworld.com/article/2486998/bio-printing-human-parts-will-spark-ethical--regulatory-debate.html [Accessed: December 1, 2022]

Chapter 3

Body Part Surrogates for Medicine, Comfort and Safety Applications

Andrey Koptyug and Mikael Bäckström

Abstract

Body part surrogates made with support from additive manufacturing (AM) technologies belong to a rapidly developing area of modeling. Although computer-based and mathematical modeling of complex processes is already an established field, these are not free from inherited problems. Surrogate modeling (physical modeling) being a subject of its own successfully complement mathematical and computer modeling and helps to cross-validate these methods and improve particular models. Present chapter provides a discussion on the general aspects of modeling relevant to the design, manufacturing and application of body part surrogates It also introduces new term 'surrogate twins' using the analogy of 'virtual twins'. It also outlines a number of known applications of body part surrogates manufactured with support of AM in medicine, safety and comfort research. Strong and weak points of particular surrogate models is discussed basing on the general concepts of modeling including defining of particular surrogate model purposes, approximations, the ways of model validation, input parameter harvesting, related measurement systems and data processing, and setups for material and product testing. Comprehensive references will allow readers getting detailed information regarding discussed issues.

Keywords: additive manufacturing, body part surrogates, physical models, digital twins, surrogate twins, safety research, medical applications

1. Introduction

Body part surrogates made with a support from additive manufacturing technologies belong to a new rapidly developing dimension in modeling, linking theories to experimental science (e.g. [1–3]). Current developments in corresponding approach and technology of physical modeling constantly bring in new applications. Present contribution is focusing at the applications of such surrogates ranging from comfort and safety research to the medicine and traumatology.

It is intuitively clear that testing or manufacturing novel safety devices without somehow proving them being safe is not possible. Two dominant ways out from this "catch 22" situation are either through developing advance computer models of human body, its parts and safety devices allowing "virtual experimentation" in close to real conditions, and experimenting with physical models having different degree of fidelity, such as for example car crash dummies and their elements. Although computer-based and mathematical (analytical) modeling of complex processes is already an established field, these are not free from problems. In relation to bio-medical, comfort and safety research, where the subjects (human) are extremely complicated dynamic systems, inability to prove the model from within a model and potential dangers in direct experimental validation cannot be easily overcome. Physical modeling helps to expand the possibilities towards real experiments with human body part surrogates allowing cross-validation of the results of both modeling approaches. Well established and standardized car crash dummies and their separate elements, such as 'head' and 'neck', find wider applications for example in protection helmet design and testing, but in many cases are not optimal for lower energy impacts. Developing physical models of human body parts (body part surrogates) with improved biofidelity is one of the fast developing trends rapidly widening possible application areas. Additive manufacturing and modern sensor application are the main enabling elements in such developments.

Modern additive manufacturing (AM) technologies often referred to as 3D printing allow for the manufacturing of the body part surrogates with high fidelity in both 3D geometry and properties. When supplemented by a number of embedded sensors-providing valuable static and dynamic data they form a basis for powerful setups and platforms for research and device testing. Such surrogates can be subjected to a 'serious abuse' replicating real world situations either dangerous or potentially leading to injuries, and allowing for the qualitative and quantitative assessment of the protection and for testing wearable devices and different apparels.

Development of the computer technology and in particular virtual representation allows for advanced modeling of extremely complicated systems [4, 5] enabling unprecedented level of details and yielding so-called 'digital twins' (for example, applied in technology [6–8], medicine [9], safety research [10–13] and garment fitting [14]). However, such computer-based representations cannot be proven from within themselves, and are in need of experimental validation. Although certain information about the performance of already existing safety devices or even unfortunate consequences of some events leading to injuries can be used for validation more experimental data is needed. Manufacturing of the advanced sensorized human body part surrogates allowing for acquisition of needed experimental data in physical modeling It is already approaching such level of complexity and fidelity that corresponding surrogates can be termed 'surrogate twins'.

Understanding the relations between 'digital twins', 'surrogate twins' and the human subjects they are designed to represent, demands some deeper analysis of pros and cons of modeling in general. So following material intends to discuss such aspects and supplement these with particular examples of applying computer and physical modeling (body part surrogates) in safety research.

2. Relevant general concepts of modeling

Although many people are using modeling in everyday practice, defining it one would commonly refer to certain specific modeling subject or approach. Given definitions commonly range from 'a representation of a system using general rules and concepts', 'a physical representation in three dimensions of an object', 'a representation of a system using mathematical concepts and formalization' to a 'simulation to reproduce a behavior of a system, and 'a formalized description of the world, its shapes, phenomena and processes happening in it' among others.

Unfortunately, the comprehensive analysis of the very basic and general modeling concepts is still not easy to find in the literature. However, one of the best concise sets of relevant definitions and basic concepts was developed by Roger D. Smith in his 'Fundamental principles of Modeling and Simulation' [15], where he is introducing basic terms, concepts, techniques, and applications of modeling from the philosophy of science perspective. He stresses that a model is designed to serve a certain specific purpose, and has certain limitations 'built in' when it is constructed and developed. Consequently, it is hard to expect a model to be universal and to cover all imaginable cases and situations. It is also inherent to all models that they are using certain approximations of the real systems and processes happening in them. Proper modeling always starts with relatively simple and in a way 'rough' approximations aiming at most important system features. When such most critical features are adequately represented, the features assumed to be of 'second order of importance' are introduced to improve the model. It is well known that *unnecessary* complicating the model contrary to the obvious assumptions can reduce its applicability: relations become obscured, conclusions become hard to comprehend, and large amount of details buries the trends. It becomes a situation when it is hard 'seeing the forest for the trees and leaves'. Thus, term *adequacy* in relation to the models means that a model uses just enough details to serve the initial purpose of its construction. When the model in its initial shape is validated, and successfully used, it is possible to expand it adding more details after finding that some of the effects, interactions, components are not represented to adequate precision or level of detail. In such expansion process 'necessity rules', exactly as it was when the basic model was constructed. Thus, well known 'Occam's razor' principle, declaring, "entities should not be multiplied beyond necessity", is equally applied in constructing and expanding/improving models.

Related to the discussed subjects, we assume that models we would like to construct are made with a certain purposes including *studying* complicated system and dynamics of the processes involved, to understand its functioning in different but well-defined conditions, to grasp interrelations of its parts in order *to make certain predictions* (e.g. [1–3, 16]). Thus, one needs to treat both physical and mathematical modeling of human body parts from the points of view of intended purpose, predictive capacity as well as strong and weak points and restrictions inherent in such models.

3. Aspects of modeling relevant to medicine, safety and comfort research

Although experiment is one of the bases for scientific research, real experimenting with human subjects has obvious limitations. One issue is due to the differences between individuals, leading to the need in collecting large amount of data from different subjects followed by statistical analysis in order to get generalized representation. In addition, in many situations experiments can be potentially dangerous or harmful, or not possible due to certain ethical reasons. Medicine and safety research are very typical in this respect. A lot of information about functioning of human body and its parts was collected through treating traumatic injuries however, it is hardly imaginable that one can subject human to the adverse conditions for studying the limits the body can tolerate, or assessing if the new protective device will be able to provide adequate protection. In a way, there is a 'catch twenty two' situation: in order to prove that safety device is providing adequate protection one should experimentally test it, but without proving it safe experiments with human subjects are impossible. Advanced models allow bypassing many of described restrictions, but along with the benefits, they bring in certain issues of their own, and it is worth looking deeper into these.

Virtual experiments, aka simulations using advanced computer models have obvious strong points. Computer models do not endanger any human test subject; can cover both static and dynamic situations; provide certain power of prediction and are very useful for research and development as well as for studying and teaching. They allow for analyzing "what if" and "ideal" cases, separating parameters "inseparable" in reality, "speeding up" or "slowing down" time and doing other useful tricks, which is not possible with real experiments. Together with advanced means of visualization provided by modern computer technology such models become quite powerful tools in research, development and teaching. However, as with any models, there are certain inherent approximations and limitations of the scope built in them. As for the *adequacy* of the used model, it should be proven in order to trust its predictions. However, it cannot be done from within the model itself, which is also one of the modeling fundamentals. For proving the adequacy of the model, its predictions should be compared with the results of previous or specially conducted experiments. It should be stressed here, that adequacy of the model is undoubtedly proven only for the tested cases, and is relative (or more correctly, statistically validated for the number of cases tested).

Mathematical and computer (numerical) models also have additional issues and potential for errors related to the limitations in algorithm stability and value quantization. Here numerically stable algorithm means that it does not magnify small errors, e.g. that small and diminishing deviation in the input parameter in turn causes steadily diminishing deviation in the computed results. Strictly speaking, each numerical algorithm should be tested, and the limits for the input parameters that guarantee stability is defined. This is a special task, and with many modeling software packages, the user commonly relies that developers have performed such tasks. Additionally, many of the commercial software packages used for modeling have at least some proprietary modules that are 'black boxes' from the user point of view. Therefore, the user may be not aware of additional limitations that were obvious for the model developers.

Yet in addition, mathematical and computer (numerical) modeling demands exact values of the input parameters, some of which can be hard or impossible to extract experimentally, especially when it concerns human body and tissues, forcing the user to 'synthesize' some of the input values. Although researchers will use 'realistic' values for such synthesized parameters, these can be differing from real ones, and it is not trivial to predict how such differences will affect the simulation results.

Experimenting with human body part surrogates (surrogate twins), or physical modeling, can complement mathematical and computer modeling also helping to cross-validate both approaches as well as underlying concepts. With additively manufactured surrogates, computer and physical modeling can use exactly same 3D shape representation of the body parts. Physical models are using synthetic materials, and necessary properties of these either are known or can be experimentally measured with adequate precision. Thus, computer models can be using exactly same parameters and same settings for the 'virtual experiments' that are used for materials and conditions of real experiments carried out using physical models (body part surrogates), simplifying result comparison.

A number of different sensors can be embedded into the body part surrogates, allowing data collection during the experiments carried out in the conditions

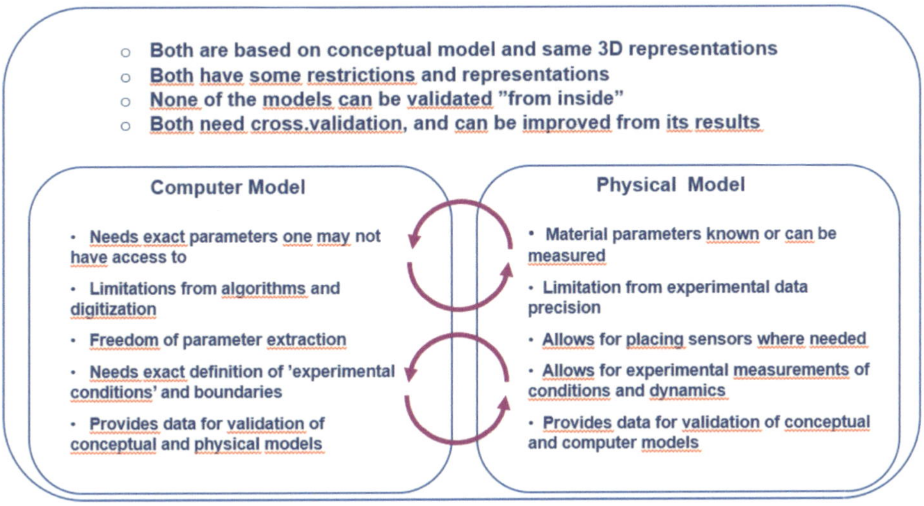

Figure 1.
"Hand in hand" development of the computer and physical models.

mimicking real cases. One can conclude that most productive is the coordinated development of both types of the models allowing for improvements in both of them and increasing the fidelity of their predictions, as schematically illustrated in **Figure 1**.

One of the specific advantages of physical modeling is in the ability of providing objective information in the cases when the data may carry subjective contributions. For example, individual differences and complications in reproducing exact test conditions have significant impact when human subjects and outdoor tests are concerned. In the cases when one needs to assess the functioning of the garments, footwear and sports, protection or wearable equipment special care is taken to separate contributions of the subjective and objective factors. In such cases, objectivization of the measurements can be a primary target of applying physical modeling even in the situations, when real experiments do not incur any risks for the test subjects and are just inconvenient or uncomfortable.

4. Additive manufacturing of body part surrogates for physical modeling

Modern additive manufacturing in a way can be regarded as a link between the 'digital world of virtual objects' and real world with real objects. Indeed, AM technology is a representative of the 'digitally empowered' way of modern design and manufacturing providing unprecedented flexibility in the object shapes and allowing for many different materials [17–20]. Files encoding the shape representations of the parts are commonly used for finite element modeling and simulating key properties of the future components. Modern software is already fit to have special inter-platform file formats also used in additive manufacturing. In relation to the biomedicine, it is used for the advanced design of the implants allowing for the case- and patient-specific fitting and implant shape optimization for the chosen functionality [9, 21–25]. Corresponding methods of human body scanning developed for biomedical applications, including optical methods, X-ray computed tomography

(CT) and magnetic resonance imaging (MRI) capable of providing high precision spatial information are successfully utilized for constructing high fidelity 3D body part representation for finite element simulations and additive manufacturing of the surrogates replicating real body parts. Additive manufacturing of body part surrogates successfully exploits known advantages of the AM method such as:

- extreme flexibility of the produced 3D shapes;
- easy linking to body part dimensions using files from 3D medical imaging;
- ability to work with different materials; possibility of combining 'direct' and 'indirect printing';
- cost-effective manufacturing of unique parts and small series;
- easy and well-controlled modification of the existing design;
- easy sharing of the production files with the manufactures localized in any part of the world.

Here 'direct printing' means that the additive machines can directly use corresponding material. In cases, where such material cannot be accepted by the machines or does not provide adequate spatial resolution in AM, one can print the mold from hard material and cast the desired material (which is a common case for soft polymers and gels, and this is often referred to as 'indirect' or 'negative' printing).

Examples of the particular applications of using additive manufacturing for making body part surrogates using both 'direct' and 'indirect' printing are given below.

4.1 AM polymer bone and tissue surrogates

Additive manufacturing of human bone and soft tissue replicas is one of the earliest applications for the physical modeling of the human body parts [9–13, 21–28]. Historically, such surrogates were primarily targeting replicating exact 3D-shapes of the selected objects and have found multiple applications as 'phantoms' used for the calibration and fine tuning of different imaging devices, and as 'bone replicas' used in preoperational planning (see, for example, [29–31]). Preoperational planning using polymer replicas of the fractured or damaged bones today is appreciated by practitioners, often preferring solid 3D models to computer images. Such replicas also allow to perform 'training runs' using surrogates aligning and fixing bone fragments of complicated fractures, applying fixation elements and placing screws, all of it without extra stress to the surgeons and any risk to the patient. As a result such 'pre-operations' allow decreasing the time patient spends under anesthetics, minimizing post-operational complications and reducing recovery times.

It was also realized, that polymer replicas of the human body parts made using additive manufacturing and basing on the real case 3D reconstructions is a powerful supplement for the teaching [32–34]. Initially these were used mainly for teaching anatomy, but recently such surrogates are used as the mean of teaching practical skills for surgeons. Indeed, bone replicas additively manufactured from acrylonitrile butadiene styrene (ABS, very common polymer used in AM) are not only looking like bones (color, shape), but also the efforts needed to insert the screws into them quite well replicate that needed in real life cases. Hence, additive manufacturing of bone surrogates allows for compiling collections providing bone replicas for 'standard cases' reflecting corresponding clinical cases for the most realistic practical student training. Files used for manufacturing such replicas can be easily shared, and bone surrogates can be manufactured following real demand.

Digital nature of the shape files used for AM allows linking shape modeling to functionality modeling and optimization through finite element simulation adding the value to the pre-operational models of the defect bones. Modern software also allows for complex manipulating of the bone shape files 'mending the bone models in virtual space', generating shape-optimized implants and fixation plates, and providing functional optimization of case- and patient specific implants and special surgery tooling (e.g. [35–37]). As a result, clinicians get a full kit including bone replicas, functionally optimized implants and special surgical tools (for example, templates for precision cutting and screw positioning) together with bone surrogates for full scale preoperative planning and training.

An interesting extension of the AM bone surrogate application is recently developed, combining finite element modeling and physical modeling [38–40]. It was realized that spatial fidelity additively manufactured bone replicas based on the CT or MRI scans of a human subject is very high, and that mechanical properties of some polymers are similar to that of natural bone. Hence, it is not only possible to predict bone fracture modalities using finite element analysis, but also possible to perform experiments with monitoring the behavior and fracturing of the bone surrogates under real loading conditions. This case also provides a good example of needed cross-validation of the computer and physical models through the comparison of the predicted and experimentally acquired results.

4.2 AM body part surrogates for studying garment and footwear performance

Assessment of the garment and footwear performance and comfort is quite a complicated task. One can design a test protocol, select a group of human subjects, provide them with apparels to be tested, and send them out for the outdoor testing. After that, subjects fill in special questioners and corresponding answers are analyzed in order to draw conclusions. In many situations like this, it is quite hard to exclude subjective elements from the final assessment. Different wearable sensors acquiring data are often used to make assessment more objective. However, the variations in weather conditions during and between different test sessions can mask the trends and affect the assessment. Moving the tests indoors even when using climate chambers and climate controlled aerodynamic tubes although improve the reproducibility of test conditions still cannot completely remove the subjective components related to human subjects. This substantially affects the ability of objective comparison of the performance and comfort of different garment and footwear designs and models, including the ones coming from different manufacturers.

Full size mannequins are quite helpful for objective assessment of the apparel performance at different conditions including different outer temperatures and even providing 'perspiration' and body heat equivalents (see, for example, [41–45]). Such complex setups are unique and not easily affordable. However, in many cases one can use body part surrogates, such as the ones of hand and foot for assessing the performance of the gloves and footwear. Additionally, it is quite possible to include good possibilities for objectively comparing apparels and wearable devices designed for example, for able-bodied or disabled subjects, male, female, children, as well as the ones reflecting individual body part variability. In such case, a single 'unified' computer-based shape model can be properly scaled and easily modified for particular situation. Corresponding surrogates will be additively manufactured with a high spatial precision.

It should be noted, that application of body part surrogates carries very high potential in supporting people with disabilities. Levels of such surrogate complexity can range from simple body part shape models (used, for example, for testing prosthetic sockets and support elements), to the advanced surrogates testing the comfort of different garments, footwear and wearable devices (taking into account distortions in body thermoregulation caused by injuries or amputations).

Described approach can be exemplified by the applications of additively manufactured body part surrogates for assessing gloves and boots for cold conditions Thin-walled shell is additively manufactured in a shape of the hand (foot). Authors use the shape files acquired using optical scans from volunteers. Initial tests were performed using polymer material (ABS), but following tests will be performed using metallic shells (Ti-6Al-4 V, one of the most common titanium alloys manufactured additively). **Figure 2** provides a typical example of such surrogate arm shell.

A number of small temperature sensors is embedded into the shell. The sell is filled with water at 32 to 34°C (temperatures of the body extremities are commonly some lower than that of the body core; see for example [44, 45]). Tested glove is placed over the surrogate, or boot (shoe) is placed over the surrogate 'wearing' a sock (socks). Assembly is placed into the fridge or freezer (controlled temperature is mimicking desired outdoor conditions) and dynamic temperature reduction at different surrogate positions is recorded. Evaluation of a separate item (glove or footwear) is rather qualitative; it clearly indicates the places where the heat is lost most effectively, and giving an indication how fast does it happen. However, this method easily allows for some quantitative comparison of the gloves and boots (shoes) of different models and different design. In present state this method is already used for supporting the manufacturers in improving heat retaining properties of their products, and comparing their models with the ones supposed to be 'best in test'.

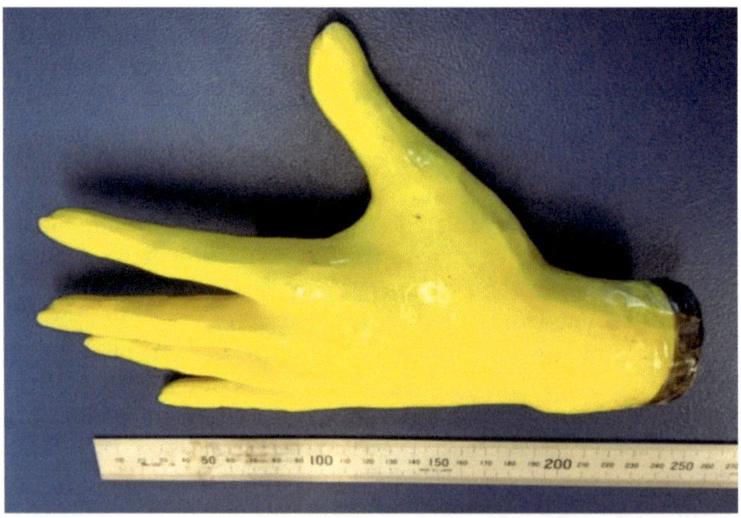

Figure 2.
Example of a 'hand surrogate' shell additively manufactured in ABS for the heat loss assessment of the gloves.

Further developments of the method are under the way. This will need coupling of the physical models to the computer ones; making surrogates better representing thermal properties of the skin and tissue; and adding thin film heaters to represent the heat generated by the body. The issue of materials properly representing the properties of the body tissues is quite complicated and needs special research. Some of the aspects related to matching the properties of synthetic materials are discussed in the following sections.

4.3 Advanced AM body part surrogates with hard and soft tissue

Additive manufacturing is also used for the manufacturing of quite complex body part surrogates combining both hard (bones) and soft (cartilage, tissue, skin) elements. These are ranging from the head and neck surrogates [10–13, 21, 22, 25, 46–50] to some simpler human thigh [51], foot [52], foot and ankle [53, 54] and wrist ones [55–57].

It should be noted, that in accordance with previously stated purpose-centered design of models, known head and neck surrogates are quite different in design and construction. Although all of them are intended to study the dynamic processes happening under the potentially traumatic conditions in the human head, some of them are targeting falls or collisions [10–13, 21, 22, 25, 46, 47], and some—blast injury conditions [48, 49]. The common automotive test dummies de facto also have head and neck surrogates of specific design. However, these are intended for studying impacts with much higher energy, as compared to the cases when head surrogates are used for assessing the falls and a level of protection of different helmets. Commonly car crash dummies have much stiffer neck surrogates, and solid head surrogates with a single sensor set placed in its center of gravity (see, for example, [58, 59]). Indeed, with such impact energy detailed head surrogates with higher biofidelity most probably are not needed.

Although initial tests with the head surrogate developed by the authors together with colleagues, were using Hybrid III crash dummy neck [10–13], after analyzing the data it was concluded that a dedicated neck surrogate should be constructed for better representing the human neck properties. Latest head-neck surrogate developed by

Figure 3.
Head surrogate with the hybrid III neck and ski helmet on (a), separate view of hybrid III neck (b), the overall view of the new biofidelic neck surrogate (c). Reference white bar is 10 cm.

Mid Sweden University and University of Padua is more biofidelic, and the analysis of the data acquired in the corresponding experiments will be published soon.

Figure 3 presents overall view of the head surrogate together with the Hybrid III neck and ski helmet on prepared for the assessment of the pendulum impact (a), separate view of Hybrid III neck with the connection flanges (b), and the overall view of the new biofidelic neck surrogate undergoing compression tests (c). Dimensions and properties of the corresponding body part surrogates are chosen to be as close to that of real ones as possible. Cranium skin and brain models are developed using a CT scan of a volunteer in order to provide realistic dimensions and fit commercial helmets. Brain surrogate weight is chosen close to the statistically average of 1300 g of a human brain accepted in modern anthropology. Hybrid III neck (**Figure 3a**) is commercial one, a part of the standard Hybrid III 50th Male crash dummy set by Humanetics.

Human head and neck surrogates developed to the date by the authors and colleagues are designed for studying dynamic processes happening in the human head subjected to the impact of heavy object (in particular case-heavy pendulum), and to assess the protective properties of different helmets. Head surrogates consist of the cranium additively manufactured from ABS; soft rubber brain surrogate with soft fabric mimicking dura matter over it; low viscosity liquid mimicking cerebral fluid; and rubber skin and head tissue surrogate. To provide needed biofidelity, 3D shapes of all components are based upon the processed CT scan of a real subject. Rubber of needed consistence (hardness) today cannot be additively manufactured, so ABS casting forms are prepared for the brain, outer skin and head tissue surrogates, and two-component silicone rubber is molded into them.

A number of sensors is embedded into the system. Three-axis accelerometers and gyroscopes are embedded at different locations of the brain surrogate. To achieve this casting of the 'brain' is carried out in steps, embedding sensors at certain level, and casting another layer of the rubber over it. During the tests, additional accelerometers and gyroscopes are attached over the cranium, over the skin surface, on the helmet and on the neck surrogate. Head-neck surrogate is placed over the force platform recording dynamic forces and torques at the base of assembly during and after the impact. Analysis of the signals from the accelerometers and gyroscopes allows for the assessment of the relative motion of the 'brain', 'cranium' 'helmet' and 'neck' during and after the impact. A number of the sensors monitoring the cranial fluid pressure is embedded into the 'scull' adding to the understanding of interactions between the surrogate elements. Novel sensor allowing measuring the tri-axial compression and strain in the 'brain' was developed and embedded into the latest version of the head surrogate [60]. Sampling of the signals during the tests is fast enough to adequately represent the dynamics of the studied processes (3–10 thousand samples of all measured values per second). A number of ski safety helmets including the ones with novel design were assessed using the setup with the described head-neck surrogate. Along with the large amount of data used for the cross-validation of the system, certain mechanisms related to traumatic brain injuries were also clarified.

There is a large number of particular mechanisms leading to human head trauma commonly summarized by the term 'concussion'. One of the large problems for treating such cases is that the events leading to it are very fast, and often either the patient or surrounding people have no clear recollection about the particular things leading to the trauma. Consequently, doctors can only back-reconstruct possible details (type of impact, direction of the forces etc.) basing on the consequences, which in many cases may show up days after the event. Using experiments with the head and neck surrogates of the described type mechanisms of some particular consequences can be clarified. For example, appearance of the hemorrhaging at the back of the scull resulting from some frontal head impacts was initially puzzling for us as researchers. Experimental data indicate that with the frontal impacts of certain energy and duration, compression of the cranial fluid at the frontal part of the head causes brain moving back without impacting the frontal part of the cranium. However, at the later stages, inertia of the brain motion causes its contact with the back of the cranium, which starts its return motion supported by the flexion of the neck. Another effect that can be clearly observed with the impact point offset from the vertical symmetry plane of the head-neck system is the cerebellum rotating around its 'neck', and two brain lobes moving separately in relation to each other. Such motion can be expected mainly from the rotation of the head, and non-central impact results in both translational motion and strong enough rotational component. In real life event, such impacts can lead to the damage of the axons in the areas where the brain tissue is twisted and is subjected to shear forces. Detailed information about the construction, data analysis and conclusions from the head and neck surrogate experiments can be found elsewhere [10–13, 21, 22, 25, 46–50].

5. Approximations inherent to the additively manufactured body part surrogates

As mentioned in the discussion on general modeling concepts, all models are approximate to some or another degree. Such approximations should be analyzed clarifying possible sources of misrepresentation and errors, and providing the ways for improving fidelity of the body part surrogates as physical models. It is quite clear that two main approximations are coming from the representation of 3D shapes of body parts, and representation of the natural tissues by synthetic materials. It is also worth analyzing what kind of uncertainties is added due to the different aspects of sensors and measurement systems used together with body part surrogates.

5.1 Shape approximations

Although the CT scans or MRI images have high spatial resolution, defining the exact boundaries for the 'selected body parts' is not a trivial exercise. Soft tissues often have properties gradually changing even across the body parts regarded by medicine as anatomically 'well-defined'. Although bone properties are commonly quite different from that of soft(er) tissues, in some cases osteoporotic bones are hard to correctly outline in the X-ray images or in the scan files. Consequently, 3D shape model files resulting from either automated or manual deconvolution based on the image density are already approximate. However, these shape files commonly contain excessive amount of details complicating the situation for the computer modeling software, and in some cases for the software generating the slice-definition files accepted by the production systems using layer-by-layer additive manufacturing. Although additive manufacturing systems in many cases can handle large files with scores of small details in 3D-shape files, finite element modeling and other simulating software either would need too much computing time, or simply would completely reject such files. As a result, 3D shape files are smoothed to be acceptable for both physical and mathematical modeling. For example, in the case of described head-neck surrogate, the 'wrinkled' surface of the brain, and both inner and outer surfaces of the cranium

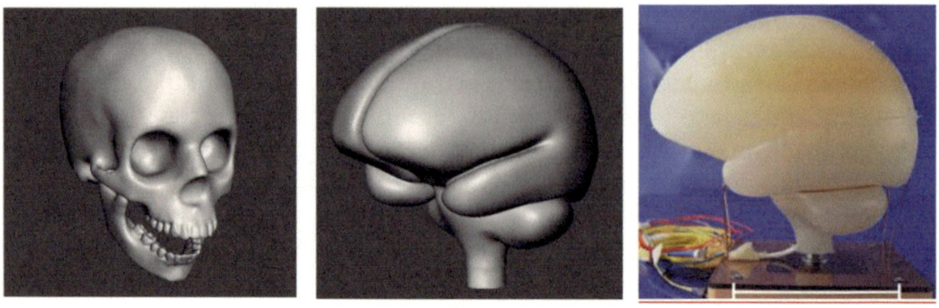

Figure 4.
Smoothed 3D representations of the scull (a) and brain (b), and resulting soft polymer brain surrogate (c). Reference white bar is 10 cm.

were approximated by the smooth envelopes. **Figure 4** presents the images of the smoothed 3D representations of the scull (a) and brain (b), and a photograph of the brain surrogate prepared by casting using molds additively manufactured basing on the smoothed brain shape file.

5.2 Approximations related to synthetic material properties

Another source of approximation relates to the fact that mechanical properties of the synthetic materials do not exactly replicate the ones of natural body tissues. This brings in a number of approximating steps. For example, ABS polymer is not as strong as natural cranium bone and has lower hardness, consequently the temporal sides of the cranium surrogate were made some thicker to provide adequate strength. Although this approximates directly relates to the difference in material properties (human Vs synthetic 'materials'), it further contributes to the shape approximations. In addition, properties in different parts of the tissue (for example, brain or skin) vary depending on the exact location (see, for example [61]), can be different from subject to subject, or even change for the same subject depending on the level of tissue hydration etc. At present majority of the body part surrogates use different materials only for the tissues with significant differences in the tissue properties (bone, brain, skin and head tissue) and disregard the property variations using same synthetic material in defined sections. In relation to the discussed head and neck surrogate, cranial bone is modeled by ABS polymer, brain- by soft two-component silicone rubber, skin and head tissue- by the two-component silicone rubber harder than that of the brain disregarding any porosity and density variations of natural tissues.

Even bigger source of approximations is due to selecting synthetic materials with properties adequately replicating the ones of natural tissues. As it was mentioned earlier, in order to increase the fidelity of the models and adequate cross-validation, both models should be based on materials with identical properties. However, using synthetic materials with known properties strongly deflecting from that of the natural tissues is of no use. Thus, a choice of materials with 'good enough' properties most closely matching that of natural tissues becomes a separate issue.

Studying mechanical properties of natural tissues in itself is a very complex research subject. There exists a number of publications where corresponding parameter values can be found. However, it is not possible to review all different values reported and considerations of variable conditions under which such values were measured within present chapter. Interested reader can be referred to some relevant publications, and scores of references given therein (see for example [26, 61–65]). Here we should just point out to some of the aspects related to the measuring of corresponding properties.

First of all, properties of the tissues *in vivo* and *ex vivo* are different, as the tissue extracted from the body rapidly degrades even if dehydration or drying is prevented by performing the experiments in the buffer solution (see, for example, [64, 65]). Secondly, in many cases mechanical properties of natural tissues are nonlinear [66], and can differ for static and dynamic conditions. The latter is quite critical for adequate computer modeling, but also means that properties of the soft synthetic materials should be measured both under static and dynamic conditions [60]. Additional issues in referring to the measured properties of natural tissues is the spread of the reported values. Indeed, measuring such properties is quite complicated, and often is performed in differing conditions. In addition, some of the publications do not specify measurement conditions exactly, and using such values results in relatively arbitrary decision, which one should be used.

It is also worth briefly discussing if so-called ballistic gel can be successfully used for the brain surrogates. Indeed, ballistic gel of different types (both gelatin-based [67] and synthetic [68]) is quite widely used in different applications. Referring to the *purpose of the modeling*, main applications of such materials are within modeling of the penetration of different hard elements (including projectiles) into the human body. Gelatin-based gels are hardly acceptable in the cases where long term use of the surrogates is intended, as they do not have reasonable longevity and will degrade being placed for example into a sealed head surrogate. In addition, validation of ballistic gels as model materials was performed with this specific purpose in mind, and would need additional validation if used for different purposes (e.g. human brain surrogate for the impact tests) as compared to what they were designed for. At the moment main medicine-related applications for such ballistic gels is with variety of 'phantoms' used for testing or calibrating different scanning devices (CT, MRI or ultrasonic). Novel synthetic ballistic gels can also be rather expensive. Two-component silicone rubber mainly used for manufacturing artistic masks appeared to be an adequate choice in many cases.

Due to the complexity of the subject under the study, validation of the body surrogate models is carried out not only against the computer ones, but also against the measurements acquired using laboratory animals and human cadavers. Corresponding literature on these is quite abundant, but one should be careful with the parameter values measured in animal experiments. Although it is commonly assumed that pigs and primates are relatively good *model subjects* (e.g. acquired results are again approximate), access to such experiments is quite restricted. Thus, results acquired win the human cadaver studies in many cases are regarded as a 'golden standard'. Interested readers can be suggested to start from the papers on cadaver studies authored by Nahum et al. [69, 70] and Hardy et al. [71]. In addition, one can start from the papers on the computer models used in head trauma research and protection development authored by Hardy et al. [72], Kleiven et al. [73–76] and Gilchrist et al. [77, 78]. These papers are authored by the multidisciplinary groups with the specialists in medicine, engineering, physics and computer technology.

5.3 Extra uncertainties added by measurement-related errors

There are certain issues related to the measurements contributing to the approximate nature of the body part surrogate systems. As mentioned earlier, a great advantage of such systems is the possibility to have multiple embedded sensors of different types. However, application of the sensors and integration of them into the measurement system is a separate task demanding special care.

Basing on the common purposes of the body surrogates as models, most commonly used sensors are for the temperature, acceleration (linear and angular), pressure and tension. For the first three types of measurements, one can use modern chip-like components having very small volume (down to 1 mm thick, and few mm footprint, making them almost noninvasive. Pressure sensors are almost equally small, and when used for monitoring cranium liquid dynamics embedding them into the 'bone' does not disturb the system. Tension in the bone surrogates can be successfully monitored by the strain gauges. However, the issue of measuring of the tension in soft tissue is not trivial. Strain gauges are very sensitive to bending, their application in soft tissues was tested using silicone rubber samples and disregarded. There are some suggestions that optical fibers can be used for such measurements (e.g. [79]), but so far no body surrogate-based systems using them were reported.

Many modern pressure sensors can detect both positive and negative values, which allows using them inside the soft tissue surrogates to monitor both compression and tension. However, if one needs to make corresponding 'triaxial' element, P-sensor tripod is constructed [60]. It is some larger than any of the previously discussed sensors, and its distortion of the soft media should be carefully considered, and resulting sensor element should be carefully calibrated.

Next issue is related to the cables connecting sensors to the measurement system. Each of the sensors needs up to 6–8 connecting wires, and using commercial cabling is too disturbing when embedded into soft tissue surrogates. Authors were forced to make own wire bundles containing enameled wires with outer diameter of 0.10–0.15 mm. These bundles were deformed near the sensors (the part which is resides in the tissue surrogate surrogate) to produce a meander section, which does not prevent free compression-expansion of the tissue surrogate material. Unfortunately, this does not prevent such wires to be broken, and first head surrogate versions were replaced as many of the connections to the embedded sensors were lost. Corresponding bundles of connecting wires coming out of the head and brain surrogates can be seen in the images of **Figures 3a** and **4c**.

Additional issues are related to the sensors, their calibration and ways of data acquisition. In the described head and neck surrogate authors are using triaxial MEMS (micro electromechanical sensors) accelerometers and gyroscopes with large dynamic range (±200 g in linear and ±3000 °/s in angular acceleration) and very small footprint. The manufacturer does not calibrate such commercial chips, and their sensitivity is commonly declared in the datasheets as 'typical' with possible deflections within 5–10%. It is possible to calibrate them, but this needs very special complicated and costly equipment. Consequently, authors have accepted the levels of uncertainty declared in the datasheets, and potential errors in measured values are all the time accounted for.

Described head and neck surrogate design is using all sensors with analogue output. One of the basic reasons is the need to monitor multiple sensors with quite high sampling rate. It is possible to have commercial data acquisition systems with even higher than necessary sampling rate and quite large number of analogue input channels. Therefore, limitation to the time resolution of such measurement systems with analogue-output sensors is limited by the sensing elements. Unfortunately, acceleration and gyroscope MEMS sensors with analogue outputs and adequate sensitivity are not manufactured any longer, forcing the migration to the digital output versions. Such digital-output sensors with serial type interfaces are commercially available, commonly have less connecting wires and allow using bus-type connection (multiple sensors are connected to the same data lines). However, they do not allow for easy arrangement of the synchronized sampling and high acquisition rates with long connecting cables. In many systems with the body part surrogates, such data collection is adequate. However, when fast acquisition rates and large numbers of sensors are needed digital data interfacing becomes problematic. Consequently, it forces reconsidering whole data acquisition concept for the body part surrogates containing large numbers of embedded sensors. To our knowledge, such systems were not yet constructed, and most probably, they will need intermediate small-size microcontroller-based data collection modules embedded into or directly attached to the body part surrogates. These modules should perform synchronized data collection with high sampling rates and storing data into onboard memory, and acquired data transfer into the mainframe computer after the experiment.

Described head-neck surrogate setup using cable connections between the sensors and data acquisition system. This seriously limits its possibilities to the impacts from the falling weights or pendulum strikes. The same time, validation of the protection helmets (including the standard methods) is commonly using drop tests from significant height (one can be referred to good and not extremely specialized sources such as [80–85]). Cable-based data acquisition in such systems is at least inconvenient, and hardly possible in many cases. Potential way forward for the data collection from head surrogate drop tests is applying wireless digital data links to the data acquisition systems and mainframe computers. Such system can be quite technically demanding in design and construction, but would provide much higher flexibility as compared to the wire-link systems. To be successful, such system should overcome potential issues with digital data link stability, and most probably, it should present some compromise between the desire of having multiple sensor channels and capacity of the data rate transfer. Another possibility can be provided by a 'hybrid' data transfer scheme using a single optical fiber link between the head-neck surrogates and data acquisition modules. It would be definitely advantageous for the systems, where the 'surrogate twin' is stationary, but considering low cost of the modern optical fibers, it can be used for the drop tests, from time to time replacing the damaged fiber link.

Further developments of the complicated 'surrogate twin' systems will definitely need alignment with the developments in the purposes for such modeling, including the development of the new standards for the safety equipment testing (see, for example, [84, 85]).

6. Conclusions

Surrogate-based physical models became a potent addition to the computer modeling allowing mutual cross-validation of both approaches and leading to higher result reliability. In many cases, experimental setups using human body part surrogates aiming for testing safety equipment being more intuitively understandable and in some cases providing faster analysis are better appreciated by the industry.

Although additive manufacturing can be regarded as just an enabling tool, body part surrogate-based modeling using AM starts to form own research and development field with fast penetration from comfort and safety applications into the field of medicine and traumatology. Together with embedded sensors, body part surrogates provide data collection in controlled environment and conditions closely reflecting real life situations.

Although different surrogate-based systems are constructed with different target purposes, all of them have similar advantages:

- performing experiments with 'surrogate twins' that approximate real humans without endangering any human subject,
- allowing for monitoring and clarification of the dynamics of the processes inside the 'body parts' with the sensors placed in positions not even imagined with alive subjects;
- providing data for the cross-validation and improvement of both themselves as physical models, and computer models and 'virtual twins',
- providing the means for testing and assessment of the protection devices before they are allowed for any real life use;
- allowing for the objective comparison of the wearables and wearable devices.

Designing and constructing body part surrogate systems should start from clear understanding of the purpose, which will strongly influence the complexity of the surrogates themselves and measurement systems linked to them. Early analysis of the approximations, which are inevitably 'built into' each model, including body part surrogates and the overall setup, will prevent many possible mistakes and help avoiding potential disillusions. Thus, two major approaches in further development of the human body part surrogates can be outlined: research-centered and development-centered. Both approaches are actively developed but have different aims.

Research-centered approach commonly aims at quite complex, flexible setups, allowing studies within a number of differing applications under a wide variety of experimental conditions. Such systems commonly have large number of sensor channels providing fast sampling and resulting in large volumes of static and dynamic data. Running such systems, and performing data analysis with such systems is quite complicated and often time consuming.

Although more testing and analysis commonly needed for the industry can be performed by research systems, they are not optimal for many industry-demanded applications. Corresponding development-centered body part surrogate systems should be focusing at certain applications, allowing for faster data analysis and provide the results in the way common for industrial users. Such systems are also easier to manufacture and are less expensive. Our experience shows, that well-designed application-oriented body part surrogate systems that can be used without involving special research personnel have a good potential of becoming commercial products.

Acknowledgements

Authors are acknowledging the input from Professor Nicola Petrone from Padua University, Italy, for conceiving many of the described ideas and participating in the 'head-neck' surrogate project. We also acknowledge the contribution of students from Padua University, who were performing their Master Thesis work at SportsTech research Center, Östersund, Sweden, aligned with this project: Ludovico Riello

(2021), Leonardo Marin (2020), Gianluca Candiotto (2019), Eduardo Marzella (2018), Federico Uriati (2018), Luca Broggio (2017) and Giovanni Carraro (2016).

Conflict of interest

The authors declare no conflict of interest.

Author details

Andrey Koptyug* and Mikael Bäckström
SportsTech Research Center, Mid Sweden University, Östersund, Sweden

*Address all correspondence to: andrey.koptyug@miun.se

References

[1] Frigg R, Hartmann S. Models in science. In: Zalta E, editor. The Stanford Encyclopedia of Philosophy. Stanford; 2020. Available from: https://plato.stanford.edu/archives/spr2006/entries/models-science/. [Accessed: November 16, 2022]

[2] Morgan M, Morrison M, editors. Models as Mediators: Perspectives on Natural and Social Science (Ideas in Context). Cambridge, UK: Cambridge University Press; 1999. DOI: 10.1017/ CBO9780511660108. ISBN: 9780511660108. Available from: http://strangebeautiful.com/other-texts/morgan-morrison-mods-mediators.pdf. [Accessed: November 16, 2022]

[3] Hartmann S. Modeling in philosophy of science. Representation, evidence, and justification. In: Frauchiger M, Essler WK, editors. Themes from Suppes (Lauener Library of Analytical Philosophy). Berlin, Boston: De Gruyter; 2013. pp. 95-122. DOI: 10.1515/9783110323566.95. Available from: https://www.modelbenders.com/pom.html

[4] LaValle SM. Virtual Reality. Cambridge, UK: Cambbridge University Press; 2016. p. 418. Available from: http://msl.cs.uiuc.edu/vr/vrbookbig.pdf. [Accessed: November 16, 2022]

[5] Onyesolu MO, Eze FU. Understanding virtual reality technology: Advances and applications. In: Schmitt M, editor. Advances in Computer Science and Engineering. London, UK: IntechOpen; 2011. DOI: 10.5772/15529. Available from: https://www.intechopen.com/chapters/14397 [Accessed: November 16, 2022]

[6] Segovia M, Garcia-Alfaro J. Design, modeling and implementation of digital twins. Sensors. 2022;**22**:5396. DOI: 10.3390/s22145396

[7] Fuller A, Fan Z, Day C, Barlow C. Digital twin: Enabling technologies, challenges and open research. IEEE Access. 2020;**8**:108952-108971. DOI: 10.1109/ACCESS.2020.2998358

[8] Weifei H, Tongzhou Z, Xiaoyu D, Zhenyu L, Jianrong T. Digital twin: A state-of-the-art review of its enabling technologies, applications and challenges. Journal of Intelligent Manufacturing and Special Equipment. 2021;**2**(1):1-34. DOI: 10.1108/jimse-12-2020-010

[9] Cronskär M, Rännar L-E, Bäckström M. Implementation of digital design and solid free-form fabrication for customization of implants in trauma orthopaedics. Journal of Medical and Biological Engineering. 2012;**32**(2):91-96. DOI: 10.5405/jmbe.883

[10] Koptyug A, Petrone N, Bäckström M. Body part surrogates for safety research. Transactions on Additive Manufacturing Meets Medicine. 2021;**3**(1):559. DOI: 10.18416/AMMM.2021.2109559

[11] Petrone N, Candiotto G, Marzella E, Uriati F, Carraro G, Bäckström M, et al. Feasibility of using a novel instrumented human head surrogate to measure helmet, head and brain kinematics and intracranial pressure during multidirectional impact tests. Journal of Science and Medicine in Sport. 2019;**Suppl. 1**:S78-S84. DOI: 10.1016/j.jsams.2019.05.015

[12] Petrone N, Carraro G, Castello SD, Broggio L, Koptyug A, Bäckström M. A novel instrumented human head surrogate for the impact evaluation of helmets. MDPI Proceedings. 2018;**2**(6):269. DOI: 10.3390/proceedings2060269

[13] Koptyug A, Bäckström M, Petrone N. Developing an instrumented physical model of the human head for studying concussion mechanisms. In: Proc. iSSs (International Ski Safety Society): The Ski Trauma and Ski Safety Congress; April 17-22, 2017; Innsbruck Austria

[14] Goldstein Y. From 3D design to 2D patterns involving realistic drape/fit and comfort simulation. In: Walter L, Kartsounis G-A, Carosio S, editors. Transforming Clothing Production into a Demand-Driven, Knowledge-Based, High-Tech Industry. London: Springer; 2009. pp. 95-105. DOI: 10.1007/978-1-84882-608-3

[15] Smith RD. Principles of Modeling Real World Systems. Model Benders LLC; 2010. Available from: https://www.modelbenders.com/pom.html. [Accessed: November 30, 2022]

[16] Spirtes P, Glymour C, Scheines R. Causation, Prediction, and Search. Adaptive Computation and Machine Learning Series. 2nd ed. Cambridge: MA: MIT Press; 2001. p. 568

[17] Crandall JR, Bose D, Forman J, Untaroiu CD, Arregui-Dalmases C, Shaw CG, et al. Human surrogates for injury biomechanics research. Clinical Anatomy. 2011;**24**(3):362-371. DOI: 10.1002/ca.21152

[18] Vafadar A, Guzzomi F, Rassau A, Hayward K. Advances in metal additive manufacturing: A review of common processes, industrial applications, and current challenges. Applied Sciences. 2021;**11**:1213. DOI: 10.3390/app11031213

[19] Jiménez M, Romero L, Domínguez LA, del Mar Espinosa M, Domínguez M. Additive manufacturing technologies: An overview about 3D printing methods and future prospects. Complexity, Special issue on Complexity in Manufacturing Processes and Systems. 2019;**2019**:9656938. DOI: 10.1155/2019/9656938

[20] Popov VV, Grilli ML, Koptyug A, Jaworska L, Katz-Demyanetz A, Klobčar D, et al. Powder bed fusion additive manufacturing using critical raw materials: A review. Materials. 2021;**14**:909. DOI: 10.3390/ma14040909

[21] Koptyug A, Rännar L-E, Bäckström M, Franzén S, Dérand P. Additive manufacturing technology applications targeting practical surgery. International Journal of Life Science and Medical Research. 2013;**3**(1):15-24. DOI: 10.5963/LSMR0301003

[22] Koptyug A, Rännar L-E, Bäckström M, Cronskär M. Additive manufacturing for medical and biomedical applications: Advances and challenges. Materials Science Forum. 2014;**783-786**:1286-1291. DOI: 10.4028/www.scientific.net/MSF.783-786.1286

[23] Bibb R, Eggbeer D, Paterson A. Medical Modelling: The Application of Advanced Design and Rapid Prototyping Techniques in Medicine. Cambridge, UK: Woodland Publishing Series in Biomaterials; 2015;**91**:516. DOI: 10.1016/C2014-0-01365-2

[24] Ganesan B, Al-Jumaily A, Luximon A. 3D printing technology applications in occupational therapy. Physical Medicine and Rehabilitation International. 2016;**3**(3):1085-1088. Available from: https://austinpublishinggroup.com/physical-medicine/fulltext/pmr-v3-id1085.pdf. [Accessed: November 16, 2022]

[25] Koptyug A, Bäckström M, Rännar L-E. 3D-printing: A future "Magic Wand" for global manufacturing— How can we benefit from it today for sports and health care? In: Vilas-Boas JP, Correia PP, Cabri J, editors. Proceedings

of the 5th International Congress on Sport Sciences Research and Technology Support, icSPORTS 2017, Funchal, Madeira, Portugal, October 30-31, 2017. Lda, Setúbal, Portugal: SciTePress, Science and Technology Publications; 2017. DOI: 10.5220/0006810300010001. [Accessed: November 16, 2022]

[26] Okkalidis N, Marinakis G. Technical note: Accurate replication of soft and bone tissues with 3D printing. Medical Physics. 2020;**47**(5):2206-2211. DOI: 10.1002/mp.14100

[27] Hongjian L, Fa W, Xiao Z. Three-dimensional printing: The potential technology widely used in medical fields. Journal of Biomedical Materials Research Part A. 2020;**108**(11):2217-2229. DOI: 10.1002/jbm.a.36979

[28] O'Reilly M, Hoff M, Friedman SD, Jones JFX, Cross NM. Simulating tissues with 3D-printed and castable materials. Journal of Digital Imaging. 2020;**33**(5):1280-1291. DOI: 10.1007/s10278-020-00358-6

[29] Segaran N, Saini G, Mayer JL, Naidu S, Patel I, Alzubaidi S, et al. Application of 3D printing in preoperative planning. Journal of Clinical Medicine. 2021;**10**(5):917. DOI: 10.1007/s10278-020-00358-6

[30] Tejo-Otero A, Buj-Corral I, Fenollosa-Artés F. 3D printing in medicine for preoperative surgical planning: A review. Annals of Biomedical Engineering. 2020;**48**(2):536-555. DOI: 10.1007/s10439-019-02411-0

[31] Manmadhachary A, Malyala SK, Alwala A. Medical Applications of Additive Manufacturing. Pandian D, Fernando X, Baig Z, Shi F, editors. Proceedings of the International Conference on ISMAC in Computational Vision and Bio-Engineering 2018 (ISMAC-CVB), ISMAC. Lecture Notes in Computational Vision and Biomechanics. Cham: Springer; 2019; LNCVB. 2018;**30**:1643-1653. DOI: 10.1007/978-3-030-00665-5_152

[32] Wu AM, Wang K, Wang JS, Chen CH, Yang XD, Ni WF, et al. The addition of 3D printed models to enhance the teaching and learning of bone spatial anatomy and fractures for undergraduate students: A randomized controlled study. Annals of Translational Medicine. 2018;**6**(20):403. DOI: 10.21037/atm.2018.09.59

[33] Ye Z, Dun A, Jiang H, Nie C, Zhao S, Wang T, et al. The role of 3D printed models in the teaching of human anatomy: A systematic review and meta-analysis. BMC Medical Education. 2020;**20**:335. DOI: 10.1186/s12909-020-02242-x

[34] Ramirez M, Nurmukhametov R, Musa G, Barrientos Castillo RE, Arno Encarnacion VL, Soriano Sanchez JA, et al. Three-dimensional plastic Modeling on bone frames for cost-effective neuroanatomy teaching. Cureus. 2022;**14**(7):e27472. DOI: 10.7759/cureus.27472

[35] Cronskär M, Rännar L-E, Bäckström M, Nilsson KG, Samuelsson B. Patient-specific clavicle reconstruction using digital design and additive manufacturing. Journal of Mechanical Design. 2015;**137**(11):111418. DOI: 10.1115/1.4030992

[36] Cronskär M, Bäckström M. Modeling of fractured clavicles and reconstruction plates using CAD, finite element analysis and real musculoskeletal forces input. WIT Transactions on Biomedicine and Health. 2013;**17**:235-243. DOI: 10.2495/BIO130211

[37] Cronskär M, Rasmussen J, Tinnsten M. Combined finite element and multibody musculoskeletal

investigation of a fractured clavicle with reconstruction plate. Computer Methods in Biomechanics and Biomedical Engineering. 2015;**18**(7):740-748. DOI: 10.1080/10255842.2013.845175

[38] Emerson N, Carré M, Reilly G, Offiah A. Geometrically accurate 3D FE models from medical scans created to analyse the causes of sports injuries. Procedia Engineering. 2015;**13**:422-427

[39] Emerson N, Carré M, Reilly G, Offiah A. Simulation based upon medical data offers a fast and robust method for the prediction of fracture risk. Procedia Engineering. 2013;**2011**(60):459-466. DOI: 10.1016/j.proeng.2013.07.051

[40] Emerson N, Offiah A, Reilly G, Carré M. Patient-specific finite element modelling and validation of porcine femora in torsion. Strain. 2013;**49**(3):212-220. DOI: 10.1111/str.12029

[41] Fan J, Chen Y, Zhang WA. Perspiring fabric thermal Manikin: Its development and use. In: Proceedings of the Fourth International Meeting on Thermal Manikins (4IMM). Hong Kong: The Hong Kong Polytechnic University; 2006. pp. 7-12 . Available from: https://www.lboro.ac.uk/microsites/lds/EEC/ICEE/textsearch/manikin%20meeting%202006/2nd%20publish%20Proceeding_all-in-one%5B061005%5Drev02.pdf. [Accessed: November 16, 2022]

[42] Watson C, Nawaz N, Troynikov O. Design and evaluation of sport garments for cold conditions using human thermoregulation modeling paradigm. Procedia Engineering. 2013;**60**:151-156. DOI: 10.1016/j.proeng.2013.07.053

[43] Anttonen H, Niskanen J, Meinander H, Bartels V, Kuklane K, Reinertsen RE, et al. Thermal manikin measurements—exact or not? International Journal of Occupational Safety and Ergonomics. 2004;**10**(3):291-300. DOI: 10.1080/10803548.2004.11076616

[44] Tamura T. Development of a two-layer movable sweating thermal manikin. Industrial Health. 2006;**44**(3):441-444. DOI: 10.2486/indhealth.44.441

[45] Sakoi T, Tsuzuki K, Kato S, Ooka R, Song D, Zhu S. Thermal comfort, skin temperature distribution, and sensible heat loss distribution in the sitting posture in various asymmetric radiant fields. Building and Environment. 2007;**42**(12):3984-3999. DOI: 10.1016/j.buildenv.2006.10.050

[46] Awad N, El-Dakhakhni W, Gilani A. A physical head and neck surrogate model to investigate blast-induced mild traumatic brain injury. Arabian Journal for Science and Engineering. 2015;**40**:945-958. DOI: 10.1007/s13369-015-1583-3

[47] Freitas CJ, Mathis JT, Scott N, Bigger RP, MacKiewicz J. Dynamic response due to behind helmet blunt trauma measured with a human head surrogate. International Journal of Medical Sciences. 2014;**11**(5):409-425. DOI: 10.7150/ijms.8079

[48] Zhang J, Pintar FA, Yoganandan N, Gennarelli TA, Son SF. Experimental study of blast-induced traumatic brain injury using a physical head model. Stapp Car Crash Journal. 2009;**53**:215-227. DOI: 10.4271/2009-22-0008

[49] Zhu F, Wagner C, Leonardi ADC, Jin X, VandeVord P, Chou C, et al. Using a gel/plastic surrogate to study the biomechanical response of the head under air shock loading: A combined experimental and numerical investigation. Biomechanics and Modeling in Mechanobiology. 2012;**11**:341-353. DOI: 10.1007/s10237-011-0314-2

[50] Awad IA, Carter BS. Surrogate measures of outcome in clinical trials: Clues worth examining? Neurosurgery. 2008;**63**(2):N5. DOI: 10.1227/01.NEU.0000335793.88007.CE

[51] Payne T, Mitchell S, Halkon B, Bibb R, Waters M. Development of a synthetic human thigh impact surrogate for sports personal protective equipment testing. Proceedings of the Institution of Mechanical Engineers, Part P: Journal of Sports Engineering and Technology. 2016;**230**(1):5-16. DOI: 10.1177/1754337115582294

[52] Tao Z, Ahn HJ, Lian C, Lee K-H, Lee C-H. Design and optimization of prosthetic foot by using polylactic acid 3D printing. Journal of Mechanical Science and Technology. 2017;**31**(5):2393-2398. DOI: 10.1007/s12206-017-0436-2

[53] Telfer S, Pallari J, Munguia J, Dalgarno K, McGeough M, Woodburn J. Embracing additive manufacture: Implications for foot and ankle orthosis design. BMC Musculoskeletal Disorders. 2012;**13**(1):84. DOI: 10.1186/1471-2474-13-84

[54] Thibodeau A. 3D-printed Surrogate of Lower Limb for testing ankle-foot [thesis]. Ottawa, Canada: Ottawa-Carleton Institute for Mechanical and Aerospace Engineering; 2021 . Available from: https://ruor.uottawa.ca/bitstream/10393/42755/1/Thibodeau_Alexis_2021_thesis.pdf. [Accessed: November 16, 2022]

[55] Adams C, James D, Senior T, Allen T, Hamilton N. Development of a method for measuring quasi-static stiffness of snowboard wrist protectors. Procedia Engineering. 2016;**147**:378-383. DOI: 10.1016/j.proeng.2016.06.320

[56] Adams C, Allen T, Senior T, James D, Hamilton N. Impact testing of snowboarding wrist protectors. Proceedings of the Institution of Mechanical Engineers, Part P: Journal of Sports Engineering and Technology. 2016;**230**(1):5-16. DOI: 10.1177/1754337115582294

[57] Adams C, James D, Senior T, Allen T, Hamilton N. Effect of surrogate design on the measured stiffness of snowboarding wrist protectors. Sports Engineering. 2018;**21**:217-225. DOI: 10.1007/s12283-018-0266-1

[58] Foster JK, Kortge JO, Wolanin MJ. Hybrid III-A BiomechanicallyBased crash test dummy. SAE Technical Paper 770938. 1977. DOI: 10.4271/770938

[59] Petrone N, Tamburlin L, Panizzolo F, Atzoria B. Development of an instrumented anthropomorphic dummy for the study of impacts and falls in skiing. Procedia Engineering. 2010;**2**:2587-2592. DOI: 10.1016/j.proeng.2010.04.036

[60] Zullo G, Leidy Silvestroni A, Candiotto G, Koptyug A, Petrone N. A novel multi-axial pressure sensor probe for measuring Triaxial stress states inside soft materials. Sensors. 2021;**21**(10):3487. DOI: 10.3390/s21103487

[61] van Dommelen JAW, van der Sande TPJ, Hrapko M, Peters GWM. Mechanical properties of brain tissue by indentation: Interregional variation. Journal of the Mechanical Behavior of Biomedical Materials. 2010;**3**(2):158-166. DOI: 10.1016/j.jmbbm.2009.09.001

[62] Dąbrowska AK, Rotaru G-M, Derler S, Spano F, Camenzind M, Annaheim S, et al. Materials used to simulate physical properties of human skin. Skin Research and Technology. 2016;**22**(1):3-14. DOI: 10.1111/srt.12235

[63] Pal S. Mechanical properties of biological materials. In: Pal S, editor.

Design of Artificial Human Joints & Organs. Boston, MA: Springer US; 2014. pp. 23-40. DOI: 10.1007/978-1-4614-6255-2_2

[64] Gefen A, Margulies SS. Are in vivo and in situ brain tissues mechanically similar? Journal of Biomechanics. 2004;**37**(9):1339-1352. DOI: 10.1016/j.jbiomech.2003.12.032

[65] Walton ME, Mars RB. Probing human and monkey anterior cingulate cortex in variable environments. Cognitive, Affective, & Behavioral Neuroscience. 2007;**7**(4):413-422. DOI: 10.3758/cabn.7.4.413

[66] Chen H, Zhao X, Lu X, Kassab G. Non-linear micromechanics of soft tissues. International Journal of Non-Linear Mechanics. 2013;**58**:79-85. DOI: 10.1016/j.ijnonlinmec.2013.03.002

[67] Ravikumar N, Noble C, Cramphorn E, Taylor ZA. A constitutive model for ballistic gelatin at surgical strain rates. Journal of the Mechanical Behavior of Biomedical Materials. 2015;**47**:87-94. DOI: 10.1016/j.jmbbm.2015.03.011

[68] Jason A. Synthetic Gelatins as soft tissue simulants. AFTE Journal. 2020;**52**(2):68-84

[69] Nahum AM, Smith RW. An experimental model for closed head impact injury. In: Proceedings of the 20th Stapp Car Crash Conference, Society of Automotive Engineers; October 18-20, 1976; Warrendale. 1976. DOI: 10.4271/760825

[70] Nahum AM, Smith RW, Ward CC. Intracranial pressure dynamics during head impact (No. 770922); SAE Technical Paper. In: NahumAM SR, Raasch F, editors. Twenty-First Stapp Car Crash Conference-P-073, Biomechanics of Impact Injury and Injury Tolerances of the Head-Neck Complex-PT-43. Warrendale, PA, USA: SAE International; 1977. DOI: 10.4271/770922

[71] Hardy WN, Mason MJ, Foster CD, Shah CS, Kopacz JM, Yang KH, et al. A study of the response of the human cadaver head to impact. Stapp Car Crash Journal. 2007;**51**:17-80. DOI: 10.4271/2007-22-0002

[72] Kleiven S. Predictors for traumatic brain injuries evaluated through accident reconstruction. Stapp Car Crash Journal. 2007;**51**:81-114. DOI: 10.4271/2007-22-0003

[73] Kleiven S, Hardy W. Correlation of an FE model of the human head with local brain motion – Consequences for injury prediction. Stapp Car Crash Journal. 2002;**46**:123-144. DOI: 10.4271/2002-22-0007

[74] Kleiven S. Head Injury Biomechanics & Criteria. Lecture notes for SD2450, Biomechanics and Neuronics. Available from: https://canvas.kth.se/courses/17965/files/2174379/download?verifier=ZAiCqV2OzQA9VRXZh5Ugj9kzH9HWcaWfyzmJNSA0&wrap=1

[75] Giordano C, Kleiven S. Evaluation of axonal strain as a predictor for mild traumatic brain injuries using finite element modeling. Stapp Car Crash Journal. 2014;**58**:29-61. DOI: 10.4271/2014-22-0002

[76] Giordano C, Kleiven S. Development of an unbiased validation protocol to assess the biofidelity of finite element head models used in prediction of traumatic brain injury. SAE Technical Paper Series, Stapp Car Crash Journal. 2016;**60**:363-471. DOI: 10.4271/2016-22-0013

[77] Horgan TJ, Gilchrist MD. The creation of three-dimensional finite

element models for simulating head impact biomechanics. International Journal of Crashworthiness. 2003;**8**(4):353-366. DOI: 10.1533/ijcr.2003.0243

[78] Rueda F, Cui L, Gilchrist MD. Finite element modelling of equestrian helmet impacts exposes the need to address rotational kinematics in future helmet designs. Computer Methods in Biomechanics and Biomedical Engineering. 2011;**4**(12):1021-1031. DOI: 10.1080/10255842.2010.504922

[79] Miller JL. Optical measurements probe the pressure and density of water under tension. Physics Today. 2011;**64**(1):14. DOI: 10.1063/1.3541933

[80] UK Department of Transport. SHARP helmet safety scheme. The Safety Helmet Assessment and Rating Programme. Laboratory test procedures. Available from: https://sharp.dft.gov.uk/wp-content/themes/sharp2017/pdfs/SHARP-laboratory-test-procedures.pdf. [Accessed: November 16, 2022]

[81] ASTM F1446. Standard Test Methods for Equipment and Procedures Used in Evaluating the Performance Characteristics of Protective Headgear

[82] Helmets: Testing Procedures & Standards. Available from: https://www.team-bhp.com/forum/ride-safe/147027-helmets-testing-procedures-standards-2.html. [Accessed: November 16, 2022]

[83] Haldin P, Kleiven S. The development of next generation test standards for helmets. In: Proceedings of the 12t International Conference on Helmet Performance and Design. London, UK; 2013. Available from: http://kth.diva-portal.org/smash/get/diva2:1276564/FULLTEXT01.pdf. [Accessed: November 16, 2022]

[84] MIPS (Multi-directional Impact Protection) system. Available from: https://mipsprotection.com/science-technology/test-lab/. [Accessed: November 16, 2022]

[85] Bishop PJ, Wells RP. The inappropriateness of helmet drop tests in assessing neck protection in head-first impacts. The American Journal of Sports Medicine. 1990;**18**(2):201-205. DOI: 10.1177/036354659001800215

Chapter 4

Technical Challenges and Future Environmentally Sustainable Applications for Multi-Material Additive Manufacturing for Metals

Valentina Pusateri, Constantinos Goulas and Stig Irving Olsen

Abstract

Through additive manufacturing (AM), it is now possible to produce functionally gradient materials (FGM) by depositing different metal alloys at a specific location to locally improve mechanical properties and enhance product performance. Despite recent developments, however, there are still some important trade-offs to consider and inherent challenges that must be addressed. These include limitations to the volume, size, and range of materials used and a data-driven strategy to drive decision-making and automation. Additionally, many potential advantages exist in environmentally sustainable terms of multi-material additive manufacturing (MM-AM). In particular, for products that require a complex design, high value, and low production volume, material and energy use can be reduced significantly. However, there are significant uncertainties in terms of environmental impact and applications of MM-AM that need to be addressed during the initial stage of the technology development to understand its potential future environmental performance improvements.

Keywords: functionally gradient materials, topology optimization, alloys, scale-up, sustainability

1. Introduction

There is agreement among all the countries of the world that we need to become more sustainable. The Paris Agreement on climate change provides the basis for an urgent need for global response to the threat of climate change and the 2030 Agenda for Sustainable Development addresses sustainability more broadly. A core aim is to increase the ability of countries to reduce and tackle the impacts of climate change, by supporting developing and the most vulnerable countries in joining the global effort. Manufacturing has an elevated environmental impact as it contributes with roughly 98% of the annual total direct carbon dioxide (CO_2) emissions, and the industrial sector alone (among energy, transportation, and building sectors) accounts for approximately a quarter of the global carbon emissions [1]. One of the major contributors of the industry sector is the steel production that represents 8% of the global CO_2 emission share [2]. In particular, in 2018 about 1.8 Gt steel was produced worldwide [3], this corresponds to roughly 2.1 Gt direct CO_2 emissions worldwide [4] and represents 8% share of the global CO_2 emissions [2]. Additionally, the increasing demand of ore mining for manufacturing [3, 5] has posed attention to the sustainable extraction and management of abiotic resources [5–7]. In a global perspective, the supply horizon and scarcity of elements have been considered an important indicators for the severity of increased extraction although newer methods focusing on environmental dissipation are being developed [8]. Nevertheless, it has also been argued that resource availability is very dependent on sociopolitical issues, opportunity costs, fixed stocks, etc., and therefore not well evaluated using an environmental tool such as life cycle assessment (LCA) [9]. From the perspective of the European economy, the European Commission monitors and evaluates the critical resources for the European economy based on their economic importance and supply risk [10]. In that report, they also evaluate the potential significance of supply risks in different industries, that is, 3D printing. They recommend diversifying the materials supply, especially titanium, and minor alloying elements such as scandium and niobium should be in focus. Additionally, the possibilities of recycling and reuse should be further investigated [10].

In the literature, the overall sustainability of conventional or additive manufacturing appeared to be product application or production technique dependent [1, 11–17]. In this context, the European project, Grade2XL, aims to investigate the potential of multi-material wire arc additive manufacturing (MM-WAAM) for large objects relative to strength, durability, and sustainability of engineering structures [18].

MM-WAAM is a variant of the wire arc additive manufacturing process (WAAM) that allows for the fabrication of complex parts with multiple materials. In WAAM, an arc is used to melt a wire consumable, which is then deposited layer by layer to build up the part [14, 19]. In multi-material WAAM, multiple wire consumables are used in the same part, allowing for the creation of multi-material parts with unique properties and improved performance.

From both a sustainability or recycling, and technical point of views, MM-WAAM still faces various challenges. The primary reason is that the materials are not easily differentiated into the different waste streams, thus reducing the recycling quality, and layering different alloys constitute a challenge [20, 21].

2. Current state of the technology for multi-material additive manufacturing

Multi-material additive manufacturing can produce highly complex products with improved functional behavior [20, 22], often at a reduced total cost. There are several available additive manufacturing methods able to achieve multi-material additive manufacturing with metals, each with its own set of advantages and limitations. The most common methods include:

- Laser Powder Bed Fusion: This method involves using a high-powered laser or electron beam to melt and fuse metal powders on a powder bed layer by layer. Even though typically suited for single material builds, certain modules exist that enable the selective distribution of different types of material powder at different locations in a given layer (e.g., Aerosint multi-material printing bundle [23]).

- Binder Jetting: Binder Jetting involves a printer head which deposits a liquid binder onto layers of powder. The binder binds the powder together, forming a solid part. The type of binder or powder being used, can be changed locally and in this way it is possible to create multi-material parts.

- Sheet Lamination: Sheet Lamination is an additive manufacturing method in which thin sheets of material (usually supplied via a system of feed rollers) are bonded together layer-by-layer to form a single piece that is cut into a 3D object using a CNC machine. Ultrasonic Consolidation is an AM method belonging to the sheet lamination family that enables multi-material printing by joining dissimilar metal sheets in solid state using ultrasonic vibrations.

- Directed Energy Deposition: This method involves using a high-powered laser or other heat source (Arc, electron beam) to deposit and fuse metallic material in powder or wire form directly onto a build platform. By moving the heat source and changing the type of material being deposited, it is possible to create multi-material parts. A recent commercialized multi-material DED solution was presented by Meltio [24], which enables the coaxial feeding of two different wire feedstock into a focus point of six individual lasers. MM-WAAM, the main technology studied in the Grade2XL project belongs to the DED family, and for this reason, we will analyze MM-WAAM further.

DED (and MM-WAAM) has some unique advantages [25], which make it a very attractive manufacturing method for large-multi-material components. These advantages include:

- High rate of material deposition, reaching up to 5 kg/h deposition rate per deposition unit.

- Ability to use widely available feedstock in the form of conventional welding wire and/or coarse metal powder, which is low cost compared to LPBF powders.

- Ability to produce components of large size without significant increase in equipment investment cost, since there is no need for a defined chamber in protected atmosphere. As a motion system, an industrial robot can be used, which enables to reach dimensions greater than 1 m.

- Ability to deposit different materials where it is needed in the component, without major process disruptions. The way the second material is delivered is versatile; it can be introduced by a second robot, tandem torch, or cold wire/powder feeder.

- Using multiple types of wire or blown powder as feedstock allows for the production of complex functional gradients, multi-material layers, and even composite materials.

3. Technical challenges and future applications for multi-material additive manufacturing for metals

For all Metal Additive Manufacturing (MAM) methods, their technical challenges currently hinder broad adoption of the technology in the industry. Some challenges are specific to each MAM method and some are common. The common challenges are related to the fact that a monolithic component is built with different materials, which is rather uncommon in the industry. Mixing and melting different materials together in a single component causes the risk of creating intermetallic phases and interfaces in the material that may exhibit undesired behavior, which needs to be studied case-by-case for every combination and mixing level. This of course means that in case the multi-material part needs to be certified, new protocols will need to be developed to account for these variations. Additionally, the alloys currently used are not designed to be mixed, so the elements present in each alloy are not necessarily designed to be compatible with other alloys in the same component. Non-material related but still very important is the limitation that MAM has compared to the single material counterpart in terms of scalability. Since MAM processes become increasingly complicated to accommodate multiple materials, the production workflows also become complicated, raising the costs.

The MAM techniques have some extra challenges. In LPBF, for example, the use of multiple materials becomes a complex matter that influences the whole process chain. From separating single material powders, reclaiming them, as well recycling the multi-material component remain challenging. However, there are more difficulties such as complexity in selective material deposition, issues with co-processing and material interface formation and developing materials and process modeling [26, 27]. For DED, there are some specific challenges as well, including the low resolution in terms of where the materials are mixed, which also results in extensive post-processing requirements. Additionally, in the case blown powder is used, a large proportion of the material used is not ending up in the part, but is lost or contaminated and not reused, which reduces the efficiency of the method.

Multi-material wire arc additive manufacturing (WAAM) can be used to fabricate a wide range of complex parts with unique properties and improved performance. Because it allows for the use of multiple materials in the same part, it opens new possibilities for the design and manufacture of parts with functional and performance characteristics that cannot be achieved using traditional manufacturing processes. Some examples of parts that can be made with multi-material WAAM include:

- Structural components with tailored mechanical properties: by combining materials with different mechanical properties, such as stiffness, strength, and toughness, it is possible to create structural components with tailored properties that are optimized for specific applications. For example, multi-material WAAM could be used to create aircraft components with high strength and low weight, or automotive or railway components with high stiffness and good fatigue resistance. For example, in **Figure 1**, the MM-WAAM produced bogie consists of a combination of steels with different strengths, which optimizes the mechanical behavior of the component while it minimizes the material that is used for its production.

Figure 1.
Topology optimized, MM-WAAM produced bogie for the railway industry. Combination of different steels yields superior mechanical behavior of the component. Image courtesy of RAMLAB BV.

- Functionally graded materials: by carefully controlling the composition and location of the different materials in a part, it is possible to create functionally graded materials with graded properties. For example, a part could be designed with a stiff core and a compliant outer layer, or with a high-conductivity region and a low-conductivity region. This can improve the performance and functional properties of the part.

- Complex shapes with internal features: because the WAAM process allows for the deposition of material layer by layer, it is possible to create complex parts with internal features that would be difficult or impossible to fabricate using other manufacturing processes. For example, multi-material WAAM could be used to create parts with internal channels, passages, or voids that are required for specific applications.

The potential applications of multi-material WAAM are vast and will continue to expand as the technology matures and advances. Since the potential of the technology is high, it is very important to assess the environmental consequences of producing parts with MM-WAAM and other MAM methods instead of traditional methods.

4. Life cycle assessment for multi-material additive manufacturing for metals

It is important to assess the sustainability impact of anthropogenic activities and technologies in order to understand what contributes most and how to reduce the impacts. To this purpose, life cycle assessment (LCA) was applied to WAAM, which is an emerging technology still under development and optimization. LCA is a state-of-the-art methodology for assessing multiple environmental impacts of a system over time and space throughout its lifecycle from cradle to grave, that is, from extraction of the materials through production and use or operation of the system till its end-of-life (EoL) [28]. It is further an ISO-standardized methodology [29, 30].

Many claim several benefits of additive manufacturing compared to conventional manufacturing [14, 15, 17, 19]. For instance, reduction of waste, energy, or fuel consumption during the use of product and reduction of cost due to optimization of shapes, lightweight design, and shorter time from ordering a product until you receive it. However, to what extent are these claims valid? Which quantifiable trade-offs are important? As explained, LCA takes a life cycle thinking perspective; this means that all the processes required to deliver the function of a product or activity from the raw material production to the disposal of it are included in the assessment (see **Figure 2**) [31].

Figure 3 represents a generic overview of all life cycle processes considered for Grade2XL products produced with WAAM. The two boxes with dotted lines represent avoided activities as incineration provides heat and power additionally to burning of waste, and with recycling of metal scrap is possible to avoid the production of a certain amount of virgin material. In this study, the modeling approach taken was attributional and system expansion was used to assess processes that provided a second service in addition to the main one (e.g., incineration) [31].

A further advantage of LCA is the holistic perspective of the comprehensive coverage of environmental issues. Rather than focusing exclusively on climate change, which currently receives generally most attention, LCA covers a broad range of environmental issues and impacts. For instance, it usually includes among others: freshwater use, land occupation and transformation, toxic impacts on human health, and depletion of non-renewable resources. In this way, the major trade-offs between impacts are addressed and burden-shifting can be avoided.

For all the reasons previously mentioned, the LCA framework is currently used to compare the sustainability of all Grade2XL products to the same objects fabricated with the traditional manufacturing processes. **Figure 4** illustrates the preliminary results of the cradle-to-grave LCA of the production of a holding ring for spherical turbine inlet valves in hydraulic power plant produced with WAAM or casting. The software and database used were SimaPro 9.4.0.2 and ecoinvent 3.8, respectively. The functional unit was defined as "enabling the production of an average amount of GWh for 10 years in France." The specific value of the average amount is confidential and

DOI: http://dx.doi.org/10.5772/intechopen.109788

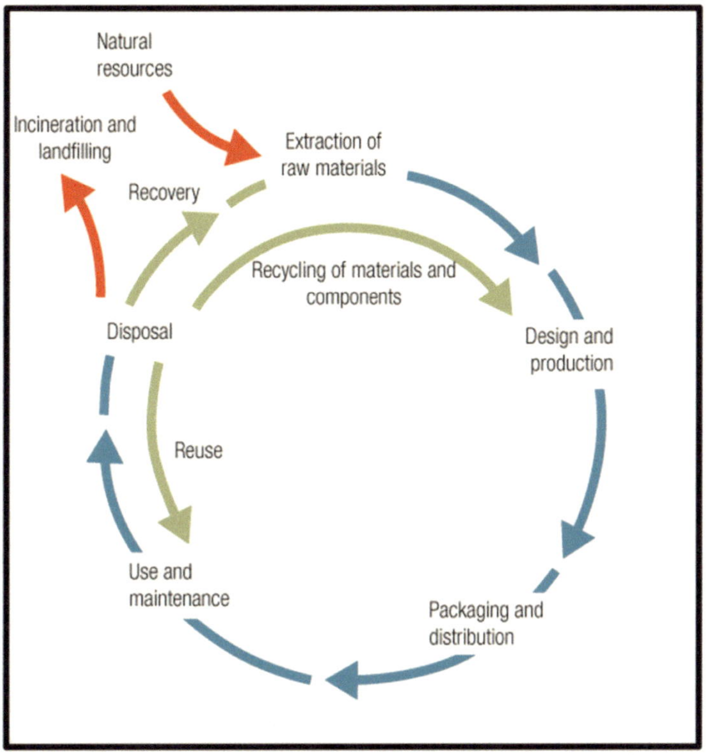

Figure 2.
Schematic representation of life cycle assessment (LCA).

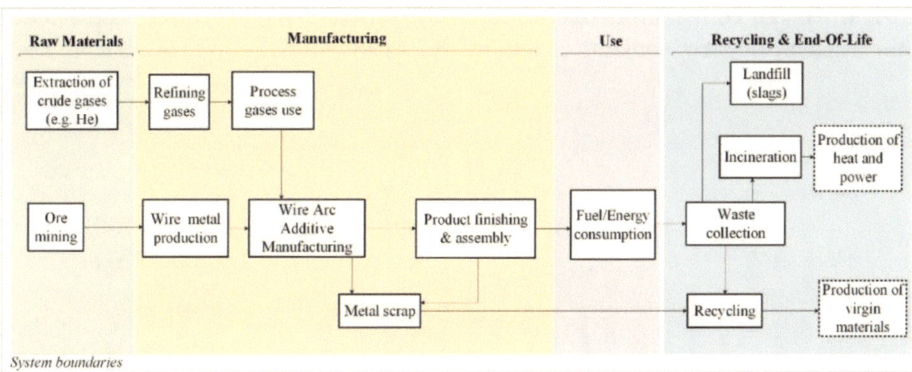

Figure 3.
System boundaries considered in cradle-to-grave LCA of Grade2XL products. Processes related to energy consumption, transportation, and capital goods were included in the model, but they are not represented here. The two boxes with the dotted line are avoided products.

thus not disclosed here. All the equivalent processes were excluded from the calculation of the environmental impact. **Figure 4** illustrates the impact score for multiple environmental issues considered in the assessment (see also **Table A3** in the Appendix).

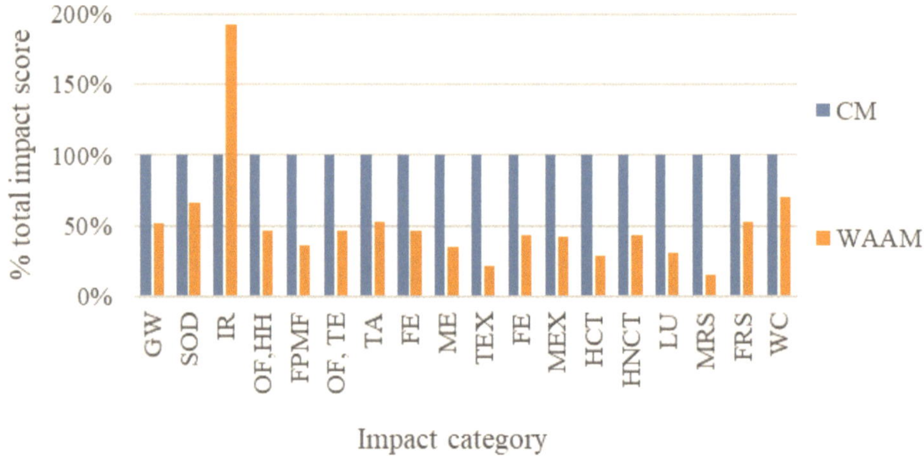

Figure 4.
Internally normalized impact score with ReCiPe2016 (H) midpoint for holding ring produced by conventional manufacturing (CM, columns on the left) and WAAM (columns on the right). The full name of the impact categories in the x-axis is reported in Appendix A.2.

It is clear the quite better environmental performance of WAAM over conventional processes, except for the impact category "ionizing radiation". In order to understand the reason why the impact score is visibly higher for WAAM for this impact category, a simple hot-spot analysis was developed by doing a process contribution analysis for WAAM. **Figure 5** illustrates the processes that contribute to the impact score for each impact category of the holding ring produced with WAAM. In general, the steel and bronze wires and the energy used during manufacturing are clearly the major contributors to both type of manufacturing processes. In particular, in the impact category "ionizing radiation" the major contribution comes from the

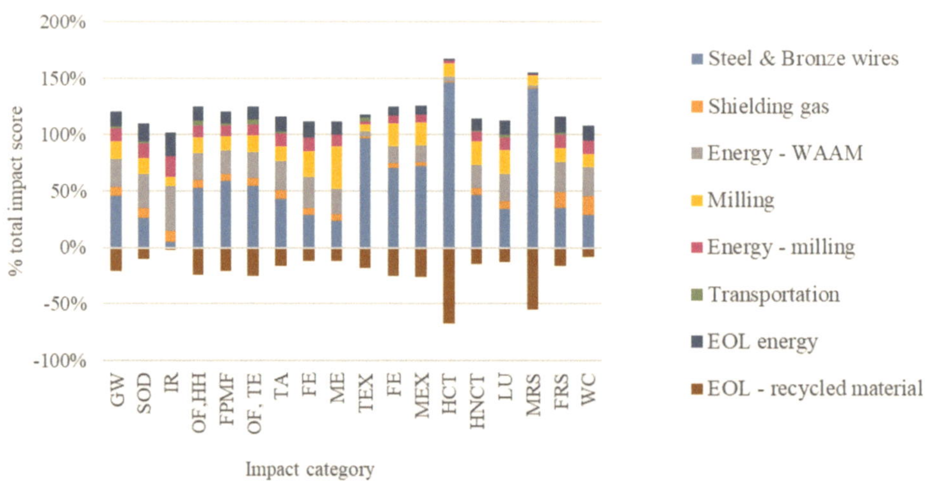

Figure 5.
Process contribution analysis with ReCiPe2016 (H) midpoint for holding ring fabrication by WAAM. The full name of the impact categories in the x-axis is reported in Appendix A.2.

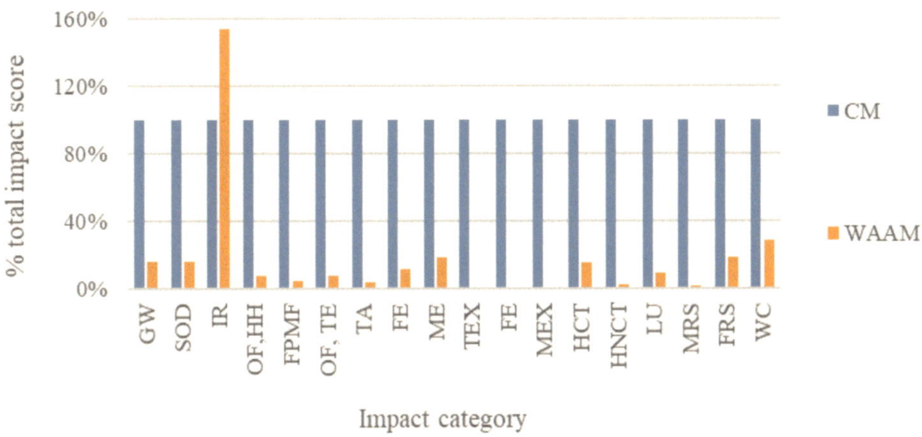

Figure 6.
Internally normalized impact score with ReCiPe2016 (H) midpoint for bathtub mold fabricated by conventional manufacturing (CM, columns on the left) and WAAM (columns on the right). The full name of the impact categories in the x-axis is reported in Appendix A.2.

energy consumption during WAAM operation. The negative contribution of the recycled material to the total environmental impact can be explained by the chosen modeling method, in which the recycled material is avoiding the extraction and production of primary material [28, 31].

Similarly, a Life Cycle Assessment was done for a bathtub mold. **Figure 6** shows the internally normalized results of the LCA of the production applying WAAM or conventional manufacturing (i.e., casting and Nickel vapor deposition in sealed vessel). In this case, the functional unit was defined as "enabling the production of 10,000 polyurethane bathtubs without surface defects in the Netherlands." All the equivalent processes between the two systems were excluded from the calculation of the environmental impact. **Figure 6** clearly shows a lower impact score of WAAM over conventional processes, except for the impact category "ionizing radiation". It is worth to mention that for "Terrestrial/Freshwater/Marine ecotoxicity" the WAAM alternative's impact score is 1% of the conventional alternative explaining why it does not show in **Figure 6**. It is possible to see the numerical results more precisely at **Table A2** in the Appendix.

In this case, as well, a simplified hot-spot analysis was done for WAAM by analyzing its process contribution for the bathtub mold. **Figure 7** presents the processes that contribute to the impact score for each environmental impact category of the bathtub mold manufactured with WAAM. It is clear that, in general, the steel wires are the major contributors. In particular, in the impact category "ionizing radiation" the major contribution comes from the energy consumption during WAAM operation. The negative contribution of the recycled material to the total environmental impact as for the holding ring is linked to the fact that system expansion is considered [28, 31].

Overall, for both products considered their LCA showed the highest impact score for the impact category "ionizing radiation" due to the elevated electricity use during WAAM. However, except for this impact category WAAM appears to be a better alternative than the conventional manufacturing route.

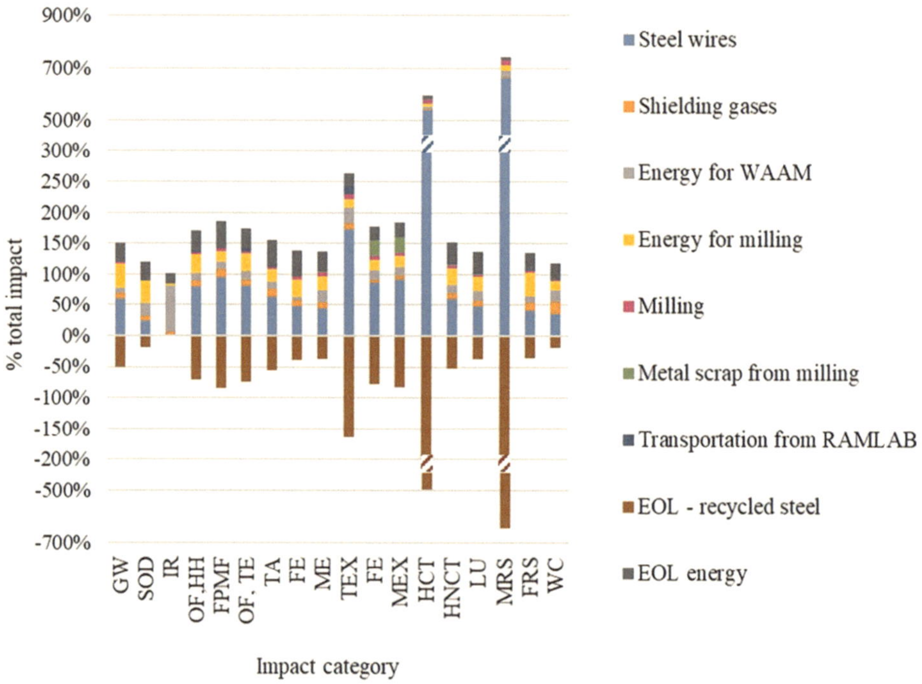

Figure 7.
Process contribution analysis with ReCiPe2016 (H) midpoint for the fabrication of a bathtub mold by WAAM. The full name of the impact categories in the x-axis is reported in Appendix A.2.

5. Expected changes on the environmental performance by upscaling and optimizing a metal additive manufacturing (MAM)

Considering sustainable development of new and future technologies is essential. However, future applications of different technologies involve many uncertainties both related to the markets, to technology upscaling, etc. [32]. Assessing this with LCA in a so-called prospective LCA, entails issues of unknown future applications (i.e., aim, functionality, system boundaries), industrial scales (compared to lab scale), and large inventory data gaps comprising issues of data availability and data quality, altogether increasing the level of LCA uncertainty [32, 33]. However, it is important to address those aspects during an initial stage of the technology development to understand potential future environmental performance improvements of a technology [33, 34]. [32] reviewed 44 case studies of prospective LCA and developed a framework for facing the challenges mentioned and [35] developed a prospective assessment of the environmental impact of incremental sheet forming (ISF). [36] took an approach mainly looking at the expected changes of environmental performance by upscaling and process optimization. Taking this approach, an individual metal additive manufacturing (MAM) system from lab- to full-scale production was qualitatively analyzed in order to anticipate the potential influence of upscaling. In particular, it was assumed that the upscaled system would be fully optimized, automated, and continuously operating for large metal objects production with a low annual volume production demand. Below is presented a more detailed list of assumptions:

- Medium or large objects;
- Same technology different products (thanks to build-in flexibility);
- Produce many different objects every year;
- All types of metal additive manufacturing techniques are considered (e.g., directed energy deposition [DED], sheet lamination, electron beam melting [EBM], laser powder bed fusion [LPBF], binder jetting);
- Fully optimized;
- Industrial scale;
- Automated;
- Low volume production demand of the same part per year;
- Not improvement in manufacturing energy efficiency;
- Not considered downtime;
- Continuously working system;

For this qualitative forecast, upscaling factors and rule-of-thumb were discussed with MAM process operation experts, and retrieved from the literature [24, 37–41]. **Table 1** reports the expected changes in environmental performance by upscaling a metal additive manufacturing system from small to full scale.

Overall, upscaling is expected to result in a reduction of environmental impact of MAM per unit part produced. The larger manufacturing system configuration will result in an increased material consumption, but this is forecasted to not significantly affect the environmental performance. The main benefits from MAM system upscaling stem primarily from an improved process capability. This would

Model parameter	Forecasted change when up-scaled from lab-scale to full-scale		Differences induced by upscaling and expected consequences on environmental performance
	Factor	Motivation	
Process capability (yield)	Increase by a factor 2 to 10 (considering kg product/h)	For some laser-based directed energy deposition (DED) and laser powder bed fusion (LPBF) techniques, it has been already possible to increase massively the deposition rate in laboratory with industrial scale manufacturing systems (e.g., for WAAM from 0.5 to 5 kg/h, for EBM from 2 to 20 kg/h)	Larger need of metals per year is forecasted to enlarge the impacts associated to the MAM system on climate change, resource depletion, and various toxicity- and non-toxicity-related impact categories due to the need for more material. Of course, this will also depend on what the products substitute

Model parameter	Forecasted change when up-scaled from lab-scale to full-scale		Differences induced by upscaling and expected consequences on environmental performance
	Factor	**Motivation**	
Material input for construction of MAM full-scale system for unit part produced	Decrease by a factor 1 to 10 (in relation to process outputs)	The factor would depend on the upscaling approach considered since it is possible to have: a) increased electricity use and deposition rate, or b) higher number of 3D printing systems. In the former case, the amount of necessary material for MAM equipment for unit part produced would decrease proportionally to the deposition rate (see process capability above). However, this is not applicable to all MAM techniques (e.g., LPBF). In the second case, the needed material input for construction would be the same as for now. In general, since an increase in process capability is forecasted, the material for capital goods per unit part produced will likely decrease proportionally with the deposition rate	Smaller need of metals and crude oil per unit part produced is forecasted to decrease the impacts on climate change, resource depletion, and various toxicity- and non-toxicity-related impact categories due to the less need for fabrication of supplementary parts and equipment
Material input for construction of post-treatment MAM full-scale system for unit part produced	Decrease by a factor 1 to 5 (in relation to process output)	The factor depends on several aspects: a) heat treatment temperature, and b) geometry of the part(s). There is a difference if the heat treatment is done locally or the part does not fit into an available furnace and the producer uses a portable heat treatment service. Moreover, once MAM will be optimized the need for subtractive manufacturing processes (e.g., milling, grinding) will be decreased	
Electricity use for unit part produced	Decrease by a factor 2 to 10 (in relation to process output)	The manufacturing processes are assumed to not have any improvement in energy efficiency. However, the electricity use is dependent on the process capability, thus if the latter is improved for the production of one part, then also the electricity use can be reduced to the same extent	Climate change, ionizing radiation, and toxic-impacts on humans and freshwater ecosystems are projected to stay the same due to reduced emissions of fossil CO_2, NO_x, SO_2, and metals stemming from lower electricity use for unit part

Model parameter	Forecasted change when up-scaled from lab-scale to full-scale		Differences induced by upscaling and expected consequences on environmental performance
	Factor	Motivation	
Process gas use for unit part produced	Decrease by a factor 2 to 10 (in relation to process output)	The increase in efficiency of process gas use is related to a) impurities differences in the part, and b) process capability. For instance, if it is known how to manipulate well the production of a specific product, the process gas use could be reduced up to 40%	Climate change, fossil depletion, water consumption and toxic-impacts on humans and freshwater ecosystems are projected to be minimized due to lower emissions of fossil CO_2, NO_x, SO_2, and vapor due to reduced process gas use for unit part
Metal scrap for unit part produced	Increase by a factor 1.5 (base plate excluded and in relation to process output)	The bigger the part, the more a distortion would have an impact on the tolerance of the as-built part. For manufacturing one part and increase its tolerance, the amount of material which has to be grinded off later (more scrap) will also increase. Except the surface can remain in the as-built state, then we get an increased safety by thicker parts. Moreover, if more than one part is manufactured, the tolerance would be optimized and minimized. Since a scenario with large product is assumed, a factor of 1.5 in relation to the part weight was assumed. This means if the part weight is doubled, the scrap is three times as high	Reduced metals scraps production per unit of manufacturing system is forecasted to relatively decrease the impacts on climate change, resource depletion, and various toxicity- and non-toxicity-related impact categories due to the need for less waste treatment processes and less need for raw materials
Fumes and powder dispersion during manufacturing for unit part produced	No change (in relation to process output)	Generally, it is expected that machines would be encased, as standardized machines in the market, with the possibility of an effective fume extraction. Thus, the emission outside the machine would be minimized (near zero)	Fumes and powder dispersion during manufacturing for unit part is forecasted to not change so the impacts would be on climate change, resource depletion, and various toxicity- and non-toxicity-related impact categories, also thanks to special working garments and an improved filtration system (e.g., hoods)

Table 1.
Expected changes established by upscaling from lab-scale to full-scale MAM system and likely consequences of its environmental performance.

influence positively also the input/consumption of electricity and process gas use for unit part produced. Moreover, it is expected that an advancement of MAM system optimization can achieve a lower production of metal scrap and an improved handling of welding fumes or powder dispersion. On the whole, this could reduce the current environmental impact associated with a MAM system from 1.5 to 10 times. The numbers would depend on several factors (e.g., MAM technology, product shape, etc.). Overall, these numbers should be considered carefully, also because they were obtained through a qualitative investigation of future expected in environmental performance due to improvement of the technology, and there are several underlying assumptions within this analysis and uncertainties were not quantified.

6. Future environmentally sustainable applications for multi-material additive manufacturing for metals

On one hand, when planning product fabrication for additive manufacturing, often the focus is on design shape complexity (e.g., "solid-to-cavity ratio" [1, 19, 42, 43]), and lightweight and strength (e.g., "strength-to-weight ratio" [44]). In particular, there are studies illustrating the potential of multi-material additive manufacturing to enhance product strengths and prolong its lifetime [20, 21] which are beneficial aspects in relation to circular economy. On the other hand, there is a general lack of argument relating to how product recyclability design can or should be applied to multi-material additive manufacturing. This topic has been discussed relative to the metal manufacturing sector, life cycle assessment (LCA) and circular economy but only in generic terms addressing the complexity of production of clean recycled materials [45–50]. Currently, there is a decrease in quality of metal stocks as a result of nowadays use of complex alloys and the recovery practice of those metals. That happens even though the technology and contemporary infrastructures would allow to undertake this challenge and maintain the quality. Thus, to bring forward industrial ecology concepts into businesses, the metallurgic constraints of metals recovery should complement policy development and product design taking into account costs feasibility aspects [48–50]. Indeed, all systems and technologies starting from product design to metal recovery are interconnected and could support in addressing this challenging task [47] but there is a need for tools or frameworks to understand and help with this task. Particularly interesting in this context is the concept of the metal wheel introduced by [48]. The first illustrates metals linkages in geology, showing the capacity of current metallurgical technologies for the recovery of trace elements in their (primary or secondary) feed.

Here we applied this framework to investigate design for recyclability and resource efficiency relative to alloys used for multi-material Wire-arc Additive Manufacturing (WAAM). Indeed, being aware of those during product design and process optimization can support the prevention of negligent recycling and develop product design for recycling and resource efficiency [49].

Figure 8 illustrates the Metal Wheel for the small holding ring (described earlier) made of three different types of wire: high-strength alloy and hot rolled steel, stainless steel martensitic, and bronze.

DOI: http://dx.doi.org/10.5772/intechopen.109788

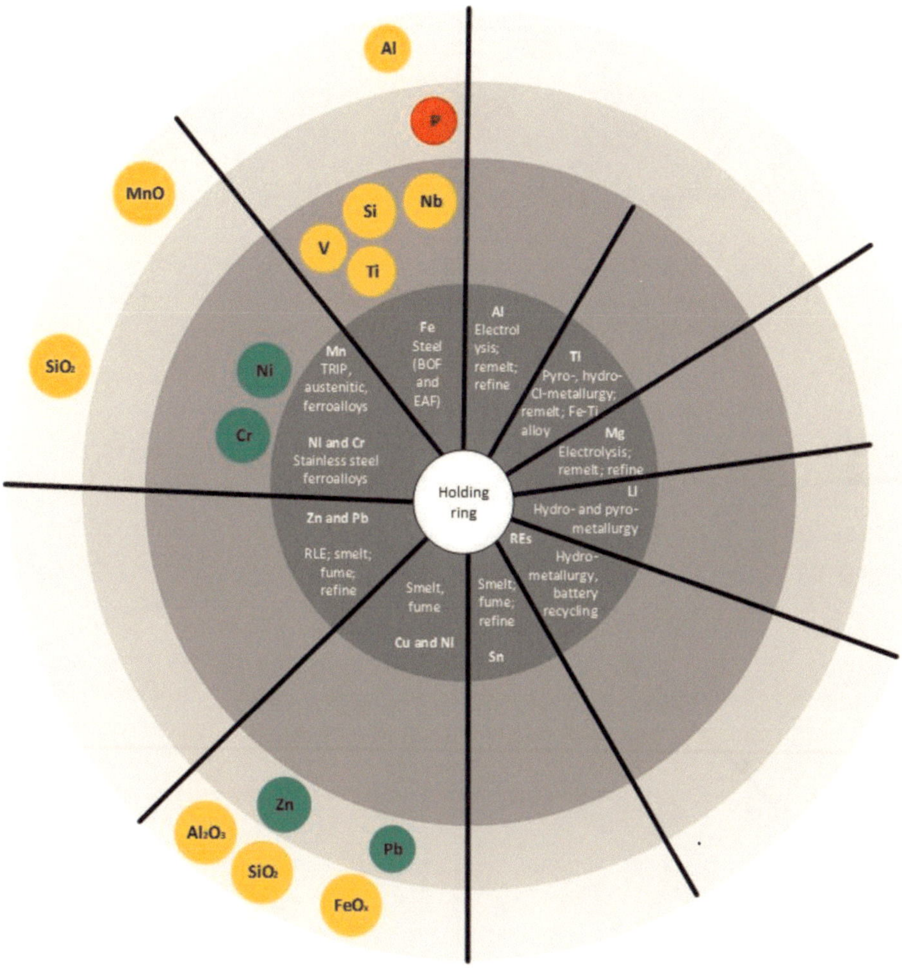

Figure 8.
Metal Wheel for a small holding ring adapted from [49]. In the small circles are represented with different colors, the trace compounds. When they are green, they can be mainly recovered; in yellow, they can be recovered in the alloys or lost if they are directed to incorrect stream or scrap; in red, the elements are lost as they are not compatible with carrier metal.

Here it is clear that the majority of elements present high chances to be lost if they end up in the wrong scrap stream (i.e., yellow circles). However, in stainless steel, Nickel and Chromium would be mainly recovered, and in bronze both zinc and lead would be primarily retrieved.

Figure 9 shows the Metal Wheel applied to investigate potential for design for recycling and resource-efficiency for a bathtub mold (described above) produced with two wires: high-alloyed ferritic martensitic stainless steel, high strength alloy, and hot rolled steel.

Figure 9 shows that also in this case, only a few elements can be recovered (see Nickel and Chromium in stainless steel alloy). The others show high probability of being dispersed if they finish in the wrong scrap stream. On the other hand, in

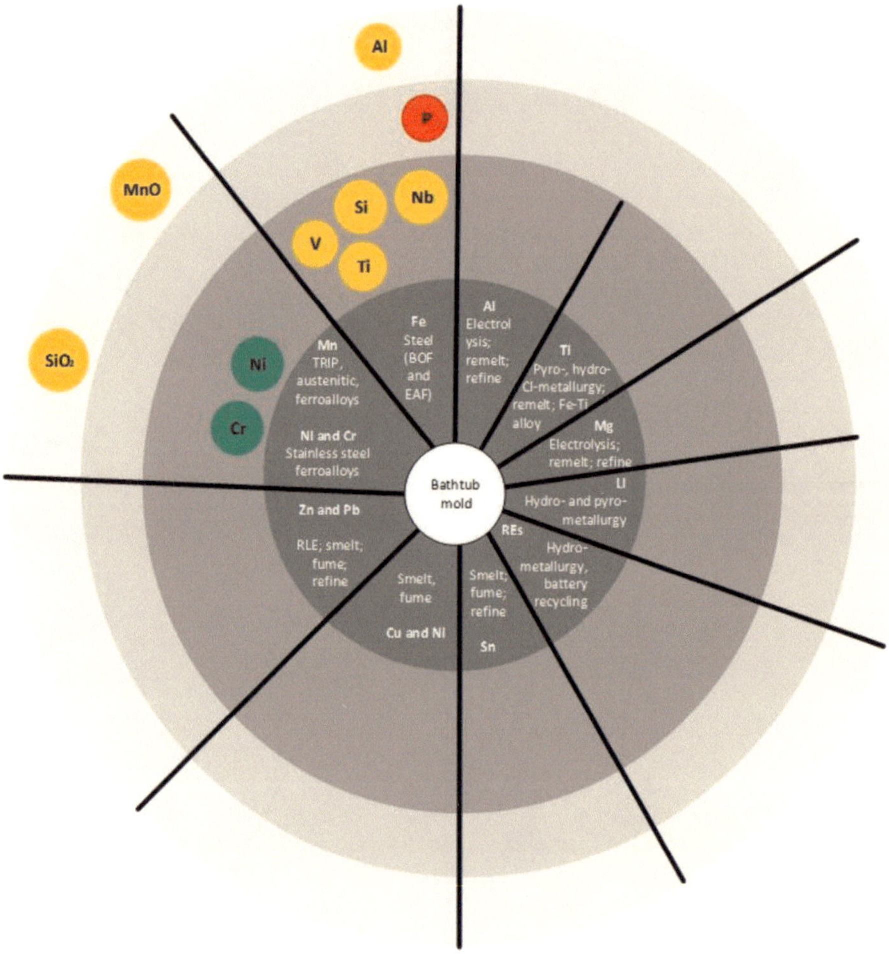

Figure 9.
Metal Wheel for a bathtub mold adapted from [50]. In the small circles are represented with different colors, the trace compounds. When they are green, they can be mainly recovered; in yellow, they can be recovered in the alloys or lost if they are directed to incorrect stream or scrap; in red, the elements are lost as they are not compatible with carrier metal.

general, the fact that multiple different alloys are melted together adds complexity to the efficiency of the named trace elements recovery.

7. Conclusions

The manufacturing sector, especially metal, as a significant contributor to greenhouse gas emissions, needs to adopt a lower emission strategy. Additionally, there is an increased demand for circular economy strategies aiming to retain the value of the material for as long as possible meaning either reducing use, prolonging lifetime, or optimizing recycling of materials. This chapter aims to address the potential future sustainability of metal additive manufacturing. At the same time, there are common technical challenges, such as the possibility of obtaining undesired behavior of parts originated by mixing several materials, which might not be compatible, in a single component due to lack of standardization.

Most MAM technologies are currently still at small scale or lab scale. Through a qualitative evaluation, upscaling to industrial scale will probably reduce the environmental impacts of the technology. The results from the environmental life cycle assessments already show an environmental benefit of multi-material additive manufacturing in relation to conventional manufacturing processes. The main impacts from MAM originate from metal production, which is why there is also significant benefit to harvesting by recycling the metals after the manufactured product's end. In particular, recycling multi-material objects may constitute challenges with reduced quality of metal stocks if individual alloying elements are dissipated due to not being recoverable from the recycled metal base. The compatibility and recoverability should be considered early in the design stage.

Acknowledgements

The authors thank for helpful comments to Sami Kara, Martin Schmitz Niederau, and Michael Zwicky Hauschild for providing useful insights on assumptions and principles for the upscaling of a metal additive manufacturing system. This work was supported by the European Union's Horizon 2020 research and innovation programme under grant agreement no. 862017. This publication reflects only the author's view and the Commission is not responsible for any use that may be made of the information it contains.

Conflict of interest

The authors declare no conflict of interest.

A. Appendix

A1. Abbreviations

Nomenclature	
LCA	**Life Cycle Assessment**
MAM	Metal Additive Manufacturing
MM-AM	Multi-material Additive Manufacturing
WAAM	Wire-arc Additive Manufacturing
FGM	Functionally gradient materials
AM	Additive Manufacturing
CM	Conventional Manufacturing
EoL	End-of-Life
DED	Directed Energy Deposition

Nomenclature	
LCA	**Life Cycle Assessment**
GW	Global warming
SOD	Stratospheric ozone depletion
IR	Ionizing radiation
OF,HH	Ozone formation, Human health
FPMF	Fine particulate matter formation
OF, TE	Ozone formation, Terrestrial ecosystems
TA	Terrestrial acidification
FE	Freshwater eutrophication
ME	Marine eutrophication
TEX	Terrestrial ecotoxicity
FE	Freshwater ecotoxicity
MEX	Marine ecotoxicity
HCT	Human carcinogenic toxicity
HNCT	Human non-carcinogenic toxicity
LU	Land use
MRS	Mineral resource scarcity
FRS	Fossil resource scarcity
WC	Water consumption

Table A1.
List of abbreviations used in the book chapter.

A2. Life cycle assessment—results

Tables A2 and **A3** illustrate the characterized midpoint results with ReCiPe2016 (H) midpoint for holding ring and bathtub mold.

Impact category	Unit	Casting	WAAM
Global warming	kg CO_2 eq	5.94E+03	3.07E+03
Stratospheric ozone depletion	kg CFC11 eq	2.16E-03	1.43E-03
Ionizing radiation	kBq Co-60 eq	5.50E+02	1.06E+03
Ozone formation, Human health	kg NO_x eq	1.33E+01	6.23E+00
Fine particulate matter formation	kg PM2.5 eq	1.55E+01	5.61E+00
Ozone formation, terrestrial ecosystems	kg NO_x eq	1.36E+01	6.37E+00
Terrestrial acidification	kg SO_2 eq	2.16E+01	1.13E+01
Freshwater eutrophication	kg P eq	6.51E+00	3.02E+00
Marine eutrophication	kg N eq	7.85E-01	2.73E-01

Impact category	Unit	Casting	WAAM
Terrestrial ecotoxicity	kg 1,4-DCB	1.06E+05	2.27E+04
Freshwater ecotoxicity	kg 1,4-DCB	5.51E+02	2.41E+02
Marine ecotoxicity	kg 1,4-DCB	7.68E+02	3.24E+02
Human carcinogenic toxicity	kg 1,4-DCB	4.71E+03	1.34E+03
Human non-carcinogenic toxicity	kg 1,4-DCB	1.20E+04	5.18E+03
Land use	m2a crop eq	3.65E+02	1.12E+02
Mineral resource scarcity	kg Cu eq	3.83E+02	5.73E+01
Fossil resource scarcity	kg oil eq	1.56E+03	8.12E+02
Water consumption	m^3	8.77E+01	6.19E+01

Table A2.
Characterized midpoint results with ReCiPe2016 (H) midpoint for holding ring.

Impact category	Unit	CM	WAAM
Global warming	kg CO_2 eq	3.40E+04	5.48E+03
Stratospheric ozone depletion	kg CFC11 eq	1.95E-02	3.22E-03
Ionizing radiation	kBq Co-60 eq	3.33E+03	5.13E+03
Ozone formation, human health	kg NO_x eq	1.22E+02	9.39E+00
Fine particulate matter formation	kg PM2.5 eq	1.20E+02	5.95E+00
Ozone formation, terrestrial ecosystems	kg NO_x eq	1.23E+02	9.55E+00
Terrestrial acidification	kg SO_2 eq	3.34E+02	1.45E+01
Freshwater eutrophication	kg P eq	3.52E+01	4.04E+00
Marine eutrophication	kg N eq	2.03E+00	3.82E-01
Terrestrial ecotoxicity	kg 1,4-DCB	1.47E+06	1.09E+04
Freshwater ecotoxicity	kg 1,4-DCB	4.63E+04	3.38E+02
Marine ecotoxicity	kg 1,4-DCB	5.67E+04	4.38E+02
Human carcinogenic toxicity	kg 1,4-DCB	5.16E+03	7.97E+02
Human non-carcinogenic toxicity	kg 1,4-DCB	2.64E+05	6.26E+03
Land use	m2a crop eq	1.73E+03	1.64E+02
Mineral resource scarcity	kg Cu eq	1.30E+03	2.13E+01
Fossil resource scarcity	kg oil eq	8.44E+03	1.57E+03
Water consumption	m^3	4.24E+02	1.20E+02

Table A3.
Characterized midpoint results with ReCiPe2016 (H) midpoint for bathtub mold.

Author details

Valentina Pusateri[1*], Constantinos Goulas[2] and Stig Irving Olsen[1*]

1 Division for Quantitative Sustainability Assessment (QSA), Department of Environmental and Resource Engineering, Technical University of Denmark, Kgs. Lyngby, Denmark

2 Faculty of Engineering Technology, Department of Design Production and Management, University of Twente, Enschede, The Netherlands

*Address all correspondence to: valpu@dtu.dk and siol@dtu.dk

References

[1] Priarone PC, Lunetto V, Atzeni E, Salmi A. Laser powder bed fusion (L-PBF) additive manufacturing: On the correlation between design choices and process sustainability. Procedia CIRP. 2018;**78**:85-90. DOI: 10.1016/j.procir.2018.09.058

[2] Krishnan M, Samandari H, Woetzel J, Smit S, Pacthod D, Pinner D, et al. The Net-Zero Transition - What it Would Cost, What it Could Bring. McKinsey Global Institute; 2022. [cited 2022 Nov 02]. Available from: https://policycommons.net/artifacts/2213603/the-net-zero-transition/2970547/. CID: 20.500.12592/0d6rdj

[3] World Steel Association. World steel in figures [Internet]. 2020 [cited 2021 Dec 23]. Available from: https://www.worldsteel.org/en/dam/jcr:f7982217-cfde-4fdc-8ba0-795ed807f513/World%2520Steel%2520in%2520Figures%25202020i.pdf

[4] IEA. Industry – topics [cited 2022 Nov 29]. Available from: https://www.iea.org/topics/industry

[5] Henckens MLCM, Driessen PPJ, Worrell E. Metal scarcity and sustainability, analyzing the necessity to reduce the extraction of scarce metals. Resources, Conservation and Recycling. 2014;**93**:1-8

[6] Yokoi R, Nansai K, Hatayama H, Motoshita M. Significance of country-specific context in metal scarcity assessment from a perspective of short-term mining capacity. Resources, Conservation and Recycling. 2021;**166**:1-9. DOI: 10.1016/j.resconrec.2020.105305

[7] Graedel TE, Erdmann L. Will metal scarcity impede routine industrial use? MRS Bulletin. 2012;**37**(4):325-331

[8] Owsianiak M, van Oers L, Drielsma J, Laurent A, Hauschild MZ. Identification of dissipative emissions for improved assessment of metal resources in life cycle assessment. Journal of Industrial Ecology. 2022;**26**(2):406-420

[9] Drielsma JA, Russell-Vaccari AJ, Drnek T, Brady T, Weihed P, Mistry M, et al. Mineral resources in life cycle impact assessment—defining the path forward. International Journal of Life Cycle Assessment. 2016;**21**(1):85-105

[10] Bobba S, Carrara S, Huisman J, Mathieux F, Pavel C. Critical Raw Materials for Strategic Technologies and Sectors in the EU. Publications Office of the European Union; 2020. [cited 2022 Nov 15]. Available from: https://ec.europa.eu/docsroom/documents/42881. ISBN: 978-92-76-15336-8

[11] Priarone PC, Ingarao G, Lorenzo R, Settineri L. Influence of material-related aspects of additive and subtractive Ti-6Al-4V manufacturing on energy demand and carbon dioxide emissions. Journal of Industrial Ecology. 2016;**21**:191-202. DOI: 10.1111/jiec.12523

[12] Böckin D, Tillman AM. Environmental assessment of additive manufacturing in the automotive industry. Journal of Cleaner Production. 2019;**226**:977-987

[13] Tang Y, Mak K, Zhao YF. A framework to reduce product environmental impact through design optimization for additive manufacturing. Journal of Cleaner Production. 2016;**20**(137):1560-1572

[14] Bekker ACM, Verlinden JC. Life cycle assessment of wire + arc additive manufacturing compared to green sand casting and CNC milling in stainless steel. Journal of Cleaner Production. 2018;**177**:438-447. DOI: 10.1016/j.jclepro.2017.12.148

[15] Gao C, Wolff S, Wang S. Eco-friendly additive manufacturing of metals: Energy efficiency and life cycle analysis. Journal of Manufacturing Systems. 2021;**60**:459-472

[16] Yusuf SM, Cutler S, Gao N. The impact of metal additive manufacturing on the aerospace industry. Metals. 2019; **9**:1-35. DOI: 10.3390/met9121286

[17] Monteiro H, Carmona-Aparicio G, Lei I, Despeisse M. Energy and material efficiency strategies enabled by metal additive manufacturing – A review for the aeronautic and aerospace sectors. Energy Reports. 2022; **8**:298-305

[18] Grade2xl Home. [cited 2022 Nov 29]. Available from: https://www.grade2xl.eu/

[19] Priarone PC, Campatelli G, Montevecchi F, Venturini G, Settineri L. A modelling framework for comparing the environmental and economic performance of WAAM-based integrated manufacturing and machining. CIRP Annals. 2019;**68**(1): 37-40. DOI: 10.1016/j.cirp.2019.04.005

[20] Bandyopadhyay A, Heer B. Additive manufacturing of multi-material structures. Materials Science & Engineering R: Reports. 2018; **129**:1-16

[21] Treutler K, Kamper S, Leicher M, Bick T, Wesling V. Multi-material design in welding arc additive manufacturing. Metals (Basel). 2019;**9**(7):1-14. DOI: 10.3390/met9070809

[22] Mehrpouya M, Tuma D, Vaneker T, Afrasiabi M, Bambach M, Gibson I. Multimaterial powder bed fusion techniques. Rapid Prototyping Journal. 2022;**28**:1-19

[23] Aerosint Multi-material 3D printing. [cited 2022 Dec 7]. Available from: https://aerosint.com/

[24] Meltio Wire-laser metal 3D printing: Multi-metal laser metal deposition [cited 2022 Dec 12]. Available from: https://meltio3d.com/

[25] Feenstra DR, Banerjee R, Fraser HL, Huang A, Molotnikov A, Birbilis N. Critical review of the state of the art in multi-material fabrication via directed energy deposition. Current Opinion in Solid State & Materials Science. 2021; **25**(4):1-12. DOI: 10.1016/j.cossms.2021.100924

[26] Wei C, Li L. Recent progress and scientific challenges in multi-material additive manufacturing via laser-based powder bed fusion. Virtual and Physical Prototyping. 2021;**16**:347-371

[27] Horn M, Prestel L, Schmitt M, Binder M, Schlick G, Seidel C, et al. Multi-Material additive manufacturing – recycling of binary metal powder mixtures by screening. Procedia CIRP. 2020;**93**:50-55. DOI: 10.1016/j.procir.2020.04.098

[28] Hauschild MZ, Rosenbaum RK, Olsen SI. Life Cycle Assessment: Theory and Practice. Springer; 2018. DOI: 10.1007/978-3-319-56475-3

[29] Iso.org. n.d. ISO 14044:2006(en) Environmental management—Life cycle assessment—Requirements and guidelines. [cited 2022 Oct 02]. Available

from: https://www.iso.org/standard/38498.html

[30] Iso.org. n.d. ISO 14040:2006(en) Environmental management—Life cycle assessment—Requirements and guidelines. [cited 2022 Oct 02]. Available from: https://www.iso.org/standard/37456.html

[31] International Reference Life Cycle Data System (ILCD) Handbook - General guide for Life Cycle Assessment - Provisions and Action Steps. EUR 24378 EN. Luxembourg (Luxembourg): Publications Office of the European Union; 2010. JRC58190. DOI: 10.2788/94987

[32] Thonemann N, Schulte A, Maga D. How to conduct prospective life cycle assessment for emerging technologies? A systematic review and methodological guidance. Sustainability. 2020;**12**(3):1-23. DOI: 10.3390/su12031192

[33] Hetherington AC, Borrion AL, Griffiths OG, McManus MC. Use of LCA as a development tool within early research: Challenges and issues across different sectors. International Journal of Life Cycle Assessment. 2014;**19**(1):130-143

[34] Wender BA, Foley RW, Prado-Lopez V, Ravikumar D, Eisenberg DA, Hottle TA, et al. Illustrating anticipatory life cycle assessment for emerging photovoltaic technologies. Environmental Science & Technology. 2014;**48**(18):10531-10538

[35] Cooper DR, Gutowski TG. Prospective environmental analyses of emerging technology: A critique, a proposed methodology, and a case study on incremental sheet forming. Journal of Industrial Ecology. 2020;**24**(1):38-51

[36] Owsianiak M, Ryberg MW, Renz M, Hitzl M, Hauschild MZ. Environmental performance of hydrothermal carbonization of four wet biomass waste streams at industry-relevant scales. ACS Sustainable Chemistry & Engineering. 2016;**4**(12):6783-6791

[37] Liao J, Cooper DR. The environmental impacts of metal powder bed additive manufacturing. Journal of Manufacturing Science and Engineering. Transactions of the ASME. 2021;**143**(3):1-11. DOI: 10.1115/1.4048435

[38] Ahn DG. Directed energy deposition (DED) process: State of the art. International Journal of Precision Engineering and Manufacturing – Green Technology. 2021;**8**:703-742. DOI: 10.1007/s40684-020-00302-7

[39] 3D Printing Industry. Fraunhofer opens "World's largest SLM facility". 2022 [cited 2022 Feb 18]. Available from: https://3dprintingindustry.com/news/fraunhofer-opens-worlds-largest-slm-facility-aachen-germany-115961/

[40] Gibson I, Rosen DW, Stucker B. Additive Manufacturing Technologies. Springer; 2010. DOI: 10.1007/978-3-030-56127-7

[41] Li G, Odum K, Yau C, Soshi M, Yamazaki K. High productivity fluence based control of directed energy deposition (DED) part geometry. Journal of Manufacturing Processes. 2021;**65**:407-417

[42] Ingarao G, Priarone PC, Deng Y, Paraskevas D. Environmental modelling of aluminium based components manufacturing routes: Additive manufacturing versus machining versus forming. Journal of Cleaner Production 2018;**176**:261–275. Available from: https://doi.org/10.1016/j.jclepro.2017.12.115

[43] Morrow WR, Qi H, Kim I, Mazumder J, Skerlos SJ. Environmental aspects of laser-based and conventional tool and die manufacturing. Journal of Cleaner Production. 2007;**15**(10): 932-943

[44] Ford S, Despeisse M. Additive manufacturing and sustainability: An exploratory study of the advantages and challenges. Journal of Cleaner Production. 2016;**137**:1573-1587

[45] Verhoef EV, Reuter MA, Scholte A, Dijkema GPJ. A dynamic LCA model for assessing the impact of lead free solder. In: Proceedings of the TMS Yazawa International Symposium: "Metallurgical and Materials Processing: Principles and Technologies". Vol. 2. 2003. pp. 1-19

[46] Reuter MA, Kojo IV. Challenges of metals recycling. Materia. 2012;**2**:50-57

[47] Reuter MA, van Schaik A. Strategic metal recycling : Adaptive metallurgical processing infrastructure and technology are essential for a Circular Economy. Annales des Mines - Responsabilité et environnement. 2016;**82**(2):62-66

[48] Verhoef EV, GPJ D, Reuter MA. Process knowledge, system dynamics, and metal ecology. Journal of Industrial Ecology. 2004;**8**:23-43

[49] Reuter MA, van Schaik A, Gutzmer J, Bartie N, Abadías-Llamas A. Challenges of the circular economy: A material, metallurgical, and product design perspective. 2019. DOI: 10.1146/annurev-matsci-070218-

[50] van Schaik A, Reuter MA. Material-centric (aluminum and copper) and product-centric (cars, WEEE, TV, lamps, batteries, catalysts) recycling and DfR rules. In: Handbook of Recycling: State-of-the-Art for Practitioners, Analysts, and Scientists. Chapter 22. Elsevier; 2014. pp. 307-378. DOI: 10.1016/B978-0-12-396459-5.00022-2

Section 2

3D Printing in Device Engineering

Chapter 5

3D Printing of Biomimetic Functional Nanocomposites *via* Vat Photopolymerization

Tengteng Tang, Dylan Joralmon and Xiangjia Li

Abstract

The complex structures and functional material systems of natural organisms effectively cope with crisis-ridden living environments such as high temperature, drought, toxicity, and predator. Behind these excellent survival strategies evolved over hundreds of millions of years is a series of effective mechanical, optical, hydraulic, and electromagnetic properties. Bionic design and manufacturing have always attracted extensive attention, but the progress has been limited by the inability of traditional manufacturing techniques to reproduce microscopically complex structures and the lack of functional materials. Therefore, there is an urgent need for a fabrication technique with a high degree of fabrication freedom and using composites derived from biological materials. Vat photopolymerization, an emerging additive manufacturing (aka 3D printing) technology, exhibits high manufacturing flexibility in the integrated manufacturing of multi-material systems and multi-scale structures. Here, biomaterial-inspired heterogeneous material systems based on polymer matrices and nanofillers, and the introduction of magnetic and electric fields on the basis of conventional 3D printing systems to spatially and programmably distribute nanofillers are summarized, which provides a new strategy for fabricating anisotropic structures. The application of this versatile 3D printing system in fabricating mechanically reinforced structures, polymer/metal structures, self-actuating, and superhydrophobic structures is also elaborated.

Keywords: 3D printing, biomimicry, nanocomposite, magnetic field, electric field

1. Introduction

Organisms in nature exhibit unique survival strategies due to their special multi-scale and multi-material structures. These biological structures are not only highly hierarchical but also highly flexible and have a clear division of labor in function. The multi-scale or multi-layer structure allows it to exhibit excellent heat conduction efficiency and energy dissipation when it is subjected to external stimuli, such as high temperature and impact force. Biomaterials make it have excellent mechanical, optical, and hydraulic performance. For example, the eggbeater-like structure on the surface of *Salvinia molesta* has excellent superhydrophobic properties, keeping the leaf surface clean all the time [1]. Nacre's bricks-and-mortar multilayer structure allows it to effectively disperse the energy shock when it is subjected to external force, showing excellent energy dissipation capability [2]. Due to the mechanical reinforcement of the arrayed nanofiller in the polymer, the limpet teeth, as the hardest structure in nature, make it firmly attached to the rock wall in the turbulence of the ocean [3, 4]. The microneedles on the surface of the cactus can efficiently absorb moisture in the air even in arid environments [5]. These structures with different functions all contribute to improving the survival rate of the natural organisms. Hierarchical structure [1, 6, 7], biomaterial matrix [8, 9], and organic/inorganic fillers [8, 10, 11] are all essential to realize highly functional structures, which also provide broad opportunities for fabricating next-generation highly integrated functional structures and materials.

Bionic design and manufacturing have always been hot spots in scientific research, and a lot of research work has been carried out in this field. Some reported preliminary results also demonstrate the feasibility of artificial biomimetic structures. For example, an anisotropic structure whose deformation can be programmably controlled is produced based on the hygroscopic properties of cellulose [9], and the mechanical reinforcement caused by the electric field arrangement of multiwalled carbon nanotubes (MWCNTs) inspired by shrimp claws [12], and the superhydrophobic structure based on *S. molesta* leaves showed excellent performance in oil separation [1]. Among the manufacturing technologies corresponding to these artificial structures, vat photopolymerization (VPP) [13], a layer-by-layer projection and selective solidification printing technology, can not only manufacture complex structures but also far outperform 3D printing methods such as fused deposition modeling (FDM) [14] and direct ink writing (DIW) [15] in terms of printing accuracy and efficiency. However, the degree of biomimeticity of current artificial structures and the development of bioinspired materials are still limited.

In order to solve the aforementioned problems, the researchers made the following new attempts based on the previous bionic manufacturing. By learning the material composition of natural structures, a series of heterogeneous hybrid systems with mixtures of polymer matrix and nanofillers have been developed. The introduction of magnetic and electric fields makes it possible to align nanofillers during printing, which further endows the structure with anisotropic behavior. Specifically, vat photopolymerization of mechanically enhanced structures inspired by Limpet teeth and magnetic field-assisted alignment of nanofillers will be described in Section 2. The heterogeneous material system inspired by the multilayer metal-containing shell of scaly-foot snail, polymer/metal structure, and electric field-assisted 3D printing system will be described in Section 3. The anisotropic gradient distribution and deformation-programmable porous structure inspired by the mechanism of Delosperma nakurense's seed compartment moisture absorption and release of seeds, and liquid crystal templating assisted vat photopolymerization will be described in Section 4. The immersed surface accumulation 3D printing of superhydrophobic structure inspired by *S. molesta* leaves will be described in Section 5. Finally, the summary and prospect of materials and methods for biomimetic manufacturing will be detailed in the conclusion section.

2. Limpet teeth inspired nanocomposite for mechanical reinforcement

Limpets are a type of aquatic gastropod mollusk with flat, cone-shaped shells that use a specialized organ called the radula to acquire food from hard ocean rocks [16, 17]. As shown in **Figure 1**, the radula is composed of rows of mineralized teeth that exhibit

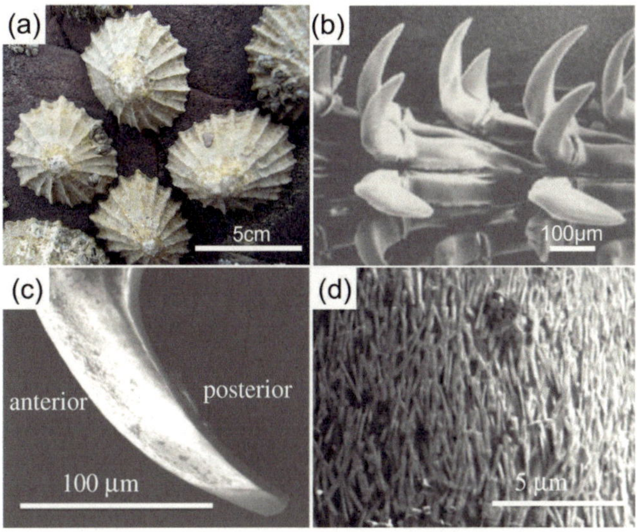

Figure 1.
Images of limpet tooth. (a) Top view of limpet shells; (b) microscale image of radula organ with exposed teeth; (c) microscale image of a single limpet tooth showing the base and cusp; (d) scanning electron microscope (SEM) image of limpet teeth microstructure with aligned mineral fibers. (a) Copyright from ref. [18]; (b) copyright from ref. [16]; (c, d) copyright from ref. [17].

the highest known mechanical strength, surpassing that of any other naturally occurring materials [16]. This excellent mechanical strength results from biomineralization, a cyclical process, that reinforces the chitin protein matrix through the distribution of aligned iron oxide minerals to create distinctive hierarchical structures [17]. These mineral-based microstructures have evolved to provide limpets with the necessary strength to graze on rough ocean surfaces, enabling them to extract nutrients from their environment without incurring damage [16]. However, the structural rigidity of mature limpet teeth eventually declines due to repeated feedings, which prompts the formation of new rows of biomineralized teeth [17]. The naturally occurring nanocomposite, a combination of soft protein and aligned mineral phase, is primarily dictated by the presence of mineral nanorods, which enhance the mechanical performance of otherwise weak biomaterials [17]. Consequently, limpet teeth provide an ideal design inspiration for the fabrication of a biomimetic nanocomposites with hierarchical microstructures and superior mechanical strength.

2.1 Functional architecture of limpet teeth

Biomaterials found in living organisms have adapted through natural selection processes to develop unique microstructures with enhanced physical properties [2, 12, 16, 17]. For example, the complex microstructures found in mantis shrimp have adapted a twisted plywood structure, known as a Bouligand structure, to enhance its mechanical properties and flexibility in order to withstand high-impact punches using its fist-like clubs [2, 12, 19]. Recent studies have determined that limpet teeth exhibit a linear elastic modulus of 120 GPa under tensile testing conditions [16]. Furthermore, the mineral phase of tooth samples was demonstrated to have a linear elastic of 180 GPa, which, compared to the overall structure, indicates that the observed mechanical strength is primarily attributed to the mineral phase [16]. This superior mechanical strength, through the integration of biomineralized reinforcement of a polymer matrix, relaxes challenges associated with manufacturing technologies to fabricate microstructures with highly complex geometries and excellent mechanical performance. Current fabrication methodologies have certain limitations that must be addressed given the high degree of difficulty in achieving microstructures with aligned mineral biomaterials. Through recent advancements in biomimetic AM technologies, these limitations can be mitigated by rigorously controlling the microstructures using 3D printing methods that were previously unachievable through traditional fabrication methods [3].

Considerable challenges arise from the capability to create uniform mineral nanofillers with adjustable dimensions, such as length and cross-sectional area. The interaction of the mineral nanofillers within the chitin-polymer matrix plays a critical role in the observed tensile strength in the natural composite material based on varying dimensions and orientation. Moreover, the ability to develop mineral nanofillers with desired dimensions, with high repeatability, is of key importance to maximizing the mechanical strength in 3D printed nanocomposite parts. Therefore, these challenges have led to strategies that incorporate physical fields to aid in the alignment of magnetic microbundles during photopolymerization [2, 3, 12].

Integrating a physical field during the 3D printing process allows for high control of the alignment direction of mineral nanofillers in order to reinforce the printed part such that the compressive, tensile, and bending are deflected by the mineral constituents. The controlled alignment of mineral nanofillers within the 3D printed part is imperative to effectively create an AM process capable of fabricating hierarchical microstructures into a fully functional 3D object using a bioinspired nanocomposite. Using methods such as physical field-assisted 3D printing, a bioinspired nanocomposite material can be created and manipulated into a fully functional 3D shape that exhibits a significantly high mechanical strength in comparison to other microscale manufacturing processes as well as highly accurate and controllable micro features.

2.2 Magnetic field-assisted vat photopolymerization

In order to reproduce the microstructures observed in limpets' teeth, an AM process, known as magnetic field assisted vat photopolymerization (MF-VPP), can be employed because of its flexibility to control the alignment direction of ferromagnetic nanofillers in polymeric resins, depicted in **Figure 2a**. For example, when a dynamic magnetic field is applied to the printing region, randomly distributed magnetic nanoparticles align along magnetic flux lines and subsequently agglomerate to form magnetic microbundles. Furthermore, the magnetic-field-assisted 3D printing process is highly advantageous because of its capability to rapidly align magnetic nanofillers in any spatial direction without direct contact between the magnet and nanocomposite material. A digital light projector (DLP) selectively cross-links the polymer resin on the printing platform, using a microscale mask image, which as a result constrains the magnetic nanofillers in the structure, as seen in **Figure 2b**. This process of alignment and photopolymerization is consecutively repeated, in a layer-by-layer fashion, to fabricate the desired 3D printed object.

Furthermore, optimization of printing parameters is highly dependent on the strength of the magnetic field and nanofiller concentrations because of light scattering effects that inhibit photopolymerization. As shown in **Figure 2c**, as the concentration of magnetic nanofiller increases, microbundle length also increases while the gap between adjacent bundles is reduced. Moreover, increasing the strength of the magnetic field significantly increases microbundle length, when compared to

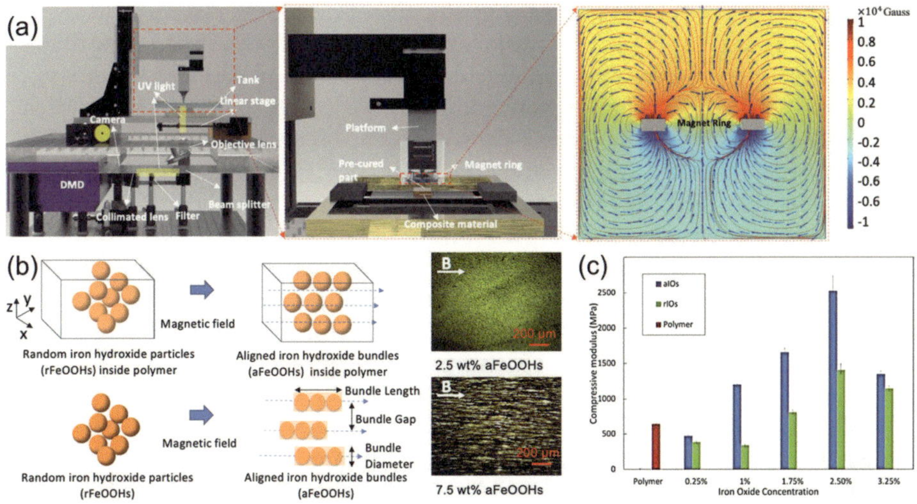

Figure 2.
Magnetic field assisted vat photopolymerization. (a) Schematic of MF-VPP printer with COMSOL Multiphysics simulation of magnetic flux lines in the printing region; (b) images showing the alignment of different concentrations of iron hydroxide nanoparticles within photopolymer matrix; (c) compression strength of reinforced iron hydroxide with different iron oxide concentrations. All figures' copyright from ref. [4].

microbundle length under weaker magnetic fields, while simultaneously decreasing the bundle diameter and gap. Consequently, nanocomposite material with high concentrations of magnetic nanofiller requires longer exposure times since it is difficult for the projected 2D light beam to penetrate the material to initialize photopolymerization. Taking these printing parameters into consideration, the process of coupling a dynamic magnetic field combined with 2D light projections during cross-linking allows for the creation of bioinspired microstructures with attractive anisotropic mechanical performance greater than that seen in limpet teeth. The 3D-printed microstructures can be fabricated at a low cost with unique features that are modifiable resulting in differing morphologies for various applications.

2.3 Biomimetic material and structures

Recent studies have revealed that the excellent mechanical performance exhibited by limpets' teeth can be attributed primarily to the reinforcement of a soft protein matrix by the controlled alignment of embedded iron-based minerals, specifically a mineral known as goethite [16]. Thus, the anisotropic mechanical strength of biomimetic hierarchical microstructures should be controlled by managing the spatial orientation of the magnetic nanofiller within the polymer resin. With the purpose of achieving high mechanical performance, a nanocomposite is prepared through the homogenous distribution of goethite nanoparticles in photocurable polymer resin. When initially distributed in the photocurable resin, the goethite nanoparticles have a random spatial orientation and must be coupled with the MF-VPP process in order to reinforce the normal weak polymer material *via* controlled nanofiller alignment.

Based on the morphologies seen in limpets' teeth, mechanical reinforcement is strongest when the mineral nanofillers are aligned parallel to the direction of the applied force. For example, this can be clearly observed when comparing the compressive strength of random and aligned magnetic nanofillers with just the pure polymer. As depicted in **Figure 2c**, the compressive strength of aligned nanofillers outperforms test specimens with random alignment and pure polymer. The maximum compressive load of aligned iron oxide particles-based composite is 80 times greater than that of the pure polymer. Anisotropic mechanical reinforcement of the microstructures is heavily influenced by the magnetic field intensity, dimension of magnetic nanofiller, and the magnetic nanofiller concentration in the 3D printable nanocomposite. However, only compression strength was enhanced due to the microbundle alignments, and the alignment of magnetic particles have no significant effect on the improvement of the tensile strength and bending strength. This is because there are no constraints between two adjacent magnetic particles, and the dissipations may lead to an early failure between each particle and polymer under the bending and tensile load.

Furthermore, bundles of magnetic nanofillers can be annealed, using a high-temperature furnace, to form long fibers that better mimic the microstructures seen in limpets' teeth. Similarly, these aligned magnetic nanofibers provide anisotropic mechanical reinforcement to the nanocomposite when a compressive load is applied. For example, the reinforced printed microstructures show superior mechanical performance when compared to the pure polymer and randomly oriented magnetic microbundles. However, high annealing temperatures, above the critical temperature for the photocurable resin, can cause cracks in the polymer material, which can lead to a reduction in the overall compressive strength. This can be easily mitigated with different polymers, such as polymer-derived ceramics, to fabricate high-strength composites with enhanced thermal performance. Thus, the reinforcement mechanism coupled with customizable alignment of magnetic nanofillers opens intriguing perspectives for designing high-strength 3D printed material based on bioinspired features with modifiable configurations.

2.4 Applications of anisotropically enhanced structure

Limpets' teeth provide a promising design inspiration for the creation of a nanocomposite material with anisotropic functions with enhanced mechanical thermal and electrical properties. The MF-VPP method has been demonstrated to have a wide range of capabilities, including the ability to fabricate unique geometries and distinctive microstructures that are difficult to manufacture using conventional manufacturing techniques. These advancements have the potential to open new possibilities for creating intricate structures at the microscale, with exceptional mechanical strength. This technology is expected to have a significant impact in various fields such as aerospace, biomedical, and electronics in the coming years [4, 20–23]. For instance, rapid prototyping of highly accurate and low-cost scale models can aid in the design and optimization of lightweight and fuel-efficient airplane components [24]. Furthermore, 3D printing of nanocomposites can be used to generate complex components for jet engines that have enhanced damage tolerance and corrosion resistance [24].

This method of 3D printing nanocomposites is also advantageous for electrical components where heat dissipation is crucial for the functionality of key components that are suspectable to overheating. Effective heat transfer in electrical components can be achieved by adjusting the orientation of aligned nanofiller materials with anisotropic thermal properties [23]. Moreover, polymers with enhanced energy capacity and anisotropic electrical conductivity can be realized through the alignment of functional nanofillers for a wide variety of electronics [25]. Additionally, high-resolution and porous biomaterials, such as scaffolds for bone regeneration, can be easily designed and fabricated using the MF-VPP method to create implants capable of withstanding shearing, compressive, and tensile loading [26]. In conclusion, creating complex structures with high mechanical strength on demand, which are both low cost and reliable, can be realized for numerous applications using the biomimetic approach described here.

3. Scaly-foot snail inspired multimaterial for property enhancement

3.1 Functional architecture of scaly-foot snail

Organisms in nature have evolved excellent survival strategies due to their special living environments. Scaly-foot snail (**Figure 3a**), a creature that lives in a deep-sea crater, is the only organism with metal in its skeleton (**Figure 3c**) [6, 29]. The multi-layer structure of its shell can effectively deal with the threat of toxic substances and high temperature in the environment (**Figure 3b**) [6, 29]. The multi-layer multi-material structure also endows shell with extremely high hardness and excellent energy dispersion capability under impact. This outstanding property has widely attracted the attention of researchers, but it is extremely difficult to reproduce this structure in an integrated manner, both in terms of materials and manufacturing. Inspired by the excellent performance of the heterogeneous material system inside the scaly-foot snail's shell, the researcher Tang et al. developed a heterogeneous mixture that can be both photocured and electroplated. The matrix of the heterogeneous material system is PEGDA with good biocompatibility, PEDOT:PSS with excellent conductivity is used as filler to improve the conductivity of the matrix, photoinitiator is used to initiate the crosslinking of PEGDA, and CuSO4 solution is used as electrolyte to adjust rheological properties to meet the viscosity requirements of DLP printing.

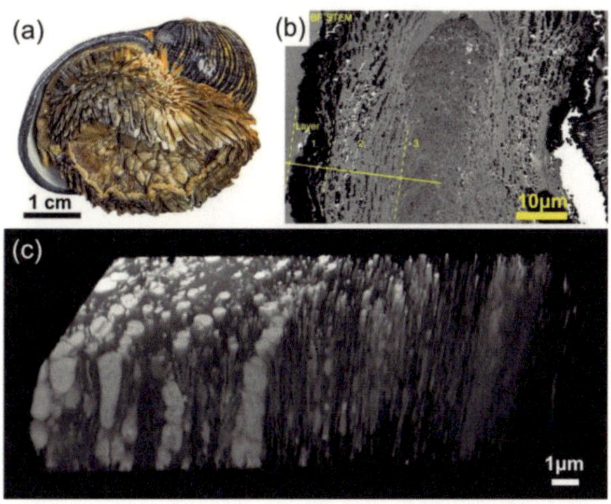

Figure 3.
Structural images of scaly-foot snail. (a) Image of the scaly-foot snail; (b) multilayered structure shown in cross-section of scaly-foot snail shell; (c) columnar channels and dispersed iron sulfide inside the shell. (a) Copyright from ref. [27]; (b, c) copyright from ref. [28].

3.2 Electrical field-assisted vat photopolymerization

The printing system in the reported work is a typical digital light processing (DLP) [30], which is mainly composed of digital micromirror device (DMD), linear stage, solution tank, printing platform, and control system (**Figure 4a**). The difference is that in order to introduce electroplating in the conventional layer-by-layer photocuring printing, two copper sheets are placed in the solution tank and connected to the negative pole of the power supply as an anode, and one corresponding copper sheet is placed on the bottom of the printing platform for both adhering the cured layer and as a cathode (**Figure 4b**). As shown in **Figure 4c**, when curing the polymer matrix, the projector projects a 2D light pattern of a specific shape to selectively cure the local area, and the conductive filler PEDOT:PSS and $CuSO_4$ electrolyte are sealed in the PEGDA polymer chain. The projection of UV light is stopped during electrodeposition, the power supply is connected to the electric field between the cathode and anode, and the copper particles migrate to the bottom of the polymer layer to obtain electrons and then reduce to copper (**Figure 4d**). So far, this article has

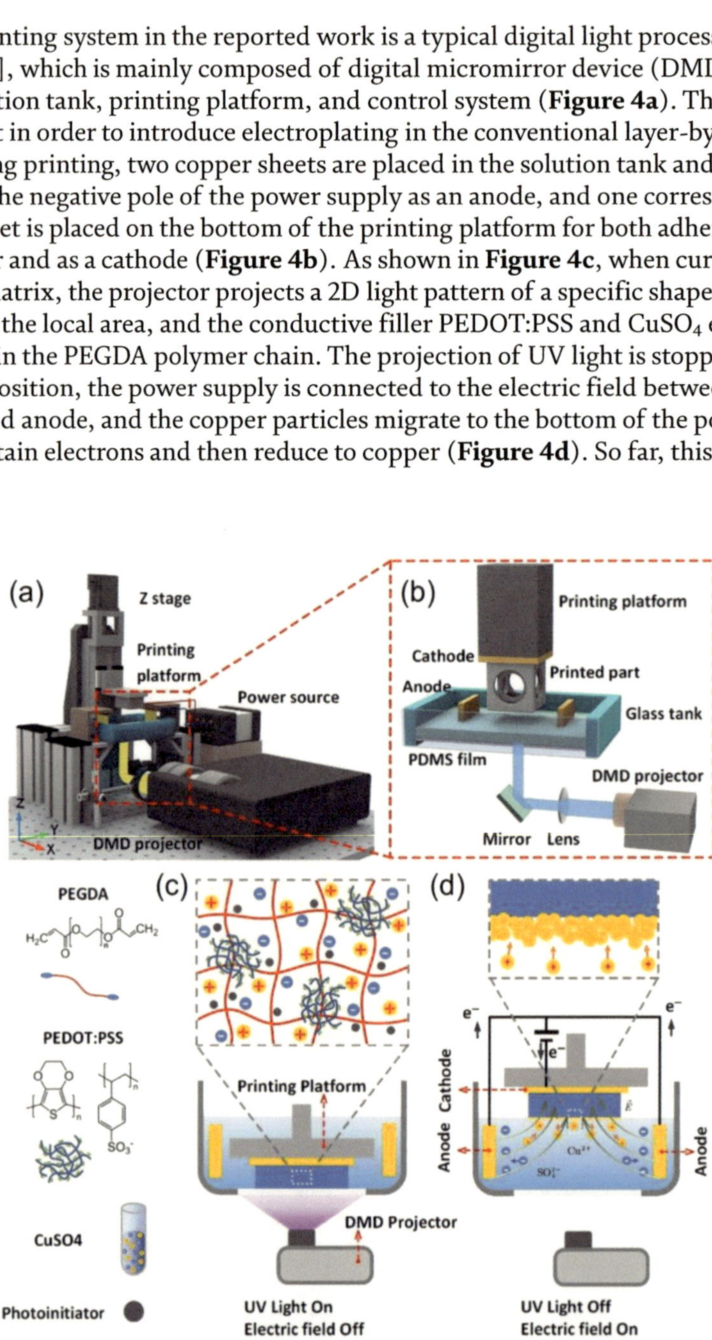

Figure 4.
Manufacturing of bioinspired polymer/metal structures using electrical field-assisted vat photopolymerization and heterogeneous material systems. (a) Schematic diagram of the electrical field-assisted vat photopolymerization set-up; (b) illustration diagram of the electric field and projection system of the printing set-up; (c) demostration diagram of photocuring process in a single printed layer; and (d) schematic diagram of electrical field assisted metal deposition process. All figures' copyrights from ref. [29].

demonstrated the composition of the heterogeneous material system and the working logic of the printing system, and the electroplating results also show the effectiveness of the materials and manufacturing methods [29].

3.3 Fabrication of polymer/metal structures

Polymer/metal structures are widely used in flexible circuits [31, 32], sensors [33, 34], and soft robots [35, 36] because of their light weight, extraordinary corrosion, and wear resistance. Most of the existing manufacturing methods are to first manufacture the polymer base by injection molding or 3D printing and then perform electroplating [37]. Since most polymers are not conductive, it is necessary to spray a conductive layer before electroplating [37]. Here, electric field-assisted 3D printing avoids the tedious and energy-wasting, time-consuming, and labor-intensive steps of traditional manufacturing [29]. It can print the polymer base and electroplate the metal surface in one step. The specific steps are as follows. The 3D model is first established by SolidWorks, and the output STL format file is imported into the self-built program for slicing. The system automatically assigns different material indexes to different layers during slicing. An index of zero corresponds to the photocuring step, and an index of one corresponds to the electroplating process. The obtained black and white mask images are then loaded into the printing program. Before printing starts, the motion system needs to be initialized, and the motion control program is used to move the printing platform to the zero point. At this time, the ground of the printing platform is in contact with the Teflon film at the bottom of the resin pool. After the printing process starts, the printing platform rises to leave a gap to solidify the base layer. Afterward, the system judges the material index corresponding to the picture. When the index is zero, the system projects UV light to cure polymer. The cured layer is then lifted by the platform to generate a new printing layer gap, and all following layers can be printed in this way. When the index is one, the system automatically progresses to the electroplating step. At this time, the polymer layer is lifted as an anode and only the bottom touches the mixture solution. After the electric field is turned on, the copper ions accept electrons on the surface of the anode and reduce to copper. As shown in **Figure 5**, the pitchfork polymer base was first printed, and then

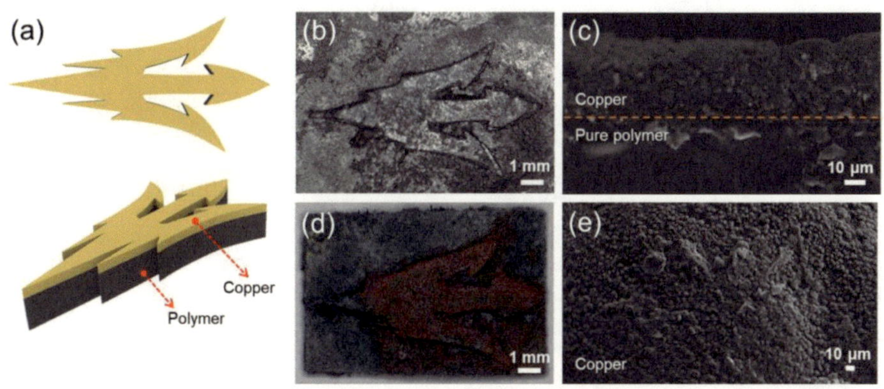

Figure 5.
Polymer/metal structure. (a) Schematic diagram of the polymer ASU pitchfork with copper surface. (b-e) Electroplating results and SEM images of the printed ASU pitchfork. All figures are copyrighted from ref. [29].

the structure was placed in the same mixture for electroplating to obtain the metal surface. The copper and polymer layer can be clearly seen in **Figure 3c**. Due to the polymer base photocuring material and the material used for electroplating being the same material, the copper grows toward the polymer layer, which strengthens the polymer/metal interlayer bonding to avoid delamination.

3.4 Applications of polymer/metal structure

Benefiting from the aforementioned photocurable heterogeneous materials that can be used as electrolytes at the same time, it is possible to manufacture polymer/metal structures in an integrated manner, making up for the time-consuming and redundant multi-step process of traditional manufacturing methods. After testing the curing properties and plating properties of the material, the authors printed a series of different structures in **Figure 6** [29]. In order to demonstrate the corresponding multi-layer printing logic, the author printed a triangular base, electroplated a layer of copper and continued to print a layer of triangular polymer cylinders, and finally obtained a polymer-metal-polymer sandwich structure (**Figure 6a, b**). Metals are used in circuit manufacturing because of their excellent conductive properties, and the metal/polymer structure not only makes up for the low conductivity of polymers but also has the advantages of lightweight and flexibility of polymers. Based on this, circuits printed with pure polymers and electroplated circuits were manufactured (**Figure 6c-e**). The experimental results show that the circuits after metal plating have better electrical conductivity. The LED lamp beads light up when the circuit is connected, but the polymer circuit cannot be lighted because of its high resistivity. As a result, the multi-layered complex polymer-metal-polymer sandwich structure and the enhanced conductivity after electroplating demonstrate the high degree of manufacturing flexibility, structural complexity, and functionality of electric field-assisted printing heterogeneous materials in integrated packaged circuits [38], flexible sensors [38], and electromagnetic interference (EMI) shielding [39, 40].

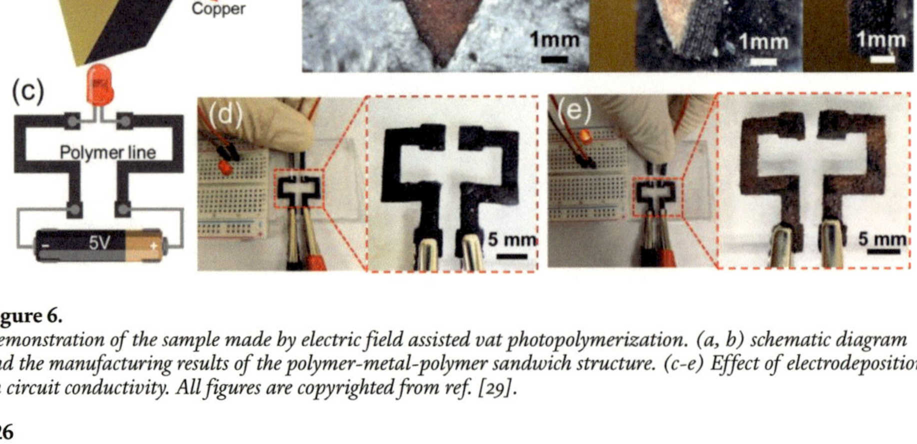

Figure 6.
Demonstration of the sample made by electric field assisted vat photopolymerization. (a, b) schematic diagram and the manufacturing results of the polymer-metal-polymer sandwich structure. (c-e) Effect of electrodeposition on circuit conductivity. All figures are copyrighted from ref. [29].

4. Delosperma nakurense inspired nanocomposite for shape changing

4.1 Functional architecture of Delosperma nakurense

Delosperma nakurense, a plant that lives in arid environments, chooses to open its seed chambers after rain or when there is high humidity to release its seeds to increase seed survival rate. This specific strategy of hygroscopic deformation is reflected in many plant species, such as pinecones [41], wheat awns [42], seedpods of orchid trees [43], and spikemoss stems [43]. These hygro-responsive structures choose to open the structure under wet or dry conditions to release seeds to complete reproduction. The spontaneous hygroscopic deformation of the seed chamber structure is of great significance to the design and manufacture of flexible devices, self-responsive sensors, and soft robots [44]. **Figure 7** describes the process of the seed chamber absorbing moisture and expanding to release the seeds. The researchers found that the three main points of this process are: porous structures are the basis of moisture absorption deformation, arranged cell wall realizes anisotropic deformation ratio, and swellable cellulose fiber is used to absorb water. It is essential to replicate these three elements in principle to bionically manufacture this structure. In order to achieve this goal, the researchers Tang et al. used liquid crystals and nano-fillers as the materials and used the phase separation of the two liquid crystals during curing to obtain a porous structure that can absorb moisture.

4.2 Liquid crystal templating assisted vat photopolymerization

Liquid crystal, a substance that is both liquid and crystal and has a specific effect on light, has a strong polarization property that enables it to align under the action of electric field. This electro-alignment property is exploited to align nanofillers to impart anisotropic deformation of the structure. Correspondingly, using an electric field to align polymers and nanofillers is a proven approach to obtain anisotropic structures. Unlike extrusion molding, which uses shear force or ultrasonic vibrations to align polymers and nanofillers, ellipsoidal liquid crystal molecules are deflected under the action of an electric field force so that the long axis aligns with the direction of the electric field. At the same time, the deflected liquid crystal molecules drive the nanofillers to align in the direction of the electric field. This indirect arrangement makes up for the deficiency that the electric field can only arrange conductive substances (such as CNT [46], graphene [47]), and makes it possible to use the electric field to arrange non-conductive substances, such as the SiC nanofiller used in this study.

Based on the characteristics of liquid crystal materials, the authors Tang et al. propose an electric field-assisted printing strategy, which places electrode pairs facing

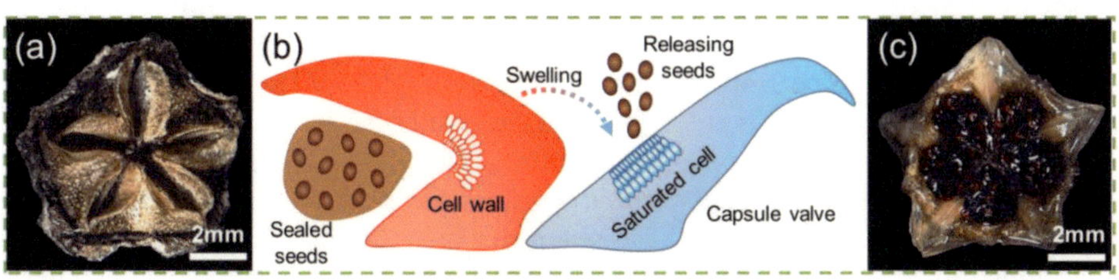

Figure 7.
Seed capsule releases the seeds after absorbing moisture. (a, c) copyright from ref. [9]; (b) copyright from ref. [45].

in different directions in the resin tank on the basis of traditional digital light processing. The printing system is mainly composed of a projector that projects a specific 2D light pattern, a stage that carries a cured layer and moves linearly, and DC power supply that is used to generate a high-voltage electric field. The electrode pair in a certain direction is first connected during printing, and after the LC and nanofillers are arranged for a period of time under the action of the electric field, the projector projects a beam of light to cure the selected area. By connecting electrode pairs in different directions and printing repeatedly layer by layer, LC and nanofillers with different alignment directions can be obtained by curing in a single layer or inside different layers.

4.3 Dynamic electro-alignment characterization

Controlling the strength and timing of the electric field is critical for aligning liquid crystals and SiC nanofillers. Even though both liquid crystal and SiC are poor conductors, applying a high-intensity electric field for a long time generates a lot of heat, which causes the liquid crystal to undergo a nematic to isotropic phase transition [48]. Once the phase transition occurs, the liquid crystal becomes disordered [49], and the SiC that was previously aligned indirectly is simultaneously driven out of the ordered alignment state by the moving liquid crystal. When the electric field strength is weak, the polarization force applied to the liquid crystal molecules is not enough to drive the liquid crystal to move or it takes a long time to complete the arrangement. Therefore, choosing an appropriate electric field strength and application time is an indispensable condition for obtaining a homogeneous unidirectional alignment. In addition, the researchers also found that the alignment results using an alternating current (AC) electric field are significantly different from those using a direct current (DC) field. Even if the AC electric field is applied for a long time, there is no significant movement once the alignment is completed. In the case of the same voltage value, although the DC electric field increases the rate of arrangement, if the electric field is still applied after the arrangement reaches the highest degree of anisotropy, the order of the arrangement changes significantly, and the mixture undergoes approximately macroscopic disordered turbulence. This is detrimental to the desired result and should be avoided.

Specifically, when an AC voltage of 1 kV is applied, the long axis of the liquid crystal is slowly deflected to the direction of the electric field while driving the dispersed SiC nanofillers to gather and arrange in a line. The Fourier transform results in **Figure 8** show that the mixture is not directional when the electric field is not applied (0 s). After 195 s of alignment, the prominent peak curve in the probability distribution curve indicates that the mixture is anisotropic. Continuing to apply electric field, the degree of alignment does not change. In order to demonstrate the sustainability of the arrangement, within 2 mins after the electric field was turned off, the directionality of the arrangement did not decrease significantly, see the relative positions of the red and green curves in the **Figure 9**. When aligning mixture with 1 kV DC compared to AC, the results were significantly different in terms of time and sustainability of the alignment. The DC electric field makes the alignment more rapid, and an application time of 12 s is sufficient to obtain the most anisotropic alignment, compared to the 195 s required for the AC electric field (**Figure 10**). However, this directionality is destroyed with the prolongation of time. When the arrangement is 120 s, the mixture again becomes chaotic and enters a turbulent state. The potential reason for this phenomenon is that the direction of the DC electric field is always constant, and the liquid crystal molecules

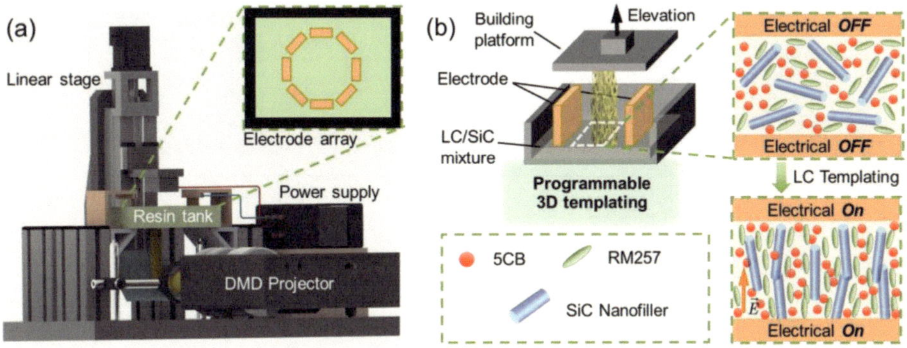

Figure 8.
Probabilistic analysis of the directionality of the alignment under alternating electric field.

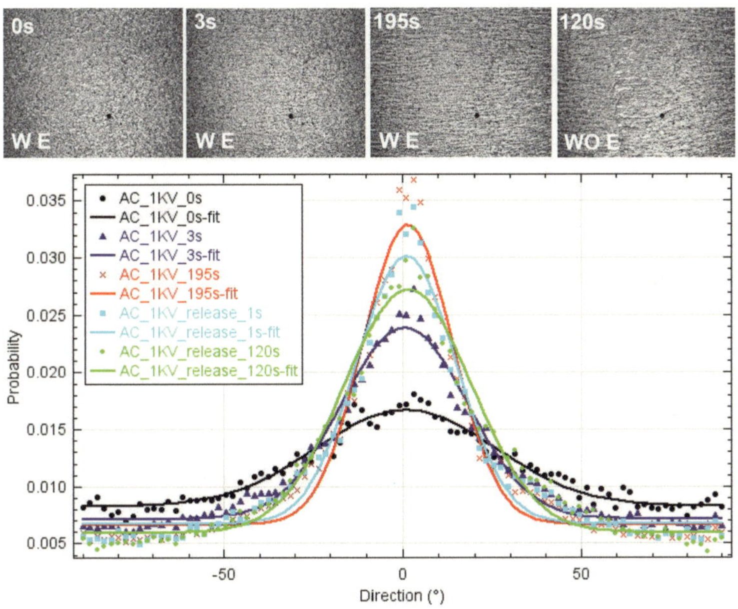

Figure 9.
Liquid crystal templating assisted vat photopolymerization. (a) Schematic illustration of the liquid crystal templating assisted vat photopolymerization; (b) schematic diagram of electric field alignment of liquid crystal monomer and SiC nanofiller. All figures are copyrighted from ref. [45].

are subjected to the electric field force of a single direction and then always migrate in the same direction. In the AC electric field, the liquid crystal molecules are subjected to alternating electric field forces, and there is no dominant force to move them in a specific direction, so the liquid crystals are only deflected *in situ*. The alignment state of the LC/SiC nanofiller does not change over time as long as the heat caused by the electrification does not cause the liquid crystal to undergo a nematic to isotropic phase transition. In summary, the AC electric field can make the arrangement more uniform and avoid entering a chaotic state, while the SiC nanofiller under the arrangement of the DC electric field has a more significant bundle, and the degree of anisotropy of the arrangement is higher but need to avoid applying an electric field for a long time to maintain the directionality of the arrangement.

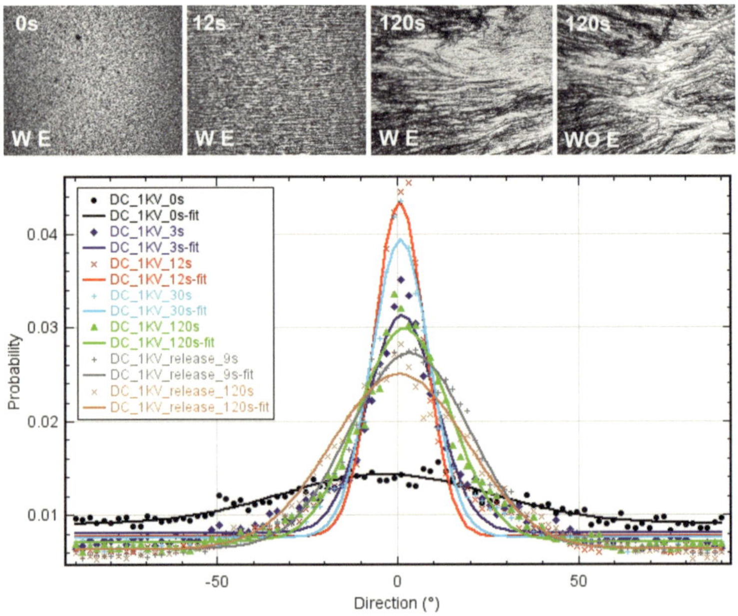

Figure 10.
Probabilistic analysis of the directionality of the alignment under direct current electric field.

4.4 Photopolymerization induced phase separation

As mentioned in Section 4.1, the core of realizing hygroscopic deformation is an anisotropic porous structure, and liquid crystal as a functional material is well suited for this purpose. The phase separation of the two liquid crystals during solidification provides the possibility to fabricate a porous structure, and the arrangement of the applied electric field allows the pores to be aligned along a specific direction, and finally, an anisotropic and gradient porous structure can be obtained [45]. As shown in **Figure 11a**, under the action of an electric field in the RM257/5CB/SiC homogeneous mixture, the long axes of the liquid crystals RM257 and 5CB are deflected to be in line with the direction of the electric field, and this microscopic movement then indirectly drives the SiC nanofillers to converge and align in a straight line. Subsequent irradiation of UV light triggers the crosslinking of RM257, and 5CB, which does not participate in the reaction as a solvent, separates from the crosslinked RM257 and gathers together, and the area occupied by it is cleaned by acetone, leaving a large number of pores. The timing of phase separation is different due to the difference in light intensity at the bottom and top of the cured layer. The bottom is rapidly cross-linked under strong light irradiation, and it is difficult for 5CB to gather locally on a large scale, so the pore size at the bottom is smaller. On the contrary, due to the lower cross-linking rate at the top, the liquid crystals phase-separated to a higher degree, eventually leaving larger-sized pores. **Figure 11b-(i, ii)** show a porous structure with a gradient distribution, and iii and iv show elongated pores with long and short axes and a SiC nanofiller whose arrangement direction is consistent with the long axis of the pores. Therefore, the anisotropic porous structure mechanically enhanced by SiC can be obtained using liquid crystal phase separation and electric field indirect alignment of SiC nanofiller, which opens the possibility of programmable deformations and hygroscopic actuation described in the following section.

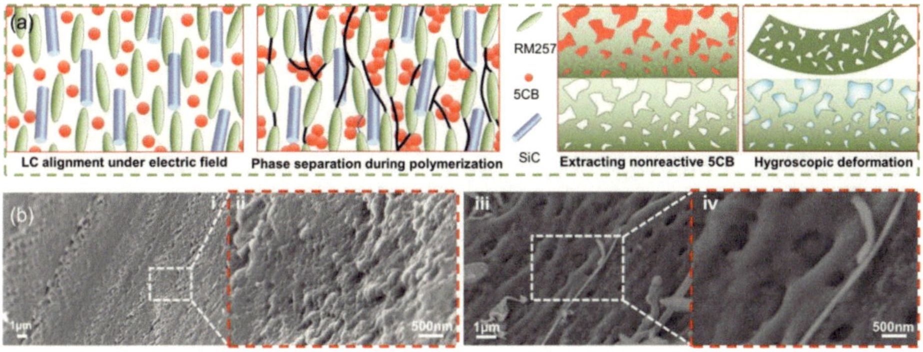

Figure 11.
Fabrication process of anisotropic porous structure. (a) Schematic illustration of aligning composite, phase separation during the photopolymerization, and the anisotropic porous structure after extracting unreacted 5CB; (b) SEM images of the gradient anisotropic porous structure with aligned SiC nanofiller. All figures are copyrighted from ref. [45].

4.5 Applications of gradient anisotropic porous structure

As previously conceived, when the LC/SiC mixture is aligned, the cured structure undergoes deformation in a specific direction after cleaning the unreacted 5CB and drying it. Since the major axis of the pores is aligned with the alignment direction, the dry structure curls up along the major axis. As shown in **Figure 12a**, the curling direction corresponding to the blue area is perpendicular to the red area. The annular structure in **Figure 12b** is divided into four regions and has a bowl shape when dry. The smaller rectangular region in **Figure 12c** corresponds to the hinge during deformation, along which the larger rectangle deflects. When two adjacent areas in the same layer are at +45° and −45°, respectively, the deformation directions of the two areas are perpendicular to each other after drying, and the deformations are mutually restrained and folded inward (**Figure 12d**). In order to demonstrate the deformation constraints between different layers due to the different alignment directions, a double-layer ring structure as shown in **Figure 12e** was printed, and the deformation directions perpendicular to each other in the two layers lead to a saddle-shaped deformation. However, the structure in **Figure 12f** is diagonally curled between the respective layers, and the opposite curling directions of the two layers are perpendicular to

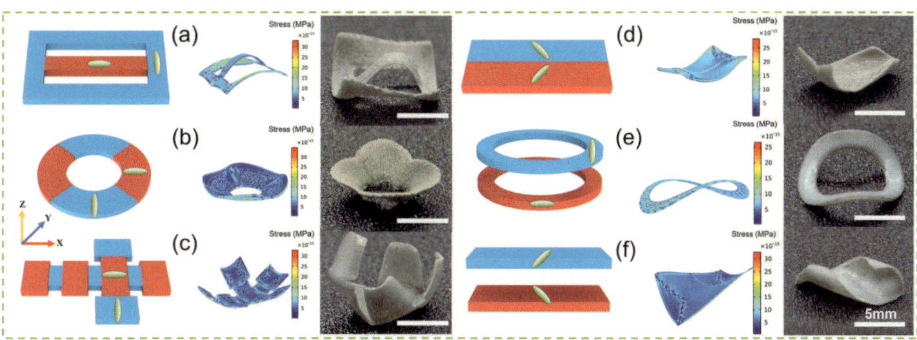

Figure 12.
Spatially programmable alignment and deformation with the prediction of simulation. All figures are copyrighted from ref. [45].

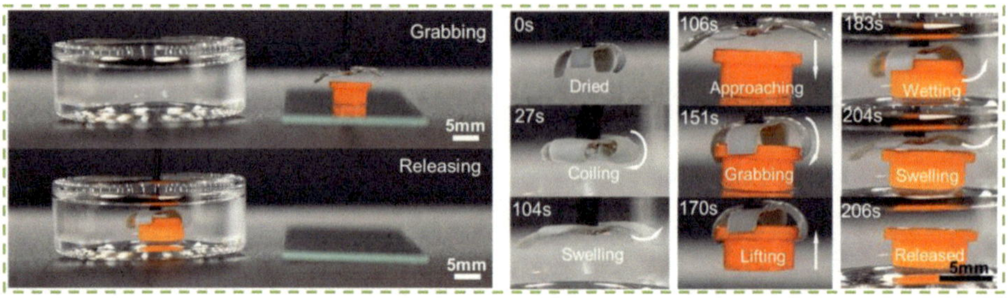

Figure 13.
Dynamic grasping process of hygroscopic gripper. All figures are copyrighted from ref. [45].

each other. So far, by setting different alignment directions in different regions in the same layer or between different layers, a series of programmable deformations can be obtained. Furthermore, the final deformation result can be predicted in advance through simulation, and the experimental results also confirm that the simulation is highly consistent with the real deformation.

Soft grippers have attracted widespread attention because of their flexibility and unique driving methods. The common ones are pneumatic drive [50, 51], electromagnetic drive [52–54], and cable chain drive [55, 56]. However, the structures corresponding to these driving methods are very complex and require the input of external energy. Therefore, designing and manufacturing a self-actuating flexible gripper is the focus of current research. The spontaneous deformation property of the porous structure upon moisture absorption and drying is a favorable way to fabricate self-propelled flexible grippers. The author designed a four-beam gripper, and the liquid crystal alignment direction is perpendicular to the direction of the beams. As shown in **Figure 13**, the beam is fully bent, and the entire gripper is closed in the dry state. After placing it in the acetone, the gripper opens after the pores absorb moisture and expand, and then moves the gripper over the yellow target. As acetone volatilizes in the air, the pore previously filled with acetone collapses, and the gripper bends and closes to grab the target. Finally, the target is moved into the acetone, and the beam continues to expand hygroscopically to release the target. The entire grasping process is about 206 s, and the deformation of the gripper is reversible, as a hygro-responsive structure stretches in acetone and contracts in the air. In summary, the programmable control of the hygroscopic deformation of the anisotropic porous structure opens up new opportunities for the fabrication of flexible grippers [57], ultrafiltration membranes [58, 59], and flexible sensors [60]. The future research direction is mainly to control the shape and distribution of pores on the micron scale, improve the actuation response rate, and enhance the mechanical properties of composites.

5. *Salvinia molesta* inspired nanocomposite for controllable wettability

5.1 Functional architecture of *S. molesta*

Since Dettre and Johnson discovered in 1964 that the superhydrophobicity of the lotus leaf is related to the nano/microscale dual-scale structure of its surface [61], a large number of biomimetic structures have been designed and manufactured by researchers, such as the eggbeater-like structure of *S. molesta* that exhibits long-term

Figure 14.
Superhydrophobic structures of Salvinia molesta in nature. All figures are copyrighted from ref. [62].

air retention capability, which enables water droplets to remain on the structure's surface (**Figure 14a, b**). The core of the superhydrophobicity exhibited by this structure is that the hair on the head of the eggbeater is coated with nano-scale wax crystals and the hair ends have hydrophilic patches (**Figure 14c**). This nano/microscale dual-scale roughness effectively reduces the surface energy of the structure. In order to bionically manufacture this structure, researchers mostly use polymer resin to print structures below the millimeter level and the smallest size is only a few microns [63]. Although this method can reproduce the natural structure excellently and has good superhydrophobic performance, the dual-scale roughness characteristics of the surface of the natural structure cannot be reflected. Therefore, Yang et al. tried to add multiwalled carbon nanotubes (MWCNTs) to the resin to increase the roughness at the nanometer level. The results showed that MWCNTs improved the superhydrophobicity of the polymer structure.

5.2 Immersed surface accumulation vat photopolymerization

Different from conventional top-down or bottom-up layer-by-layer printing, immersed surface accumulation innovatively uses optical fiber to guide the light beam, combined with computer numerical control (CNC) multi-axis printing platform to obtain greater degree of printing freedom (**Figure 15a**). The mask image obtained by slicing the three-dimensional model is projected onto the focusing objective lens by the digital micromirror device (DMD), and the shrunken optical pattern

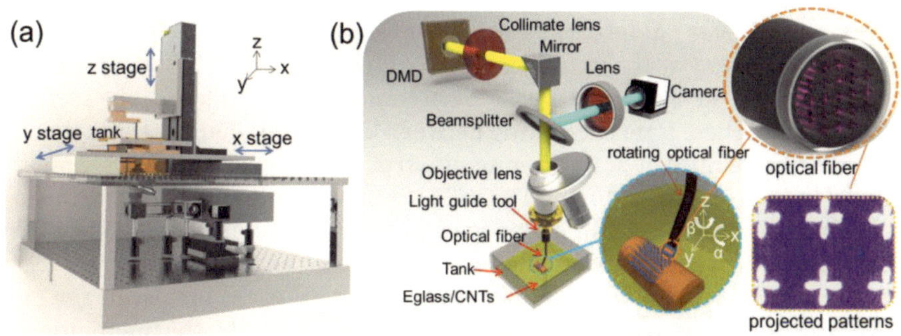

Figure 15.
Immersed surface accumulation 3D printing. (a) Schematic diagram of the vat photopolymerization set-up; (b) schematic diagram of the optical system of the immersed surface accumulation strategy. All figures are copyrighted from ref. [1].

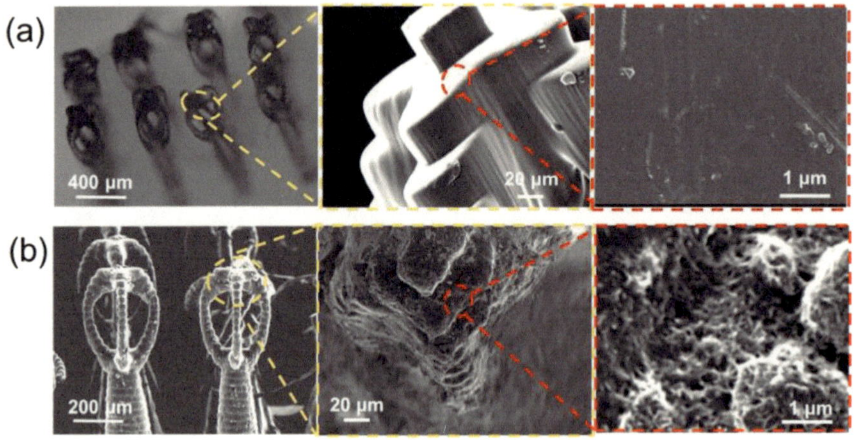

Figure 16.
Scanning electron microscope (SEM) images of the eggbeater-like structure. (a) Structures made from polymer; (b) structures made from polymer/MWCNTs mixture. All figures are copyrighted from ref. [1].

is guided into the photocurable resin solution through the optical fiber. The polymer undergoes photoinitiated polymerization under light, and the mixture is selectively cured. The MWCNTs are sealed in the polymer network, some MWCNTs protrude from the surface of the structure to form nanoscale roughness. The polydimethylsiloxane covering the end of the optical fiber makes the cured polymer easily separated from the optical fiber. At the same time, the printing result can be observed through the beam splitter, so as to detect the printing quality in real time. At the level of printing freedom, immersed surface accumulation can not only realize the layer-by-layer printing of conventional vat photopolymerization on the plane but also can print on complex curved surfaces or sides. The scanning electron microscope image in **Figure 16** shows the effectiveness of the printing system. The shape of the structure is similar to the natural structure. The structure below the micron level has a minimum size of only tens of microns. The subsequent contact angle test also fully confirmed the superhydrophobicity of the structure.

5.3 Fabrication of eggbeater-like microstructure

By using pure polymer and polymer/MWCNTs to print the eggbeater structure, the effect of MWCNTs on the surface roughness can be compared. **Figure 16a** is the top view and side view of the structure made of pure polymer. It is obvious that the entire *Salvinia molesta*-like structure has been completely reproduced, the surface is very smooth, and there is no overcuring between layers. This demonstrates that it is feasible to replicate micron-scale biomimetic structures using photocurable polymers and digital light processing. When MWCNTs were mixed into the polymer, the surface of the structure was distributed with microscale microgroove (**Figure 16b**), and the roughness increased. More importantly, part of the MWCNTs protrudes from the surface of the structure (**Figure 16b**), which constitutes the nanoscale roughness. Experiments prove that the nanoscale roughness caused by the microgrooves on the surface of the structure and the protruding MWCNTs significantly reduces the surface energy of the structure, and the two synergistically enhance the superhydrophobicity of the structure.

5.4 Applications of superhydrophobic structure

In recent years, with the increasingly frequent exploitation and transportation of marine resources, emergencies such as offshore oil spills have occurred more and more frequently, which not only caused huge economic losses but also caused serious damage to the marine ecological environment. Traditionally, spilled crude oil has been dealt with by physical, chemical, and biological methods. For example, a porous lipophilic material absorbs oil through capillary action and stores it in the pores of the material, and finally recovers the crude oil adsorbed by the porous material through extrusion or centrifugation. In addition, the oil-water separation ability of the structure can be greatly improved by changing the lipophilic and hydrophobic properties of the structure. Therefore, through the design and surface treatment of the surface micro-nano structure, the oil absorption and hydrophobicity of the oil-absorbing material can be effectively improved. At the same time, due to the capillary force caused by the surface micro-nano structure, its adsorption capacity to the oil layer is greatly enhanced. As shown in **Figure 17**, under the action of surface tension and gravity of the droplet, the droplet of the oil-water mixture remains spherical on the surface of the eggbeater-like structure. Since the structure is superhydrophobic and has a strong capillary effect on oil droplets, the oil droplets penetrate into the gaps of the columns while the water droplets remain spherical. After 6 s, the oil droplets were completely absorbed by the biomimetic structure, achieving oil-water separation. When the eggbeater-like structure holds the oil-water droplets in different relative positions, even if the droplets are suspended below the surface of the structure, the result of oil-water separation does not change. At tiny scales, oil-liquid separation will not be affected by gravity but is mainly determined by the surface tension of the droplets and the surface energy of the structure. Therefore, regardless of the relative position of the droplet and the eggbeater-like superhydrophobic structure, the result and rate of the oil-water separation did not change significantly. In summary, the bionic superhydrophobic structure will have a wide range of applications in surface self-cleaning, oil-water separation, and anti-icing.

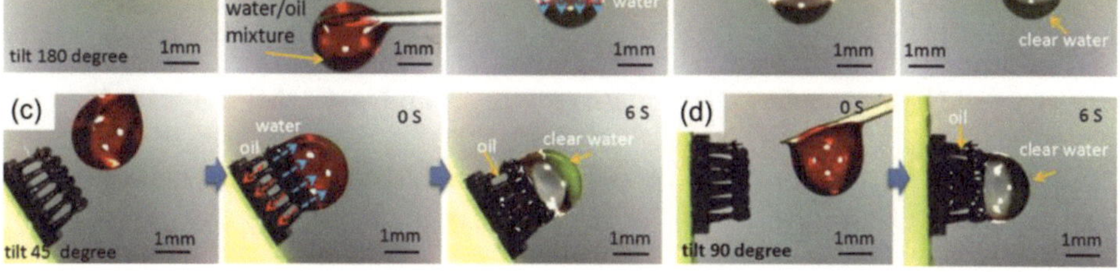

Figure 17.
Oil/water separation of the printed eggbeater-like structure. Oil/water separation performance under the tilt condition of (a) 0°, (b) 180°, (c) 45°, and (d) 90°. All figures are copyrighted from ref. [1].

6. Conclusion

The implications for biomimetic fabrication of highly functional structures found in nature are both significant and challenging. This paper summarizes the difficulties and shortcomings of existing research in terms of materials, structures, and manufacturing methods required for bionic manufacturing, and proposes a series of preparation methods for bionic materials and highly flexible additive manufacturing methods. Firstly, since a single material is not enough to make the structure highly functional, the bioinspired polymer/nanofiller composite can effectively endow the structure with better mechanical performance, optical characteristics, thermal conductivity, hygroscopic deformation, and superhydrophobicity. Secondly, in view of the fact that there are still some slight differences between the bionic structure and the natural structure, and these differences are decisive for the performance, it is necessary to replicate the natural structure both macroscopically and microscopically, which will play an important role in improving performance. Thirdly, at the level of manufacturing methods, vat photopolymerization provides an effective means for manufacturing multi-scale complex structures, and the introduction of magnetic and electric fields has created the possibility to increase the complexity of the structure, and the nanofillers spatially arranged in a specific direction have brought the structure mechanical, thermal, optical, and deformational manifestations of anisotropy. Although the current research on biomimetic manufacturing is quite effective and the performance of the structure is excellent, there are still many potential research fields worth studying in the future: 1) The study and imitation of the structure, material, function, and interaction of natural organisms must be the result of comprehensive application and cross-learning of biological principles, material chemistry, advanced manufacturing technology, and numerical simulation; 2) Bionic structures need not be limited to natural structures despite their superior performance, the next-generation functional structures can be predicted through big data analysis and numerical simulation; 3) Continue to develop advanced manufacturing methods to cope with multi-scale, multi-material, and multi-physical-field printing needs; 4) The intelligent bionic structure should not only originate from nature but also return to nature, which realizes the purpose of green manufacturing and sustainable development. Overall, deconstructing at the material and structural level and using additive manufacturing to reproduce natural structures is effective in promoting bionic design and manufacturing, and opens new doors for the manufacture of next-generation high-performance mechanical, thermal, optical, and hydraulic structures.

Acknowledgements

The authors acknowledge ASU startup funding, ASU FSE Strategic Interest Seed Funding, and National Science Foundation (NSF grant No. CMMI-2114119).

Conflict of interest

The authors declare no conflict of interest.

Author details

Tengteng Tang, Dylan Joralmon and Xiangjia Li*
Arizona State University, Tempe, USA

*Address all correspondence to: xiangjia.li@asu.edu

References

[1] Yang Y, Li X, Zheng X, Chen Z, Zhou Q, Chen Y. 3D-printed biomimetic super-hydrophobic structure for microdroplet manipulation and oil/ water separation. Advanced Materials. 2018;**30**(9):1704912. DOI: 10.1002/ adma.201704912

[2] Yang Y et al. "electrically assisted 3D printing of nacre-inspired structures with self-sensing capability," science. Advances. 2019;**5**(4):eaau9490. DOI: 10.1126/sciadv.aau9490

[3] Li X et al. Limpet tooth-inspired painless microneedles fabricated by magnetic field-assisted 3D printing. Advanced Functional Materials. 2021;**31**(5):2003725. DOI: 10.1002/ adfm.202003725

[4] Joralmon D, Amonoo E, Zhu Y, Li X. Magnetic field-assisted 3D printing of limpet teeth inspired polymer matrix composite with compression reinforcement. Journal of Manufacturing Science and Engineering. 2021;**144**(4): 041012. DOI: 10.1115/1.4052296

[5] Li X et al. 3D-printed cactus-inspired spine structures for highly efficient water collection. Advanced Materials Interfaces. 2020;**7**(3):1901752. DOI: 10.1002/admi.201901752

[6] Yao H et al. Protection mechanisms of the iron-plated armor of a deep-sea hydrothermal vent gastropod. Proceedings of the National Academy of Sciences. 2010;**107**(3):987-992. DOI: 10.1073/pnas.0912988107

[7] Zhu Y et al. Recent advancements and applications in 3D printing of functional optics. Additive Manufacturing. 2022;**52**:102682. DOI: 10.1016/j. addma.2022.102682

[8] Sydney Gladman A, Matsumoto EA, Nuzzo RG, Mahadevan L, Lewis JA. Biomimetic 4D printing. Nature Materials. 2016;**15**(4):413-418. DOI: 10.1038/nmat4544

[9] Harrington MJ et al. Origami-like unfolding of hydro-actuated ice plant seed capsules. Nature Communications. 2011;**2**(1):337. DOI: 10.1038/ ncomms1336

[10] Li X et al. 3D printing of flexible liquid sensor based on swelling behavior of hydrogel with carbon nanotubes. Advanced Materials Technologies. 2019;**4**(2):1800476. DOI: 10.1002/ admt.201800476

[11] Compton BG, Lewis JA. 3D-printing of lightweight cellular composites. Advanced Materials. 2014;**26**(34):5930-5935. DOI: 10.1002/ adma.201401804

[12] Yang Y et al. Biomimetic anisotropic reinforcement architectures by electrically assisted nanocomposite 3D printing. Advanced Materials. 2017;**29**(11):1605750. DOI: 10.1002/ adma.201605750

[13] Pagac M et al. A review of vat Photopolymerization technology: Materials, applications, challenges, and future trends of 3D printing. Polymers. 2021;**13**(4):598 [Online]. Available: https://www.mdpi. com/2073-4360/13/4/598

[14] Dhinakaran V, Manoj Kumar KP, Bupathi Ram PM, Ravichandran M, Vinayagamoorthy M. A review on recent advancements in fused deposition modeling. Materials Today: Proceedings. 2020;**27**:752-756. DOI: 10.1016/j. matpr.2019.12.036

[15] Saadi MASR et al. Direct ink writing: A 3D printing Technology for Diverse Materials. Advanced Materials. 2022;**34**(28):2108855. DOI: 10.1002/adma.202108855

[16] Barber AH, Lu D, Pugno NM. Extreme strength observed in limpet teeth. Journal of the Royal Society Interface. 2015;**12**(105):20141326. DOI: 10.1098/rsif.2014.1326

[17] Lu D, Barber AH. Optimized nanoscale composite behaviour in limpet teeth. Journal of the Royal Society Interface. 2012;**9**(71):1318-1324. DOI: 10.1098/rsif.2011.0688

[18] Tango22. Common limpets1. Wiki. Internet. 2006. [Online]. Available: https://commons.wikimedia.org/wiki/File:Common_limpets1.jpg.

[19] Tadayon M, Amini S, Wang Z, Miserez A. Biomechanical Design of the Mantis Shrimp Saddle: A biomineralized spring used for rapid raptorial strikes. iScience. 2018;**8**:271-282. DOI: 10.1016/j.isci.2018.08.022

[20] Zhu Y et al. 3D printing biomimetic materials and structures for biomedical applications. Bio-Design and Manufacturing. 2021;**4**(2):405-428. DOI: 10.1007/s42242-020-00117-0

[21] Shahrubudin N, Lee TC, Ramlan R. An overview on 3D printing technology: Technological, materials, and applications. Procedia Manufacturing. 2019;**35**:1286-1296. DOI: 10.1016/j.promfg.2019.06.089

[22] Vinayagam M, Ali A, Ranganathan K, Jeffrey J, Ravikumar M, Rajkumar S. The roles and applications of additive manufacturing in the aerospace and automobile sector. Materials Today: Proceedings. 2021;**47**:405-409. DOI: 10.1016/j.matpr.2021.04.596

[23] Wang H et al. Design of 3D printed bioinspired nacre-like structured materials with significantly enhanced thermal conductivity. Applied Physics Letters. 2021;**118**(13):131903. DOI: 10.1063/5.0042812

[24] Joshi SC, Sheikh AA. 3D printing in aerospace and its long-term sustainability. Virtual and Physical Prototyping. 2015;**10**(4):175-185. DOI: 10.1080/17452759.2015.1111519

[25] Ajjan FN, Mecerreyes D, Inganäs O. Enhancing energy storage devices with biomacromolecules in hybrid electrodes. Biotechnology Journal. 2019;**14**(12):1900062. DOI: 10.1002/biot.201900062

[26] Mehrali M et al. Nanoreinforced hydrogels for tissue engineering: Biomaterials that are compatible with load-bearing and electroactive tissues. Advanced Materials. 2017;**29**(8):1603612. DOI: 10.1002/adma.201603612

[27] Chen C, Uematsu K, Linse K, Sigwart JD. By more ways than one: Rapid convergence at hydrothermal vents shown by 3D anatomical reconstruction of Gigantopelta (Mollusca: Neomphalina). BMC Evolutionary Biology. 2017;**17**(1):62. DOI: 10.1186/s12862-017-0917-z

[28] Okada S et al. The making of natural iron sulfide nanoparticles in a hot vent snail. Proceedings of the National Academy of Sciences. 2019;**116**(41):20376-20381. DOI: 10.1073/pnas.1908533116

[29] Tang T, Ahire B, Li X. Scalable multi-material additive manufacturing of bioinspired polymeric material with metallic structures via electrically assisted Stereolithography. Journal of Manufacturing Science and Engineering. 2023;**145**(1):011004. DOI: 10.1115/1.4055793

[30] Zhao Z, Tian X, Song X. Engineering materials with light: Recent progress in digital light processing based 3D printing. Journal of Materials Chemistry C. 2020;**8**(40):13896-13917. DOI: 10.1039/D0TC03548C

[31] Kang H, Jung S, Jeong S, Kim G, Lee K. Polymer-metal hybrid transparent electrodes for flexible electronics. Nature Communications. 2015;**6**(1):6503. DOI: 10.1038/ncomms7503

[32] Decataldo F et al. Stretchable low impedance electrodes for bioelectronic recording from small peripheral nerves. Scientific Reports. 2019;**9**(1):10598. DOI: 10.1038/s41598-019-46967-2

[33] Lin Y, Bariya M, Javey A. Wearable biosensors for body computing. Advanced Functional Materials. 2021;**31**(39):2008087. DOI: 10.1002/adfm.202008087

[34] Nugroho FAA et al. Metal–polymer hybrid nanomaterials for plasmonic ultrafast hydrogen detection. Nature Materials. 2019;**18**(5):489-495. DOI: 10.1038/s41563-019-0325-4

[35] Byun J et al. Electronic skins for soft, compact, reversible assembly of wirelessly activated fully soft robots. Science robotics. 2018;**3**(18):eaas9020. DOI: 10.1126/scirobotics.aas9020

[36] Shintake J, Cacucciolo V, Shea H, Floreano D. Soft biomimetic fish robot made of dielectric elastomer actuators. Soft Robotics. 2018;**5**(4):466-474. DOI: 10.1089/soro.2017.0062

[37] Grujicic M. Injection Overmolding of Polymer–Metal Hybrid Structures. Joining of Polymer-Metal Hybrid Structures: Principles and Applications. 2018:277-305. DOI: 10.1002/9781119429807.ch10

[38] Tiwari R, Garcia E. The state of understanding of ionic polymer metal composite architecture: A review. Smart Materials and Structures. 2011;**20**(8):083001. DOI: 10.1088/0964-1726/20/8/083001

[39] Sankaran S, Deshmukh K, Ahamed MB, Khadheer Pasha SK. Recent advances in electromagnetic interference shielding properties of metal and carbon filler reinforced flexible polymer composites: A review. Composites Part A: Applied Science and Manufacturing. 2018;**114**:49-71. DOI: 10.1016/j.compositesa.2018.08.006

[40] Tiwari L, Tang T, Rong J, Shan W, Yang Y, Li X. Thermoelectric material fabrication using mask image projection based Stereolithography integrated with hot pressing. Journal of Material Science and Technology Research. 2022;**9**(1):105-113

[41] Quan H, Pirosa A, Yang W, Ritchie RO, Meyers MA. Hydration-induced reversible deformation of the pine cone. Acta Biomaterialia. 2021;**128**:370-383. DOI: 10.1016/j.actbio.2021.04.049

[42] Burgert I, Fratzl P. Actuation systems in plants as prototypes for bioinspired devices," (in eng). Philosophical Transactions A Mathematical, Physical and Engineering Science. 2009;**367**(1893):1541-1557. DOI: 10.1098/rsta.2009.0003

[43] Erb RM, Sander JS, Grisch R, Studart AR. Self-shaping composites with programmable bioinspired microstructures. Nature Communications. 2013;**4**(1):1712. DOI: 10.1038/ncomms2666

[44] Zhang Y et al. An asymmetric hygroscopic structure for moisture-driven Hygro-ionic electricity generation and storage. Advanced Materials. 2022;**34**(21):2201228. DOI: 10.1002/adma.202201228

[45] Tang T, Alfarhan S, Jin K, Li X. 4D Printing of Seed Capsule-Inspired

Hygro-Responsive Structures via Liquid Crystal Templating-Assisted Vat Photopolymerization. Advanced Functional Materials. 2023;**33**(5):2211602. DOI: 10.1002/adfm.202211602

[46] Goh PS, Ismail AF, Ng BC. Directional alignment of carbon nanotubes in polymer matrices: Contemporary approaches and future advances. Composites Part A: Applied Science and Manufacturing. 2014;**56**:103-126. DOI: 10.1016/j.compositesa.2013.10.001

[47] Wu S et al. Aligning multilayer graphene flakes with an external electric field to improve multifunctional properties of epoxy nanocomposites. Carbon. 2015;**94**:607-618. DOI: 10.1016/j.carbon.2015.07.026

[48] Rjumtse EI, Osipov MA, Rotinyan TA, Yevlampieva NP. Electric field effect on the nematic-isotropic phase transition. Liquid Crystals. 1995;**18**(1):87-95. DOI: 10.1080/02678299508036595

[49] Joralmon D, Alfarhan S, Kim S, Tang T, Jin K, Li X. Three-dimensional printing of liquid crystals with thermal sensing capability via multimaterial vat Photopolymerization. ACS Applied Polymer Materials. 2022;**4**(4):2951-2959. DOI: 10.1021/acsapm.2c00322

[50] Su H et al. Pneumatic soft robots: Challenges and benefits. Actuators. 2022;**11**(3):92 [Online]. Available: https://www.mdpi.com/2076-0825/11/3/92

[51] Rajappan A, Jumet B, Preston DJ. Pneumatic soft robots take a step toward autonomy. Science robotics. 2021;**6**(51):eabg6994. DOI: 10.1126/scirobotics.abg6994

[52] Mao G et al. Soft electromagnetic actuators. Science Advances. 2020;**6**(26):eabc0251. DOI: 10.1126/sciadv.abc0251

[53] Wang X et al. Untethered and ultrafast soft-bodied robots. Communications Materials. 2020;**1**(1):67. DOI: 10.1038/s43246-020-00067-1

[54] Mao G et al. Ultrafast small-scale soft electromagnetic robots. Nature Communications. 2022;**13**(1):4456. DOI: 10.1038/s41467-022-32123-4

[55] Wang R, Zhang X, Zhu B, Zhang H, Chen B, Wang H. Topology optimization of a cable-driven soft robotic gripper. Structural and Multidisciplinary Optimization. 2020;**62**(5):2749-2763. DOI: 10.1007/s00158-020-02619-y

[56] Shi K, Song A, Li Y, Li H, Chen D, Zhu L. A cable-driven three-DOF wrist rehabilitation exoskeleton with improved performance," (in eng). Frontiers in Neurorobotics. 2021;**15**:664062. DOI: 10.3389/fnbot.2021.664062

[57] Shintake J, Cacucciolo V, Floreano D, Shea H. Soft robotic grippers. Advanced Materials. 2018;**30**(29):1707035. DOI: 10.1002/adma.201707035

[58] Fukuda M et al. Insights into gradient and anisotropic pore structures of Capiox® gas exchange membranes for ECMO: Theoretically verifying SARS-CoV-2 permeability. Membranes. 2022;**12**(3):314 [Online]. Available: https://www.mdpi.com/2077-0375/12/3/314

[59] Karausta A, Bukusoglu E. Liquid crystal-templated synthesis of mesoporous membranes with predetermined pore alignment. ACS Applied Materials & Interfaces. 2018;**10**(39):33484-33492. DOI: 10.1021/acsami.8b14121

[60] Butt MA, Kazanskiy NL, Khonina SN. Revolution in flexible

wearable electronics for temperature and pressure monitoring— a review. Electronics. 2022;**11**(5):716 [Online]. Available: https://www.mdpi.com/2079-9292/11/5/716

[61] Johnson RE Jr, Dettre RH. "Contact angle hysteresis," in Contact Angle, Wettability, and Adhesion, Vol. 43, (Advances in Chemistry, no. 43). Washington DC, US: American Chemical Society; 1964, ch. 7, pp. 112-135. DOI: 10.1021/ba-1964-0043.ch007. Available from: https://pubs.acs.org/doi/10.1021/ba-1964-0043.ch007

[62] Gandyra D, Walheim S, Gorb S, Ditsche P, Barthlott W, Schimmel T. Air retention under water by the floating Fern Salvinia: The crucial role of a trapped air layer as a pneumatic spring. Small. 2020;**16**(42):2003425. DOI: 10.1002/smll.202003425

[63] Tricinci O, Terencio T, Mazzolai B, Pugno NM, Greco F, Mattoli V. 3D micropatterned surface inspired by Salvinia molesta via direct laser lithography. ACS Applied Materials & Interfaces. 2015;7(46):25560-25567. DOI: 10.1021/acsami.5b07722

Chapter 6

Advances in Large-Scale Robotic 3D Printing with Plastic Pellets

Adolfo Nadal Serrano and José María Espejo Bares

Abstract

3D printing with robotics is reaching an unprecedented level of maturity both in the market and at a technological level. This paper discusses current applications of large-scale robotics applied to 3D printed real-scale final parts for the construction and product design industries, including state-of-the-art methodologies, technologies and applications. Furthermore, an in-depth view of technologies and applications developed by the author will be provided, including robot-end effector integration and the automated generation of machine code through an ad-hoc computer aided design to computer aided manufacturing (CAD-CAM) integration. This integration accounts for parametric capabilities and design-time feedback. Consequently, advances in the seamlessly integrated design-to-manufacturing workflow will be presented: (i) design (by means of employing parametric modeling software), (ii) geometry analysis (by means of ad-hoc machine manufacturing process simulations), (iii) CAD-CAM integration (by means of automated geometry processing and machine code generation), and (iv) manufacturing and testing (by combining 6-axis robots and large-scale plastic extruders).

Keywords: advanced additive manufacturing, large-scale, 3D printing, robotics, recycled plastic extrusion, pellets, automation

1. Introduction

Additive manufacturing has exponentially increased its adoption and widened its use in the industry in the last 10 years after a short but significant decay in its implementation relative to former adoption expectations [1]. Its use focused on small-scale parts and rapid prototyping mainly, while allowing a decent amount of market space aside in terms of final parts for the automotive, aerospace, product design or construction industry, to name just some. One of the main issues behind this shortfall was high, unstable production costs, deriving simultaneously from both raw printing material and manufacturing costs—including operating and entry—level ones, such as high machinery prices. Besides, most industries have traditionally operated on metal materials—mostly aluminum or steel—and Computerized Numerical Control (CNC) based techniques, such as milling, drilling or similar.

Furthermore, the main drivers at a technological level for the adoption of additive manufacturing techniques, in general, and 3D printing, in particular, showed a bottom-up pattern [2]. These were generally pushed by technically-oriented workforce but barely precepted by top-management implementations, often unable to carry out the necessary innovations due to a series of impediments, namely: (i) market constraints, (ii) misleading and conflicting market interests or (iii) lack of capacity to prove the feasibility of the investment in 3D printing technologies [3]. Finally, 3D printing technology has proven unable to scale appropriately in terms of size, production costs and time, despite the efforts made to bridge this gap [4].

On the material side, despite the fact that many metals have gained presence and microstructural stability in the 3D printing market, especially in small-scale applications related to jewelry, biomedicine or high-end parts [5] are still barely affordable for most end-use products. Besides, the precision and manufacturing tolerances of 3D printing techniques with titanium, for example, are far from those reachable by milling or CNC machining techniques. Therefore, mixed manufacturing technologies [6] have augmented their importance, thus allowing firms to gradually adhere to those in order to optimize their production processes while refining their product quality and expanding to otherwise unapproachable market opportunities [7].

As a consequence, there is plenty of room to implement other materials and large-size printing technologies with wide potential use in sectors such as the automotive industry or the construction and engineering fields, where large parts are needed to replace their more factory-like, traditionally-made counterparts at an affordable price while maintaining a well-balanced material performance and consumption [8, 9]. Moreover, automation processes are nowadays commoditized, pushed by the power of Supervisory Control and Data Acquisition and Artificial Intelligence (SCADA-AI) integrative applications [10] that allow to obtain large sets of end-effector data and produce behavioral patterns for a variety of practices and routines. For instance, data pattern analysis enables firms to control the overall performance of their production lines and tools in integrated user interfaces, favoring the use of "intelligent devices" able to provide data in real time. Robots, as fully customizable and programmable machines, gain further momentum over simpler CNC-driven architectures.

This research combines the two aforementioned opportunities into a single solution while providing affordable hardware and software solutions in an attempt to democratize technology and widen its use and implementation in predominantly the engineering and construction sectors. Thus, a solution for plastic pellet extrusion is presented, combined with a 6-axis robotic arm through seamless software and hardware integration.

2. Current research and technological development

While most large-scale 3D printing technologies rely typically on 3-axis gantry-like machines, these have proven to have relevant limitations in terms of usability and scope: (i) these require large structures and initial investments, (ii) the crane-like structures must be used off-site and present constraints when located on-site, (iii) they are restricted to a horizontally layered logical structure, which (iv) restricts the geometrical capabilities of the systems, such as its use in cantilevers or inclined geometries in general. As a result, the use of robots and thereby adapted 3D printing technologies is growing steadily, activating research projects and technological advancements worldwide. Much research has been conducted by ETH Zurich, a pioneer in the development of advanced industrialized methods through an architectural lens leading the research in the field. Numerous initiatives have been carried out by this institution in the last 10 years since the creation of the Gramazio and Kohler research group [11]. ETH's block research group presented in 2021 the first-ever pure unreinforced concrete offsite 3D printed bridge alongside Zaha Hadid Architects' computation and design group and other industrial partners, showing a unique dry-assembled construction stabilized solely by its geometry through shape-based topological optimization [12], thus minimizing material usage [13] through a compression-only, computationally pre-optimized design. This bridge was displayed at the Venice Biennale of Architecture in 2021, allowing visitors to see the lightweight, single-layered shelled assembled panels generated by a fusion of FDM 3D printing with casting methods [14].

Further mentioned worthy efforts have been realized by market players and private companies alike. Aectual, a Netherlands-based company producing furniture and architectural products, focuses on the use of terrazzo, bioplastics and plastic pellets. The firm implements an ad-hoc setup employing 6-axis robotic arms and a regular-size pellet extruder controlled by Siemens PLCs, thus externalizing a meaningful part of their printing technology [15]. Furthermore, Aectual uses a 9-meter-long track to extend its maximum buildable volume capacity to 170cubic meters in order to create interior design elements.

Also located in the Netherlands, MX3D claims to introduce the advantages of 3D metal printing to new high-impact industries. MX3D uses a robot-mounted wire arc additive manufacturing (WAAM) system [16]. WAAM allows the use of conventional welding filler materials, which may reduce material costs drastically when compared with SLS manufacturing in a ratio of 10:1 [17], despite the high energy requirements of the system. The MX3D bridge, built with that technology at an early stage of development, took 6 months to print and required extensive testing in order to test the material's mechanical properties and calibrate the calculus of the structural section. Printed steel's properties and characteristics proved to differ significantly from regular steel in the elastic modulus, which affected the overall stability and stiffness of the bridge [18].

3. 3D printing design-to-manufacture. A comprehensive software and hardware solution for plastic pellet extrusion

Optimizing the manufacturing capacity is one of the most substantial aims of every production industry. These intend to achieve mass customization without sacrificing efficiency or the benefits of economy of scale in terms of return on investment. As a result, the ability to respond to customer requirements in a quick and flexible manner while keeping high version numbers at low batch sizes must increase [19]. The implementation of a customizable, programmable and fully integrated large-scale production system is of use precisely to tackle this issue at (i) software, (ii) hardware and (iii) material levels. This concept is a great advancement in the design industry, due to the time-saving factor and the subsequent cost-effectiveness component. This design method also challenges sustainability goals. It enhances the material reduction in the manufacturing process and subsequent effect of generating less waste at the end of the product lifetime.

Therefore, a comprehensive 3D printing design-to-manufacture technology is proposed, which entails: (i) A parametric or-feature based modelling tool that grants the designer the possibility of modifying an established geometry in a matter of instances by easily generating algorithms grounded on the Rhino-based visual programming interface Grasshopper or others, which allow for single- or multi-solutions-based algorithmic design [20] and receive intense optimization attention [21]; (ii) a quick-response, integrated slicing simulation algorithm capable of dealing with complex and intricate geometries of various topologies including boundary representations (BReps) and meshes [22]; (iii) a CAD-CAM translator for a variety of robot models; (iv) a visualization interface whereby production can be simulated and robot signals set and programmed; and (v) a fully-functional end-effector comprising an extruder and low-cost, reliable control electronics for (vi) pellets obtained from thermoplastics including Polylactic Acid (PLA), polyethylene terephthalate (PET) and polyethylene terephthalate glycol PETG.

3.1 Parametric modeling and geometrical analysis

Parametric modeling is a rather restrictedly employed term that defines the ability to design parts and products based on implicit geometrical relationships rather than via explicit dimensioning. This associative way of modeling and depicting solids and other topologies—such as BReps, meshes, and others—relies on the ability of 3D CAD platforms to define geometry through either (i) a history-based hierarchical object-dependency tree, (ii) customizable parameters and equations, (iii) programmable functions or (iv) a combination of the above. Each 3D modeling software offers a variety of interfaces, including but not limited to (i) equation editors, (ii) visual programming interfaces, (iii) application programming interface (API) accessibility through programming interfaces or a (iv) combination of any of those. McNeel Rhinoceros is used for testing and programming purposes, as well as a platform to program and extend its built-in modeling capabilities. Rhinoceros is a relatively lightweight NURBS-based software able to produce and handle all sorts of geometry, which constitutes an ideal system for advanced users that like to generate their own custom scripts, create generative systems [23], or create logic models.

For the purpose of this research, a set of different geometries comprising numerous variable conditions are tested, such as those displayed in **Figures 1** and **2**. These

Figure 1.
3D printing of a Striatus bridge block. Studio Naaro.

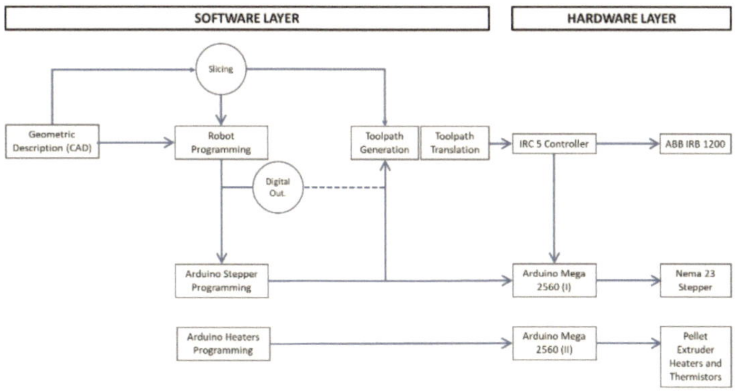

Figure 2.
Software and hardware layers. Integration of electronics.

display entirely parameterized sets of sizes, curvature settings, cantilevers and overhangs, among others, allowing the designer to modify and adapt the design according to manufacturing results and analyse optimal printing setup and material results in the printed part. Besides, three main 3D printing set-up-related parameters are implemented in the design of the parts: (i) nozzle diameter, (ii) shell overlap (which define the overall wall thickness for the part), and total part length, which relates simultaneously to printing speed and temperature cool down—the latter crucial to the successful layering and cohesion of the parts. Parts were designed hollow and single or two-cycled (this is, employing just one or two outlines per layer). **Figures 3** and **4** show the analysed printed configuration and the variable parameters used throughout the design process that defines the non-uniform rational Bezier spline (NURBS) model. Please note that the single-cycle chair design considers variable separations between contours in order to test the actual part's thickness as compared to the nominal size and adjust material flow.

Table 1 shows the relationship between material flow, layer height setup, resulting part thickness (for a single wall pass) and qualitative results at a printing speed of

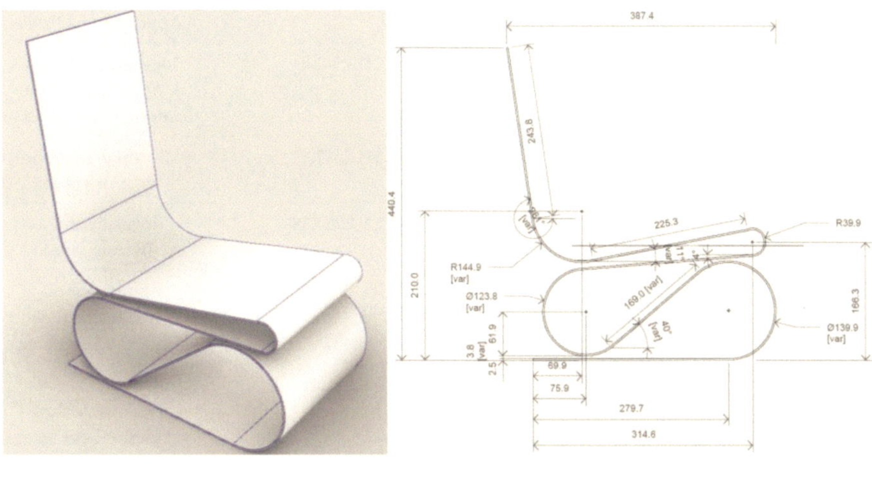

Figure 3.
Test parts (I). Single-cycled chair-like design showing dimensioning.

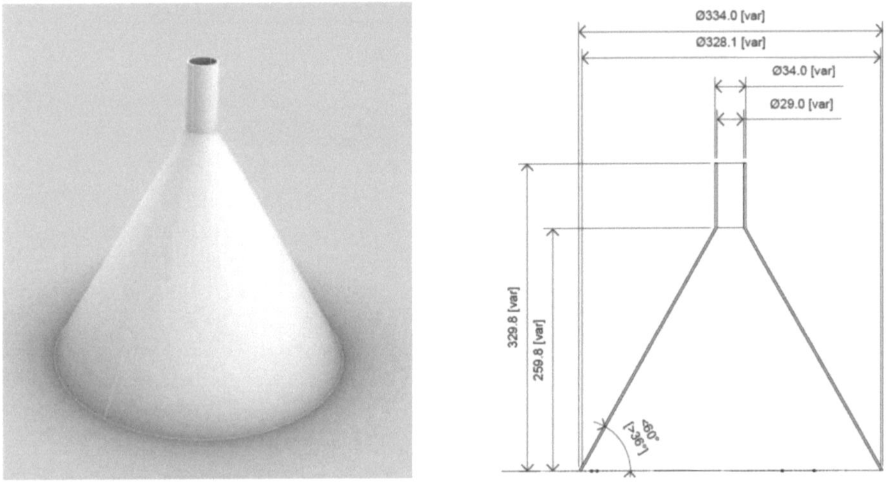

Figure 4.
Test parts (II). Hopper showing dimensioning constraints.

40 mm/s. These affected the design tolerances to a great extent, showing that speed could be adjusted to ranges of 60mm/s to 200mm/s if desired. Optimal material flow was found at 375.5 to 392 mm3/s at speed rates of 40 mm/s to 60mm/s, although further research must be conducted to reduce data dispersion. As seen in **Table 1**, layer heights could also be modified accordingly.

Material flow adjustments			
Flow (mm3/s)	**Layer height (mm)**	**Part thickness (mm) real/nominal**	**Results**
300	0.8	8.6–9.8/1.25–1.8	Increase layer height Increase flow Appropriate adhesion
325	0.8	9.5–11.2/1.25–1.8	Increase layer height
350	1	8.1–9.6/1.25–1.8	Appropriate flow Appropriate adhesion
375	1	8.8–10.1/1.25–1.8	Increase layer height Appropriate flow Appropriate adhesion
400	1	9.4–11.4/1.25–1.8	Regular adhesion Increase speed
450	1	10.56–12.2/1.25–1.8	Regular adhesion Increase speed
400	1.25	7.3–8.6/1.25–1.8	Regular adhesion Excessive flow Undesired overhangs
450	1.25	8.4–9.7/1.25–1.8	Regular adhesion Excessive flow Undesired overhangs

Table 1.
Material flow adjustments and results.

4. Pre-processing for manufacturing. CAD-CAM integration

Geometrical processing entails a three-step software-based workflow: (i) robot model and overall setup, (ii) model slicing and (iii) target generation. The software, an extension of Rhinoceros' Grasshopper visual interface, as indicated above, is capable of processing both mesh-like and surface-like topologies, including solids. From a mesh point of view, a wide range of extensions are allowed, including but not limited to *.stl, *.igs, *.obj, *.ply, *.msh and similar mesh-compliant file formats. This functionality is provided by the platform itself. Surface-like and solids can also be processed as BReps, which support ASCII encoding; some formats allowed are *.3 dm (native Rhinoceros format), *.brep, *rle, *.step and *.sat, widely used in engineering and product design. Solids or surface-like geometries may be translated into meshes to the user's intent, who will decide fabrication tolerances or further geometrical affections (**Figure 5**).

The workflow comprises the following stages:

1. Robot model and overall setup. Employing the built-in functionality of the robot's extension on Grasshopper, a robot model is provided. **Figure 6** depicts the selected ABB IRB 1200 for the development and experiment's case studies. The setup includes also: (i) the design and input of an end-effector, including its gripper or adaptor to the robotic arm and the (ii) configuration of printing parameters, namely target's speed and precision zones. For the purpose of the present experiment, a range of 40 mm/s to 200 mm/s was used as maximum linear printing speeds and a redefined z0 as a precision zone for the pass-by toolpath points. Further robot parameters are also set up, such as elbow configuration and digital output (DO) naming.

2. Model slicing. The imported or natively generated model is sliced by equally distanced planes according to the manufacturing requirements. A range of 1 to 2 mm is used. The slicing operation results in polylines, which are then divided

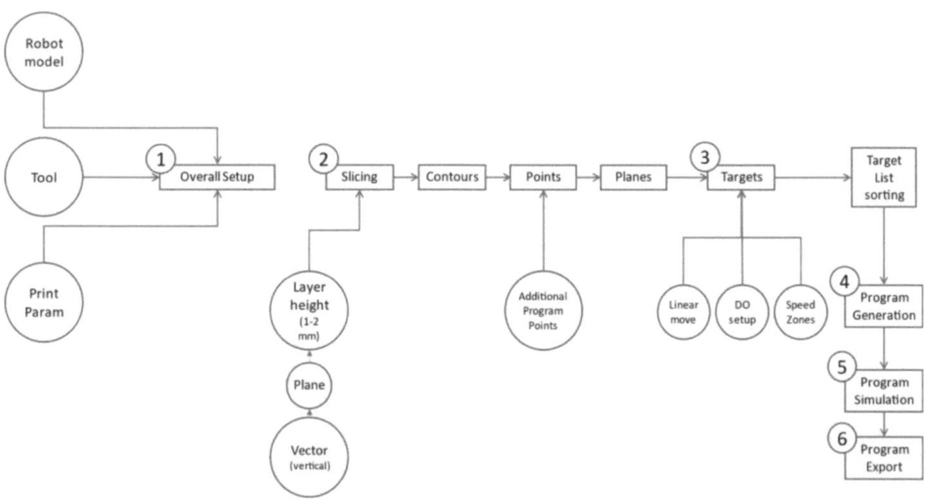

Figure 5.
Programming workflow diagram.

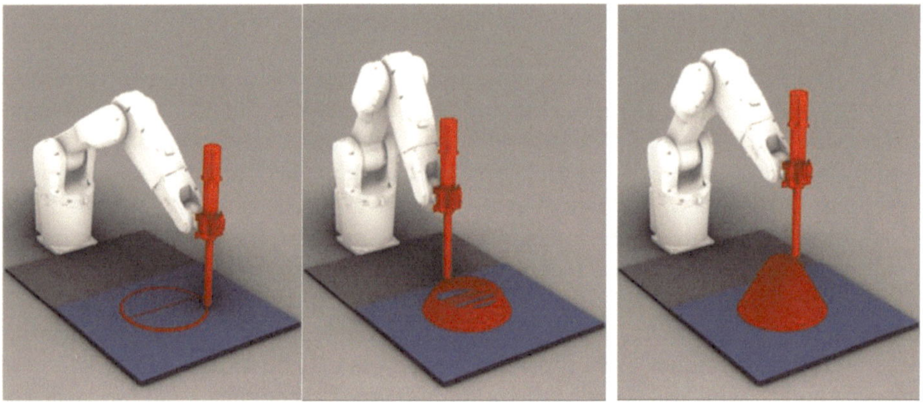

Figure 6.
Test part (II). Hopper in a simplified simulation. Printing bed width 700 mm.

either by length or by curvature to obtain the points that will become the targets' origin by assigning them an orientation plane.

3. Target generation. A horizontal plane in each point is generated, providing the actual orientation for the tool (in this case, vertical). The movement interpolation type for the targets is set to linear. Additional home or safe positions are programmed for the purpose of safety and further program control. Finally, the electronic control output is fed into the desired targets, allowing to turn the extruder either on or off when needed. In this case, continuous printing is tested, thus demanding a single turn-on and turn-off cycle after completion of the part's printing.

4. Program generation. Targets are combined through simple list-management python-written routines. These targets are stored in a linear array and then fed as a single data stream into the program generator routine.

5. Program simulation. The program is simulated showing errors or warnings in execution time. Warnings may include robot configuration issues, while errors prevent the program from running, and might refer to (i) out-of-boundaries errors, (ii) "close to singularity" instructions and (iii) collisions or similar faults related to the mechanical unit.

6. Program export. The program is exported and saved to a user-defined folder. In the case of the current example, a RAPID program (*.pgf) is generated, comprising XML-written data that refers: (i) a MAIN module (*.mod file) including variables and overall setup and (ii) additional modules containing movement instructions and other commands (additional *.mod files). Due to the used IRC5 limited RAM size, modules are split into 32,000-long instruction modules. These contain as well dynamic memory loading and unloading. Further considerations must be implemented in order to control the extruder's behavior while the movement modules are loaded or unloaded from memory, in order to prevent material overflow at stop points.

7. The program is loaded onto the controller and runs normally.

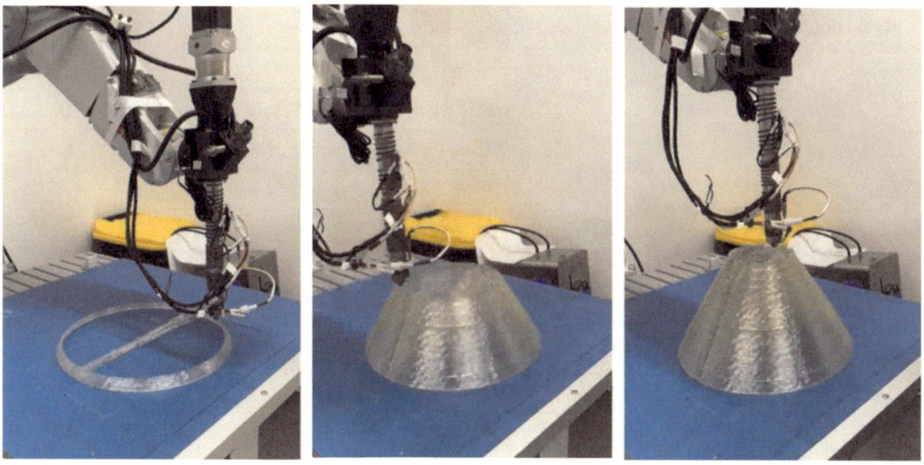

Figure 7.
Test part (II). Hopper in actual printing process. Printing bed width 700 mm.

Figure 6 above demonstrate the software's flexibility and accuracy in depicting the system's behavior via an ad-hoc geometry processing algorithm. **Figure 7** depicts the corresponding printing stages on the 700 mm-wide printing bed proving the workflow's feasibility.

5. Manufacturing and testing

As introduced above, the manufacturing test setup conveys a series of components, entailing mechanical, logical and ad-hoc control units. The main components are shown in **Table 2**, including robot, end-effector, feeding system and printing bed.

The main component for the present setup is the robotic arm ABB IRB 1200 with the IRC 5 compact controller. Secondly, an adapted pellet extruder is used. Heating is delivered by three heating coils that work at 220 volts and are controlled by 3 PT100

Main components	
Robot	
Mechanical arm	ABB IRB 1200 H 7–70
Controller	ABB IRC 5 Compact
End effector	Adapted pellet extruder
Size	150 mm x 100 mm x 495 mm
Weight	5.25 kg
Extruder motor type	Stepper (Nema)
Extruder motor torque	3 Nm
Material flow rate	<500 mm3/s
Transmission type	Planetary gear 1:10 ratio
Heating	Direct coil heating

Main components	
Robot	
Coil number/specs	3220V coils w/ PT 100thermistors
Electronics	Ad-hoc designed control, Arduino MEGA boards, SSRs
Adaptor	Ad-hoc designed, waterjet cut adaptor
Cooling	Liquid cooling
Feeding system	Venturi valve with pressurized inlet
Air pressure	7 bars
Printing bed	Glass, 700 mm x 700 mm x 4 mm on aluminium profiles
Type	Non-heated

Table 2.
Main components of manufacturing unit.

type thermistors. An Arduino code developed ad-hoc for the purpose runs on an Arduino Mega 2560 board with an added ramps 1.4 to control the three solid state relays (SSR) that regulate the three 220 V heating coils. Thirdly, with the objective of controlling the extruded material flow a Nema 23 stepper is used, which supplies a maximum torque of 3 Nm. Fourthly, an air-refrigerating system is mounted on the nozzle to cool down the printed material, comprising a 4 mm aluminum pipe curved around the tip of the nozzle. The extruder body is cooled by liquid refrigerant pumped from a tank placed aside the robotic arm. The printing bed consists of a 4 mm thick glass surface covered with a single layer of Scotch 3D printing tape, which enhances the adhesion of the part to the bed (**Table 3**).

Whereas the robot program includes the necessary DO-related commands, the wiring was set up to minimize the need for electronic components and coding. An Arduino MEGA board is used to control the overall behavior of the end-effector (this is, to turn the stepper motor on or off according to needs) while keeping track of the temperature regulation in each coil, as mentioned previously.

ABS was not tested due to its thermal instability while printing and in order to prevent warping. Additionally, ABS is not a suitable printing material in open spaces and in the absence of properly heated printing beds. Solid-state relays are employed as gates to power the coils, which cool down under regular atmospheric circumstances.

Table 4 shows the test part specifications including operating time for both specimens. Note that the single-cycled design presents tighter layer heights and a lower printing speed at the same material flow rate, thus expelling more material per length

General printing parameters	
Printing speed	**Per part and layer (below/above 10 initial layers)**
Printing type	Continuous
Nozzle diameter	1.5 mm
Layer height	Per part and layer (below/above 10 initial layers)
Support required	No, geometrically feasible
Cooling	

General printing parameters	
Printing speed	**Per part and layer (below/above 10 initial layers)**
Extruder refrigeration	Refrigeration liquid cooling
Layer cooling air pressure	3 bars
Layer cooling air temperature	18°C
Layer cooling air humidity	40–55%
Adhesion	
Adhesion type	None, direct layering
Adhesion material	Adhesive tape on printing bed

Table 3.
Printing parameters.

Figure 8.
Closeup of single-cycled design showing irregular geometry, material burns (left) and warping (right), scale bar shows a mark every 10 mm.

unit. A close-up view of the part's layers is displayed in **Figure 8**, depicting geometry inaccuracies caused by intricate geometry and low precision zones (**Figure 8** left). Furthermore, material burns can be seen in intermediate layers (**Figure 8** left).

In addition to material burns, an excess of remaining latent heat in a layer can result in undesired material fluidity, causing the part to collapse partially or completely. More importantly, this may cause the nozzle to overheat, melting already printed parts or layers, thus affecting the overall stability and geometry in undesired ways. This effect may be seen in **Figure 8** right above.

In addition to adjusting printing speed per layer perimeter, a layer cooling system was developed and mounted (**Figure 9**). Also, geometry needs to be reviewed to minimize the amount of material at that height (**Figure 10**).

Tested materials include PLA and PETG presented in spherical pellet form and flake shape. Sizes are constrained to a maximum diameter of 6 mm for the former and up to 6.25 mm in any direction for the latter. The main characteristics of the used materials are displayed in **Table 5** for clarity purposes.

Two main material arrangements are employed: (i) the use of purely spherical pellets alone and (ii) a combination of spherical and flake-like pellets. These tested material combinations include a series of mixes that allow checking for the feasibility

Figure 9.
Closeup of single-cycled design showing test part (left) and regular layering (right). Scale bar shows a mark every 10 mm.

Printing parts specifications		
	Single-cycled chair-like design	**Hopper**
Size LxWxH	440.4 mm x 387.4 mm x 200 mm	334 mm x 334 mm x 329.8 mm
Layer type	Single continuous, 2 walls	Single discontinuous, 2 walls
Layer height (<10)	0.6 mm	0.65 mm
(>10)	0.8 mm	1 mm
Layers	252	333
Printing speed (1)	40 mm/s	60 mm/s
Printing speed (2)	40 mm/s	80 mm/s
Printing time	18377.72 s	3932.08 s
Printing length	735109.04 mm	311119.67 mm
Material	PLA	PLA
Material temp.	200 °C / 220 °C / 220°C	200°C / 220°C / 220°C

Table 4.
Test part specifications.

of employing recycled and reused materials alongside new ones. Spherical pellets are provided as new, while flake ones are recycled from a variety of untraceable sources.

Therefore, their specifications and applications may only be deemed as approximate values. **Table 5** specifies the results and issues tested on different material mixes. As shown, most reliable results are obtained when combining new pellets and recycled ones in a proportion of 80–20% or higher due predominantly to two factors: (i) the substantial decrease in the reliability of the venturi feeding system while using flakes even at high air pressures and (ii) the affection on printing temperature requirements when combining different material sources, some of which might include unknown origins and previous physical or chemical treatments or pre-processing (**Figure 11** right).

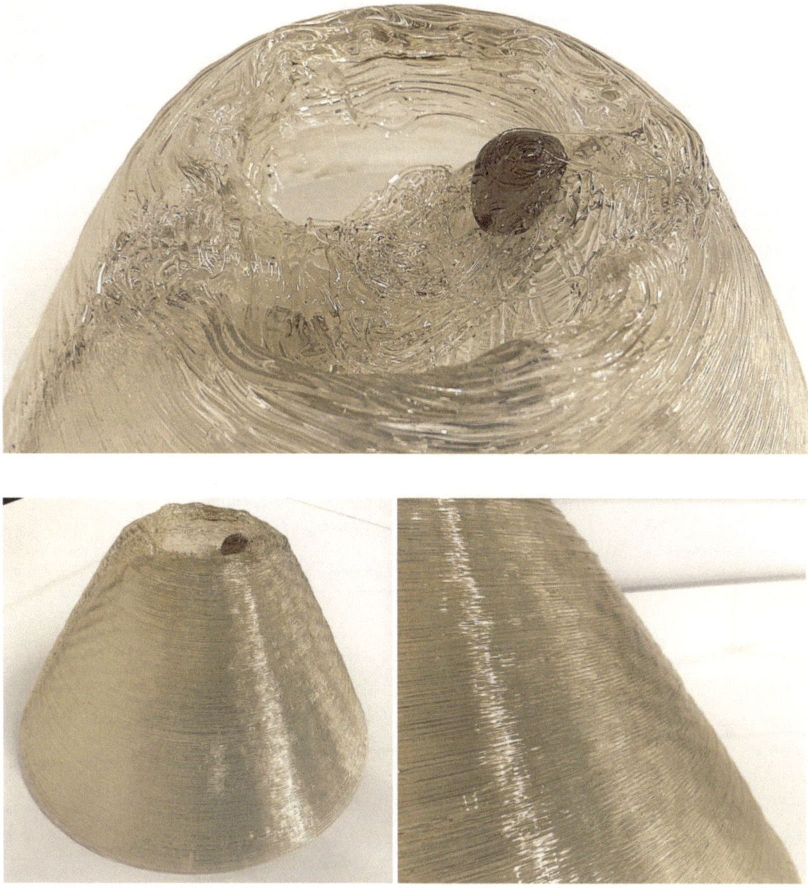

Figure 10.
Collapse of hopper's tip due to heat accumulation and part's layer quality (detail and overall).

Material specifications (spherical and flake pellets)				
	Polylactic acid (PLA)		**Polyethylene terephthalate glycol (PETG)**	
	Typical value	**Test method**	**Typical value**	**Test method**
Material density	1.24 g/cm^3	ISO 1183	1.27 g/cm^3	ASTM D792
Glass transition temp.	60 °C	D3418	85°C	D3418
Tensile strength at break	50 MPa	D882		
Tension yield strength	60 MPa	D882	50 MPa	ASTM D638
Rockwell hardness (R Sc.)			108 R	ASTM D785
Tensile modulus	3.5 GPa	D882		
Tensile elongation	6%	D882		
Notched Charpy Impact	5 KJ/m^2	ISO-179-1eA		
Notched Izod Impact			105 J/m	ASTM D256
Flexural strength	83 MPa	D790	69 MPa	ASTM D790

Material specifications (spherical and flake pellets)				
	Polylactic acid (PLA)		**Polyethylene terephthalate glycol (PETG)**	
	Typical value	**Test method**	**Typical value**	**Test method**
Flexural modulus	3.8 GPa	D790	2100 MPa	ASTM D790
Heat distortion temp.	55°*C*	E2092	70°*C*	ASTM D648
Pellet (spherical)—new				
	Nominal	Actual	Nominal	Actual
Shape	Spherical	Irregular ellipsoid	Spherical	Irregular ellipsoid
Size	ϕ 2.5 mm	Variable ϕ 2 -ϕ 4 mm	ϕ 2.5 mm	Variable ϕ 1.5 -ϕ 3 mm
Pellet (flake)—reused				
Shape	Irregular planar polygonal flake	Irregular non-planar polygon	Irregular planar polygonal flake	Irregular non-planar polygon
Size	< 10 mm	<7.5 mm	<10 mm	<7 mm

Table 5.
Printing material parameters.

Figure 11.
Photography of ϕ2–ϕ4 mm PLA spherical and ϕ1.5–ϕ3 mm flake-shaped pellets used in the experiment.

6. Results and conclusions

In terms of size, the presented robot-based 3D-printing system is easily scalable to produce as-large-as-required parts simply by modifying or selecting robot models with a higher movement range, such as an IRB 1600, IRB 2600 or higher. Above mentioned parameters such as printing speed or layer air-cooling might need further

DOI: http://dx.doi.org/10.5772/intechopen.109438

adjustment, despite the fact that larger parts allow more time for layers to cool down, which plays a crucial role in the stability of the parts while printing. In other words, larger sizes should be easier to print.

Special attention should be paid to warping, which highly depends on the geometry of the parts. The single-cycle chair-like design presents warping, in every instance of the process, exactly where the geometry is more acute and the part's printing length is higher, causing the part to be less condensed or present a lower level of adhesion to the printing bed. Warping, as opposed to layer cooling, will become a challenge with bigger parts.

Results show that a proper material mix combining new-to-used PLA and PETG is suitable for large-scale printing as shown in **Table 6**. Proportions of 80% new spherical pellets to 20% reused flake-shaped pellets and higher work properly. Whereas printing with pellets from equal thermoplastic types is achievable, mixing PLA and PETG has proven pointedly a more delicate task. Nonetheless, manufactured parts have proven to be stable and are well preserved through time, thus evidencing the feasibility and reliability of the entire system. The implications of this technology on the circular economy and environmental impact are left aside for future discussion.

Material combination results				
	Spherical polylactic acid (PLA) pellets		**Spherical polyethylene terephthalate glycol (PETG) pellets**	
	Proportion spherical*	Proportion PLA flakes*	Proportion spherical*	Proportion PLA flakes*
Proportion 1	100%	0%	100%	0%
Result	Satisfactory printing Not satisfactory Not printing		Satisfactory printing	
Issue	None		None	
Proportion 1	60%	40%	60%	40%
Result	Not printing		Intermittent printing	
Issue	• Feeder not fed • Material not sucked into feeding pipe • Venturi system failure		• Material feeding interrupted • Unpredictable feeding behavior • Interrupted layer printing • Unstable mixture-temp	
Proportion 2	65%	35%	65%	35%
Result	Not printing		Intermittent printing	
Issue	• Feeder not fed • Material not sucked into feeding pipe • Venturi system failure		• Material feeding interrupted • Unpredictable feeding behavior • Interrupted layer printing • Stepper motor misses steps • Unstable mixture -temp	
Proportion 3	70%	30%	70%	30%
Result	Irregular printing		Intermittent/irregular printing	
Issue	• Material feeding interrupted • Partially unpredictable feeding behavior • Irregular layer printing		• Material feeding interrupted • Partially unpredictable feeding behavior • Irregular layer printing • Stepper motor misses steps • Unstable mixture—temp	

Material combination results				
	Spherical polylactic acid (PLA) pellets		Spherical polyethylene terephthalate glycol (PETG) pellets	
Proportion 4	80%	20%	80%	20%
Result	Regular, medium-quality printing		Low-quality printing	
Issue	• Flakes are dragged by spherical pellets • Warping • Occasional material defects		• Unstable mixture—temp • Increased warping • Unreliable printing	
In terms of weight.				

Table 6.
Material combination results.

Further research and funding will be required to challenge the thesis hereby presented and to be able to print bigger, more intricate geometries. Also, a significant part of future work will focus on the development of a more stable extruder, despite efforts made in that direction.

Acknowledgements

Special thanks are given to Archiologics and its staff for their invaluable support and for providing funding, materials, hardware and valuable knowledge. This research would have been impossible without them.

Conflict of interest

The authors declare no conflict of interest.

Author details

Adolfo Nadal Serrano[1*] and José María Espejo Bares[2]

1 Universidad Francisco de Vitoria, Archiologics, Madrid, Spain

2 Universidad Francisco de Vitoria, Madrid, Spain

*Address all correspondence to: adolfo.nadal@ufv.es; adolfo.nadal@archiologics.com

References

[1] Schniederjans DG. "Adoption of 3D-printing technologies in manufacturing: A survey analysis." International Journal of Production Economics. January 2017; 183(Part A): 287-298

[2] Mellor S, Hao L, Zhang D. Additive manufacturing: A framework for implementation. International Journal of Production Economics. 2014;**149**: 194-201

[3] Petrovic V, Haro González JV, Jordá Ferrando O, Delgado Gordillo J, Blasco Puchades JR. Additive layered manufacturing: Sectors of industrial application shown through case studies. International Journal of Production Research. 2010;**49**(4):1061-1079

[4] Yuan PF. Feasibility study of large-scale mass customization 3D printing framework system with a case study on Nanjing Happy Valley east gate. Frontiers of Architectural Research. 2022;**11**:670-680

[5] Korium M, Roozbahani D, Alizajeh M, Perepelkina S, Handroos H. Direct metal laser sintering of precious metals for jewelry applications: Process parameter selection and microstructure analysis. IEEE. 2021;**9**(99):126530

[6] Gibson IRDSB. Post-processing. In: Additive Manufacturing Technologies. New York, NY: Springer; 2015. pp. 329-350

[7] Gibson IRDSB. The impact of low-cost AM systems. In: Additive Manufacturing Technologies. New York, NY: Springer; 2015. pp. 293-301

[8] Ashraf M, Gibson I, Rashed MG. Challenges and prospects of 3D printing in structural engineering. In: 13th International Conference on Steel, Space and Composite Structures at Perth, Australia; 2018. Available from: https://www.researchgate.net/publication/320943125_CHALLENGES_AND_PROSPECTS_OF_3D_PRINTING_IN_STRUCTURAL_ENGINEERING

[9] Nadal A, Cifré H, Pavón J, Liébana Ó. "Material use optimization in 3D printing through a physical simulation algorithm". Automation in Construction. June 2017;**78**:24-33

[10] Narayanan LK, Subbiah P, Mary SA, Muralidharan RA, Gururajan I, Sampathkumar R, et al. Chapter 4 - A smart and efficient IoT-AI and ML-based multifunctional system for multilevel power distribution management. In: Padmanaban S, Holm-Nielsen JB, Padmanandam K, Dhanaraj RK, Balusamy B, editors. Smart Energy and Electric Power Systems. Elsevier; 2023. pp. 49-63. ISBN 9780323916646. DOI: 10.1016/B978-0-323-91664-6.00008-5. Available from: https://www.sciencedirect.com/science/article/pii/B9780323916646000085

[11] ETH Zürich. "Gramazio Kohler Research, Professur für Architektur und Digitale Fabrikation, ETH Zürich" [Online]. Available from: https://gramaziokohler.arch.ethz.ch/ [Accessed: November 21, 2022]

[12] Nowak A. Architectural and structural optimization research of structural. In: 1st International Conference on Optimization-Driven Architectural Design (PROTARCH). Jordan: Amman; 2019

[13] Walther M. ETH Zürich News [Online]. 2021. Available from: https://ethz.ch/en/news-and-events/eth-news/news/2021/07/3d-printed-and-unreinforced.html. [Accessed: June 14, 2022]

[14] Everett H. ETH Zurich creates first 3D printed concrete bridge without reinforcement, 3D Printing Industry [Online]. 2021. Available from: https://3dprintingindustry.com/news/eth-zurich-creates-first-3d-printed-concrete-bridge-without-reinforcement-193090/ [Accessed: January 13, 2022]

[15] Aectual BV. Aectual Systems [Online]. 2022. Available from: https://blog.aectual.com/3d-printing-technology [Accessed: November 11, 2022]

[16] MX3D. WAAM System [Online]. 2022. Available from: https://mx3d.com/technology/robotic-waam/

[17] Arana M, Ukar E, Rodríguez I, Iturrioz A, Alvarez P. Strategies to reduce porosity in Al-Mg WAAM parts and their impact in their mechanical properties. Metals. 2021;**11**:524

[18] Joosten S, Ren S, van Horn P, Vola M. MX3D bridge. The Arup Journal. 2022;**2022**(1):14-21

[19] Pasetti Monizza G, Bendetti C, Matt DT. Parametric and generative design techniques in mass-production. Automation in Construction. 2018;**92**:270-285

[20] García Dominguez A, Claver J, Sebastián MA. Methodology for the optimization of work pieces for additive manufacturing by 3D printing. In: Procedia Manufacturing. Vigo, Pontevedra: Elsevier B.V.; 2017

[21] Saremi S, Mirjalili S, Lewis A. Grasshopper optimisation algorithm: Theory and application. Advanced in Engineering Software. 2017;**105**:30-47

[22] Nadal Serrano A. Towards the standardization of real-scale, robotic 3D printing. Software integration. In: Beyond Rapid Prototyping: Automation of Robotic 3D Printing for Construction [Thesis]. Madrid: Universidad Complutense; 2018. pp. 75-102

[23] Sass L, Oxman R. Materializing design: The implications of rapid prototyping in digital. Design Studies. 2006;**27**(3):325-355

Chapter 7

Can the DryLyte® Technology Polish 3D Printed Ceramic/Metal Samples and in Particular WC-Co?

Guiomar Riu Perdrix and Joan Josep Roa Rovira

Abstract

DryLyte® Technology is an effective surface finish technique, which follows the same traditional electrolytic cell principle, but uses an electrolytic solid non-conductive medium rather than a liquid one. For the last 10 years, this technology has been attracting a lot of attention compared to conventional ones due to the selective smoothing of the surface technique, interacting only with the roughness peaks and not with the valleys, etc. In this book's chapter, for 3D-printed cemented carbides (WC-Co) polished with DryLyte® Technology, it is shown the correlation between the microstructure and the surface integrity, in terms of mechanical properties, at submicrometric length scale. Also, a particular case study is presented of 3D-printed WC-Co as a function of the testing temperature, ranging from room temperature up to service-like working conditions. Finally, the mechanical properties are correlated as function of the chemical nature and/or crystallographic phase.

Keywords: DryLyte® technology, roughness evolution, surface integrity, ceramic-metal materials, cemented carbides

1. Introduction

The tribological and mechanical behavior of a tool under service-like working conditions depends not only on intrinsic properties of the constitutive phases but also on the material surface and subsurface properties—such as topography and residual stress state. During the last 10 years, extensive research has been dedicated to investigating the relation between chemistry, microstructure, and the resulting mechanical properties [1–3]. In addition to those bulk-related features, surface characteristics are also crucial in determining the functional response of a given material. Commonly, materials used for structural components are shaped into final dimensions and geometry changed as the material goes through different surface modification processes as mechanical working operations, material removal methods, heat treatment, and other finishing practices.

In this regard, surface alterations and/or modifications become critical in controlling the properties and performance of final products, particularly if service conditions involve contact loading (e.g., wear, impact, fatigue, etc.) and/or environmental (e.g., corrosion, oxidation, etc.) interaction.

In the case of manufacturing stages involving material removal (e.g., grinding, lapping, etc.), a complex surface interaction exists between the tool and workpiece. In this sense, the temperature near the surface increases which could produce microstructural changes, including oxidation and also possible local melting at the surface level. Furthermore, plastic deformation, tearing, and fracture also occur. For all the aforementioned information, mechanical and thermal changes at the surface level take place along thermal stages which may induce relevant residual stresses [4].

The existence of a pronounced influence of manufacturing methods on mechanical properties and service performance, as a result of the type of surface produced, is well-establish. Within this framework, the concept of surface integrity in terms of roughness has been introduced as a key parameter. In this sense, the surface integrity, which contained not only the geometry consideration, including surface roughness and accuracy, but also other microstructural aspects at the surface and/or subsurface level.

Within this framework, this topic demands synergic interdisciplinary expertise in different fields: materials science, machining and shaping technology, as well as surface integrity in terms of mechanical properties [5, 6].

Ceramic/metallic materials and in particular the WC-Co cemented carbides (also known as hardmetals), are the key materials for tooling industry. These materials present an excellent combination of properties (e.g., hardness, strength, fracture toughness, wear, and abrasion resistance) [7–12]. Hardmetal tools are produced through a powder metallurgical (PM) route, where mixed WC and Co powders are sintered at high temperatures to consolidate the composite material [4, 13]. WC-Co grades are classified according to their Co content and WC grain size. The proportion of WC phase is generally between 70 and 97% of the total weight of the composite and its grain size averages between 0.4 and 10 μm. This range of cemented carbides can be subdivided into its major application area as described below [14, 15]:

- **Nano, ultrafine, and submicron grades**, with a metallic Co binder content in the range of 3–10 wt.% and WC grain size below 1 μm. WC-Co grades are widely used in a wide range of wear part applications as well as cutting tools designed, where the resulting tool needs high strength, wear resistance and sharp cutting edges are essential under service-like working conditions.

- **Fine and medium grades**, with a metallic Co binder ranging between 6 and 30 wt. % and WC grain size ranging between 1 and 3 μm. WC-Co material is used mainly in wear parts and cutting tools is employed in applications where strength and shock resistance is required.

- **Medium coarse, coarse, and extra coarse grades**, with a metallic Co binder content ranging between 6 and 15 wt.% and a WC grain size above 3 μm. WC-Co grades are employed in oil and gas, and mining applications in which resistance to high impact and abrasive wear is required.

However, during the last decade 10 years, the Additive Manufacturing (AM) routes are gaining more importance in this field. However, due to the complex shapes as well as closed cavities, conventional post-processing routes do not give out the desired quality in terms of surface finish.

DOI: http://dx.doi.org/10.5772/intechopen.110299

2. Surface modification techniques on WC-Co grades

In the production of conventional WC-Co components, it is occasionally necessary to carry out a number of shaping operations before final sintering. Green compacts can then be produced in simple shapes, such as rectangular and round blanks, by means of conventional methods, such as turning, drilling, and grinding.

After sintering, the corresponding blank has achieved its fully density and final mechanical properties, and it is ready to be dispatched. In this regard, most blanks need to be further post-processed in order to get the desired shape, size, flatness, and surface finish by using the traditional post-processing routes on WC-Co grades; being the most common techniques: diamond wheel grinding, diamond lapping, and grinding and electrical discharge machining (EDM) [16]. All these traditional post-processing technologies will be briefly described in Section 2.1.

During the post-processing process, two aspects need to be taken into account: (a) the functionality of the machined workpiece and (b) the economic efficiency. According to different applications, the functionality of the workpiece after the post-processing process can be divided into different groups as follows [17]:

- **Mechanical functions**, defined as the capability of carrying mechanical loads; even monotonically and/or cyclically.

- **Thermal functions**, considering the heat resistance and/or conductivity as a function of the testing temperature.

- **Tribological functions**, defined as the surface interaction with other media

- **Optical functions**, considering visible appearance and/or light reflection behavior

- **Flow functions**, which consider the influence on the flow of fluids.

Each of the aforementioned steps in a manufacturing chain influences the workpiece properties, which directly link to its functionality. Along this chapter of the book, we will focus attention to changes induced at the surface level changing the resulting roughness. In particular, the traditional post-processing techniques will be briefly explained (see Section 2.1) and in particular compared with a disruptive dry-electropolished technology (see Section 2.2) focusing the attention to these technologies on WC-Co AM specimens.

2.1 Traditional polishing techniques

The main requirements which lead to use post-processing routes on WC-Co cemented carbides are:

- to reduce superficial roughness

- to keep geometry and preserve the specimen tolerances

- to polish complex shapes

- to reduce the polishing times.

The most widely employed post-processing techniques on WC-Co cemented carbides are:

- **Diamond wheel grinding** contains abrasive compounds for grinding and abrasive machining operation. The wheels generally are made with composite material. This consists of coarse-particle aggregate pressed and bonded together by a cementing matrix to form a solid, circular shape. Grinding wheels are consumables, although the life span can vary widely depending on the use case, from less than a day to man years. Mainly, the abrasive aggregate is selected primarily according to the hardness of the material being cured.

- **Lapping** also known as mechanical lapping, is widely applied in diamond polishing for its simple process, low cost, and high efficiency [18]. This technique consists on a machining process in which two surfaces are rubbed together with an abrasive between them, by hand movement or using a machine. This technique often follows other subtractive processes with more aggressive material removal as a first step, such as milling and/or grinding. Lapping can take two forms: (a) *traditionally often called grinding,* which involves rubbing a brittle material, producing a microscopic conchoidal fractures as the abrasive rolls about between the two surfaces and removes material from both and (b) involves a softer material such as pitch or a ceramic for the lap, which is charged with the abrasive. The lap is then used for finishing of WC-Co cutting tools, have proven to be economical for this typology of materials.

- **Grinding** is virtually unchallenged for machining of materials which, due to their extreme hardness or brittleness, cannot be efficiently shaped by other methods. In the particular case of study, machining is almost exclusively dependent on this process [19, 20]. This post-processing technique is a representative abrasive process in which a grinding wheel is used. These wheels are composed of two materials—*tiny abrasive particles called grains* and a *softer bonding agent to hold the countless grits together in a solid mass.* During this process, each protruding abrasive grain interacts with the workpiece surface and a local stress field upon each contact point causes irreversible material deformation in the form of dislocation, cracks, and voids [21].

- **EDM** is a post-processing technique, which is increasingly used on WC-Co suppliers. This is a thermal process where a workpiece electrode is shaped through the action of a succession of discrete electrical discharges which local erode (melt or vaporize) the material surface. Mainly, this technology is used for close tolerances [10].

Mainly, the most common post-processing technology to process the workpieces of WC-Co is the grinding (**Figure 1a**) and EDM process (**Figure 1b**), as shown in the SEM micrographs presented on **Figure 1**.

After that, it is necessary to reduce the surface roughness induced by these techniques by using chemo-mechanical process and/or chemical polishing process until the WC-Co cemented carbides workpieces present a precise dimension and shape that fulfill requirements in real applications [22–25]. However, the media used to reduce the surface roughness induced by the ground and chemo-mechanical polishing

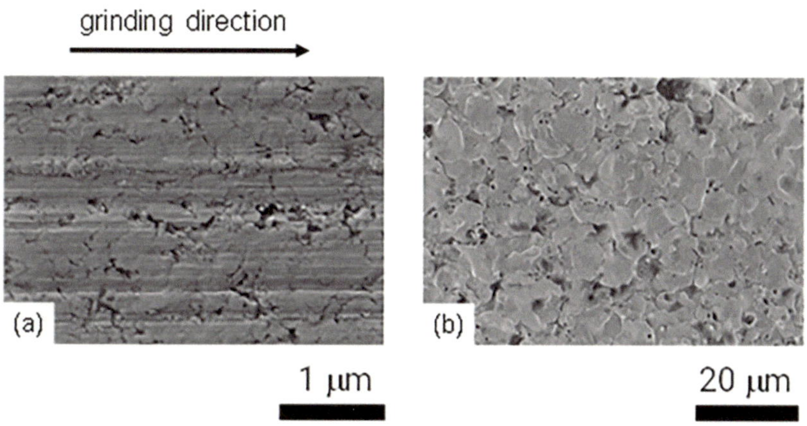

Figure 1.
Scanning electron microscopy (SEM) micrographs of the microstructural quality for WC-Co post-processed by (a) grinding and (b) EDM technologies.

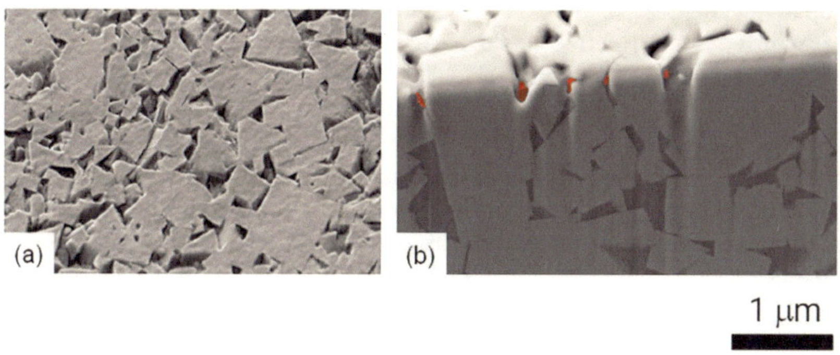

Figure 2.
Scanning electron microscopy (SEM) micrographs of the microstructural quality for the WC-Co chemo-mechanical polished. (a) Surface micrograph and (b) focused ion beam (FIB)—cross section showing the leaching induced (painted in red in the SEM micrograph) between both constitutive phases.

process can induce localized leaching at the metallic Co binder as shown in **Figure 2a**. As reported in Refs. [26–28], when the pH media is below 7, the media induces a galvanic couple, where the ceramic reinforcement particles (WC) act as the cathode while the metallic Co binder as an anode. Thus, the potential difference between both parts generates a microgalvanic couples that induce a corrosion effect generating a difference in height between both constitutive phases of around 100 nm as shown in **Figure 2b**. This difference in height may produce a reduction of the workpiece under service-like working conditions.

From all the aforementioned information, the conventional post-processing techniques present some drawbacks for the WC-Co workpieces when these have been superficially treated by a ground and/or EDM process. Furthermore, these drawbacks are more evident when the geometry of the specimens' have complex shapes. In this sense, the AM specimens cannot be post-processed using conventional techniques as these technologies do not lead to the final workpiece:

(1) keep the geometry and preserve the tolerances and

(2) be able to polish complex shapes in a short polishing time.

For all the aforementioned information, recently the DryLyte® Technology increases their applicability in the market.

2.2 DryLyte® technology

This technology was developed to overcome the limitations of classic polishing technologies (e.g., mechanical process, chemo-mechanical process, electropolishing, etc.). In this sense, DryLyte® (contraction of the words dry- and electrolyte) Technology is based on a dry-electrolyte and dry-suspension electrolyte made up of an ionic exchange resin.

In this sense, the structural characteristics, together with the functional group's nature and degree of the resin, has an important effect on its exchange-ion behavior. Morphologically, they can be classified as macroreticular or gel-type resins and macroreticular or macroporous resins:

1. **Macroporous resins** have stable macropores in the dry state and to fabricate them, a minimum of around 10% cross-link degree is required. Their appearance is opaque. The beads have a porous multi-channeled structure which provides them with a high effective surface area. Nonetheless, they are more fragile than gel-type resins as reported in [29].

2. **Gel-type resins** are a three-dimensional (3D) swollen structure with a solvent evenly dispersed through it. When the solvent is removed, the gel shrinks and its porosity is not appreciable in the dry state. This type of resins presents heterogeneous micropores (ranging between 0.7 and 2 nm) and a cross-link degree between 4 and 10%. Visually, they have a translucid aspect [30].

The electrolyte presents a multi-modal particle size distribution (see **Figure 3a**) with an interconnected porous microstructure (see **Figure 3b**) and is chemically constituted of PolyStyrene DiVinylBenzene (PS-DVB). In this sense, the electrolyte particles are arranged in such a way that they make electrical contact with the negative pole of the power supply (cathode) and with the metal part to be polished (anode), generating a close circuit as schematically depicted in **Figure 3b**.

From all the aforementioned information, unlike traditional polishing systems, DryLyte® Technology achieves a uniform surface finish, avoiding surface marks patterns such as those generated by machining, and it is able to process complex geometries without generating micro-scratches on the surfaces, preserving the edges of complex geometries as the one produced by AM routes. Furthermore, this technology leads to polish close cavities by using internal cathodes. This is the main advantage to employ this technology to polish AM specimens in comparison with the traditional post-processing routes.

As it does not generate any polishing and/or surface modification texture (e.g., grinding patterns, etc.), this technology improves surface integrity in terms of mechanical properties (e.g., fatigue and wear) and chemical properties (e.g., corrosion resistance, aging, etc.). In this sense, along the entire chapter, this technology will be applied to ceramic/metal samples and in particular, to WC-Co cemented carbides produced by AM routes.

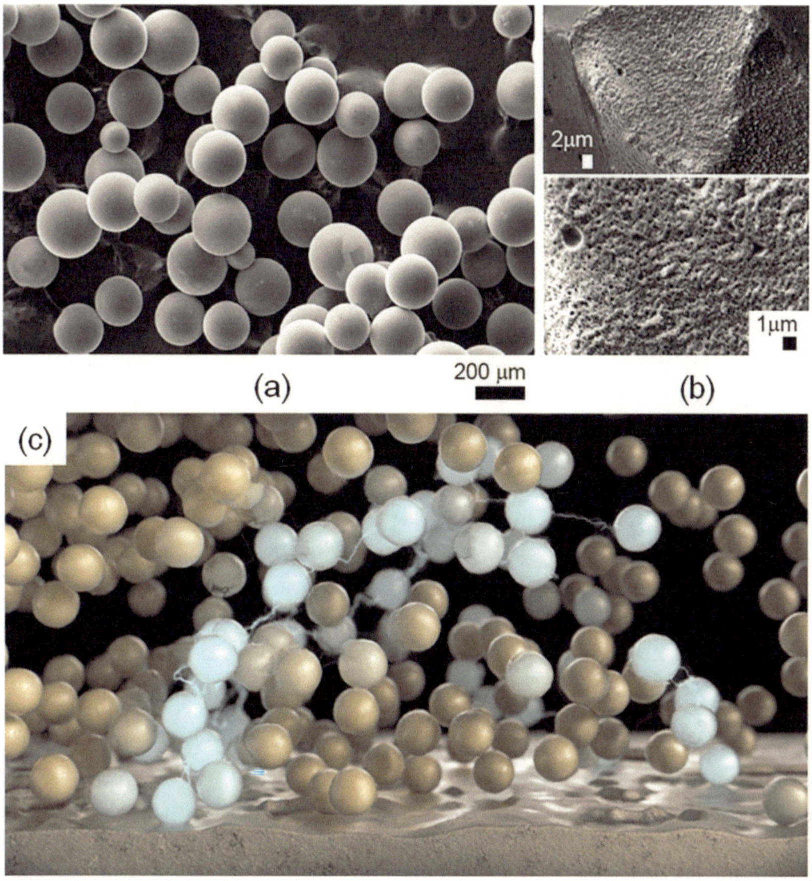

Figure 3.
(a) SEM micrograph of the PS-DVB electrolyte showing a multi-modal particle size distribution; (b) SEM magnification micrograph showing the internal porosity; and (c) schematic representation of how the electricity passes through the PS-DVB particles during the dry-electropolishing process.

3. AM of WC-Co hardmetals

3.1 Theoretical background: bibliometric review

AM technologies process has emerged as an alternative to traditional manufacturing process that can fabricate very complicated geometries with high efficiency and reduce post-processing [31–37].

AM technology is increasingly considered as a key manufacturing technology of tomorrow's society due to the fact, it makes possible the production of complex near-set shaped parts which are not feasible with conventional technologies (e.g., PM and/or subtractive). With this bottom-up approach, material is added layer-by-layer where necessary and the complex shape obtained. This technology offers several advantages, like:

- greater design freedom compared to conventional routes.
- the possibility of creating close cavities and/or channels to reduce the density and optimize the cooling process, respectively.

- the possibility of saving on raw material consumption.
- to reduce the production time.
- no need of using customized tools.

From all these advantages, AM technology is being employed in many industries, like biomedical [38] and electronic [39] ones, as well as within scientific fields, such as mechanical devices [40, 41], periodic microstructures [42], and ceramics [43]. In this regard, the increasing interest in this technology is driven by its ability to produce custom devices, its simplicity and speed.

During the last 10 years, the AM of WC-Co has been the subject of several research efforts, and it is still in an early stage of development. AM technologies offer attractive advantages in terms of producing WC-Co hardmetal cutting tools with complex geometries, such as U-shaped or helical cooling channels inside. These internal, contour-adapted cooling channels allow higher cutting speeds and, consequently, a remarkable increase in the productivity of the machining process. The main AM technologies suitable for metal are selective laser melting (SLM), selective electron beam melting (SEBM), laser powder deposition, binder jet AM (BJAM), and wire arc AM (WAAM) [44–54]. So far, AM technologies have been successfully applied to stainless steels [55–59], Ni alloys [60, 61], Ti alloys [62, 63], refractory metals [64, 65], Al alloys [66, 67], etc.

For WC-Co, it remains very challenging to use AM due to its very high melting temperature. SLM and BJAM are the most attempted AM processes for manufacturing WC-Co hardmetals. Besides, a few studies on SEBM [46], 3D gel-printing (3DGP) [68], and fused filament fabrication (FFF) [69] of WC-Co hardmetal samples were deeply investigated.

The AM techniques used for fabrication of WC-Co hardmetals and the main research groups working on WC-Co AM were summarized in **Tables 1** and 2.

AM process	Other names	Advantages	Disadvantages
SLM	Laser powder bed fusion, L-PBF	(1) High dimensional accuracy (2) High geometric freedom (3) Less performing steps (4) High hardness	(1) High residual stress (2) Uneven microstructure (3) Carbon loss (4) Evaporation of metallic Co binder
SEBM	Electron beam powder bed fusion, E-PBF	(1) High dimensional accuracy (2) High geometric freedom (3) Less performing steps (4) High hardness (5) High scan speed	(1) High residual stress (2) Uneven microstructure (3) Expensive equipment (4) Needs vacuum to avoid the oxidation of the metallic Co binder
BJAM	Binder jet 3D printing, BJ3DP	(1) Uniform microstructure (2) High toughness (3) Low cost (4) Low residual stress	(1) Complicated processes (2) Large shrinkage (3) Low hardness (4) Moderate strength
3DGP	N/A	(1) Low residual stress (2) Uniform microstructure (3) Low powder requirements (4) No waw material loss	(1) Complicate processes (2) Large shrinkage

AM process	Other names	Advantages	Disadvantages
FFF	N/A	(1) Low residual stress (2) Uniform microstructure (3) Low powder requirements (4) No raw material loss	(1) Complicated processes (2) Large shrinkage (3) Needs filament fabrication of the desired material (4) Rough surface

Table 1.
Main research groups working on AM of WC-Co hardmetals [46, 68, 70–84].

Author	Printing technology	Material	Research group/country	Year
Wang [85]	SLS	WC-Co	Catholic University of Leuven/Belgium	2002
Maeda [86]	SLS	WC-4Co, WC-5Co, WC-10Co, WC-30Co	University of Leeds/UK	2004
Gu [87]	DMLS	WC-10Co with Cu matrix	Nanjing University of Aeronautics and Astronautics/China	2006
Gu [88]	DMLS	WC-10Co with Cu matrix	Nanjing University of Aeronautics and Astronautics/China	2006
Kernan [89]	3DP™ (SFF)	WC-10Co	Massachusetts Institute of Technology/USA	2007
Gu [90]	DMLS	WC-10Co	Nanjing University of Aeronautics/China	2007
Gu [91]	DMLS	WC-10Co with Cu matrix	Nanjing University of Aeronautics and Astronautics/China	2007
Gu [92]	DMLS	WC-10Co/Cu	Nanjing University of Aeronautics and Astronautics/China	2008
Kumar [93]	SLS	WC-9Co, WC-12Co, WC-18Co, WC-50Co	Utah State University/USA	2008
Xiong [94]	LENS®	WC-10Co	University of California /USA	2008
Kumar [95]	SLS	WC-10Co	Utah State University/USA	2009
Kyogoku [96]	DMLS	WC-10Co, WC-10Co with Cu-Sn, WC-10Co with Cu	Kinki University Hihashihiroshima/Japan	2012
Ghosh [97]	SLS	WC-15Co, WC-20Co	National Institute of technology Agartala/India	2015
Scheithauer [98]	T3DP	WC-10Co (+ grain size inhibitors VC and Cr_3C_2)	Fraunhofer IKTS/Germany	2017
Khmyrov [99]	SLM	WC-8Co	Moscow State University of Technology/Russia	2017
Kumar [100]	SLS	WC-17Co	York University/Canada	2017
Kumar [101]	SLS	WC-17Co	York University/Canada	2018
Uhlmann [73]	SLM	WC-Co	Fraunhofer Institute for Production Systems and Design Technology/Germany	2018
Enneti [102]	BJ3DP	WC-12Co	Global Tungsten and Powder Corp/USA	2019

Author	Printing technology	Material	Research group/country	Year
Cramer [103]	BJAM + Infiltration	WC-19Co	Oak Ridge National Laboratory/USA	2020
Padmakumar [104]	SLM, SLS, BJ3DP	WC-Co	Technology Center/India	2020
Yang [105]	SLM, SEBM, BJAM, 3DGP, FFF	WC-Co	Shanghai Jiao Tong University/China	2020
Al-Thamir [106]	LPBF	WC-12Co	University of Nottingham	2020
Bricín [107]	SLM	WC-13Co	University of West Bohemia/Czech Republic	2020
Agyapong [108]	SLS	WC-17Co	York University/Canada	2021
Liu [109]	SLM, LENS, SEBM, BJ3DP, SLS	WC-Co	North University of China/China	2021
Mariani [110]	BJ	WC-12Co	Politecnico di Milano	2021
Rieger [111]	Lithography-base	WC-(88-12)Co	Aalen University/Germany	2021
Ibe [112]	LPBF	WC-17Co, WC-25Co	Fujimi Incorporated/Japan	2021
Jucan [113]	SLS	WC-12Co/PA12	Technical University of Cluj-Napoca/Romania	2021
Lakovakis [114]	EBM	WC-Co with CrC-rich	The University of Manchester/UK	2021
Kim [115]	DED	WC-12Co	Inha University/Republic of Korea	2021
Kim [116]	MEX (Similar to MJT)	WC-10Co	Korea Institute of Ceramic Engineering and Technology/Korea	2022
Xing [117]	LPBF	WC-12Co	Beijing University of Technology/China	2022
Zhao [118]	SLM	WC-12Co	North University of China/China	2022
Li [119]	LMD	WC-20Co	North University of China/China	2022
Wang [120]	SEBM	WC-12Co	North University of China/China	2022
Tang [121]	BJ3DP	WC-18Co, WC-35Co	Hefei University of Technology/China	2022
Papy [122]	LPBF	WC-17Co	Technogenia/France	2022
Schwanekamp [123]	LPBF	WC-12Co, WC-25Co	University of Applied Sciences Cologne	2022
Wolfe [124]	BJ3DP	WC-12Co, WC-10Co	Global Tungsten and Powers Corp/USA	2023
Fries [125]	LPBF, SLM	WC-12Co	Institute for Materials Applications in Mechanical Engineering RWTH Aachen University/Germany	2023

Table 2.
State of the art of the AM techniques used for printing WC-Co hardmetals.

3.2 Can the DryLyte® technology act as a post-processing technique for the AM specimens?

However, laser-based AM technologies for manufacturing WC-Co hardmetals have encountered several issues as thermally induced cracks, heterogeneous microstructures, and embrittlement due to the formation of undesirable phases [71, 126]. On the other hand, sinter-based AM technologies offer the highest potential to process complex WC-Co parts with properties similar to commercial grades [127]. For these techniques, at this point in time limitations are found in regard to layer deposition quality, part size,

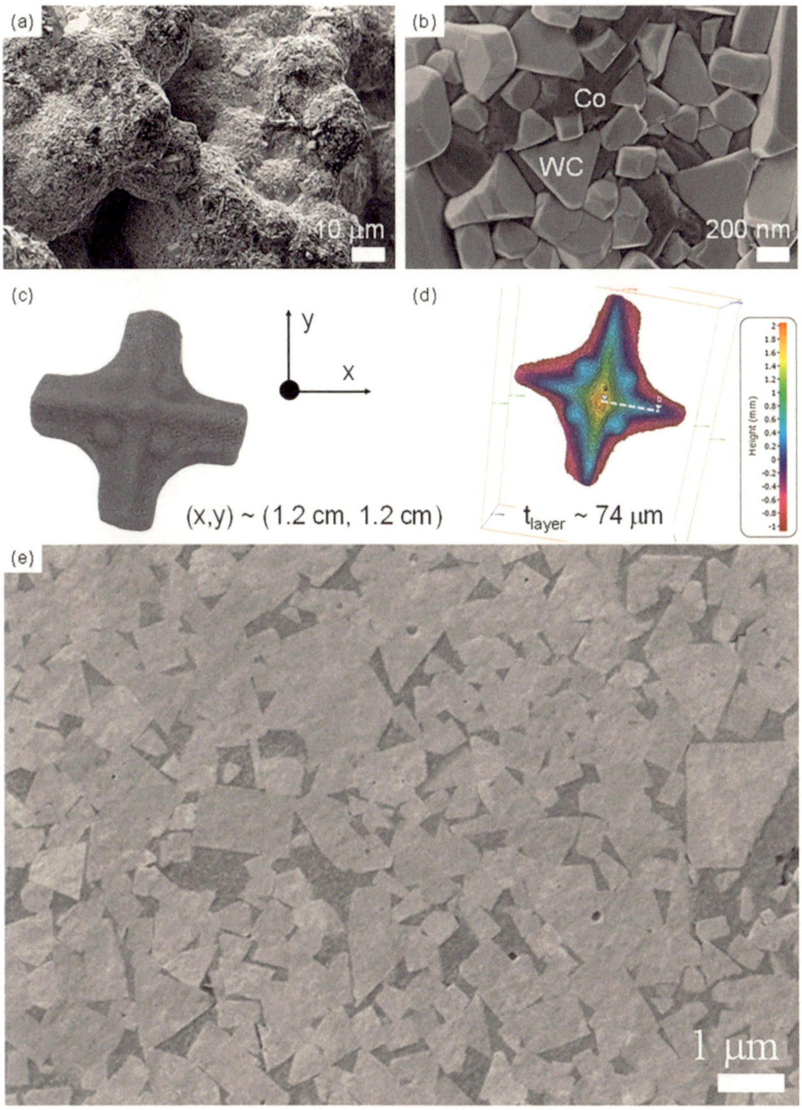

Figure 4.
(a) Low SEM micrograph of the WC-Co sintered granules; (b) high-magnification SEM micrograph showing the internal part of the sintered WC-Co granules where it is evident the low amount of metallic Co content; (c) 3DP WC-Co drill bit head; (d) confocal laser scanning microscopy (CLSM) 3D reconstruction of the AM WC-Co drill bit investigated [40]; and (e) SEM micrograph after being polished by using the DryLyte® Technology [128].

and batch size. Recently, solved on granules 3D-Printing (SG-3DP) has been successfully used to process fully dense WC-Co complex parts. In this chapter, the feasibility to polish SG-3DP by means of the DryLyte® Technology with lower Co content is explored. This technique consists in spreading powder-binder granules produced by a spry-drier technique (see **Figure 4a**) which were deposited layer-by-layer. **Figure 4b** shows the WC-Co granules microstructure where low Co content is clearly visible. As it is shown in **Figure 4c** by using the SG-3DP technology a drill bit head was printed with a thickness layer of around 74 μm as determined by using the Confocal Scanning Laser Microscopy, CSLM (see **Figure 4d**). As it is clearly evidenced in the SEM and CLSM micrographs, the sintered drill bit produced by the AM technology, presents a complex shape with high roughness values. To reduce the roughness of this complex specimen is not possible using conventional post-processing technology as the AM specimen will not preserve the desired geometry. In this sense, by using the DryLyte® Technology, it is possible to reduce considerably the superficial average roughness (S_a) until reaching a value of around 20 nm as shown in **Figure 4e**. This S_a reduction implies a roughness reduction of around 33% of the initial roughness.

After using the post-processing technologies, it is possible to superficially change the mechanical integrity in terms of hardness and elastic modulus. In this sense, preliminary results highlight that the surface mechanical integrity for the WC-Co specimens polished with DryLyte® Technology remains constant [128, 129] and equals as the once reported in the literature after being polished using conventional post-processing techniques [2]. Then, by using the DryLyte® technology leads to polish complex shapes processing from the AM routes keeping constant the mechanical surface integrity at the superficial level.

4. Conclusions

The conventional post-processing techniques do not lead to homogeneously polish the AM specimens due to the fact these techniques do not keep constant the geometry, preserve tolerances, etc.

DryLyte® Technology presents several advantages compared with the conventional post-processing techniques as it allows to homogeneously polish complex specimens. Keeping constant the surface integrity in terms of mechanical properties under service-like working conditions.

Acknowledgements

The authors are grateful to HILTI AG (Schaan, Liechtenstein) to provide us with the different samples investigated here. Furthermore, the authors are also grateful to Zeppelin 3D methodology company (especially to Javier Ledesma and María Gil) and the *Centro de Fabricacción Avanzada Aeronáutica* (especially to Guillermo González) to conduct the 3D measurements by using the InfiniteFocus G5 plus from Bruker Alicona. Finally, G.R. acknowledges the Ministerio de Ciencia e Innovación for the industrial PhD (Grant number: DIN2021-011846).

Conflict of interest

The authors declare no conflict of interest.

Appendices and nomenclature

AM	additive manufacturing
BJAM	binder jet additive manufacturing
CLSM	confocal laser scanning microscopy
DED	direct energy deposition
DMLS	direct metal laser sintering
EBM	electron beam melting
EDM	electrical discharge machining
FFF	fused filament fabrication
LENS	laser engineered net shaping
LMD	laser melting deposition
L-PBF	laser-based powder bed fusion
MEX	material extrusion
MJT	MultiJet technology
PS-DVB	PolyStyrene DiVinylBenzene
PM	powder metallurgical
SEBM	selective electron beam melting
SEM	scanning electron microscopy
SLM	selective laser melting
SLS	selective laser sintering
SG-3DP	solved on granules 3D-printing
S_a	superficial average roughness
T3DP	thermoplastic 3D printing
WAAM	wire arc additive manufacturing
WC-Co	cemented carbide or hardmetal
3D	three-dimensional

Author details

Guiomar Riu Perdrix and Joan Josep Roa Rovira*
Steros GPA Innovative S.L., Barcelona, Spain

*Address all correspondence to: jj.roa@gpainnova.com

References

[1] Mozetic M. Surface modification to improve properties of materials. Materials. 2019;**12**:441/1-441/8. DOI: 10.3390/ma12030441

[2] Roa JJ, Jiménez-Piqué E, Verge C, Tarragó JM, Mateo A, Fair J, et al. Intrinsic hardness of constitutive phases in WC-Co composites: Nanoindentation testing, statistical analysis, WC crystal orientation effects and flow stress for the constrained metallic binder. Journal of the European Ceramic Society. 2015;**35**: 3419-3425. DOI: 10.1016/j.jeurceramsoc.2015.04.021

[3] Ravichandran KS. Fracture toughness of two phase WC-Co cermets. Acta Metallugica et Materialia. 1994;**43**: 143-150. DOI: 10.1016/0956-7151(94) 90057-4

[4] Astakhov VP. Surface integrity—definition and importance in functional performance. In: Paulo Davim J, editor. Surface Integrity in Machining. London: Springer; 2010. pp. 1-35

[5] Field M, Kahles JF, Cammett JT. A review of measuring methods for surface integrity. CIRP Annals. 1972;**21**:219-238

[6] Jawair IS, Brinksmeier E, M'Saoubi R, Aspinwall DK, Outeiro JC, Meyer D, et al. Surface integrity in material removal processes: Recent advances. CIRP Annals. 2011;**60**:603-626. DOI: 10.1016/j.cirp.2011.05.002

[7] Prakash L. 1.02—Fundamentals and general applications of hardmetals. In: Sarin VK, Mari D, Llanes L, editors. Comprehensive Hard Materials. Vol. 1. Oxford: Elsevier; 2014. pp. 29-90

[8] Roebuck B, Almond EA. Deformation and fracture processes and the physical metallurgy of WC-Co hardmetals. International Materials Reviews. 1988;**33**:90-112. DOI: 10.1179/imr.1988.33.1.90

[9] Shatov AV, Ponomarev SS, Firstov SA. 1.10—Fracture and strength of hardmetals at room temperature. In: Sarin VK, Mari D, Llanes L, editors. Comprehensive Hard Materials. Vol. 1. Oxford: Elsevier; 2014. pp. 301-343

[10] Sigl LS, Exner HE. Experimental study of the mechanics of fracture in WC-Co alloys. Metallurgical and Materials Transactions A. 1987;**18**: 1299-1308. DOI: 10.1007/bf02647199

[11] Llanes L, Torres Y, Anglada M. On the fatigue crack growth behavior of WC-Co cemented carbides: Kinetics description, microstructural effects and fatigue sensitivity. Acta Materialia. 2002; **50**:2381-2393. DOI: 10.1016/s1359-6454 (02)00071-X

[12] Van der Berg H. Hardmetals: Trends in devilment and application. Powder Metallurgy. 2007;**50**:7-10. DOI: 10.1179/174329007X186822

[13] Byrne G, Dornfeld D, Denkena B. Advancing cutting technology. CIRP Annals. 2003;**53**:483-507. DOI: 10.1016/S0007-8506(07)60200-5

[14] Bouzakis KD, Michailidis N, Skordaris G, Bouzakis E, Biermann D, M'Saoubi R. Cutting with coated tools: Coating technologies, characterization methods and performance optimization. CIRP Annals. 2012;**61**:703-723. DOI: 10.1016/j.cirp.2012.05.006

[15] Hogmark S, Jaobson S, Larsoon M. Design and evaluation of tribological coatings. Wear. 2000;**246**:20-33. DOI: 10.1016/S0043-1648(00) 00505-6

[16] Schubert W-D, Lassner E, B¨hlke W. Cemented carbide—a success story. ITIA Newsletter. June 2010

[17] Griffiths B. Manufacturing Surface Technology: Surface Integrity and Functional Performance. London: Penton Press; 2001

[18] Yang N, Zong W, Li Z, Sun T. The dependency of diamond lapping surface morphology on crystal orientation. The International Journal of Advanced Manufacturing Technology. 2014;**77**: 1029-1034. DOI: 10.1007/s00170-014-6516-x

[19] Malkin S, Guo C. Grinding Technology: Theory and Application of Machining with Abrasives. 2nd ed. New York: Industrial Press Inc.; 2008

[20] Marinescu ID, Tonshoff HK, Inaski I. Handbook of Ceramic Grinding and Polishing. New York: Noyes Publications/William Andre Publishing, L.L.C; 1998

[21] Zhang B, Howes TD. Material-removal mechanisms in grinding ceramics. CIRP Annals. 1994;**43**: 305-308. DOI: 10.1016/S0007-8506(07)62219-7

[22] Swain MV. Microfracture about scratches in brittle solids. Proceedings of the Royal Society A-Mathematical, Physical and Engineering Science. 1979; **366**:575-597. DOI: 10.1098/rspa.1979.0070

[23] Marshall DB, Evans AG, Khuri Yakub BT, Tien JW, Kino GS. The nature of machining damage in brittle materials. Proceedings of the Royal Society A-Mathematical, Physical and Engineering Science. 1983;**385**:461-475. DOI: 10.1098/rspa.1983.0023

[24] Marshall DB. Failure from contact-induced surface flaws. In: Freiman SW, Hudson CM, editors. Methods for Assessing the Structural Reliability of Brittle Materials. Philadelphia, PA: American Society for Testing and Materials; 1984. pp. 3-21

[25] Kirchner HP. Damage penetration at elongated machining grooves in hot-pressed Si_3N_4. Journal of the American Ceramic Society. 1984;**67**:127-132. DOI: 10.1111/j.1151-2916.1984.tb09629

[26] Engqvist H, Beste U, Axén N. Influence of pH on sliding wear of WC-based Materials Science. International Journal of Refractory Metals & Hard Materials. 2000;**18**:103-109. DOI: 10.1016/S0263-4368(00)00007-X

[27] Pugsley VA, Sockel HG. Corrosion fatigue of cemented carbide cutting tool materials. Materials Science and Engineering A. 2004;**366**:87-95. DOI: 10.1016/j.msea.2003.08.057

[28] Human AM, Exner HE. The relationship between electrochemical behavior and in-service corrosion of WC based cemented carbides. International Journal of Refractory Metals and Hard Materials. 1997;**15**:65-71. DOI: 10.1016/S0263-4368(96)00014-5

[29] Crittenden JC, Truseel RR, Hand DW, Howe KJ, Tchobanoglous G. MWH's Water Treatment: Principles and Design. 3rd ed. John Wiley & Sons; 2012. DOI: 10.1002/9781118131473. Print ISBN: 9780470405390

[30] Fathi MB, Rezai B, Alamdari EK, Alorro RD. Mechanisms and equilibrium modeling of Re and Mo adsorption on a gel type strong base anion resin. Russian Journal of Applied Chemistry. 2017;**90**: 1504-1513. DOI: 10.1134/s1070427217080208

[31] Simonelli M, Tse YY, Tuck C. Effect of the build orientation on the mechanical properties and fracture modes of SLM Ti-6Al-4V. Materials Science and Engineering A. 2014; **616**:1-11. DOI: 10.1016/j.msea.2014.07.086

[32] Liu Y, Yang Y, Mai S, Wang D, Song C. Investigation into spatter behavior during selective laser melting of AISI 316K stainless steel powder. Materials & Design. 2015;**87**:797-806. DOI: 10.1016/j.matdes.2015.08.086

[33] Grigoriev S, Tarasova T, Gusarov A, Khmyrov R, Egorov S. Possibilities of manufacturing products from cermet compositions using nanoscale powders by additive manufacturing methods. Materials. 2019;**12**:3425/1-3425/16. DOI: 10.3390/ma12203425

[34] Hou H, Simsek E, Ma T, Johnson NS, Qian S, Cissé C, et al. Fatigue-resistant high-performance elastocaloric materials made by additive manufacturing. Science. 2019;**366**:1116-1121. DOI: 10.1126/science.aax7616

[35] Todaro CJ, Easton MA, Qiu D, Zhang D, Bermingham MJ, Lui EW, et al. Grain structure control during metal 3D printing by high-intensity ultrasound. Nature Communications. 2020;**11**:142. DOI: 10.1038/s41467-019-13874-z

[36] Lin T-C, Cao C, Sokoluk M, Jiang L, Wang X, Schoenung JM, et al. Aluminum with dispersed nanoparticles by laser additive manufacturing. Nature Communications. 2019;**10**:4124. DOI: 10.1038/s41467-019-12047-2

[37] DebRoy T, Mukherjee T, Milewsk JO, Elmer JW, Ribic B, Blecher JJ, et al. Scientific, technological and economic issues in metal printing and their solutions. Nature Materials. 2019;**18**:1026-1032. DOI: 10.1038/s41563-019-0408-2

[38] Barba A, Diez-Escudero A, Maazouz Y, Rappe K, Espanol M, Montufar EB, et al. Osteoinduction by foamed and 3D printed calcium phosphate scaffolds: Effect of nanostructure and pore architecture. ACS Applied Matererials & Interfaces. 2017;**9**:41722-41736. DOI: 10.1016/j.actbio.2018.09.003

[39] Ahn BY, Duoss EB, Motala MJ, Guo X, Park SI, Xiong Y, et al. Omnidirectional printing of flexible, stretchable, and spanning silver microelectrodes. Science. 2009;**323**:1590-1593. DOI: 10.1126/science.1168375

[40] Au AK, Huynh W, Horowitz LF, Folch A. 3D-printed microfluidics. Angewandte Chemie, International Edition. 2016;**55**:3862-3881. DOI: 10.1002/anie.201504382

[41] Ho CMB, Huan Ng S, Li KHH, Yoon YJ. 3D printed microfluidics for biological applications. Lab on a Chip. 2015;**15**:3627-3637. DOI: 10.1039/C5LC00685F

[42] Smay JE, Gratson GM, Shepherd RF, Cesarano J, Lewis JA. Directed colloidal assembly of 3D periodic structures. Advanced Materials. 2002;**14**:1279-1283. DOI: 10.1002/1521-4095(20020916)14:18<1279::AID-ADMA1279>3.0.CO;2-A

[43] Álvarez F, Cifuentes A, Serrano I, Franco L, Fargas G, Fenollosa F, et al. Optimization of the sintering thermal treatment and the ceramic ink used in direct ink writing of α-Al_2O_3: Characterization and catalytic application. Journal of the European Ceramic Society. 2022;**42**:2921-2930. DOI: 10.1016/j.eurceramsoc.2022.01.032

[44] Herzog D, Seyda V, Wycisk E, Emmelmann C. Additive manufacturing of metals. Acta Materialia. 2016;**117**: 371-392. DOI: 10.106/j.ctamat.2016. 07.019

[45] Gu DD, Meiners W, Wissenback K, Poprawe R. Laser additive manufacturing of metallic components: Materials, processes and mechanisms. International Materials Reviews. 2013;**57**:133-164. DOI: 10.1179/1743280411y.0000000014

[46] Konyashin I, Hinners H, Ries B, Kirchner A, Klöden B, Kieback B, et al. Additive manufacturing of WC-13%Co by selective electron beam melting: Achievements and challenges. International Journal of Refractory Metals and Hard Materials. 2019;**84**: 105028. DOI: 10.1016/j. ijrmhm.2019.105028

[47] DebRoy T, Wei HL, Zuback JS, Mukherjee T, Elmer JW, Milewski JO, et al. Additive manufacturing of metallic components—process, structure and properties. Progress in Materials Science. 2018;**92**:112-224. DOI: 10.1016/j. pmatsci.2017.10.001

[48] Drescher P, Sarhan M, Seitz H. An investigation of sintering parameters on titanium powder for electron beam melting processing optimization. Materials. 2016;**9**:974. DOI: 10.3390/ ma9120974

[49] Hernández-Nava E, Tammas-Williams S, Smith C, Leonard F, Withers P, Todd I, et al. X-Ray tomography characterization of lattice structures processed by selective electron beam melting. Metals. 2017;**7**: 300. DOI: 10.3390/met7080300

[50] Maizza G, Caporale A, Polley C, Seitz H. Micro-macro relationship between microstructure, porosity, mechanical properties, and build mode parameters of a selective-electron-beam-melted Ti-6Al-4V alloy. Metals. 2019;**9**: 786. DOI: 10.339/met9070786

[51] Veiga A, Artaza S. Experimental investigation of the influence of wire arc additive manufacturing on the machinability of titanium parts. Metals. 2019;**10**:24. DOI: 10.3390/met10010024

[52] Wang F, Williams S, Colegrove P, Antonysamy AA. Microstructure and mechanical properties of wire and arc additive manufactured Ti-6Al-4V. Metallurgical and Materials Transactions A. 2013;**44**:968-977. DOI: 10.1007/s11661-012-1444-6

[53] Marinelli G, Martina F, Ganguly S, Williams S. Grain refinement in an unalloyed tantalum structure by combining wire + arc additive manufacturing and vertical cold rolling. Additie Manufacturing. 2019;**32**:101009. DOI: 10.1016/j.addma.2019.101009

[54] Marinelli G, Martine F, Ganguly S, Williams S. Dvelopment of wire + arc additive manufacture for the production of large-scale unalloyed tungsten components. International Journal of Refractory Metals and Hard Materials. 2019;**82**:329-335. DOI: 10.1016/j. ijrmhm.2019.05.009

[55] Tucho WM, Cuvillier P, Sjolyst-kverneland A, Hansen V. Microstructure and hardness studies of Inconel 718 manufactured by selective laser melting before and after solution heat treatment. Materials Science and Engineering A. 2017;**689**:220-232. DOI: 10.1016/j.msea. 2017.02.062

[56] Mazur M, Leary M, McMillan M, Elambasseril J, Brandt M. SLM additive manufacture of H13 tool steel with conformal cooling and structural lattices. Rapid Prototyping Journal. 2016;**22**:

504-518. DOI: 10.1108/RPJ-06-2014-0075

[57] Seyedkashi A, Kang K, Moon W. Analysis of melt-pool behaviors during selective laser melting of AISI 304 stainless-steel composites. Metals. 2019; **9**:876. DOI: 10.3390/met9080876

[58] Jin W, Zhang C, Jin S, Tian Y, Wellmann D, Liu W. Wire arc additive manufacturing of stainless steels: A review. Additive Manufacturing in Industry. 2020;**10**:1563. DOI: 10.3390/app10051563

[59] Zai L, Zhang C, Wang Y, Guo W, Wellmann D, Tong X, et al. Selective laser melting of precipitation-hardened martensitic stainless steels: A review. Metals. 2020;**10**:255/1-255/25. DOI: 10.3390/met10020255

[60] Arison YM, Criales LE, Özel T, Lane B, Moylan S, Donmez A. Influence of scan strategy and process parameters on microstructure and its optimization in additively manufactured nickel alloy 625 via laser powder bed fusion. The International Journal of Advanced Manufacturing Technology. 2017;**90**: 1393-1417. DOI: 10.1007/s00170-016-9429-z

[61] Chauvet E, Kontis P, Jägle EA, Gault B, Raabe D, Tassin C, et al. Hot cracking mechanisms affecting a non-weldable Ni-based superalloy produced by selective electron beam melting. Acta Materialia. 2018;**142**:82-94. DOI: 10.1016/j.actamat.2017.09.047

[62] Promoppatum P, Onler R, Yao S-C. Numerical and experimental investigtions of micro and macro characteristics of direct metal laser sintered Ti-6Al-4V products. Journal of Materials Processing Technology. 2017; **240**:262-273. DOI: 10.1016/j.jmatprotec.2016.10.005

[63] Aithilingam J, Goodridge RD, RJM H, Christie SDR, Edmondson S. The effect of laser remelting on the surface chemistry of Ti6Al4V components fabricated by selective laser melting. Journal of Materials Processing Technology. 2016;**232**:1-8. DOI: 10.1016/j.matprotec.2016.01.022

[64] Sidambe AT, Tian Y, Prangnell PB, Fox P. Effect of processing parameters on the densification, microstructure and crystallographic texture during the laser powder bed fusion of pure tungsten. International Journal of Refractory Metals and Hard Materials. 2019;**78**: 254-263. DOI: 10.1016/j.ijrmhm.2018.10.004

[65] Sidambe AT, Judson DS, Colosimo SJ, Fox P. Laser power bed fusión of a pure tungsten ultra-fine single pinhole collimator for use in gamma ray detector characterization. International Journal of Refractory Metals and Hard Materials. 2019;**84**: 104998. DOI: 10.1016/j.ijrmhm.2019.10.4998

[66] Schmidtke K, Palm F, Hawkins A, Emmelmann C. Process and mechanical properties: Applicability of a scandium modified Al-alloy for laser additive manufacturing. Physics Procedia. 2011; **12**:369-374. DOI: 10.1016/j.phpro.2011.03.047

[67] Chou R, Milligan J, Paliwal M, Brochu M. Additive manufacturing of Al-12Si alloy via pulsed selective laser melting. Journal of Metals. 2015;**67**: 590-596. DOI: 10.1007/s11837-014-1272-9

[68] Zhang X, Guo Z, Chen C, Yang W. Additive manufacturing of WC-20Co components by 3D gel-printing. International Journal of Refractory Metals and Hard Materials. 2018;**70**:

215-223. DOI: 10.1016/j.ijrmhm.2017.10.005

[69] Lengauer W, Duretek I, Fürst M, Schwarz V, Gonzalez-Gutierrez J, Schuschnigg S, et al. Fabrication and properties of extrusion-based 3D-printed hardmetal and cermet components. International Journal of Refractory Metals and Hard Materials. 2019;**82**:141-149. DOI: 10.1016/j.ijrmhm.2019.04.011

[70] Chen J, Huang M, Fang ZZ, Koopman M, Liu W, Deng X, et al. Microstructure analysis of high density WC-Co composite prepared by one step selective laser melting. International Journal of Refractory Metals and Hard Materials. 2019;**84**:104980. DOI: 10.1016/j.ijrmhm.2019.104980

[71] Uhlmann E, Bergmann A, Gridin W. Investigation on additive manufacturing of tungsten carbide-cobalt by selective laser melting. Procedia CIRP. 2015;**35**:8-15. DOI: 10.1016/j.procir.2015.08.060

[72] Gusarov AV, Pavlov M, Smurov I. Residual stress at laser surface remelting and additive manufacturing. Physics Procedia. 2011;**12**:248-254. DOI: 10.1016/j.phpro.2011.03.032

[73] Uhlmann E, Bergmann A, Bolz R. Manufacturing of carbide tools by selective laser melting. Procedia Manufacturing. 2018;**21**:765-773. DOI: 10.1016/j.promfg.2018.02.182

[74] Kear BH, Skandan G, Sadangi RK. Factors controlling decarburization in HVOF sprayed nano-WC/Co hardcoatings. Scripta Materialia. 2001;**44**:1703-1707. DOI: 10.1016/s1359-6462(01)00867-3

[75] Körner C. Additive manufacturing of metallic components by selective electron beam melting—a review. International Materials Reviews. 2016;**61**:361-377. DOI: 10.1080/09506608.2016.1176289

[76] Cramer CL, Nandwana P, Lowden RA, Elliott AM. Infiltration studies of additive manufacture of WC with Co using binder jetting and pressureless melt method. Additive Manufacturing. 2019;**28**:333-343. DOI: 10.1016/j.addma.2019.04.009

[77] Cramer CL, Wieber NR, Aguirre TG, Lowden RA, Elliott AM. Shape retention and infiltration height in complex WC-Co parts made via binder jet of WC with subsequent Co melt infiltration. Additive Manufacturing. 2019;**29**:100828. DOI: 10.1016/j.addma.2019.100828

[78] Enneti RK, Prough KC, Wolfe TA, Kelin A, Studley N, Trasorras JL. Sintering of WC-12%Co processed by binder jet 3D printing (BJ3DP) technology. International Journal of Refractory Metals and Hard Materials. 2018;**71**:28-35

[79] Enneti RK, Prough KC. Wear properties of sintered WC-12%Co processed via binder Jet 3D printing (BJ3DP). International Journal of Refractory Metals and Hard Materials. 2019;**78**:228-232. DOI: 10.1016/j.ijrmhm.2018.10.003

[80] Berman B. 3-D printing: The new industrial revolution. Business Horizons. 2012;**55**:155-162. DOI: 10.1016/j.bushor.2011.11.003

[81] Spierings AB, Voegtlin M, Bauer T, Wegener K. Powder flowability characterization methodology for powder-bed-based metal additive manufacturing. Progress in Additive Manufacturing. 2016;**1**:9-20. DOI: 10.1007/s40964-015-0001-4

[82] Ren X, Shao H, Lin T, Zheng H. 3D gel-printing-an additive manufacturing method for producing complex shape parts. Materials & Design. 2016;**101**: 80-87. DOI: 10.1016/j.matdes.2016.03.152

[83] Kukla C, Gonzalez-Gutierrez J, Burkhardt C, Weber O, Holzer C: The production of magnets by FFF-fused filament fabrication. In Proceedings of Proceedings of the Euro PM2017 Congress&Exhibition. Milan, Italy. 2017;1-5

[84] Gonzalez-Gutierrez J, Cano S, Schuschnigg S, Kukla C, Sapkota J, Holzer C. Additive manufacturing of metallic and ceramic components by the material extrusion of highly-filled polymers: A review and future perspectives. Materials. 2018;**11**:840. DOI: 10.3390/ma11050840

[85] Wang XC, Laoui T, Bonse J, Kruth JP, Lauwers B, Froyen L. Direct selective laser sintering of hard metal powders: Experimental study and simulation. International Journal of Advanced Manufacturing Technology. 2002;**19**:351-357. DOI: 10.1007/s001700200024

[86] Maeda K, Childs THC. Laser sintering (SLS) of hard metal powder for abrasion resistant coatings. Journal of Materials Processing Technology. 2004; **149**:609-615. DOI: 10.1016/j.jmatprotec.2004.02.024

[87] Gu D, Shen Y. WC-Co particulate reinforcing Cu matrix composites produced by direct laser sintering. Materials Letters. 2006;**60**:3664-3668. DOI: 10.1016/j.matlet.2006.03.103

[88] Gu D, Shen Y. Processing and microstructure of submicron WC-Co particulate reinforced Cu matrix composites prepared by direct laser sintering. Materials Sciencie and Engineering A. 2006;**435-436**:54-61. DOI: 10.1016/j.msea.2006.07.105

[89] Kernan BD, Sachs EM, Oliveira MA, Cima MJ. Three-dimensional printing of tungsten carbide-10 wt% cobalt using a cobalt oxide precursor. International Journal of Refractory Metals and Hard Materials. 2007;**25**:82-94. DOI: 10.1016/j.ijrmhm.2006.02.002

[90] Gu D, Shen Y. Influence of reinforcement weight fraction on microstructure and properties of submicron WC-Cop/Cu MMCs prepared by direct laser sintering. Journal of Alloys and Compounds. 2007;**431**(1-2): 112-120. DOI: 10.1016/j.jallcom.2006.05.044

[91] Gu D, Shen Y, Zhao L, Xiao J, Wu P, Zhu Y. Effect of rare earth oxide addition on microstructures of ultra-fine WC-Co particulate reinforced Cu matrix composites prepared by direct laser sintering. Materials Science and Engineering A. 2007;**455-446**:316-322. DOI: 10.1016/j.msea.2006.09.057

[92] Gu D, Shen Y. Direct laser sintered WC-10Co/Cu nanocomposites. Applied Surface Science. 2008;**254**(13):3971-3978. DOI: 10.1016/j.apsusc.2007.12.028

[93] Kumar S, Kruth JP, Froyen L: Wear behaviour of SLS WC-Co composites. Annual International Solid Freeform Fabrication Symposium SFF. 2008:543-557. DOI: 10.26153/tsw/14982

[94] Xiong Y, Smugeresky JE, Ajdelsztajn L, Schoenung JM. Fabrication of WC-Co cermets by laser engineered net shaping. Materials Science and Engineering A. 2008;**493**: 261-266. DOI: 10.1016/j.msea.2007.05.125

[95] Kumar S. Manufacturing of WC-Co moulds using SLS machine. Journal of Materials Processing Technology. 2009; **209**:3840-3848. DOI: 10.1016/j.jmatprotec.2008.08.037

[96] Kyogoku H, Uemori T, Ikuta A, Yoshikawa K, Ohmori H. Direct selective laser sintering of WC-Co cemented carbide by premixing of additives. ASME/ISCIE International Symposium on Flexible Automation, Conference Proceeding. 2012:465-468. DOI: 10.1115/ISFA2012-7182

[97] Ghosh SK, Das AK, Saha P. Selective laser sintering: A case study of tungsten carbide and cobalt powder sintering by pulsed Nd:YAG Laser. In: Joshi SN, Dixit US, editor. Materials Engineering, Chapter 4. Springer; 2015. pp. 441-459. DOI: 10.1007/978-81-322-2352-8_22. ISSN 23643293

[98] Scheithauer U, Pötschke J, Weingarten S, Schwarzer E, Vornberger A, Mortiz T, et al. Droplet-based additive manufacturing of hard metal components by thermoplastic 3D Printing (T3DP). Journal of Ceramic Science and Technology. 2017;**8**:155-160. DOI: 10.4416/JCST2016-00104

[99] Khmyrov RS, Shevchukov AP, Gussarov AV, Tarasova TV. Phase composition and microstructure of WC-Co alloys obtained by selective laser melting. Mechanics and Industry. 2017; **18**(1):714. DOI: 10.1051/meca/2017059

[100] Kumar S, Czekanski A. Optimization of parameters for SLS of WC-Co. Rapid Prototyping Journal. 2017;**23**(6):1202-1211. DOI: 10.1108/RPJ-10-2016-0168

[101] Kumar S. Process chain development for additive manufacturing of cemented carbide. Journal of Manufacturing Processes. 2018;**34** (PartA):121-130. DOI: 10.1016/j.jmapro.2018.05.036

[102] Enneti RK, Prough KC. Effect of binder saturation and powder layer thickness on the green strength of the binder jet 3D printing (BJ3DP) WC-12% Co powders. International Journal of Refractory Metals and Hard Materials. 2019;**84**:104991. DOI: 10.1016/j.ijrmhm.2019.104991

[103] Cramer CL, Aguirre TG, Wieber NR, Lowden RA, Trofimov AA, Wang H, et al. Binder jet printed WC infiltrated with pre-made melt of WC-Co. International Journal of Refractory Metals and Hard Materials. 2020;**87**: 105137. DOI: 10.1016/j.ijrmhm.2019.105137

[104] Padmakumar M. Additive manufacturing of tungsten carbide hardmetal parts by selective laser melting (SLM), selective laser sintering (SLS) and binder jet 3D printing (BJ3DP) techniques. Laser in Manufacturing and Materials Processing. 2020;**7**(3):338-371. DOI: 10.1007/s40516-020-00124-0

[105] Yang Y, Zhang C, Wang D, Nie L, Wellmann D, Tian Y. Additive manufacturing of WC-Co hardmetals: A review. International Journal of Advanced Manufacturing Technology. 2020;**108**:1653-1673. DOI: 10.1007/s00170-020-05389-5

[106] Al-Thamir M, McCartney DG, Simonelli M, Hague R, Clare A. Processability of atypical WC-Co composite feedstock by laser powder-bed fusion. Materials. 2020;**13**:50. DOI: 10.3390/ma13010050

[107] Bricín D, Ackermann M, Jansa Z, Kubátová D, Kříž A, Špirit Z, et al. Development of the structure of cemented carbides during their

processing by SLM and HIP. Metals. 2020;**10**:1477. DOI: 10.3390/met10111477

[108] Agyapong J, Czekanski A, Boakye-Yiadom S. Effect of heat treatment on microstructural evolution and properties of cemented carbides (WC-17Co) reinforced with 3% volume hexagonal-boron nitride (h-BN) and processed by selective laser sintering (SLS). Materials Characterization. 2021;**174**:110968. DOI: 10.1016/j.matchar.2021.110968

[109] Liu Y, Guo Z, Wang H, Zhang J, Li X. Research progress of 3D printing cemented carbide. Chinese Journal of Rare Metals. 2021;**45**:484-492. DOI: 10.13373/j.cnki.cjrm.XY19090015

[110] Mariani M, Goncharov I, Mariani D, De Gaudenzi GP, Popovich A, Lecis N, et al. Mechanical and microstructural characterization of WC-Co consolidated by binder jetting additive manufacturing. International Journal of Refractory Metals and Hard Materials. 2021;**100**:105639. DOI: 10.1016./j.jrmhm.2021.105639

[111] Rieger T, Schubert T, Schurr J, Butschle M, Schwenkel M, Bernthaler T, et al. Slurry development for lithography-based additive manufacturing of cemented carbide components. Powder Technology. 2021;**383**:498-508. DOI: 10.1016/j.powtec.2021.01.049

[112] Ibe H, Kato Y, Yamada J, Kato M, Suzuki A, Takata N, et al. Controlling WC/Co two-phase microstructure of cemented carbides additive-manufactures by laser power bed fusion: Effect of powder composition and post heat-treatment. Materials and Design. 2021;**210**:110034. DOI: 10.1016/j.matdes.2021.110034

[113] Jucan OD, Gădălean RV, Chicinaş HF, Hering M, Bâlc N, Popa CO. Study on the indirect selective laser sintering (SLS) of WC-Co/PA12 powders for the manufacturing of cemented carbide parts. International Journal of Refractory Metals and Hard Materials. 2021;**96**:105498. DOI: 10.1016/j.ijmhm.2021.105498

[114] Lakovakis E, Avcu E, Roy MJ, Gee M, Matthews A. Dry sliding wear behaviour of additive manufactured CrC-rich WC-Co cemented carbides. Wear. 2021;**486-487**:204127. DOI: 10.1016/j.wear.2021.204127

[115] Kim KW, Ham GS, Park SH, Cho JW, Lee KA. Direct energy deposition of ultrastrong WC-12Co cemented carbide: Fabrication, microstructure and compressive properties. International Journal of Refractory Metals and Hard Materials. 2021;**99**:105591. DOI: 10.1016/j.ijrmhm.2021.105591

[116] Hakhyun K, Jong-Il K, Young DK, Hyeondeok J, Sung-Soo R. Material extrusion-based three-dimensional printing of WC-Co alloy with a paste prepared by powder coating. Additive Manufacturing. 2022;**52**:102679. DOI: 10.1016/j.addma.2022.102679

[117] Xing M, Wang H, Zhao Z, Lu H, Liu C, Lin L, et al. Additive manufacturing of cemented carbides inserts with high mechanical performance. Materials Science and Engineering A. 2022;**861**:144350. DOI: 10.1016/j.msea.2022.144350

[118] Zhao Y, Wang H, Zhang L, Li X, Guo Z, Zhang J, et al. Study on the microstructure and properties of WC-12Co cemented carbide fabricated by selective laser melting. Journal of Materials Research and Technology.

2022;**20**:3512-3521. DOI: 10.1016/j.jmrt.2022.08.082

[119] Li X, Zhao Y, Guo Z, Liu Y, Wang H, Zhang J, et al. Influence of different substrates on the microstructure and mechanical properties of WC-12Co cemented carbide fabricated via laser melting deposition. International Journal of Refractory Metals and Hard Materials. 2022;**104**:105787. DOI: 10.1016/j.ijrmhm.2022.105787

[120] Wang J, Han Y, Zhao Y, Li X, Yi D, Guo Z, et al. Microstructure and properties of WC-12Co cemented carbide fabricated via selective electron beam melting. International Journal of Refractory Metals and Hard Materials. 2022;**106**:105847. DOI: 10.1016/j.ijrmhm.2022.105847

[121] Tang JY, Luo LM, Liu Z, Zan X, Wu YC. Shape retention of cemented carbide prepared by Co melt infiltration into un-sintered WC green parts made via BJ3DP. International Journal of Refractory Metals and Hard Materials. 2022;**107**:105904. DOI: 10.1016/j.ijrmhm.2022.105904

[122] Papy K, Jean-Marc S, Alexey S, Andras B. Additive manufacturing feasibility of WC-17Co cermet parts by laser powder bed fusion. Procedia CIRP. 2022;**111**:153-157. DOI: 10.1016/j.procir.2022.08.049

[123] Schwanekamp T, Marginean G, Reuber M, Ostendorf A. Impact of cobalt content and grain growth inhibitors in laser-based powder bed fusion of WC-Co. International Journal of Refractory Metals and Hard Materials. 2022;**105**: 105814. DOI: 10.1016/j.ijrmhm.2022.105814

[124] Wolfe T, Shah R, Prough K, Trasorras JL. Coarse cemented carbide produced via binder jet 3D printing. International Journal of Refractory Metals and Hard Materials. 2023;**110**: 106016. DOI: 10.1016/j.ijrmhm.2022.106016

[125] Fries S, Vogelpoth A, Kaletsch A, Broeckmann C. Influence of post heat treatment on microstructure and fracture strength of cemented carbide manufactured using laser-based additive manufacturing. International Journal of Refractory Metals and Hard Materials. 2023;**111**:106085. DOI: 10.1016/j.ijrmhm.2022.106085

[126] Mostafaei A, Elliott AM, Barnes JE, Cramer CL, Nandwna P, Chmielus M. Binder jet 3D printing—process parameters, materials, properties, modeling, and challenges. Progress in Materials Science. 2020;**119**:100707/1-100707/138. DOI: 10.1016/j.pmatsci.2020.100707

[127] Carreño-Morelli E, Alveen P, Moseley S, Rodriguez-Arbaizar M, Cardoso K. A comparative study of cemented carbide parts produced by solvent on granules 3D-printing (SG-3DP) versus press and sinter. International Journal of Refractory Metals and Hard Materials. 2021;**97**: 105515/1-105515/6. DOI: 10.1016/j.ijrmhm.2021.105515

[128] Riu G, Roa JJ. Small-scale mechanical response at intermediate/high temperature of 3D printed WC-Co. Procedia CIRP. 2022;**108**:507-512. DOI: 10.1016/j.procir.2022.03.079

[129] Riu G, Weil D, Llanes L, Johanns KE, Oliver WC, Roa JJ. Surface integrity of new dry electropolishing technology on WC-Co cemented carbides. Procedia CIRP. 2022;**108**: 543-548. DOI: 10.1016/j.procir.2022.03.085

Section 3

3D Bioprinting

Chapter 8

Rheological Model of Materials for 3D Printing by Material Extrusion

Jorge Mauricio Fuentes Fuentes

Abstract

In this chapter, the viscoelastic model of Maxwell and Kevin-Voigt and the rheological model are described. The operation and characteristic equations of a capillary rheometer are explained, as well as the Bagley and Rabinowitch corrections. Next, the method used to determine the viscosity of semicrystalline polymer is explained, using the capillary rheometer. Finally, the Rabinowitch is explained to define a rheological model that determines the viscosity of materials using a capillary rheometer.

Keywords: rheology, additive manufacturing, viscosity, manufacturing by extrusion, Bagley correction, Rabinowitch correction, cross WLF model

1. Introduction

Semicrystalline materials, such as polypropylene, are difficult to print by MEX (material extrusion additive manufacturing), due to their lack of adhesion to the print bed and their high shrinkage [1]. The problem increases when the pieces have a large surface in relation to their height, so solutions such as cellophane adhesive tapes, adhesive sprays, and other types of solutions are used. The specimens suffer evident deformations (**Figure 1**), which might affect your application. This chapter shows how a rheological study of the material is carried out to select the optimal properties that would help to mitigate the warping problem.

In this chapter, an introduction to the viscoelasticity models and rheological models is made to later carry out the study of the rheological parameters of semicrystalline polymers, which would serve as a basis for carrying out similar studies on any other polymeric material for printing by MEX.

2. Viscoelasticity

Viscoelasticity is the property that polymers must behave like an elastic solid and a viscous fluid [2]. Elastic deformation is instantaneous and independent of time. Viscous materials, such as water, resist shear flow, and they relax linearly with time when applying a strain. Elastic materials become taut when stretched and immediately return to their original state once the tension is removed. In a polymer, deformations occur with a delay in relation to the applied stresses. Polymers also have a plastic-type

 IntechOpen

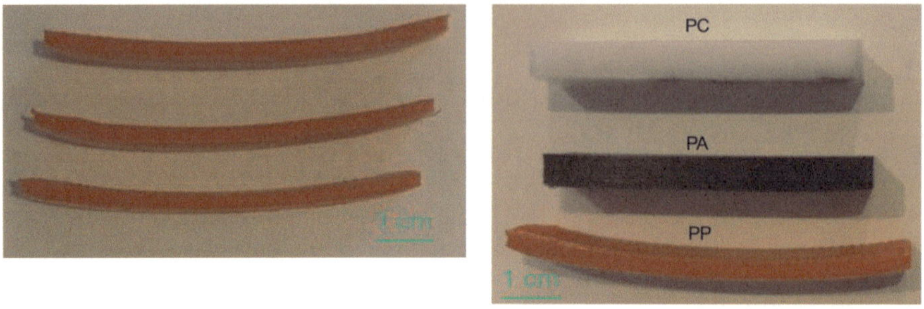

(a) (b)

Figure 1.
a) PP specimens deformed after the printing process by MEX. b) Comparison of PP, PA, and PC specimens after the printing process by MEX.

component related to non-recoverable deformation, linked to immediate permanent deformation.

In polymers, stress and time play important parameters in mechanical behavior because they help determine if it will be strong enough, if it will be tough to withstand blows without breaking, and how the polymer will deform under load.

The response of the material is conditioned by its viscoelastic nature. The analysis of the response of these materials can be carried out from mathematical models that try to explain the behaviors, among which are:

- Maxwell's model
- Kelvin-Voight model
- Combined models.

A fully viscous response is that of a Newtonian fluid, whose deformation is linear with time while stress is applied and is completely unrecoverable.

2.1 Basic models of viscoelasticity

When there is an elastic material, the deformation (ε) is instantaneous and proportional to the applied stress (σ), which is governed by Hooke's law, shown in Eq. (1), this behavior is represented by a spring [3, 4], which represents the stiffness modulus (ξ).

$$\varepsilon = \frac{\sigma}{\xi} \tag{1}$$

For a viscous fluid, the deformation is not instantaneous and will depend on time; and this deformation is not recoverable. This deformation is represented by a plunger or piston with a fluid inside, and its behavior is given by Newton's law, according to the Eq.(2). This equation indicates that the stress or stress applied is proportional to the strain rate, who's constant of proportionality is the viscous constant of the fluid (η)

$$\sigma = \eta \frac{d\varepsilon}{dt} \tag{2}$$

2.1.1 Maxwell viscoelasticity model

This model considers that the viscoelastic model of a polymer is given by the series union of a spring and a piston. An ideal elastic element is represented by a spring that obeys Hooke's law [5, 6], with a modulus of elasticity ξ.

Since there is a series coupling of both elements, the total deformation of the assembly will be the sum of the elastic deformation independent of time (ε_1) and the viscous component (ε_2) that is dependent on time, according to the Eq. (3), see **Figure 2a** and **b**.

$$\varepsilon = \varepsilon_1 + \varepsilon_2 \tag{3}$$

On the other hand, the stresses when connected in series are equal according to the Eq. (4).

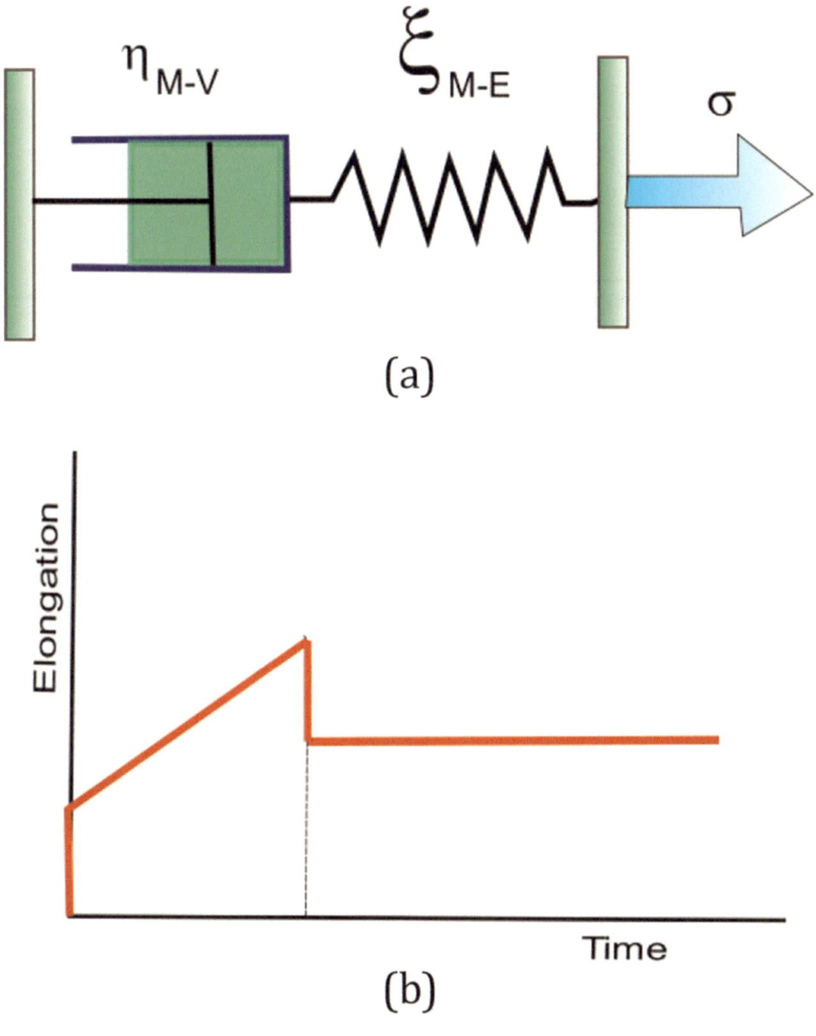

Figure 2.
(a) Diagram of Maxwell's viscoelastic model. (b) Series coupling of the elastic element (spring) and viscous element (plunger).

$$\sigma = \sigma_1 = \sigma_2 \tag{4}$$

If the time variable for the deformations is considered, the Eqs. (3) and (4) become:

$$\frac{d\varepsilon_1}{dt} = \frac{1}{\xi}\frac{d\sigma_1}{dt} \tag{5}$$

$$\frac{d\varepsilon_2}{dt} = \frac{1}{\eta}\sigma_2 \tag{6}$$

Deriving the Eq.(3) respect to time:

$$\frac{d\varepsilon}{dt} = \frac{d\varepsilon_1}{dt} + \frac{d\varepsilon_2}{dt} \tag{7}$$

$$\frac{d\varepsilon}{dt} = \frac{1}{\xi}\frac{d\sigma_1}{dt} + \frac{1}{\eta}\sigma_2 = \frac{d\varepsilon}{dt} = \frac{1}{\xi}\frac{d\sigma}{dt} + \frac{\sigma}{\eta} \tag{8}$$

If it is considered that a constant effort is applied, the expression is as indicated in Eq. (9):

$$\frac{d\varepsilon}{dt} = \frac{\sigma_0}{\eta} \tag{9}$$

Integrating the previous Eq. (9), considering that the immediate response in the elastic spring corresponds to the constant tension, we have:

$$\varepsilon = \frac{\sigma_0}{\eta}t + \frac{\sigma_0}{\xi} \tag{10}$$

It can be seen in **Figure 3** that a viscoelastic element that works at constant tension will have an immediate deformation due to its elastic component and an increasing linear deformation, due to the viscous response of the polymer.

The Maxwell model is useful when it comes to predicting instantaneous elastic deformation; however, when it comes to viscous deformation over time, it does not fit reality, since this curve is not linear.

2.1.2 Kelvin-Voigt model

In this model of viscoelasticity, the behavior of a polymer is considered as the parallel union of a piston and a spring [3, 4], as shown in the **Figure 4**.

In this model, when applying the tension, part of the energy will be stored by the spring and the rest will be slowly dissipated when the viscous element (piston) moves, resulting in a total deformation that depends on time [5, 6]. When the load is no longer applied, the original shape of the spring will recover, but not the piston, see diagram of **Figure 5**.

According to this model, the total tension applied will be equal to the sum of the tensions of the spring and that of the piston, see Eq. (11). The total deformation will be equal to the deformation of the spring and that of the piston, see Eq. (12), since they are in parallel.

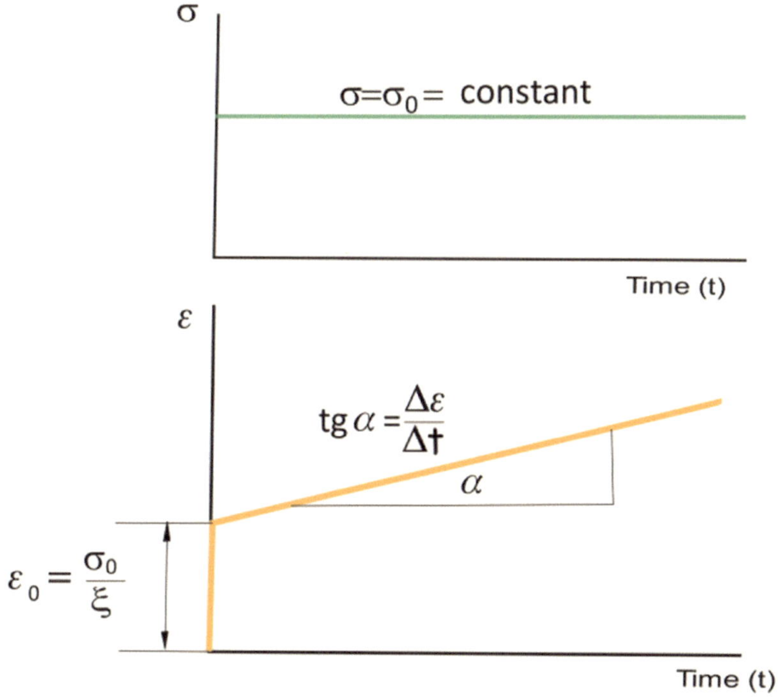

Figure 3.
Representation of the deformation with respect to time according to the Maxwell model under the action of a constant stress.

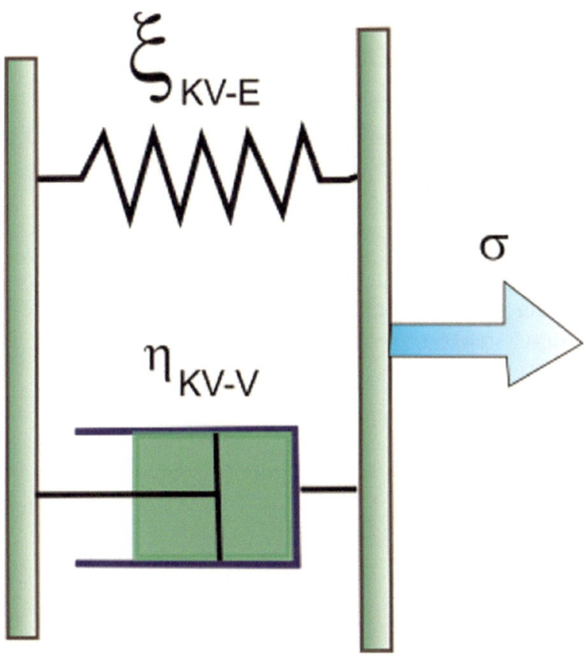

Figure 4.
Diagram of the kelvin-Voigt viscoelastic model in which there is a parallel coupling of the elastic element and the viscous element.

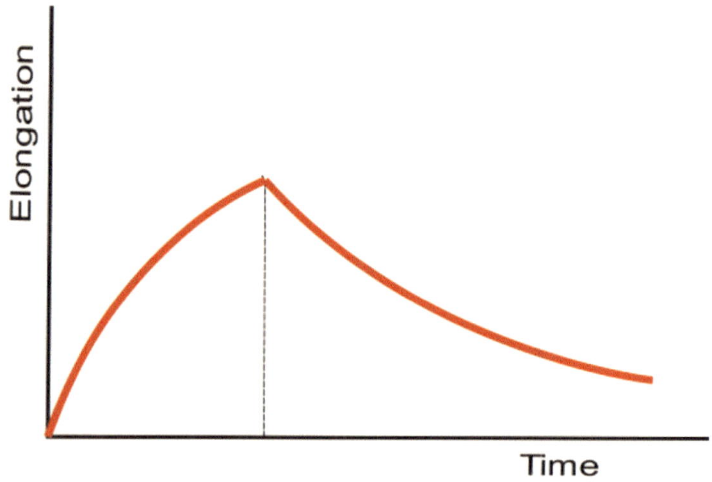

Figure 5.
Diagram of the elongation that a viscoelastic element undergoes over time that complies with the kelvin-Voigt model.

$$\sigma = \sigma_1 + \sigma_2 \tag{11}$$

$$\varepsilon = \varepsilon_1 = \varepsilon_2 \tag{12}$$

Considering the addition of stresses, we have the general expression of the Kelvin-Voigt model:

$$\varepsilon = \frac{\sigma_0}{\eta} t + \frac{\sigma_0}{\xi} \tag{13}$$

$$\sigma = \xi\varepsilon_1 + \eta\frac{d\epsilon_2}{dt} = \xi\varepsilon + \frac{d\varepsilon}{dt} \tag{14}$$

When dealing with a long-term phenomenon, we have that: $\sigma = \sigma_1\sigma_0$, and the solution of the differential Eq. is given by the form:

$$\varepsilon = \frac{\sigma_0}{\xi}\left[1 - e^{\frac{-\xi}{\eta}t}\right] \tag{15}$$

The behavior of the polymer applying this model is shown in the **Figure 6**, in this it is observed that when a constant tension is applied to the material, it experiences a

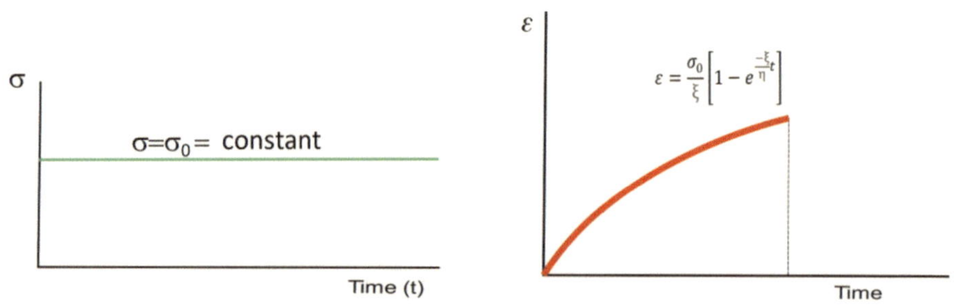

Figure 6.
Representation of the exponentially deformation with respect to time according to Maxwell model.

progressive elongation exponentially. However, this does not square with reality, since at time 0, there is an instantaneous deformation.

2.1.3 Combined model (burgers)

In general terms, the Kelvin-Voight model Eq. satisfactorily explains a real behavior such as creep, which the Maxwell model does not consider. On the other hand, the Kelvin-Voight model does not explain instantaneous deformation (**Figure** 7), which the Maxwell model does and with a good approximation.

A good, more correct approximation that has been obtained to simulate the viscoelastic behavior of polymers is to use a combined model that considers the Kelvin-Voigt and Maxwell models and adjusts more to the real behavior, which is applicable from the point of view engineering [7]. In this model, the two models are coupled in series, according to the **Figure 8**, where the first component corresponds to the Maxwell model and the second component corresponds to the Kelvin-Voigt model.

Identifying the subscripts M to the Maxwell model, KV to the Kelvin-Voigt model, E to the elastic component of the spring, and V to the viscous component of the piston, the above equations become:

Maxwell's equation:

$$\varepsilon_M = \frac{\sigma_0}{\eta_{M-V}} t + \frac{\sigma_0}{\xi_{M-E}} \tag{16}$$

Kelvin-Voigt equation:

$$\varepsilon_{KV} = \frac{\sigma_0}{\xi_{KV-E}} \left[1 - e^{\frac{-\xi_{KV-E}}{\eta_{KV-V}}.t}\right] \tag{17}$$

If the two models are considered in series, the total elongation of the system is equal to the sum of each of the Maxwell and Kelvin-Voigt systems individually. Therefore, the expression that defines the yield in the combined model is:

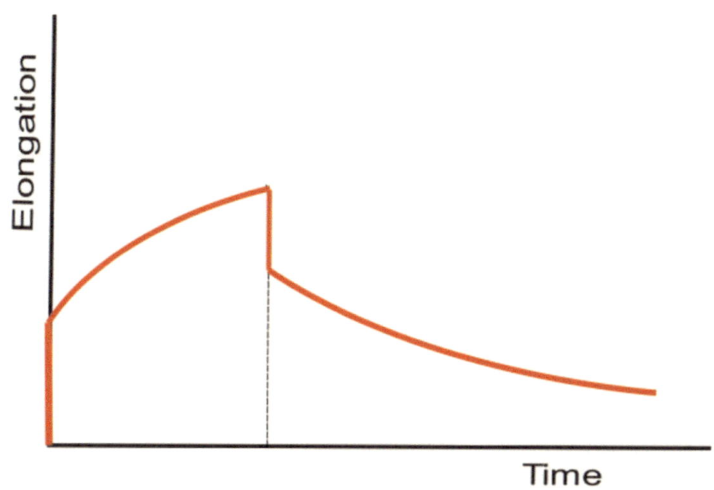

Figure 7.
Schematic representation of the actual deformation that a polymer undergoes, in which the instantaneous deformation, the deformation and the viscous recovery of the material are observed.

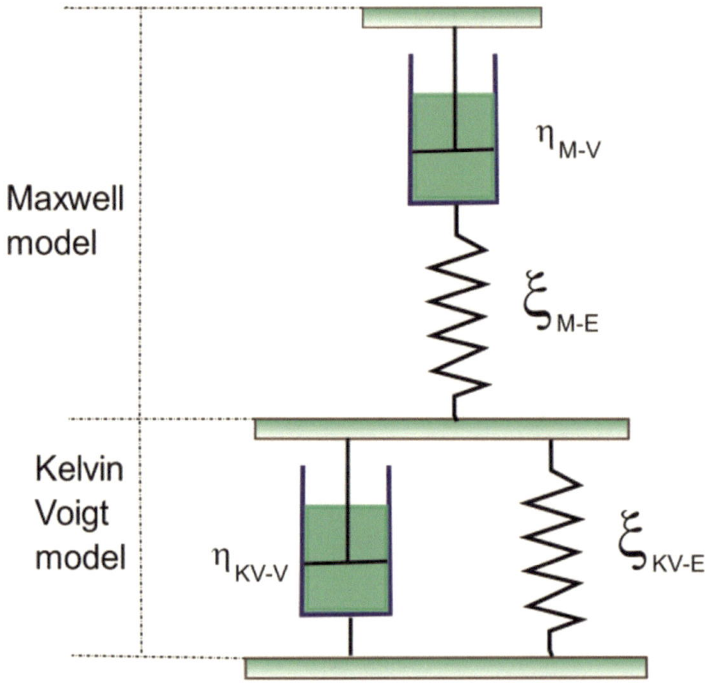

Figure 8.
Scheme of the combined viscoelastic model (Burguers).

$$\varepsilon_{\mathrm{TOTAL}} = \varepsilon_{\mathrm{M}} + \varepsilon_{\mathrm{KV}} \tag{18}$$

$$\varepsilon_{\mathrm{TOTAL}} = \frac{\sigma_0}{\eta_{\mathrm{M-V}}}\mathrm{t} + \frac{\sigma_0}{\xi_{\mathrm{M-E}}} + \frac{\sigma_0}{\xi_{\mathrm{KV-E}}}\left[1 - \mathrm{e}^{\frac{-\xi_{\mathrm{KV-E}}}{\eta_{\mathrm{KV-V}}}.\mathrm{t}}\right] \tag{19}$$

Each of the terms is defined as:
ξ_{M-E}: Elastic constant of the spring in Maxwell's element.
η_{M-V}: Viscous constant of the piston in Maxwell's element.
ξ_{KV-E}: Elastic constant of the spring in the Kelvin-Voigt element.
η_{KV-V}: Viscous constant of the piston in the Kelvin-Voigt element.
See **Table 1**.

The combination of simple moldings such as Maxwell and Kelvin-Voigt allows to obtain models that are more adjusted to the real viscoelastic behavior of polymeric materials.

$$\varepsilon_{\mathrm{M}} = \frac{\sigma_0}{\xi_{\mathrm{M-E}}} + \frac{\sigma_0.\mathrm{t}}{\eta_{\mathrm{M-V}}} \tag{20}$$

$$\varepsilon_{\mathrm{KV}} = \frac{\sigma_0}{\xi_{\mathrm{KV-E}}}\left[1 - \mathrm{e}^{\frac{-\xi_{\mathrm{KV-E}}}{\eta_{\mathrm{KV-V}}}}\right] \tag{21}$$

3. Polymer rheology

To carry out the correct modeling and simulation of the extrusion process through a nozzle for printing by MEX, it is necessary to precisely know the viscosity of the material, which depends on the physical parameters to which it is subjected [8, 9].

DOI: http://dx.doi.org/10.5772/intechopen.109630

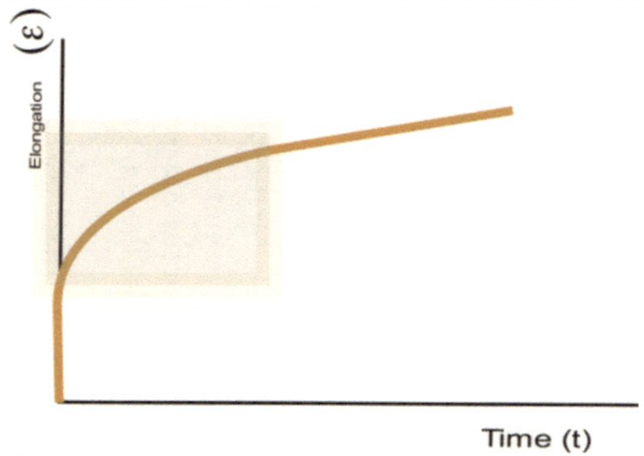 ξ_{M-E}: Elastic constant of the spring in Maxwell's element. Interpretation: Defines the initial deformation in the response of the combined model	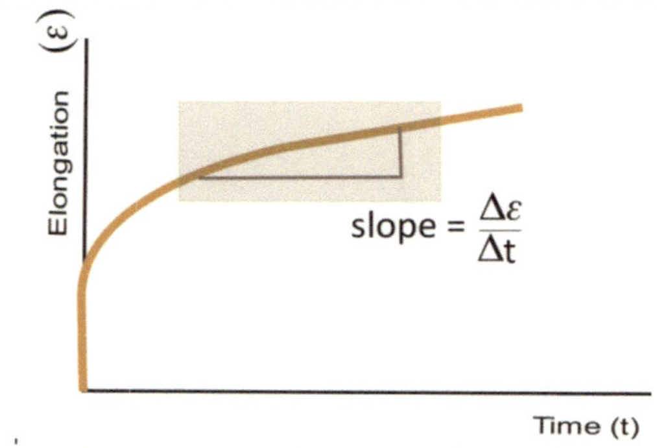 η_{M-V}: Viscous constant of the piston in Maxwell's element Interpretation: Defines the slope of the linear yield zone after stabilization
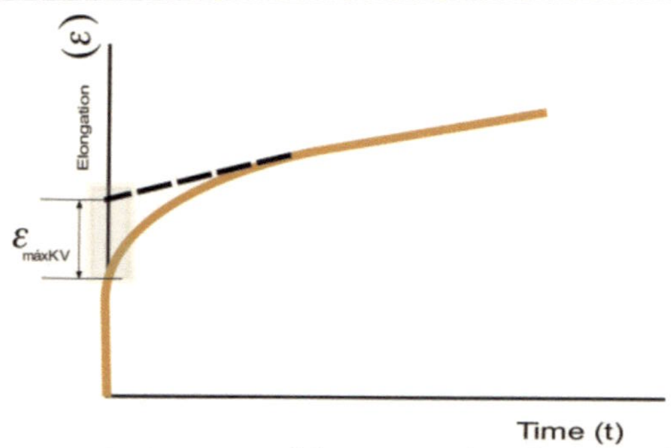 ξ_{KV-E}: Elastic constant of the spring in the Kelvin-Voigt element Interpretation: Defines the slope of the linear yield zone after stabilization	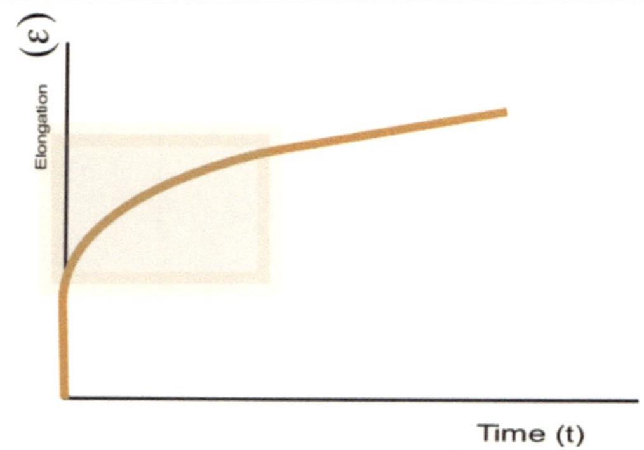 ξ_{KV-E}: Elastic constant of the spring in the Kelvin-Voigt element Interpretation: Defines the growth rate of the exponential

Table 1.
Interpretation of the parameters of the kelvin-Voigt model in the creep curve of the material.

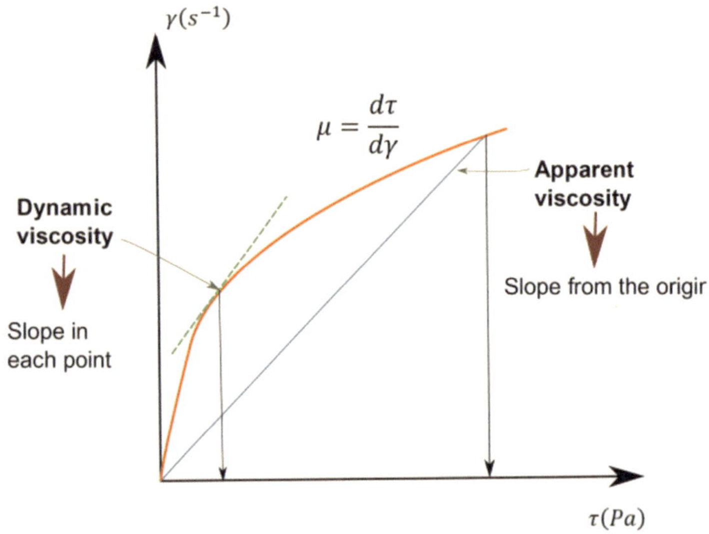

Figure 9.
Diagram of the relationship between dynamic viscosity and kinematic viscosity.

There are three types of viscosity: dynamic viscosity, kinematic viscosity, and apparent viscosity. The dynamic or absolute viscosity (μ) represents the internal resistance between the molecules of a moving fluid and determines the forces that move and deform it. The kinematic viscosity (ν) relates the dynamic viscosity to the density of the fluid used and represents the resistance of a fluid to sliding. Instead, the apparent viscosity "η" is defined as the ratio between the shear stress (τ) and the strain rate (γ) for fluids with nonlinear behavior (non-Newtonian). If the fluidity curve is plotted (shear force vs. strain rate), it is also defined as the slope at each point on the curve, see **Figure 9**.

In **Figure 10**, the required shear stress is represented as a function of the shear speed to be reached for different liquids. In a Newtonian fluid, the slope of the curve (n) is 1, while in a dilatant fluid ($n > 1$), the viscosity increases with the shear strain rate. For some fluids such as polymer melts, some paints, and fluids with suspended particles, the viscosity decreases with increasing shear stress ($n < 1$), these fluids are called pseudoplastic or shear-thinning.

Thermoplastic polymers under low or no shear conditions behave like a Newtonian fluid, while under high shear conditions they behave like a pseudoplastic, decreasing rapidly with increasing shear rate [10], as can be seen in the example of polypropylene shown in **Figure 11**.

3.1 Rheology for 3D printing

It is necessary to know if the designed plastic will be extrudable by the MEX method, to avoid performing many time-consuming and expensive empirical tests [12], as well as avoiding nozzle clogging, setting the parameters of the 3D printing machine, as well as avoiding phenomena such as contraction and adhesion to the printing table.

To simulate plastic extrusion through the die, you can use the rheology data and compare it with a plastic that extrudes correctly and compare the rheological curves, which will show, for example, shear rate and shear stress [13, 14].

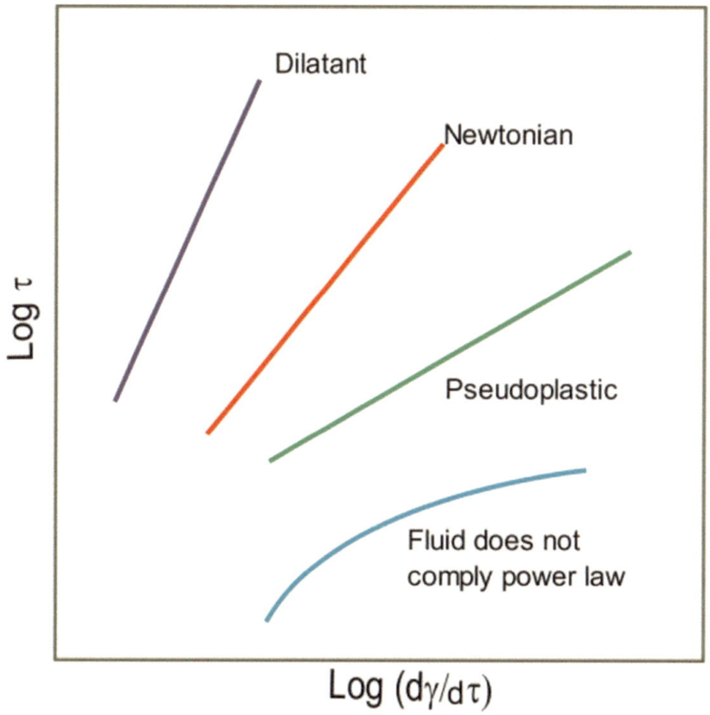

Figure 10.
Power law schematic showing logτ versus log(dγ/dτ) for different types of fluids.

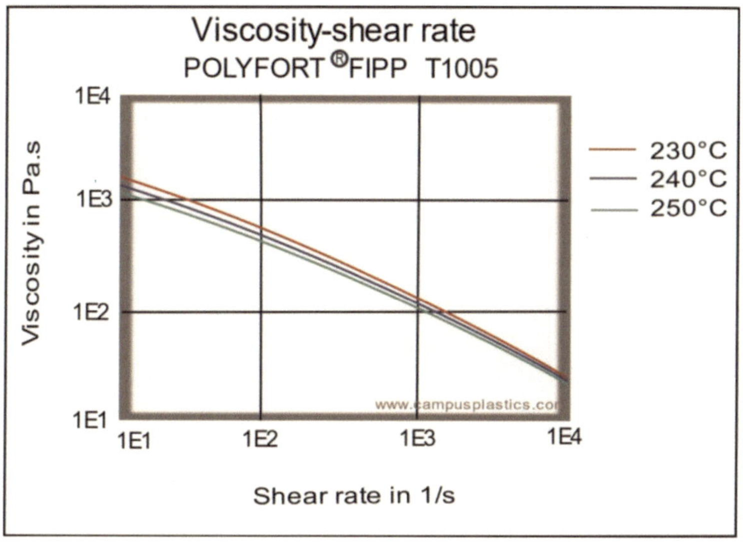

Figure 11.
Viscosity graph of PP, POLYFORT® FIPP 30 T K1005 at 3 temperatures [11].

The cutting rate can be determined by:

- making an initial shear rate measurement of the melt viscosity to obtain the power law index.

- know the printing speed and nozzle diameter.
- apply the rheological Eqs. (1)–(4).

3.2 Viscosity measurement

One of the most important characteristics to predict the behavior of an extrusion or injection process is the viscosity of the melted material. There are several methods for obtaining viscosity as a function of shear rate [15]. Instruments used to measure shear rates must shear the fluid at measurable rates, and the stress developed must be known, for which a rotational viscometer or capillary rheometer can be used [16]. For the tests of this work, a capillary rheometer was used. Extrusion pressure or volumetric flow rate can be controlled as the independent variable, and the other is the measured dependent variable [17] .

3.2.1 Rotational rheometer

In this rheometer, the fluid is sheared at a given temperature between an annular space due to the rotation of a coaxial internal cylinder inside another cylinder or by the rotation of a conical plate on a stationary plate or vice versa [18]. Rotational viscometers using two coaxial cylinders measure low viscosities of liquids, see **Figure 12a**. In a plate-and-cone viscometer, the polymer is contained between the lower plate and the cone, which rotates at a constant speed (Ω), see **Figure 12b**. These viscometers are used for viscosities less than 10 s^{-1}. This viscometer is expensive equipment, it gives information at the molecular level, and it only serves to give information about the fluid in the Newtonian regime.

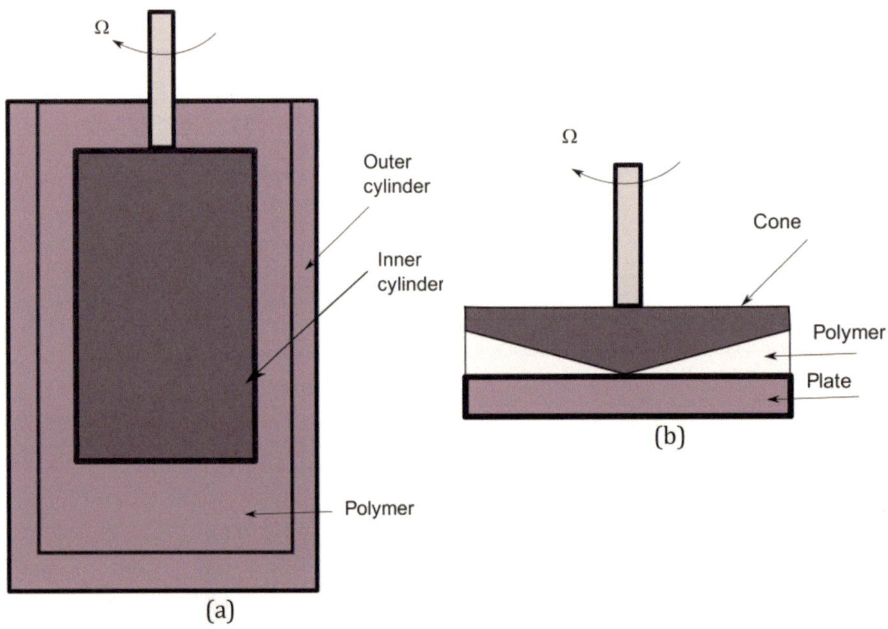

Figure 12.
a) Diagram of a coaxial cylinder viscometer. b) Diagram of a plate and cone viscometer.

3.2.2 Capillary tube rheometer

There are three main reasons why the capillary rheometer is widely used in the plastics industry [19]:

the shear rate and flow geometry in the capillary rheometer are very similar to the conditions actually encountered in extrusion and injection modeling;

a capillary rheometer typically covers the wider shear rate ranges ($10^{-6}\ s^{-1}$–$10^{6}\ s^{-1}$);

a capillary rheometer provides good practical data and information on matrix swelling, melt instability, and extrudate defects.

In this equipment, the fluid is forced to pass from a container through a small diameter hole or capillary in a nozzle, by mechanical or pneumatic actuators or pistons [12]. The fluid is kept at a constant temperature due to the use of electric heating resistances (**Figure 13**).

Under constant flow and isothermal conditions for an incompressible fluid, the viscous force resisting the movement of a column of fluid in the capillary is equal to the applied force tending to move the column in the direction of flow, then:

$$\tau = \frac{R\Delta P}{2L} \tag{22}$$

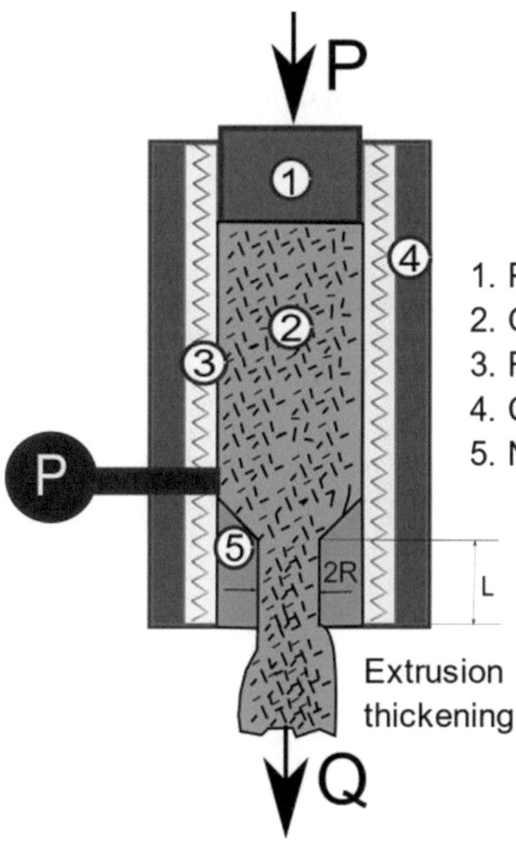

Figure 13.
Diagram of a capillary rheometer.

where,
R is the radius of the column.
Lis the length of the column.
ΔP is the pressure drop across the capillary.
τ is the shear stress.

According to the above, the shear stress τ is maximum on the cylinder walls and is zero in the center. For the calculations the maximum shear stress will be used.

In normal capillary rheometry, the molten material vents to the atmosphere, and the driving static pressure in the reservoir is taken as ΔP. In such cases, end effects involving viscous and elastic deformations at the capillary inlet and outlet must be considered when calculating the actual shear stress in the capillary wall, particularly if the ratio of capillary length to radius (L/R) is small.

For a fluid that exhibits Newtonian behavior, the shear rate at the wall is given by:$\dot{\gamma}$

$$\dot{\gamma} = \frac{4Q}{\pi R^3} \tag{23}$$

where Q is the volumetric flow rate through the capillary due to pressure ΔP, then the viscosity of the melt can be expressed as:

$$\eta = \frac{\tau}{\dot{\gamma}} = \frac{\pi R^4 \Delta P}{8LQ} \tag{24}$$

Values measured by capillary rheometers are often presented as plots of shear stress versus shear rate at certain temperatures. These values are called the apparent shear stress and the apparent shear rate at the tube wall.

3.3 Bagley's correction

This correction is applied due to the overpressure that occurs when going from a large cylinder (container or charge cylinder) to a small one (nozzle). The Bagley method is considered in which the overpressure is evaluated by relating it to an apparent increase in the length of the nozzle [20]. The total pressure drop ΔP_{tot} is given by the sum of the pressure drop in the reservoir, at the inlet ΔP_e, in the capillary ΔP, and at the outlet, due to the swelling of the polymer at the outlet. The pressure drop in the capillary is required to make the calculations and the pressure drop in the reservoir and the pressure drop at the outlet due to swelling can be neglected [21].

$$\Delta P_{tot} = \Delta P_e + \Delta P \tag{25}$$

For two or three capillaries, measurements are made of the total pressure drop as a function of the volumetric flow, for which the piston moves at different speeds, so that the flow varies, with the corresponding pressure variation, see **Figure 14**. As the flow increases, the cutting speed increases. With greater capillary length, there is a greater pressure drop.

For the same flow, the pressure drop for the three capillaries is taken. For each flow value, a minimum of 2 L/D points are obtained, and straight lines are made for each flow. In our case we have three capillaries (**Figure 15**). In total, seven flow data were taken as a function of piston speed (**Figure 16**).

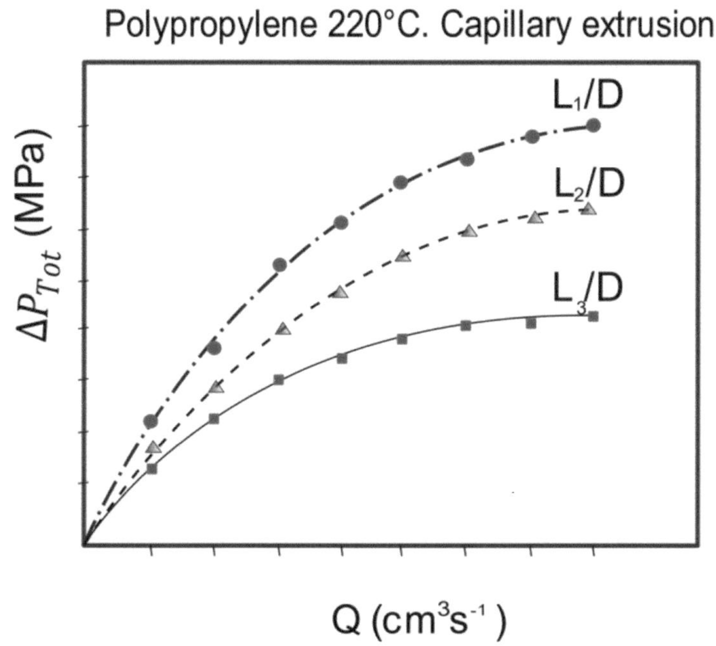

Figure 14.
Pressure drop curve (ΔP_{Tot}) versus volume flow (Q) for three nozzles (3 L/D ratios).

The determination of the correction value is carried out with a test in which at least three nozzles with different values of the L/D ratio are used.

$$\tau_{\text{corr}} = \frac{\Delta \text{P}}{2\left(\frac{\text{L}}{\text{R}} + \text{e}\right)} \tag{26}$$

where,

τ_{corr} is the corrected shear stress in [Pa].

e is the additional apparent length of capillary measured at a given shear rate measured in [mm].

In order to determine, which is an empirical constant that tries to correct the effects of exit and entrance of material in the molten state in the capillary, it is extrapolated to $\Delta P = 0$, from the representation of ΔP versus L/D at constant shear rate and for capillaries of different lengths.

Extrapolating means that the L/D ratio is 0, that is, there is no change in diameter. When it intercepts in ΔP, the value of pressure drops at the inlet (ΔP_e) is obtained, then from Eq. 9.24 we have:

$$\Delta \text{P} = \Delta \text{P}_{\text{Tot}} - \Delta \text{P}_{\text{e}} \tag{27}$$

For each of the straight lines and for the three points, the shear stress is calculated corrected $\tau_{corr.}$

The apparent shear stress is evaluated with:

$$\tau_{\text{app}} = \frac{\Delta \text{P} \cdot \text{D}}{4 * \text{L}} \tag{28}$$

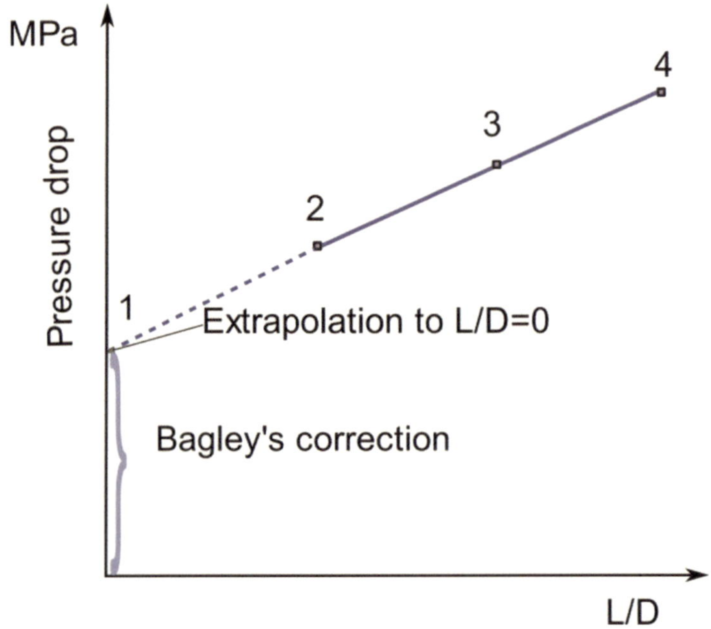

Figure 15.
Illustration of determining the value of the Bagley correction.

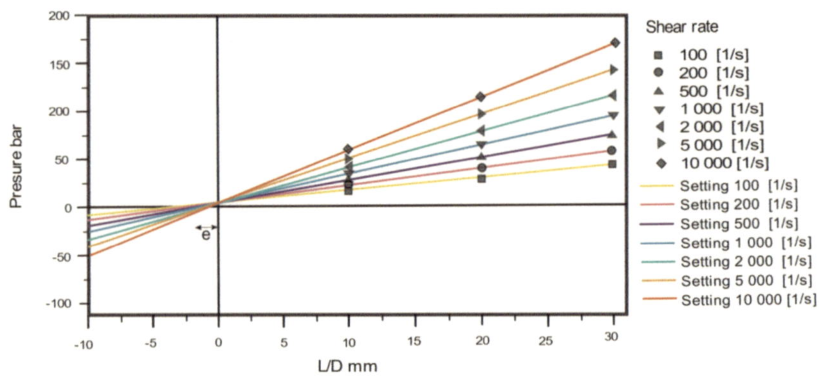

Figure 16.
Bagley plot: Pressure drop vs. L/D for different shear rates [22].

Corrected shear stress is as follows:

$$\tau_{corr} = \left(\frac{-\Delta P}{L + e \cdot D}\right) \cdot \frac{D}{4} \tag{29}$$

3.4 Weissenberg: rabinowitch correction

This correction is applied to the shear rate, since a molten plastic does not behave like a Newtonian fluid that presents a parabolic velocity distribution [23], but rather in a pseudoplastic manner, presenting a non-parabolic velocity distribution [24], this is shown in the Eq. (30)

$$\gamma_{corr} = \frac{\left(3 + \frac{1}{n}\right)\gamma_{app}}{4} \tag{30}$$

where,
γ_{corr} is the corrected shear rate in [s-1];
n is the slope of the relationship between shear rate and shear stress;
γ_{app} represents the apparent value of the shear rate on the tube wall;

$$n = \frac{d\log\tau_{app}}{d\dot{\gamma}_{app}} \tag{31}$$

n is 1 for a Newtonian fluid.

3.5 Polymer rheological models for 3D printing using Cross-WLF parameters

Cross-Williams-Landel-Ferry (WLF) model or Cross-WLF model [25] uses the results of the experimental data obtained in a capillary rheometer to describe the rheological behavior of polymeric materials, which allows the determination of their viscosity, to evaluate their processability under different conditions [26]. Within the manufacturing processes, the polymer is subjected to various shear conditions, which need to be evaluated and predicted [27]. Under conditions of low or practically zero shear, the viscosity of the material usually remains constant, presenting a Newtonian behavior. On the contrary, under high shear conditions, the viscosity decreases rapidly with the shear rate, showing a pseudoplastic behavior [28].

This model has been chosen because it can predict with a very good approximation the pseudoplastic behavior of the melt at high shear rates and the Newtonian behavior at low shear conditions [29], see **Figure 14**. This viscosity model is used in computer-aided engineering (CAE) programs such as Autodesk Moldflow™ and Ansys PolyFlow™ [30] for simulation of the plastic injection process [31, 32], because it offers the best fit to the viscosity data [33]. This model also simplifies the calculations of the pseudoplastic region, favoring the interpretation of results by considering its slope linear on a logarithmic scale [34].

This viscosity model describes the dependence of viscosity as a function of temperature (K), the shear rate $T_m(\gamma(s^{-1})$ and pressure p (Pa), which are the model dependent parameters.

The viscosity according to this model is expressed as:

$$\eta\ (\dot{\gamma}, T_m, p) = \frac{\eta_0(T_m, p)}{1 + \left(\frac{\eta_0(T_m, p)}{\tau^*}\dot{\gamma}\right)^{(1-n)}} \tag{32}$$

where,
η is the viscosity of the molten material in [Pa.s];
η_0 is the zero shear viscosity or the 'Newtonian limit' where the viscosity approaches a constant at very low shear rates [Pa.s];
τ^* is the constant of the model that indicates the shear stress from which the pseudoplastic behavior of the material begins, determined by fitting the curve [Pa] with $K = \frac{\eta_0}{\tau^*}n$ is the index of the power law in the high shear rate regime, determined by the curve fit that symbolizes the slope of the pseudoplastic behavior in the form of $(n - 1)$, see **Figure 17**.

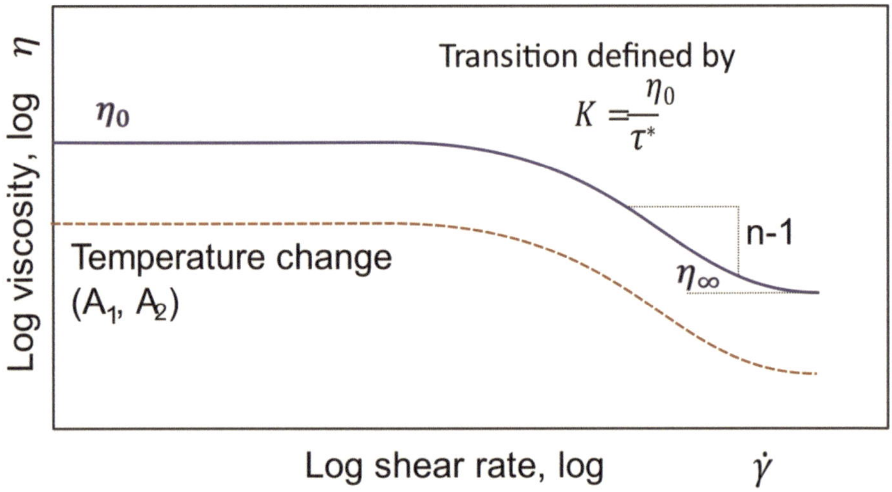

Figure 17.
Approximation of the viscosity with the Cross-WLF model in Eqs. (32) and (33) [35].

$\dot{\gamma}$ is the apparent shear rate in $[s^{-1}]$.

This model is complemented by the Williams-Landel-Ferry (WLF) model. Bagley [20] which helps to determine the behavior of the material with respect to null shear phenomena and offers reliable results. His expression is:

$$\eta_0(T_m, p) = \begin{cases} D_1 \bullet e^{\left(\frac{-A_1 \bullet (T_m - \tilde{T})}{A_2 + (T_m - \tilde{T})}\right)}, \text{ si } T_m \geq \tilde{T} \\ \infty, \text{ si } T_m < \tilde{T} \end{cases}$$

$$A_2 = \tilde{A_2} + D_3 \bullet p$$

$$\tilde{T} = D_2 + D_3 \bullet p \tag{33}$$

where:

$\tilde{T}$ is the glass transition temperature of the material in [K], which depends on the pressure.

D_1 is the model constant that indicates the viscosity under zero shear conditions at the glass transition temperature of the material and atmospheric pressure, in [Pa.s].

D_2 is a model constant indicating the glass transition temperature of the material at atmospheric pressure, in [K].

D_3, constant of the model that indicates the variation of the transition temperature of the material as a function of pressure, in [K/Pa].

$A_1, \tilde{A_2}$, are model constants, in [−], [K], respectively.

*pe*s the pressure in [Pa].

To use this model, its seven parameters can be determined based on estimates given by various authors [36] or by analysis of the characteristics of the materials.

4. Conclusions

It is important to know the viscoelastic and rheological properties of the materials for 3D printing by MEX, since this will allow to avoid performing an error trial process, which is long and expensive, being able to perform the process through the necessary parameters to simulate the process through computer-aided engineering.

The rheological parameters of a polymer or composite material for manufacturing by extrusion can be determined by using a capillary rheometer, due to its high cutting speeds and shear stresses.

Bagley and Rabinowitch corrections can be applied to the results obtained through the capillary rheometer to compensate the change in diameter in the capillary, and these results can be exported to simulation programs by computer to verify the technical feasibility of manufacturing with the composite material.

When the Cross WLF model is applied, we have a good approximation of the rheological properties of the material to any pressure and temperature condition, which allows us to extrapolate the results to adjust the properties in a 3D printing process by MEX.

Acknowledgements

We thank the Central University of Ecuador for their support with time and funds to carry out this study.

Conflict of interest

The author declares no conflict of interest.

Appendix 1

Terms and symbols

A_1 *[−]*	Constant of the WLF Cross model
A_2 *[K]*	Constant of the WLF Cross model
D [Pa·s]	Capillary diameter
D_1 *[Pa·s]*	Material viscosity, under zero shear conditions, at material transition temperature and atmospheric pressure
D_2 *[K/Pa]*	Constant of the Cross-WLF model symbolizing the variation of the transition temperature of the material as a function of pressure
ΔP*[bar]*	Constant of the Cross-WLF model symbolizing the variation of the transition temperature of the material as a function of pressure
ΔP_e*[bar]*	Inlet pressure drop (Bagley's correction)
ΔP_{tot}*[bar]*	Total pressure drop (Bagley's correction)
e [Pa·s]	Additional apparent length (Bagley's correction)
ξ_{KV-E} *[MPa]*	Elastic spring constant in the Kelvin-Voigt element
ξ_{M-E} *[MPa]*	Elastic spring constant on Maxwell's element
$\dot{\gamma}$ *[s^{-1}]*	Shear rate at which the material is being processed
γ_{app} *[s^{-1}]*	Apparent shear rate

η *[Pa/s]*	Viscosity
η_{app} *[Pa/s]*	Apparent viscosity
η_{KV-V} *[Pa/s]*	Viscous plunger constant in the Kelvin-Voigt element
η_{M-V} *[Pa/s]*	Viscous plunger constant in Maxwell's element
η_0 *[Pa/s]*	Material viscosity under zero shear conditions
L [mm]	Rheometer nozzle length
μ*[Pa·s]*	Dynamic or absolute viscosity
n [−]	Cross Model: Slope of the pseudoplastic behavior of the material
$\nu[m^2/s]$	Kinematic viscosity
P [Pa]	Pressure
$Q\ [m^3/s]$	Volumetric flow rate through the capillary due to pressure
R [mm]	Radius of capillary viscometer column
σ *[MPa]*	Applied stress
T [°C]	Temperature
T_m *[K]*	Material temperature during the Cross-WLF model process
τ *[Pa]*	Shear stress
τ_{corr} *[Pa]*	Corrected shear stress
τ^* *[Pa]*	Shear stress at which the pseudoplastic behavior of the material starts
τ_{app} *[Pa]*	Apparent shear stress
Ω *[rad/s]*	Angular velocity of the inner cylinder of a rotational viscometer

Author details

Jorge Mauricio Fuentes Fuentes
Central University of del Ecuador, Quito, Ecuador

*Address all correspondence to: jmfuentes@uce.edu.ec

References

[1] Bachhar N, Gudadhe A, Kumar A, Andrade P, Kumaraswamy G. 3D printing of semicrystalline polypropylene: Towards eliminating warpage of printed objects. Bulletin of Materials Science. 2020;1-8. DOI: 10.1007/s12034-020-02097-4

[2] Hajikarimi P, Moghadas Nejad F. Mechanical models of viscoelasticity. In: Applications of Viscoelasticity. 2021. vol. 1. pp. 27-41. DOI: 10.1016/b978-0-12-821210-3.00003-6

[3] Varna J, Pupure L. Characterization of Viscoelasticity, Viscoplasticity, and Damage in Composites. 2nd ed. London, UK: Elsevier; 2019. DOI: 10.1016/B978-0-08-102601-4.00016-3

[4] Mainardi F, Spada G. Creep, relaxation and viscosity properties for basic fractional models in rheology. European Physical Journal: Special Topics. 2011;**193**(1):133-160. DOI: 10.1140/epjst/e2011-01387-1

[5] Gutierrez-Lemini D. Engineering viscoelasticity. Engineering Viscoelasticity. Vol. 1. 2014:1-353. DOI: 10.1007/978-1-4614-8139-3

[6] Fombuena V, Boronat L, Sánchez-Nácher L, García-Sanoguera D, Balart R. "Utilidad de los modelos de viscoelasticidad en el aprendizaje de la ingeniería de materiales poliméricos," Modelling in Science Education and Learning. Jan 2017;**10**(1):137. DOI: 10.4995/msel.2017.6315

[7] Hajikarimi P, Moghadas Nejad F. Mechanical models of viscoelasticity. In: Applications of Viscoelasticity. London, UK: Elsevier; 2021. pp. 27-61. DOI: 10.1016/b978-0-12-821210-3.00003-6

[8] Mackay ME. "The importance of rheological behavior in the additive manufacturing technique material extrusion." Journal of Rheology. Nov 2018;**62**(6):1549-1561. DOI: 10.1122/1.5037687

[9] Polychronopoulos ND, Vlachopoulos J. Polymer Processing and Rheology. In: Jafar Mazumder M, Sheardown H, Al-Ahmed A, editors. Functional Polymers. Polymers and Polymeric Composites: A Reference Series. Springer, Cham. 2019. DOI: 10.1007/978-3-319-95987-0_4

[10] Irgens F. Linearly Viscoelastic Fluids. In: Rheology and Non-Newtonian Fluids. Springer, Cham. 2014. DOI: 10.1007/978-3-319-01053-3_7

[11] CAMPUSplastics | datasheet POLYFORT® FIPP 30 T K1005. https://www.campusplastics.com/campus/en/datasheet/POLYFORT%C2%AE+FIPP+30+T+K1005/LyondellBasell/103/21818245/SI?pos=2 [Accessed: June 17, 2021]

[12] Prabhu R, Devaraju A. Recent review of tribology, rheology of biodegradable and FDM compatible polymers. Materials Today: Proceedings. 2020;**39**:781-788. DOI: 10.1016/j.matpr.2020.09.509

[13] Picco S et al. Polymeric Additive Manufacturing: The Necessity and Utility of Rheology. London, UK, London, UK: IntechOpen; 2016. p. 13 no. tourism [Online]. Available: https://www.intechopen.com/books/advanced-biometric-technologies/liveness-detection-in-biometrics

[14] Carvalho C, Bom RP, Joinville C. de & Whirpool SA. Propriedades reológicas de abs e suas misturas oriundas de reciclagem primária. Anais Do 10o Congresso Brasileiro de Polímeros. Vol. 1. 2009:10

[15] Maysaa M, Rahamtalla E, Deen H. Viscosity Measurement by using Melt flow Index for Thermoplastic polymers. Sudan University of Science and Technology. 2014. Available from: http://repository.sustech.edu/handle/123456789/9099

[16] Morrison FA. Rheometry CM4650 Chapter 10: Rheometry. 2018. pp. 1–56

[17] Nikzad M, Masood SH, Sbarski I, Groth A. Rheological properties of a particulate-filled polymeric composite through fused deposition process. Materials Science Forum. 2010;**654–656**: 2471-2474. DOI: 10.4028/www.scientific.net/MSF.654-656.2471

[18] ASTM D 4402. Standard test method for measuring the viscosity of Mold powders above their melting point using a rotational viscometer. Control. 1999; **94**, no. Reapproved:5-7. DOI: 10.1520/D4402

[19] Osswald T, Rudolph N. "Rheometry," in Polymer Rheology. Munich, Germany: Carl Hanser Verlag GmbH & Co. KG; 2014. pp. 187-220. DOI: 10.3139/9781569905234.006

[20] Bagley EB. End corrections in the capillary flow of polyethylene. Journal of Applied Physics. 1957;**28**(5):624-627. DOI: 10.1063/1.1722814

[21] Dealy JM, Wang J. Methods Melt Rheology and its Applications in the Plastics Industry. Montreal, Canada: Springer; 2013. pp. 149-152 [Online]. Available: http://www.springer.com/series/4604

[22] Montanes N et al. Modelización reológica mediante Cross-WLF de un nuevo material compuesto elaborado con bioPE y Thyme. Vol. 2. pp. 22–27, [Online]. Available: http://revista.aemac.org/

[23] Pranata dkk. Study of rheological. Thermal and Mechanical Behavior of Reprocessed Polyamide. 2013;**6**(53): 679-688

[24] M. Dees, M. Mangnus, N. Hermans, W. Thaens, A. S. Hanot, and P. van Puyvelde, "On the pressure correction of capillary melt rheology data." Rheologica Acta. Feb 2011;**50**(2):117-124. DOI: 10.1007/s00397-011-0529-2.

[25] Cross MM. Rheology of Non-Newtonian Fluids: A New Flow Equation for Pseudoplastic Systems. Journal of Colloid Science. 1965;**20**:417-437. DOI: 10.1016/0095-8522(65)90022-X

[26] Ferrándiz S, Arrieta MP, López J, Navarro R. Demostració práctica de la validesa dels models matematics en elements finits. Aplicaci'o al model de Cross. Modelling in Science Education and Learning. 2013;**6**(3):67-82

[27] Osswald T. Rudolf N. Polymer Rheology: Fundamentals and Applications. Munich, Germany: Hanser, 2015

[28] Irgens F. Rheology and non-newtonian fluids. In: Rheology and Non-Newtonian Fluids. 2013;**9783319010**:1-16. DOI: 10.1007/978-3-319-01053-3

[29] Reig MJ, Segui VJ, Zamanillo JD. Rheological behavior modeling of recycled ABS/PC blends applied to injection molding process. Journal of Polymer Engineering. 2005;**25**(5): 435-457. DOI: 10.1515/POLYENG.2005.25.5.435

[30] Canonsburg TD. ANSYS POLYMAT user 's guide. Knowledge Creation Diffusion Utilization. 2012;**15317** (October):724-746

[31] Debbaut B. Rheology: from Process to Simulation. 2005;**13**:23-36

[32] Acedo J. Caracterización y simulación del comportamiento viscoelástico de materiales plásticos mediante el Método de Elementos Finitos. Valencia; Spain: Universidad Politécnica de Valencia; 2019

[33] Sanchez LC, Augusto C, Helena S, Bettini P, Costa LC. Rheological approach for an additive manufacturing printer based on material extrusion. The International Journal of Advanced Manufacturing Technology. Vol. 105. 2019. pp. 2403–2414. DOI:10.1007/s00170-019-04376-.

[34] Piotr M. Numerical and Experimental Analysis of Filament-based Material Extrusion Additive Manufacturing Ph.D. thesis, Technical University of Denmark DTU, Mechanical Engineering Section of Manufacturing Engineering. 2020

[35] Ferri D, Perolo A, Nodari M. Cross-WLF parameters to predict rheological properties of polylactic acid. Annual Transactions of the Nordic Rheology Society. 2017;**25**(1):419-426

[36] Passaglia E, Martin GM. Variation of Glass Temperature With Pressure in Polypropylene. Journal of Research of the National Bureau of Standards. Section A, Physics and Chemistry. 1964 May-Jun;**68A**(3):273-276. DOI: 10.6028/jres.068A.024. Epub 1964 Jun 1. PMID: 31834721; PMCID: PMC5327688

Chapter 9

3D Printing for Tissue Regeneration

Meghana Kasturi, Vidhi Mathur, Prachi Agarwal, Varadharajan Srinivasan and Kirthanashri S. Vasanthan

Abstract

Tissue engineering is an interdisciplinary field and 3D bioprinting has emerged to be the holy grail to fabricate artificial organs. This chapter gives an overview of the latest advances in 3D bioprinting technology in the commercial space and academic research sector. It explores the commercially available 3D bioprinters and commercially printed products that are currently available in the market. It provides a brief introduction to bioinks and the latest developments in 3D bioprinting various organs. The chapter also discusses the advancements in tissue regeneration from 3D printing to 4D printing.

Keywords: 3D bioprinting, bioink, tissue engineering, regenerative medicine, 4D bioprinting, scaffold

1. Introduction

Tissue engineering is a branch of biomedical engineering that focuses on repairing and/or replacing diseased and damaged organs. This is done primarily via developing artificial organs using natural or synthetic materials. Organ shortage is a severe problem worldwide due to the non-availability of donors and tissue engineering strategies enable to produce a scaffold that mimics the organ of interest [1]. Three-dimensional (3D) bioprinting has great potential in this field and was developed in the early 1990s, and has evolved ever since. It is an additive manufacturing (AM) technique that uses computer-aided design (CAD) models to deposit biomaterials on the substrate along with living cells, extracellular matrix (ECM) components, biochemical cues, and drugs. The three basic steps in the 3D bioprinting process are: (i) preprocessing - includes developing CAD models to develop in-vitro scaffolds or to develop organ blueprints from imaging modalities such as computer tomography (CT) and magnetic resonance imaging (ii) processing - produces a physical structure that mimics the organ/tissue of interest from the designed model (iii) postprocessing - improves the bioprinted organ model and scope for transplantation if required. Over recent years, there has been a huge demand and interest in 3D bioprinting due to its potential to produce high-throughput biomimetic organ scaffolds. Several technological advancements have come up in 3D bioprinting which are mentioned in **Figure 1.** The goal of 3D bioprinting is to provide alternative approaches to autologous and allogeneic implant treatments and avoid animal testing in drug studies and disease models. 3D bioprinting has several biological applications in the fields of

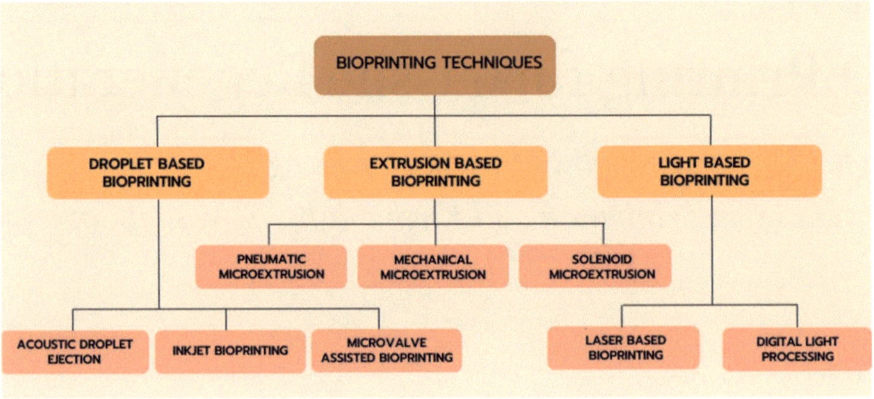

Figure 1.
Various classifications of 3D bioprinting.

tissue engineering, materials science, pharmaceutical drug development and validation, cosmetics testing, personalized medicine, regenerative medicine, cell-based biosensors, and bionics.

2. Commercial 3D bioprinters

A 3D bioprinter is an automated device that enables the development of functional tissue and organ models. The 3D bioprinting technology is generally classified into three types – (i) droplet-based bioprinters (ii) extrusion-based bioprinter (iii) light-based bioprinter (**Figure 1**). Extrusion-based bioprinters are widely used and are based on the principle of depositing the material layer by layer. Laser-based 3D bioprinters deposit the bioink drop by drop, the principle is like an inkjet 3D bioprinter. Some companies and universities have developed 3D bioprinting technologies that cannot be easily classified into widely known technologies. For example, Cyfuse Biomedics has developed a technique where cells are 3D printed on a needle array. A scaffold is not required in this method instead only a cluster of cells (not mixed with other biomaterials) are skewered onto vertical needles to fabricate 3D tissue structures. Companies like rainbow biosciences have developed a bioprinter called BiOassay where biocompatible magnetic nanoparticles are used to print the 3D structures and use the working principle of magnetic levitation [2]. There are many emerging bioprinting technologies being developed by researchers across the world to make the process more efficient and cost-effective. Currently, the 3D-printed organs can be used for research only; however, in future, they can be transplanted into human patients.

The wide range of applications has driven many companies/universities to develop bioprinting technology. The following are the types of business models utilized by these companies that exploit bioprinting technology – (i) Manufacturing bioprinters (ii) Providing bioprinting services (iii) providing cell therapies that utilize bioprinting technology. Commercially available 3D bioprinters have increased in the market over the past decade and have rapidly advanced the tissue engineering field. The 3D Bioprinting Market is expected to reach USD 3261.31 Million by 2027, from USD 796.9 Million in 2020 growing at a compound annual growth rate of 22.3% during

2021–2027 [3]. The growth of this market is due to a limited number of organ donors, and an increase in the aging population with chronic diseases. The rise in R&D investment in this sector, advancement in commercially available products, and increment in the incidence of chronic diseases are other vital factors that are likely to boost market growth during the coming years [4]. **Table 1** provides a list of commercially available bioprinters in the market.

3. Formulation of bioinks

In bioprinting, cells are placed at user-defined coordinates, along with biomaterials that are either (i) mixed with cells before printing, or (ii) printed simultaneously with one print head while the cells are deposited via the other print head (**Figure 2A** and **B**). Materials used in bioprinting that contain cells in the mixture are termed as bioink. The biomaterial is usually a polymer (natural or synthetic) that has biocompatible components and provides favorable rheological properties for the desired organ of interest. Hydrogels are the most used bioinks. However, hydrogel precursors are widely in use these days as they can be cross-linked into hydrogels post-biofabrication. Another method is to crosslink the precursor solution to obtain a viscous ink, followed by crosslinking the scaffold post-printing [5].

An ideal bioink should have the desired physicochemical properties to print mechanically stable scaffolds which mimic the organ of interest (**Figure 2C**). These properties are determined by the mechanical strength of the scaffold, viscosity of the ink, chemical structure of the polymer, and biological characteristics of the desired tissue. These properties should lead to: (i) mechanically stable scaffolds, that have the mechanical strength similar to the native tissue (ii) adjustable rheological properties (gelation, viscosity) to help in ease of bioprinting the constructs while retaining the desired structural shape (**Figure 3**); (iii) biocompatibility, biodegradability if required; (iv) not be cytotoxic to be suitable for in vivo studies and possible transplantation in future; and (v) large scale reproducibility of the ink [6]. Optimizing the bioink formulation is a vital step toward successful bioprinting and this is represented in **Figure 4** in the form of a flowchart. **Table 2** provides a few examples of different polymers that have been used in the formulation of bioinks.

4. 3D bioprinting for hard tissues

4.1 Bone

Bone is a complex tissue that has mechanical, metabolic, and hemopoietic functions. It provides structure and function to the surrounding tissues. Currently, bone defect repairs are treated by grafts: Autologous grafts, allografts, and synthetic grafts. Alternative methods like cadaver allografts and xenografts are available but they have poor biological properties like lower biocompatibility and risk of infection. Osteoconductive properties are seen in synthetic grafts but are degraded by osteoclasts and hence are suitable for small defect repairs only. Bone tissue engineering offers solutions to treat bone defects and one effective way is via 3D bioprinting. However, a major concern is to provide a solution that overcomes the challenges faced in conventional treatments by improving osteoinduction and osteoconduction. Studies have shown that 3D bioprinted bone constructs avoid the possibility of immune rejection which was

Sl. No	Company name	Model	Technology	Country	Website
1	Cellink	INKREDIBLE™ Series, BIO CELLX, BIO X™, BIO X6™	Extrusion based	Sweden	https://www.cellink.com/
		BIONOVA X, LumenX+™, Quantum X bio	Light-based		
2	Allevi	Allevi (1,2,3)	Extrusion-based	United States	https://www.allevi3d.com/
3	Advanced Solutions Life Sciences	BioBotBasic, BioAssemblyBot (200,400,500)	Extrusion-based	United States	https://www.advancedsolutions.com/
4	RegenHu	R-GEN (100,200)	Extrusion-based	Switzerland	https://www.regenhu.com/
5	Rokit Healthcare	Dr.INVIVO (4D2, 4D6)	Extrusion-based	South Korea	https://rokithealthcare.com/?ckattempt=1
6	Fluicell	Biopixlar, Biopixlar AER	Extrusion-based	Sweden	https://fluicell.com/
7	Envision Tec	3DBioplotter (started, developer, manufacturer) series	Extrusion-based	United States	https://etec.desktopmetal.com/
8	Inventia Life Science Operations	RASTRUM™	Digital bioprinting	Australia	https://inventia.life/
9	3D bioprinting Solutions	Fabion, Fabion 2	Extrusion-based	Russia	https://bioprinting.ru/
10	Poietis	NGB-R Bioprinter	Extrusion-based, laser-assisted, micro-valve bioprinting	France	https://poietis.com/
11	Organovo	NovoGen Bioprinter®	Extrusion-based	United States	https://organovo.com/
12	nScript	The BAT Series	Extrusion-based	United States	https://www.nscrypt.com/
13	Cyfuse Biomedics	Regenova	Extrusion-based	Japan	https://en.cyfusebio.com/
14	SunP biotech International	BIOMAKER, ALPHA-CPT1, ALPHA-BP11	Extrusion-based	United States	http://sunpbiotech.com/

Sl. No	Company name	Model	Technology	Country	Website
15	Next Big Innovation Labs	TRIVIMA	Extrusion-based	India	https://nextbiglab.com/
16	Axolotl Biosystems	Axo A3, Axo A6	Extrusion-based	Turkey	https://www.axolotlbio.com/
17	Brinter	Brinter®One	Extrusion-based	Finland	https://www.brinter.com/
18	GeSiM	BS5.3/E, BS5.3, BS3.3, BS3.3 Prime	Extrusion-based	Germany	https://gesim-bioinstruments-microfluidics.com/
19	Regemat 3D	BIOV1, REG4LIFE	Extrusion-based	Spain	https://www.regemat3d.com/
20	CLECELL	U-BIOLET™, U-BIOXT	Extrusion-based; laser-assisted	South Korea	https://www.clecell.co.kr/
21	UpNano	NanoOne Bio	multiphoton lithography	Austria	https://www.upnano.at/

Table 1.
List of commercially available 3D bioprinters.

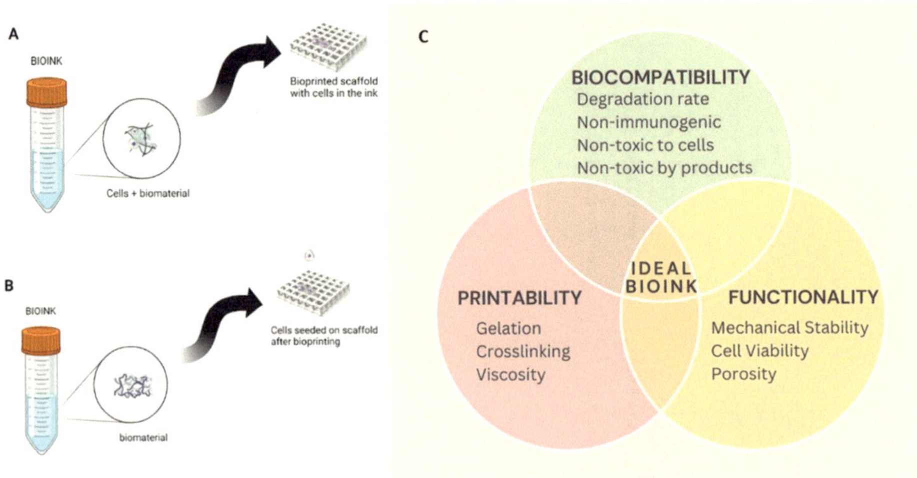

Figure 2.
(A) Bioprinted scaffolds with cells in ink (B) cells seeded on scaffolds after bioprinting (C) properties to consider for the formulation of an ideal bioink.

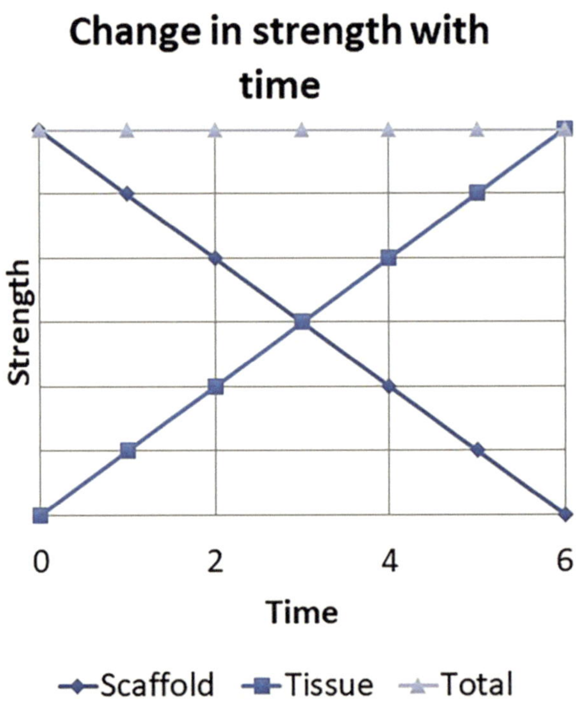

Figure 3.
Ideal strength and degradation profile of a bioprinted scaffold.

observed earlier in the use of grafts which may have otherwise led to inflammation, fibrosis, scarring, and transplant failure. The advantage of 3D bioprinting over the current grafting technique is that the cells are spatially distributed within the construct, thus optimizing tissue regeneration. 3D-bioprinted bone tissues have a huge impact

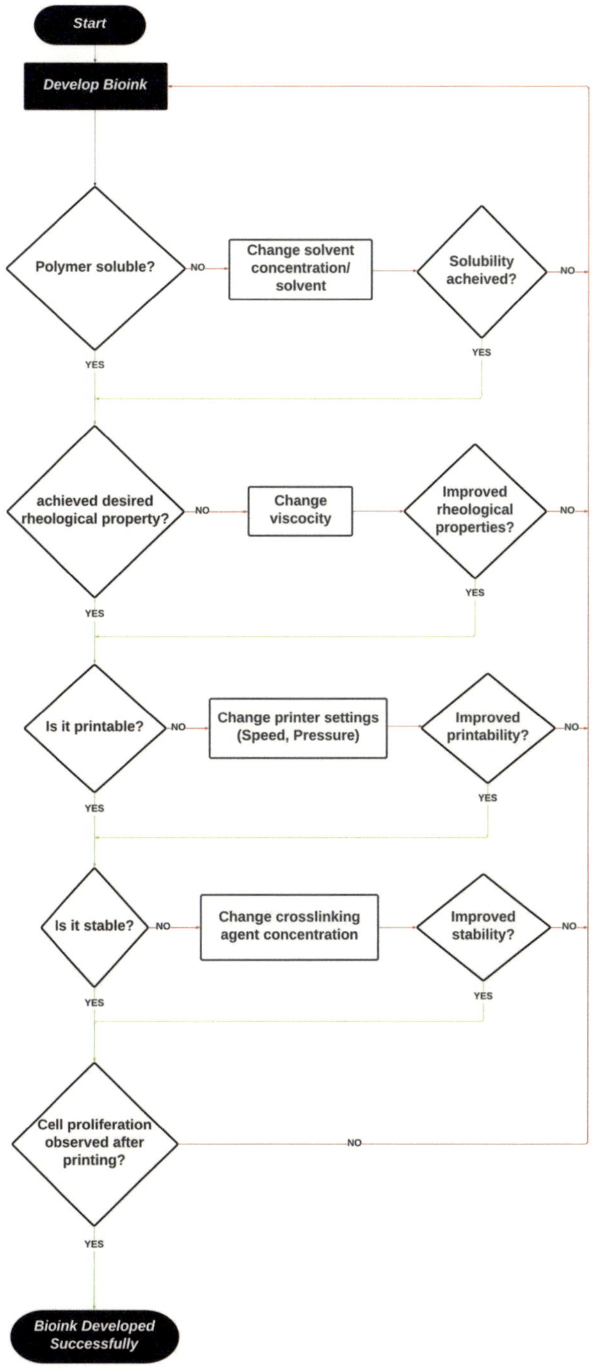

Figure 4.
Flowchart depicts the process of bioink development.

on clinical practice as it makes reconstruction of bone defects with complex shapes efficient and less time-consuming, by translating defect data from image modalities like CT to CAD designs which make it possible for patient-specific bioprinting. The ideal

Polymer origin	**Constituent materials**	**Constituents of polymers**
Natural Polymers	Polysaccharides	HA[1], dextran, chitosan, agarose, Alginate
	Decellularized Extra Cellular Matrix	dECM[2]
	Proteins	Collagen, lysozyme, silk, Matrigel™, gelatin, fibrin
Synthetic polymers	Biodegradable synthetic polymers	PCL[3], PLA[4], PLGA[5], methacrylated HA, GelMA[6]
	Non-biodegradable synthetic polymers	HEMA[7]
	Bioactive synthetic polymers	Proteins, peptides, carbohydrates, PLA
Hybrid polymers	Synthetic polymer, modified natural polymers	Pluronic® 127/carboxymethyl hexanoyl chitosan
	Synthetic peptide-modified proteins or polysaccharides	Glucose, Gluconic acid
	[9]PNIPAAm modified polymers	PNIPAAm/Collagen, PNIPAAm/Chitosan, and PNIPAAm/Alginate
	[10]PEG-modified natural polymers	Heparin, dextran, HA, fibrinogen, and albumin, HAP[8]

[1]*HA- Hyaluronic acid.*
[2]*dECM - decellularized extra cellular matrix.*
[3]*PCL-Polycaprolactone.*
[4]*PLA -poly lactic acid.*
[5]*PLGA- poly(lactide-co-glycolide).*
[6]*GelMA-methacrylated gelatin.*
[7]*HEMA- 2-Hydroxyethyl methacrylate.*
[8]*HAP- Hydroxyapatite.*
[9]*PNIPAAm- Poly(N-isopropylacrylamide).*
[10]*PEG- Polyethylene glycol.*

Table 2.
Polymers used in bioinks.

scaffold should mimic the bone structure and composition, have a good resorption rate, allow for vascularization, and have a higher bone healing/formation ability compared to ceramics and metals. Fabrication of bone constructs with various geometries, porosity, and sizes, which are specific to each patient's features is possible via 3D bioprinting. It also helps to fabricate osteoinductive scaffolds [7].

Bioink is necessary to bioprint bone and it should have good mechanical strength without losing cell viability and bioactivity. Bioinks can be classified in three categories- (i) first generation – materials that are bioinert and biocompatible. Chances of rejection are minimized in this case. The scaffold remains in vivo to provide mechanical support and does not degrade, e.g., metals (stainless steel and titanium) and polymers (ii) second generation – materials that are biocompatible and bioactive simultaneously. They allow mineralization and biodegradation over time so that the cells can replace the scaffold. (iii) third generation – bioresponsive materials. They contain growth factors and stimulatory molecules that trigger osteoblast differentiation including bone morphogenetic proteins (BMP) and fibroblast growth factors (FGF). A composite bioink is most beneficial for use since it combines the best of all three generation of bioinks i.e., a balance between mechanical and functional properties is obtained to meet the needs of the desired tissue [8].

Bone regeneration requires osteoinductive cues which include growth factors such as vascular endothelial growth factor (VEGF), fibroblast growth factors (FGF), bone morphogenetic proteins (BMPs), parathyroid hormone, and platelet-derived growth factor (PDGF). Growth factors are delivered in 3D-printed bone scaffolds during or after the printing process. The biochemical factors can be delivered on already printed scaffolds by adding them on the top surface after printing or within the micropores during printing. These osteogenic factors help in enhanced cell proliferation, differentiation, and angiogenesis. Bioinks were formulated using BMP-2 loaded PLGA nanoparticles, alginate, and mesenchymal stem cell (MSC). Composite bioink showed enhanced printability and yielded stable constructs post-printing. Sustained in vitro release for up to two weeks was observed in BMP2-loaded PLGA nanoparticles and was also noted to induce osteogenesis of the MSCs [9]. Bone healing is linked to the relationship between blood vessels and bone cells. It is known that VEGF is released during fracture healing. VEGF inhibition has been shown to interfere with fracture repairs and bone defects. However, it was not sufficient to heal large defects. Furthermore, it was observed that VEGF did not drive progenitor cells toward the chondrogenic or osteogenic lineage. Hence, combination therapies with BMPs are being developed to advance the regeneration of large bone defects. VEGF and BMP-2 were delivered to enhance the regeneration of large bone defects. The release of these growth factors was studied by 3D bioprinting alginate-based bioinks with nanoparticles. The bone formation was enhanced in vivo by slowing the release of BMP-2. Enhanced vascularization was found in vivo when VEGF was introduced in the study. Accelerated large bone defect healing was observed in this case using 3D-printed implants containing VEGF and BMP-2 [10].

Osteoinductive materials are preferred for studies in bone regeneration as they enhance regenerative properties. A study has shown that the NICE bioink (7.5% methacrylated gelatin, 1% kappa carrageenan, 2% nano silicates) showed both high print performance and enzymatic degradability. This bioink provided cell friendly microenvironment for bone bioprinting [11]. Jakus et al. developed a bioink composed of 90% HAP and 10% PCL or PLGA. It was found that the bioprinted scaffolds promoted osteogenic differentiation without additional biochemical factors. Higher biocompatibility, tissue integration, vascularization, and mineralization were observed in animal studies with no immune rejection [12].

Cell adhesion, viability, and metabolism are influenced by a scaffold's internal architecture like pore size and shape which in turn affect the bone regeneration capacity. A major challenge in developing a scaffold is to obtain a balance between the mechanical strength of the scaffold and mimicking the native strength of the tissue. A comparative study was done in the internal architecture of the scaffold- (i) continuous pattern (ii) zigzag-Spiral pattern, in treating bone defects treatments. It was found that the printed scaffolds showed characteristics like that of a native bone – permeability, porosity, and mechanical properties which are owed to the microarchitecture of the scaffold design. Human mesenchymal stem cells seeded scaffolds determined the effects of geometrical microstructure on cell attachment and morphology. The cells in the scaffold with a zigzag pattern infilled pores quickly in comparison with the other pattern [13]. **Tables 3** and **4** outline the latest developments in 3D bioprinting of bone in-vivo and in-vitro studies respectively.

4.2 Teeth

Teeth are the hardest part of the human body and have limited capacity for repair and regeneration. The gold standard treatment for permanent tooth loss by disease, defect, or injury has been dental implants. Titanium alloys are widely used to

Sl. No	Animal model	Material	Printing method	Cells	Growth factors	Time (weeks)	References
1	Rat	Fibrin, GelMA, Y-irradiated RGD modified Alginate	Extrusion based	Human umbilical vein endothelial cells, human bone marrow stem cells	VEGF	2	[14]
2	Rat	HAP, PLGA	Extrusion based	[1]C3H/10 T1/2 cells, human patient-derived osteoblasts	—	12	[15]
3	Rat	GelMA, Calcium silicate nanowires	Extrusion based	Bone mesenchymal stem cells, Schwann cells	—	8	[16]
4	Rabbit	Silk fibroin	Extrusion based	Bone marrow-derived mesenchymal stem cells	BMP 2, [2]TGF-β	12	[17]
5	Rat	Calcium phosphate-based materials	Extrusion based	Rat mesenchymal stromal cells	—	12	[18]
6	Rat	β-tricalcium phosphate and osteogenic peptide (OP) containing water/PLGA/dichlo romethane emulsion inks	Extrusion based	Rat endothelial cells and rat bone marrow-derived mesenchymal stem cells	—	12	[19]
7	Rabbit	Sodium alginate, Gelatin, HAP	Extrusion based	Bone marrow mesenchymal stem cells	—	24	[20]
8	Rabbit	GelMA	Extrusion based	Bone marrow mesenchymal stem cells	—	12	[21]
9	Rabbit	Bioactive glass	Extrusion based	Bone marrow mesenchymal stem cells	—	12	[22]
10	Rat	GelMA/Gelatin/PEG/[3]MSN	Extrusion based	Bone marrow mesenchymal stem cells	BMP 4	3	[23]

[1]*C3H/10 T1/2 cells – Mouse embryo fibroblast cells.*
[2]*TGF- transforming growth factor.*
[3]*MSN- mesoporous.*

Table 3.
3D bioprinting for bone: in vivo studies.

Sl. No	Model	Material	Printing method	Cells	Growth factors	Time (weeks)	References
1	Square scaffold	Gelatin-nano HAP	Extrusion-based	Human mesenchymal stem cells, human umbilical vein endothelial cells	—	2	[24]
2	Square scaffold	Graphene oxide/alginate/gelatin	Extrusion-based	Human mesenchymal stem cells	—	6	[25]
3	Square scaffold	Y-irradiated RGD modified Alginate, GelMA, [1]PEGMA	Extrusion-based	Mesenchymal stem cells	—	4	[26]
4	Round scaffold	Alginate-[2]PVA-HAP hydrogel	Extrusion-based	[3]MC3T3-E1	—	2	[27]
5	Square scaffold	[4]PCL/bioactive borate glass composite	Extrusion-based	Human Adipose stem cells	—	1	[28]
6	Square scaffold	Wood-based nanocellulose and bioactive glass modified Gelatin/ Alginate	Extrusion based	Human bone marrow-derived mesenchymal stem cells	—	2	[29]
7	Round scaffold	Polypropylene fumarate-PEG-PCL	Extrusion based	—	—	—	[30]
8	Round scaffold	Octacalcium phosphate, GelMA	Digital Light Processing	C3H/10 T1/2cells, Human Umbilical vein endothelial cell spheroids	—	2	[31]
9	Square scaffold	PCL/polydopamine/HA	Extrusion based	MC3T3-E1	BMP 2	3	[32]
10	Square scaffold	Silk Fibroin-MA	Digital Light Processing	MC3T3-E1	—	2	[33]

[1]*PEGMA-Poly (ethylene Glycol) Methyl Ether Methacrylate.*
[2]*PVA- Poly Vinyl Alcohol.*
[3]*MC3T3- mouse preosteoblasts.*
[4]*PCL- polycaprolactone.*

Table 4.
3D bioprinting for bone: in-vitro studies.

manufacture dental implants. The tooth has multiple internal organs and its connection to surrounding tissues (periodontal ligament and alveolar bone) is important for its function. Hence, the entire tooth unit i.e., tooth root and the adjacent connecting tissues should be considered for tooth regeneration. The 3D printing technique is beneficial for tooth regeneration and building patient-specific supporting structures for teeth (e.g., dentures, dental implants, aligners etc). Dentistry applications of scaffolds are in the limelight of research with the aim of enhancing the regeneration of dental tissues. Bioprinting dental and periodontal tissues is a primary focus of research in dental regeneration [34].

In the native tooth, two mineralized tissues are present – dentin and enamel. Dentin provides strength and toughness while enamel is hard and resistant to both fracture and wear. Titanium alloys are commonly used implants, and they exceed the required strength and stiffness that is normally found in native teeth. This leads to alveolar bone resorption post-implantation. A study used collagen, agarose, and fibrin-based bioink to bioprint dental pulp. The study was successful in vascular tube formation at the root. Human dental pulp cells and Human primary umbilical vein endothelial cells were used in the bioink. Injecting the prepared bioink by the hand-held bioprinter in-vitro showed vascularization and proved to be effective in comparison to filling up the canals with inert materials and sacrificing the tooth [35]. Yi-Ting Lin et al., developed calcium silicate-reinforced gelatin methacrylate bioink and bioprinted dental pulp stem cells along the scaffolds for dental regeneration. The release of silicon ions from the scaffolds contributed to enhanced regenerative properties by upregulating the expression of various odontogenic-related biomarkers. It was also found that these increased calcium mineralization. The developed bioink enhanced the mechanical properties of the scaffold and contributed to increased regenerative properties [36]. Jonghyeuk Han et al., developed a new Demineralized Dentin Matrix particle (DDMp) bio-ink. This bioink is composed of human tooth-derived DDMp, fibrinogen–gelatin mixture, and dental cells. It was found that the DDMp bio-ink improved odontogenic differentiation [37]. **Tables 5** and **6** outline the latest developments in 3D bioprinting of teeth in-vivo and in-vitro studies, respectively.

5. 3D bioprinting for liver

Liver disorders like acute liver failure, chronic liver disease, liver fibrosis, viral hepatitis, and carcinoma have led to high mortality which requires liver transplantation [58]. 3D printing has been used as an alternative strategy to generate organs in vitro as being a shortage of organ donors [59]. 3D printed patient-specific liver models are being used and are showing great potential in disease treatment while the constructs having scaffolds and cells (bioprinted) are being used for the fabrication of liver tissue-like constructs and whole artificial livers [60].

It is very important to select the appropriate kind of cells and scaffold when considering 3D printing of liver tissues [61]. The viability of the hepatocytes reduces in vitro and there is a loss of hepatic phenotype [62]. Many studies have focused on liver regeneration using patient-specific functional cells and pluripotent stem cells. Valve-based inkjet bioprinting has been used by Faulker-jones et al. to print human induced pluripotent stem cells and human embryonic stem cells. The cells were able to differentiate into hepatocyte-like cells post-printing, and there were positive results for nuclear factor 4-alpha and albumin secretion. The cells were compatible to fabricate mini livers as drug testing models [63]. Primary

DOI: http://dx.doi.org/10.5772/intechopen.109141

Sl. No	Animal model	Material	Printing method	Cells	Growth factors	Time (weeks)	References
1	Rat	PCL/HA	Extrusion-based	—	[1]SDF-1, BMP-7	9	[38]
2	Rat	[2]PU, [3]POSS	Extrusion-based	MC3T3-E1	—	6	[39]
3	Rat	GelMA	Extrusion-based	Dental papilla cells, Hertwig's epithelial root sheath	—	8	[40]
4	Mouse	[4]PEGDA and sodium alginate composite	Stereolithography	Human dental pulp stem cells	FGF	4	[41]
5	Dog	HAP/PLA	Extrusion-based	Human dental pulp stem cells		40	[42]
6	Mouse	Alginate, Gelatin	Extrusion-based	Gingival fibroblasts	platelet-rich fibrin	8	[43]
7	Mouse	PCL/PGA	3D wax printing	Human primary gingival fibroblast cells	BMP-7	6	[44]
8	Rat	PCL	Inkjet-based	Human Periodontal Ligament Cells	—	6	[45]
9	Mouse	PCL/HA	Extrusion-based	Human dental pulp stem cells	—	6	[46]
10	Rat	PCL	Solid-free form fabrication method	Primary human periodontal ligament cells	BMP-7	3	[47]

[1]*PEGMA- Poly (ethylene Glycol) Methyl Ether Methacrylate*
[2]*PVA- Poly Vinyl Alcohol*
[3]*MC3T3- mouse preosteoblasts*
[4]*PCL- polycaprolactone.*

Table 5.
3D bioprinting for teeth: in vivo studies.

Sl. No	Model	Material	Printing method	Cells	Growth factors	Time (weeks)	References
1	Patient-specific	Bioglass, HAP, porcelain	Digital Light Processing	—	—	—	[48]
2	Patient-specific	PCL, Fibrin	Extrusion-based	Human dental pulp stem cells	—	2	[49]
3	Round scaffold	PCL/45S5 bioglass composite and PCL/hyaluronic acid	Extrusion-based	Human gingival fibroblast cells	—	3	[50]
4	Square scaffold	GelMA	Extrusion-based	Human dental pulp stem cells	BMP	3	[51]
5	Square scaffold	GelMA and HAP-magnetic iron oxide nanoparticles	Extrusion-based	Human osteoblasts and human periodontal ligament fibroblasts	—	1	[52]
6	Square scaffold	Alginate, dentin matrix	Extrusion-based	[1]OD 21 cells	—	1	[53]
7	Round scaffold	Poloxamer-407	Extrusion-based	Apical papilla stem cells	—	2	[54]
8	Round scaffold	PLGA, HAP and β-tricalcium phosphate	Extrusion-based	—	—	—	[55]
9	Square scaffold	Alginate, gelatin, nano-HAP	Extrusion-based	Bone marrow-derived mesenchymal stem cells, gingival fibroblast cells	—	1	[56]
10	Square scaffold	GelMA	Extrusion-based	Primary human periodontal ligament cells	—	2	[57]

[1]OD 21 cells – undifferentiated dental pulp cells.

Table 6.
3D bioprinting for teeth: in-vitro studies.

rat hepatocytes, HUVECs, and human lung fibroblasts were bioprinted by Lee et al. using multiple nozzle-based extrusion printing. Collagen-based bioink was mixed with the cells and a 3D construct was fabricated by infusing the bioink into PCL framework. There was enhanced survival and functionality of the HCs in the printed liver construct due to the 3D environment-induced interaction among cells. This study showed potential in the liver tissue regeneration field for the capillary-like networked 3D constructs [64].

Robbins et al., used iPSC-derived HLCs, and endothelial and hepatic stellate cells to fabricate highly reproducible 3D liver constructs. These constructs had high viability, multi-layered architecture, tissue-like cell density. There was improved reproducibility, durability, and biological complexity in a study conducted by Nguyen et al. the liver constructs were able to show more biological functions including storing lipids and glycogens and retaining their viability, and compartmentalized structure. Kim et al., used an alginate scaffold and primary mouse hepatocytes to fabricate liver constructs [65]. The cells were viable for 14 days and there was an increase in albumin, HNF-α. Zhong et al., fabricated 3D-printed hydrogel and they were implanted in mice in different groups acting as a control, hydrogel, hydrogel with cells, and hydrogel with hepatocyte growth factor. The viability of the cells was not affected by the hydrogel. The group implanted with cells showed significant improvement in levels of albumin, bilirubin, and the group with HGF, had the longest survival time [66].

6. 3D bioprinting for tubular organs

Tubular organs like Esophagus, blood vessels, urethra, etc. are very prone to infection and can be treated via surgery, stent insertion, and organ transplant that is dependent on suitable donors and autologous organs [67]. Tissue engineering has emerged as an alternative approach for developing grafts and scaffolds.

6.1 Blood vessels

Vascular systems are the most common tissue-engineered structures in the body. Development and discoveries have happened in the past years toward the fabrication of vascular networks in all organ systems. An arterial scaffold consisting of three layers of polydioxanone, fibrin, and gelatin was fabricated by Thomas et al. [68]. The Polydioxanone (PDS) layer provided mechanical integrity and the protein layers had a similar functional extracellular matrix as blood vessels. Nguyen et al., fabricated a tubular scaffold made up of PCL/PU using electrospinning for artificial blood vessels, which demonstrated improved cell attachment and proliferation [69].

6.2 Trachea

Tracheal disorders are rare but still life-threatening like tracheal stenosis and narrowing, and such disorders require immediate treatment. 3D printing of trachea constructs is gaining popularity in the field of regenerative medicine [70]. 3D bio-print tracheal constructs were fabricated by Taniguhi et al. using chondrocytes and mesenchymal cells [71]. Spheroids were fabricated and matured in a bioreactor; then as tracheal grafts transplanted in rats. Silicone stents were used as a framework to provide support and prevent collapsing of the stent. Vascular and epithelium networks were observed over the grafts thus successfully making a way in tracheal engineering. Goa et al., fabricated a porous PCL construct that would mimic the native trachea of rabbits [72]. The graft was cytocompatible as it was observed when seeded with chondrocytes. There was successful formation of cartilage tissue in the subcutaneous spaces of the mice. Later, it was transplanted into the rabbit, and the survival time was observed as 10 weeks. The fabricated scaffolds can be used for tracheal replacement therapies and for repairing whole-segment tracheal defects.

6.3 Excretory organs

Many organs in the excretory systems are hollow and tubular in morphology including Bowman's capsule, tubules in renal nephron urethra, etc. 3D printing has been utilized and applied in the fabrication of tissues and organs in this organ system. Zhang et al., fabricated cell-laden urethra using PCL and poly (lactide-co-ε-caprolactone) (PLCL) polymers having spiral scaffold design that could mimic the native properties of the urethra of rabbits [73]. Urothelial cells and smooth muscle cells of the bladder were added to the hydrogel comprising gelatin, Dulbecco's Modified Eagle Medium (DMEM) and hyaluronic acid, and the cell-laden hydrogel was fabricated. The urethra was 3D printed by adding PCL/PLCL polymers blend in one syringe and cell-laden hydrogels in another. The polymers provided with the structural framework and the cell-laden hydrogels contributed to mimic the microenvironment. It was observed that the scaffold had the mechanical properties equivalent to native rabbit urethra and the hydrogel was able to maintain a suitable microenvironment and the results set up a strong foundation for future studies on 3d bioprinting of urethra. Pi et al., used a coaxial extrusion-based printing technique to fabricate complex tubular hollow fibers which were made up of blend bioink consisting of PEG, and GelMA/alginate hydrogel [74]. The main objective of this study was to avoid the pre/post-processing step as the coaxial nozzle allows the printing of multiple layers in one step. The team was successfully able to print cannular urothelial tissue constructs using human urothelial cells and human bladder smooth muscle cells. This kind of fabrication is a fundamental step toward creating human cannular tissues.

6.4 Gastrointestinal tract

The esophagus is the tubular tube connecting throat to stomach. Many congenital and acquired disorders of GI tract have only esophageal replacement as the treatment option. 3D printed scaffolds are being considered to repair damaged esophagus. Esophageal reconstruction has been done using resorbable materials, acellular matrices, decellularized patches, and implants of synthetic polymers [70]. Pisani et al., fabricated a biodegradable patch using PLA-PCL polymer. Two different techniques- electrospinning and temperature-induced precipitation were used to develop the cellularized patch. The protocol was repeatable, reproducible, and simple [75]. Haghdel et al., fabricated a flexible esophageal stent to treat esophageal strictures using PLA, polyurethane, and Polyvinyl alcohol (PVA) [76]. The stent was assessed in vitro and in vivo canine esophagus. The stent was implanted in a 16-year-old boy who had esophageal stricture, and it was observed for 2 months. No major inflammatory effects and cytotoxicity were observed, and the mechanical tests revealed that the nature and behavior did not change significantly. This biocompatible polymeric stent can be used as an individualized treatment for treating esophageal structures.

7. Commercial 3D bioprinted products

Manufacturing companies have been using 3D printing for years, mostly to create product prototypes. Models and molds are produced by several manufacturers using huge and quick 3D printers referred to as "rapid prototyping machines" [77]. There are many.stl files that may be used for business. Many of these printed goods are comparable to those that are made traditionally [78]. There are now businesses that employ 3D printing for industrial medical purposes [79]. These include Organovo, Helisys, and Ultimateker, a business that creates living human tissue through 3D printing. The use of 3D printing in medicine, however, is still relatively new. The market value of 3D printing is $700 million out of which only 1.6% of it is devoted to medical uses. If we look at the numbers, it is anticipated that the market value of 3D printing will expand to be a sector of $8.9 billion in the next 10 years out of which 21% of it is estimated to go toward its usage in medical applications [78].

The democratization of product design and production is another advantageous aspect of 3D printing [80].

A significant shift has been made in the manner hearing aids are made, currently 99% of hearing aids that fit in the human ear are fabricated via 3D printers. Every individual has a unique ear canal shape, and 3D printers make it possible to build custom-shaped devices quickly and affordably [81]. Another productive commercial use of 3D printing is the production of 50,000 Invisalign braces per day. Each user's set of these transparent, removable, 3D-printed orthodontic braces is unique and is created to order. This item serves as an excellent illustration of how 3D printing can be utilized effectively and commercially to create unique, personalized, complex items [80].

In 2010, Organovo made its first noteworthy business using just primary human cells to successfully bioprint entirely functional blood arteries. The year 2014 saw the introduction of Organovo's ExVive™ 3D bioprinted human liver tissue models. There were histological and functional resemblances to the natural liver, and albumin, ATP, and CYP3A4 activity are consistently expressed for up to 28 days. Drugs like Valproic acid and Monocrotaline have their therapeutic effects demonstrated using tissue models [79].

Organovo released ExVive™ Human Kidney Tissue in 2016, a complete three-dimensional bioprinted human tissue made of primary renal fibroblasts and endothelial cells at the tubule-interstitial interface, which is rich in collagen IV, and polarized primary renal proximal tubule epithelial cells (RPTECs) in the apical layer [77]. ExVive™ Human Kidney Tissue displays in vivo-like renal transporter expression, barrier function, and the production of the crucial enzyme gamma glutamyl transferase (GGT). When subjected to the chemical Cisplatin, this bioprinted kidney tissue produces kidney damage indicators and shows transporter-dependent (OCT2) drug uptake [82]. The world's first animal thyroid gland was successfully 3D printed by 3D Bioprinting Solutions (3dbio) in March 2015 and then implanted into the mouse when it was alive. In addition, to create artificial tissues in the International Space Station using a magnetic 3D bioprinter, 3dbio has been collaborating with Russia's national space agency, United Rocket and Space Corporation (URSC). The company hopes to fabricate synthetic thyroid and kidney tissue using this technology [83].

The most recent RX1TM bioprinting from Aspect Biosystems makes use of their exclusive Lab-on-a-Printer™ microfluidic technology. Contains a coaxial flow-focusing system that guarantees the direct extrusion of biological fibers in a range of diameters. The device was utilized to show how to fabricate the 3DBioRingTM artificial airway. Primary human airway smooth muscle cells make up contractile smooth muscle tissue that lines the airway. When histamine is present, the airway tissue responds with proper and repeatable contractions, and when pharmacological stimuli are present, it dilates (B2-agonist) [84].

A new biotech company called BIOLIFE4D was established in 2015. The business hopes to 3D bioprint patient-specific, perfectly operational hearts for secure and reasonably priced organ transplantation. They are a strong group of biomedical researchers and businesspeople that are now supporting their research through equity crowdfunding. The goal of the BIOLIFE4D technique is to 3D bioprint a human heart using adult induced pluripotent stem cells (iPSCs), following a complete MRI (Magnetic Resonance Imaging) scan to determine the precise dimensions needed for its production [85].

Poietis makes use of INSERM and the University of Bordeaux technology. The business focuses on D laser-assisted bioprinting technology and collaborates with BASF and L'Oréal to develop bioprinted skin models and hair follicles, respectively [86]. Their NGB 17.03 bioprinting machine, which has an eight-axis motion, can print 3D models down to the level of a single cell. Early in 2018, Poietis introduced the first bioprinted human full-skin model made with their NGB bioprinter, called Poieskin® [87].

In collaboration with scientists at Sichuan University's West China Hospital, Revotek has had success implanting 3D-printed arteries within simian test subjects. In 30 rhesus monkeys, a replacement of a 2-centimeter portion of the abdominal artery was done with a 3D-printed blood conduit, and the stem cell bioink was created using the monkeys' own autologous adipose mesenchymal stem cells (ADSCs) [88]. Using a print head with two nozzles, the printer can presently manufacture 10-centimeter blood arteries in about two minutes [89].

TeVido biodevices make use of patented Clemson University technology. TeVido's initial product is a bioprinted nippular-areola implant for breast reconstructive surgery. In two to three years, clinical trials are expected to begin. The second product from TeVido is intended for Vitiligo sufferers who desire to print skin tissues to lessen the contrast in colors [90].

8. Advancements in 4D bioprinting

The fourth dimension (4D) which is 4D printing incorporates time, and it is an improved production method based on 3D printing. With this technique, external stimulation can cause the printed constructions to alter form over time. 4D bioprinting refers to the recent expansion of 4D printing to include the printing of complex constructions from biocompatible materials or even live cells. If one of the following criteria is met, 4D printing can be referred to as 4D bioprinting. 1) Biomedical engineering may make use of printed products, such as biomedical gadgets. 2) The printed materials are transplantable into the human body and are biocompatible. 3) The printed materials are loaded with living cells. When using 4D bioprinting, the bio constructs size, form, and/or functionality might vary over time [91].

The benefits that 4D printing has over 3D printing might prove to be the necessary proof of concept and accelerate wider adoption. More precisely, 4D printing enables the implementation of micro or nano actuators by providing sensation, knowledge of the movement, and programmability embedded into the material without any requirement for an external source or system like the batteries, wires, engines etc. Additional advantages of these systems include decreased installation time, expense, human effort, mistakes, storage, and the number of components in a prototype or system.

There have been reports of 4D printing applications in several industries, including medicinal devices, security, the creation of precisely patterned surfaces for optics, electrical devices, constructions with multidirectional capabilities, and soft actuators. Recent years have seen a huge increase in the popularity of soft robotics, which attempts to emulate biology by building flexible and rigid controlled objects, notably for the medical industry. Researchers have recently been more interested in the usage of Shape memory alloys and electroactive polymers which are the materials that change their shape and size according to the temperature and electric field respectively [92], pressurized fluid or gas-operable elastomers, chemical stimuli, and light-sensitive materials with a focus on soft robotics and the biomedical area. As a result, many new opportunities and chances are anticipated to materialize soon as the development of 4D printing technology would open several new possibilities [93].

Zhang et al. modified cellulose with stearoyl moieties to create a material that responds to moisture. They created a film out of this material that, when exposed to an environment with a moisture gradient, would bend because of the non-uniform absorption of moisture [94]. To expand its biological uses or improve the control of printing accuracy, certain novel techniques are emerging. In certain ways, recent advances in 4D bioprinting have resolved issues that were once seen as obstacles, such as the development of microscale vascular models and medication delivery systems for the stomach and muscular actuators. Now that 4D bioprinting is more understood, it has drawn a lot of attention to the research of tissue regeneration and biomedical devices [95]. The fact that 4D bioprinting can better suit the physiological aspects of the body is now widely acknowledged by experts. Instead of being in a static environment like 3D printing, the 4D bioprinted devices may integrate dynamic modification. It has been demonstrated that 4D bioprinting has enormous potential to change tissue engineering, medication delivery, and other sectors. It offers up a new path for bio fabrication. We have a thorough grasp of the biomedical area thanks to the innovative features of 4D bioprinting, not only in terms of tissue and organ regeneration but also in terms of illness therapy. It totally advances the idea of biomedicine while innovating traditional industrial techniques. The tissues in the human body are exceedingly malleable, non-static, and have special roles that are ideal for dynamic alterations. Conventional 3D-printed objects may have certain forms, topologies, or cells, but they are unable to demonstrate dynamic processes. Given this, 4D bioprinting effectively satisfies the need for biomedicine. To the greatest degree possible, 4D bioprinting aims to emulate biological functions in vivo. The bodily reaction cues that trigger the shift should be secure and simple to manage [96].

9. Conclusion

In this chapter, we have reviewed the basics of 3D printing, the various types of bioprinters available like droplet-based, extrusion-based, light-based bioprinters. There are many commercially available bioprinters discussed, developed by various companies fulfilling the requisites of bioprinting. Bioinks are the core part of bioprinting, and it is important to formulate them properly to get the constructs that can be stable, biodegradable, biocompatible, and able to mimic the native microenvironment of the tissue. Numerous studies have been mentioned where 3D printing has been used to fabricate bone grafts, dental implants, liver disease models, liver tissue constructs, vascular structures, tracheal constructs, cell-laden urethra, and esophageal stent. Some of the mentionable commercially available 3D printed products available include ExVive™ 3D bioprinted human liver tissue model, ExVive™ human kidney tissue, animal thyroid gland by 3dbio, etc. The future of 3D printing is 4D printing which utilizes time as a fourth dimension. The smart materials used for 4D printing change the shape and size under the influence of an external stimulus. 4D printing will open several new possibilities in the field of biomedicine.

10. Future scope

3D printing is the latest technology creating a buzz in all fields including artificial intelligence, advanced simulations, biomedicine, and engineering. The scope embraces objects like human organs, aircraft components, and much more. The technique is being widely accepted due to the several advantages its offers including patient-specific design, high complexity, cost-effective fabrication, and high productivity. The possible uses of 3D printing are endless now, from decreasing the cost of health care to the construction of houses. The cost of production of the prosthetic limb has been reduced to 75% by using 3D printing by the company Mercuris. Rice university has developed a 3D bioprinter that can print narrow blood vessels and which led to developing lung model. 3D printing is being explored by various researchers but now many are working around 4D printing as it is the upcoming technique that is beginning to establish. There are still many challenges and hurdles that must be addressed including the lack of multi-material printers, lack of low-cost printers, and smart materials. The area is still new and unexplored.

Acknowledgements

The authors wish to acknowledge the Department of Science and Technology-Science Engineering Research Board, Early Career Research Award (ECR/2018/000709) and Indian Council for Medical Research (ICMR Adhoc -5/3/8/53/2020 – ITR) for funding the research.

Conflict of interest

The authors declare no conflict of interest.

Author details

Meghana Kasturi[1], Vidhi Mathur[1], Prachi Agarwal[1], Varadharajan Srinivasan[2] and Kirthanashri S. Vasanthan[1*]

1 Manipal Centre for Biotherapeutics Research, Manipal Academy of Higher Education, Manipal, Karnataka, India

2 Department of Civil Engineering, JSS Academy of Higher Education, Noida, Uttar Pradesh, India

*Address all correspondence to: kirthanashri.sv@manipal.edu; kirthanasv@gmail.com

References

[1] Mhanna R, Hasan A. Introduction to tissue engineering. In: Tissue Engineering for Artificial Organs: Regenerative Medicine, Smart Diagnostics and Personalized Medicine. Vol. 1. Germany. 8 May 2017. pp. 1-34

[2] Magnetic Levitation Aids 3D Bioprinting [Internet]. Engineering.com. Available from: https://www.engineering.com/story/magnetic-levitation-aids-3d-bioprinting. [Accessed: October 10, 2022]

[3] $3.26 Billion 3D Bioprinting Market, 2027 by Technology (Inkjet, Magnetic Levitation, Laser-assisted), Component (3D Bioprinters, Biomaterials, Scaffolds), Material, Application-ResearchAndMarkets.com [Internet]. 2021. Available from: https://www.businesswire.com/news/home/20210422005702/en/3.26-Billion-3D-Bioprinting-Market-2027-by-Technology-Inkjet-Magnetic-Levitation-Laser-assisted-Component-3D-Bioprinters-Biomaterials-Scaffolds-Material-Application---ResearchAndMarkets.com. [Accessed: October 10, 2022]

[4] 3D Bioprinting Market Size & Share Report, 2022-2030 [Internet]. Available from: https://www.grandviewresearch.com/industry-analysis/3d-bioprinting-market. [Accessed: October 10, 2022]

[5] Groll J, Burdick JA, Cho DW, Derby B, Gelinsky M, Heilshorn SC, et al. A definition of bioinks and their distinction from biomaterial inks. Biofabrication. 2018;**11**(1):013001

[6] Gungor-Ozkerim PS, Inci I, Zhang YS, Khademhosseini A, Dokmeci MR. Bioinks for 3D bioprinting: An overview. Biomaterials Science. 2018;**6**(5):915-946

[7] Genova T, Roato I, Carossa M, Motta C, Cavagnetto D, Mussano F. Advances on bone substitutes through 3D bioprinting. International Journal of Molecular Sciences. 2020;**21**(19):7012

[8] Salah M, Tayebi L, Moharamzadeh K, Naini FB. Three-dimensional bio-printing and bone tissue engineering: Technical innovations and potential applications in maxillofacial reconstructive surgery. Maxillofacial Plastic and Reconstructive Surgery. 2020;**42**(1):18

[9] Choe G, Lee M, Oh S, Seok JM, Kim J, Im S, et al. Three-dimensional bioprinting of mesenchymal stem cells using an osteoinductive bioink containing alginate and BMP-2-loaded PLGA nanoparticles for bone tissue engineering. Biomaterial Advanced. 2022;**136**:212789

[10] Freeman FE, Pitacco P, van Dommelen LHA, Nulty J, Browe DC, Shin JY, et al. 3D bioprinting spatiotemporally defined patterns of growth factors to tightly control tissue regeneration. Science Advances. 2020;**6**(33):eabb5093

[11] Chimene D, Miller L, Cross LM, Jaiswal MK, Singh I, Gaharwar AK. Nanoengineered Osteoinductive bioink for 3D bioprinting bone tissue. ACS Applied Materials & Interfaces. 2020;**12**(14):15976-15988

[12] Jakus AE, Rutz AL, Jordan SW, Kannan A, Mitchell SM, Yun C, et al. Hyperelastic "bone": A highly versatile, growth factor–free, osteoregenerative, scalable, and surgically friendly biomaterial. Science Translational Medicine. 2016;**8**(358):358ra127-358ra127

[13] Fallah A, Altunbek M, Bartolo P, Cooper G, Weightman A, Blunn G, et al. 3D printed scaffold design for bone defects with improved mechanical and biological properties. Journal of the Mechanical Behavior of Biomedical Materials. 2022;**134**:105418

[14] Nulty J, Freeman FE, Browe DC, Burdis R, Ahern DP, Pitacco P, et al. 3D bioprinting of prevascularised implants for the repair of critically-sized bone defects. Acta Biomaterialia. 2021;**126**:154-169

[15] Shokouhimehr M, Theus AS, Kamalakar A, Ning L, Cao C, Tomov ML, et al. 3D bioprinted bacteriostatic Hyperelastic bone scaffold for damage-specific bone regeneration. Polymers. 2021;**13**(7):1099

[16] Calcium silicate nanowires-containing multicellular bioinks for 3D bioprinting of neural-bone constructs - ScienceDirect [Internet]. Available from: https://www.sciencedirect.com/science/article/abs/pii/S1748013222002122?via%3Dihub [Accessed: November 13, 2022]

[17] Zhang X, Liu Y, Zuo Q, Wang Q, Li Z, Yan K, et al. 3D bioprinting of biomimetic Bilayered scaffold consisting of Decellularized extracellular matrix and silk fibroin for Osteochondral repair. International Journal of Bioprinting. 2021;7(4):401

[18] Korn P, Ahlfeld T, Lahmeyer F, Kilian D, Sembdner P, Stelzer R, et al. 3D printing of bone grafts for cleft alveolar Osteoplasty–In vivo evaluation in a preclinical model. Frontiers in Bioengineering and Biotechnology. 2020;**8**:217

[19] Wang C, Lai J, Li K, Zhu S, Lu B, Liu J, et al. Cryogenic 3D printing of dual-delivery scaffolds for improved bone regeneration with enhanced vascularization. Bioactive Material. 2021;**6**(1):137-145

[20] 3DBioprinted Integrated Osteochondral Scaffold-Mediated Repair of Articular Cartilage Defects in the Rabbit Knee, SpringerLink. Available from: https://link.springer.com/article/10.1007/s40846-019-00481-y

[21] Long Zhisheng XL. Effect of artificial bone with multi-scale hydroxyapatite/chitosan microtubule structure on rabbit bone defect repair and angiogenesis. Chinese Journal of Tissue Engineering Research. 2022;**26**(34):5436

[22] Zhao L, Luo Y, Wang Y, Zhao F, Chen X, Cai D. Three-dimensional printed BGS treat a large bone defect in a rabbit model. Doklady. Biochemistry and Biophysics. 2021;**497**(1):123-129

[23] Sun X, Ma Z, Zhao X, Jin W, Zhang C, Ma J, et al. Three-dimensional bioprinting of multicell-laden scaffolds containing bone morphogenic protein-4 for promoting M2 macrophage polarization and accelerating bone defect repair in diabetes mellitus. Bioactive Material. 2021;**6**(3):757-769

[24] Chiesa I, Maria CD, Lapomarda A, Fortunato GM, Montemurro F, Gesù RD, et al. Endothelial cells support osteogenesis in an in vitro vascularized bone model developed by 3D bioprinting. Biofabrication. 2020;**12**(2):025013

[25] Zhang J, Eyisoylu H, Qin XH, Rubert M, Müller R. 3D bioprinting of graphene oxide-incorporated cell-laden bone mimicking scaffolds for promoting scaffold fidelity, osteogenic differentiation and mineralization. Acta Biomaterialia. 2021;**121**:637-652

[26] Daly AC, Cunniffe GM, Sathy BN, Jeon O, Alsberg E, Kelly DJ. 3D bioprinting

of developmentally inspired templates for whole bone organ engineering. Advanced Healthcare Materials. 2016;**5**(18):2353-2362

[27] Wei M, Quinnell SP, Bendtsen ST. Development of a novel alginate-polyvinyl alcohol-hydroxyapatite hydrogel for 3D bioprinting bone tissue engineered scaffolds. Journal of Biomedical Materials Research. 2017;**105**:1457-1468

[28] Murphy C, Kolan K, Li W, Semon J, Day D, Leu M. 3D bioprinting of stem cells and polymer/bioactive glass composite scaffolds for bone tissue engineering. International Journal of Bioprinting. 2017;**3**(1):005

[29] Ojansivu M, Rashad A, Ahlinder A, Massera J, Mishra A, Syverud K, et al. Wood-based nanocellulose and bioactive glass modified gelatin–alginate bioinks for 3D bioprinting of bone cells. Biofabrication. 2019;**11**(3):035010

[30] Kondiah PJ. A 3D bioprinted pseudo-bone drug delivery scaffold for bone tissue engineering. Pharmaceutics. 2020;**12**:166

[31] Anada T, Pan CC, Stahl AM, Mori S, Fukuda J, Suzuki O, et al. Vascularized bone-mimetic hydrogel constructs by 3D bioprinting to promote Osteogenesis and angiogenesis. International Journal of Molecular Sciences. 2019;**20**(5):1096

[32] Park J, Lee SJ, Jung TG, Lee JH, Kim WD, Lee JY, et al. Surface modification of a three-dimensional polycaprolactone scaffold by polydopamine, biomineralization, and BMP-2 immobilization for potential bone tissue applications. Colloids and Surfaces. B, Biointerfaces. 2021;**199**:111528

[33] Rajput M, Mondal P, Yadav P, Chatterjee K. Light-based 3D bioprinting of bone tissue scaffolds with tunable mechanical properties and architecture from photocurable silk fibroin. International Journal of Biological Macromolecules. 2022;**202**:644-656

[34] Morrison DG, Tomlinson RE. Leveraging advancements in tissue engineering for bioprinting dental tissues. Bioprinting. 2021;**23**:e00153

[35] Duarte Campos DF, Zhang S, Kreimendahl F, Köpf M, Fischer H, Vogt M, et al. Hand-held bioprinting for de novo vascular formation applicable to dental pulp regeneration. Connective Tissue Research. 2020;**61**(2):205-215

[36] Lin YT, Hsu TT, Liu YW, Kao CT, Huang TH. Bidirectional differentiation of human-derived stem cells induced by biomimetic calcium silicate-reinforced gelatin methacrylate bioink for odontogenic regeneration. Biomedicine. 2021;**9**(8):929

[37] Han J, Jeong W, Kim MK, Nam SH, Park EK, Kang HW. Demineralized dentin matrix particle-based bio-ink for patient-specific shaped 3D dental tissue regeneration. Polymers. 2021;**13**(8):1294

[38] Kim K, Lee CH, Kim BK, Mao JJ. Anatomically shaped tooth and periodontal regeneration by cell homing. Journal of Dental Research. 2010;**89**(8):842-847

[39] Gong H, Zhao Y, Chen Q, Wang Y, Zhao H, Zhong J, et al. 3D bio-printing of photocrosslinked anatomically tooth-shaped scaffolds for alveolar ridge preservation after tooth extraction. Journal of Materials Chemistry B. 2022;**10**(41):8502-8513

[40] Tang H, Bi F, Chen G, Zhang S, Huang Y, Chen J, et al. 3D-bioprinted recombination structure of Hertwig's epithelial root sheath cells and

dental papilla cells for alveolar bone regeneration. International Journal of Bioprinting. 2022;**8**(3):512

[41] Feng S, Liu J, Ramalingam M. 3D printing of stem cell responsive Ionically-Crosslinked polyethylene glycol Diacrylate/alginate composite hydrogels loaded with basic fibroblast growth factor for dental pulp tissue engineering: A preclinical evaluation in animal model. Journal of Biomaterial and Tissue Engineering. 2019;**9**(12):1635-1643

[42] Chen RS, Hsu SH, Chang HH, Chen MH. Challenge tooth regeneration in adult dogs with dental pulp stem cells on 3D-printed hydroxyapatite/Polylactic acid scaffolds. Cell. 2021;**10**(12):3277

[43] Yi K, Li Q, Lian X, Wang Y, Tang Z. Utilizing 3D bioprinted platelet-rich fibrin-based materials to promote the regeneration of oral soft tissue. Regenerative Biomaterial. 2022;**9**:rbac021

[44] Park CH, Rios HF, Jin Q, Bland ME, Flanagan CL, Hollister SJ, et al. Biomimetic hybrid scaffolds for engineering human tooth-ligament interfaces. Biomaterials. 2010;**31**(23):5945-5952

[45] Park CH et al. Image-based, fiber guiding scaffolds: A platform for regenerating tissue interfaces. Tissue Engineering Part C: Methods. 2013;**20**:533-542

[46] Lee CH et al. Three-dimensional printed multiphase scaffolds for regeneration of periodontium complex. Tissue Engineering Part A. 2014;**20**:1342-1351

[47] Park CH, Rios HF, Jin Q, Sugai JV, Padial-Molina M, Taut AD, et al. Tissue engineering bone-ligament complexes using fiber-guiding scaffolds. Biomaterials. 2012;**33**(1):137-145

[48] Cresswell Boyes AJ et al. Composite 3D printing of biomimetic human teeth. Scientific Reports. 2022;**12**:7830

[49] Han J, Kim DS, Jang H, Kim H-R, Kang H-W. Bioprinting of three-dimensional dentin–pulp complex with local differentiation of human dental pulp stem cells. Journal of Tissue Engineering. 2019:3-7

[50] Mousavi Nejad Z, Zamanian A, Saeidifar M, Vanaei HR, Salar AM. 3D bioprinting of Polycaprolactone-based scaffolds for pulp-dentin regeneration: Investigation of physicochemical and biological behavior. Polymers. 2021;**13**(24):4442

[51] Park JH, Gillispie GJ, Copus JS, Zhang W, Atala A, Yoo JJ, et al. The effect of BMP-mimetic peptide tethering bioinks on the differentiation of dental pulp stem cells (DPSCs) in 3D bioprinted dental constructs. Biofabrication. 2020;**12**(3):035029

[52] Vurat MT, Şeker Ş, Lalegül-Ülker Ö, Parmaksiz M, Elçin AE, Elçin YM. Development of a multicellular 3D-bioprinted microtissue model of human periodontal ligament-alveolar bone biointerface: Towards a pre-clinical model of periodontal diseases and personalized periodontal tissue engineering. Genes and Diseases. 2022;**9**(4):1008-1023

[53] Athirasala A, Tahayeri A, Thrivikraman G, França CM, Monteiro N, Tran V, et al. A dentin-derived hydrogel bioink for 3D bioprinting of cell laden scaffolds for regenerative dentistry. Biofabrication. 2018;**10**(2):024101

[54] Dutta SD, Bin J, Ganguly K, Patel DK, Lim KT. Electromagnetic field-assisted cell-laden 3D printed poloxamer-407 hydrogel for enhanced

osteogenesis. RSC Advances. 2021;**11**(33):20342-20354

[55] Zhang C, Chen Z, Liu J, Wu M, Yang J, Zhu Y, et al. 3D-printed pre-tapped-hole scaffolds facilitate one-step surgery of predictable alveolar bone augmentation and simultaneous dental implantation. Composites. Part B, Engineering. 2022;**229**:109461

[56] Zhang S et al. Three-dimensional cell printed lock-key structure for Oral soft and hard tissue regeneration. Tissue Engineering Part A. 2022;**28**:13-26

[57] Thattaruparambil Raveendran N, Vaquette C, Meinert C, Samuel Ipe D, Ivanovski S. Optimization of 3D bioprinting of periodontal ligament cells. Dental Materials. 2019;**35**(12):1683-1694

[58] Picon RV, Bertol FS, Tovo CV, de Mattos ÂZ. Chronic liver failure-consortium acute-on-chronic liver failure and acute decompensation scores predict mortality in Brazilian cirrhotic patients. World Journal of Gastroenterology. 2017;**23**(28):5237

[59] Tanaka K. Resection versus transplantation for hepatocellular carcinoma exceeding Milan criteria within increasing donor shortage. Hepatobiliary Surgery and Nutrition. 2017;**6**(4):280

[60] Mitsouras D, Liacouras P, Imanzadeh A, Giannopoulos AA, Cai T, Kumamaru KK, et al. Medical 3D printing for the radiologist. Radiographics. 2015;**35**(7):1965

[61] Lewis PL, Green RM, Shah RN. 3D-printed gelatin scaffolds of differing pore geometry modulate hepatocyte function and gene expression. Acta Biomaterialia. 2018;**69**:63-70

[62] Wang JZ, Xiong NY, Zhao LZ, Hu JT, Kong DC, Yuan JY. Review fantastic medical implications of 3D-printing in liver surgeries, liver regeneration, liver transplantation and drug hepatotoxicity testing: A review. International journal of surgery. 2018;**56**:1-6

[63] Faulkner-Jones A, Fyfe C, Cornelissen DJ, Gardner J, King J, Courtney A, et al. Bioprinting of human pluripotent stem cells and their directed differentiation into hepatocyte-like cells for the generation of mini-livers in 3D. Biofabrication. 2015;**7**(4):044102

[64] Cho JW, Choi YJ, Yong WJ, Pati F, Shim JH, Kang KS, et al. Development of a 3D cell printed construct considering angiogenesis for liver tissue engineering. Biofabrication. 2016;**8**(1):015007

[65] Kim Y, Kang K, Jeong J, Paik SS, Kim JS, Park SA, et al. Three-dimensional (3D) printing of mouse primary hepatocytes to generate 3D hepatic structure. Annals of surgical treatment and research. 2017;**92**(2):67-72

[66] Zhong C, Xie HY, Zhou L, Xu X, Zheng SS. Human hepatocytes loaded in 3D bioprinting generate mini-liver. Hepatobiliary & Pancreatic Diseases International. 2016;**15**(5):512-518

[67] Góra A, Pliszka D, Mukherjee S, Ramakrishna S. Tubular tissues and organs of human body—Challenges in regenerative medicine. Journal of Nanoscience and Nanotechnology. 2016;**16**(1):19-39

[68] Thomas V, Zhang X, Vohra YK. A biomimetic tubular scaffold with spatially designed nanofibers of protein/PDS® bio-blends. Biotechnology and Bioengineering. 2009;**104**(5):1025-1033

[69] Nguyen TH, Padalhin AR, Seo HS, Lee BT. A hybrid electrospun PU/PCL scaffold satisfied the requirements of blood vessel prosthesis in terms

of mechanical properties, pore size, and biocompatibility. Journal of Biomaterials Science, Polymer Edition. 2013;**24**(14):1692-1706

[70] Farhat W, Chatelain F, Marret A, Faivre L, Arakelian L, Cattan P, et al. Trends in 3D bioprinting for esophageal tissue repair and reconstruction. Biomaterials. 2021;**267**:120465

[71] Taniguchi D, Matsumoto K, Tsuchiya T, Machino R, Takeoka Y, Elgalad A, et al. Scaffold-free trachea regeneration by tissue engineering with bio-3D printing. Interactive cardiovascular and thoracic surgery. 2018;**26**(5):745-752

[72] Gao M, Zhang H, Dong W, Bai J, Gao B, Xia D, et al. Tissue-engineered trachea from a 3D-printed scaffold enhances whole-segment tracheal repair. Scientific Reports. 2017;**7**(1):1-2

[73] Zhang K, Fu Q, Yoo J, Chen X, Chandra P, Mo X, et al. 3D bioprinting of urethra with PCL/PLCL blend and dual autologous cells in fibrin hydrogel: An in vitro evaluation of biomimetic mechanical property and cell growth environment. Acta Biomaterialia. 2017;**50**:154-164

[74] Pi Q, Maharjan S, Yan X, Liu X, Singh B, van Genderen AM, et al. Digitally tunable microfluidic bioprinting of multilayered cannular tissues. Advanced Materials. 2018;**30**(43):1706913

[75] Pisani S, Dorati R, Conti B, Modena T, Bruni G, Genta I. Design of copolymer PLA-PCL electrospun matrix for biomedical applications. Reactive and Functional Polymers. 2018;**124**:77-89

[76] Haghdel M, Alizadeh AA, Ghasemi Y, Hosseinpour H, Foroutan H, Shahriarirad S, et al. Utilization of 3D-printed polymer stents for benign esophageal strictures in patients with caustic ingestion. Journal of 3D Printing in Medicine. Mar 2021;**5**(1):11-21

[77] Hoy MB. 3D printing: Making things at the library. Medical reference services quarterly. 2013;**32**(1):93-99

[78] Schubert C, Van Langeveld MC, Donoso LA. Innovations in 3D printing: A 3D overview from optics to organs. British Journal of Ophthalmology. 2014;**98**(2):159-161

[79] Klein GT, Lu Y, Wang MY. 3D printing and neurosurgery--ready for prime time? World Neurosurgery. 2013;**80**(3-4):233-235

[80] Mertz L. Dream it, design it, print it in 3-D: What can 3-D printing do for you? IEEE Pulse. 2013;**4**(6):15-21

[81] Banks J. Adding value in additive manufacturing: Researchers in the United Kingdom and Europe look to 3D printing for customization. IEEE Pulse. 2013;**4**(6):22-26

[82] King SM, Higgins JW, Nino CR, Smith TR, Paffenroth EH, Fairbairn CE, et al. 3D proximal tubule tissues recapitulate key aspects of renal physiology to enable nephrotoxicity testing. Frontiers in Physiology. 2017;**8**:123

[83] Choudhury D, Anand S, Naing MW. The arrival of commercial bioprinters–towards 3D bioprinting revolution! International Journal of Bioprinting. 2018;**4**(2):139

[84] Ravanbakhsh H, Karamzadeh V, Bao G, Mongeau L, Juncker D, Zhang YS. Emerging technologies in multi-material bioprinting. Advanced Materials. 2021;**33**(49):2104730

[85] Biolife4D. About BIOLIFE4D. 2018. Available from: https://biolife4d.com/about/

[86] L'Oréal and Poietis sign an exclusive research partnership to develop bioprinting of hair. 2016. Available from: https://www.poietis.com/en/post-release.php?id=6&view=0.

[87] Poieskin. 2018. Available from: http://poietis.com/fr/poieskin/welcome.php.

[88] Wang S, Hunt K. Chinese Company Implants 3-D Printed Blood Vessels into Monkeys. Hong Kong: CNN; 2017

[89] Davies S. Chinese medical researchers create natural blood vessels using 3D bio-printer. 2016. Available from: https://www.tctmagazine.com/api/content/55948f80-c07b-11e6b59b-0aea2a882f79/

[90] TeVido Vitiligo. 2018. Available from: http://tevidobiodevices.com/vitiligo/

[91] Zolfagharian A, Kouzani AZ, Khoo SY, Moghadam AA, Gibson I, Kaynak A. Evolution of 3D printed soft actuators. Sensors and Actuators A: Physical. 2016;**250**:258-272

[92] Jani JM, Leary M, Subic A, Gibson MA. A review of shape memory alloy research, applications and opportunities. Materials & Design. 2014;**56**:1078-1113

[93] Ahmed K, Shiblee MN, Khosla A, Nagahara L, Thundat T, Furukawa H. Recent progresses in 4D printing of gel materials. Journal of The Electrochemical Society. 2020;**167**(3):037563

[94] Zhang K, Geissler A, Standhardt M, Mehlhase S, Gallei M, Chen L, et al. Moisture-responsive films of cellulose stearoyl esters showing reversible shape transitions. Scientific Reports. 2015;**5**(1):1-3

[95] Melocchi A, Uboldi M, Inverardi N, Briatico-Vangosa F, Baldi F, Pandini S, et al. Expandable drug delivery system for gastric retention based on shape memory polymers: Development via 4D printing and extrusion. International Journal of Pharmaceutics. 2019;**571**:118700

[96] Mathur V, Agarwal P, Srinivasan V, Panwar A, Vasanthan KS. Facet of 4D printing in biomedicine. Journal of Materials Research. 15 Nov 2022:1-7

Chapter 10

Structure and Properties of Biodegradable Polymer Materials for Fused Deposition Modeling 3D Printing

Jing Tian, Yanyan Zheng, Qing Ouyang, Ping Xue, Baohua Guo and Jun Xu

Abstract

The properties of 3D printed products are closely related to the raw materials and the processes by which they are made. The processes of melting, depositing, and cooling of polymers affect the orientation, crystallinity, and microstructure of the product. These in turn influence the thermal, mechanical, optical, and other properties of the printed part. Among various 3D printing methods, filament and pellet extrusion-based fused deposition modeling (FDM) 3D printing is the cheapest and mostly adopted. In this chapter, the devices and some biodegradable polymer materials applicable in FDM 3D printing are briefly introduced. In the first part, preparation and the structure-property relationship of polylactic acid/polybutylene succinate blend filaments are discussed. Rheological, thermal properties of the raw materials and the properties of the printed parts were characterized. In the second part, a pellet extrusion 3D printer with a micro-screw was designed for using pellets of polyhydroxyalkanoate composites, which are difficult to produce filaments. The relationship between the screw parameters of the micro-screw extrusion 3D printer, rheological properties of the composites, and the printed product performance has been investigated. Combining theory and practical application will provide guidance for formulating biodegradable polymer materials and designing equipment for FDM 3D printing.

Keywords: fused deposition modeling, biodegradable polymer materials, structure, rheological properties, mechanical properties

1. Introduction

3D printing, also known as additive manufacturing (AM), refers to a series of processes used to stack into a three-dimensional solid. AM has unique advantages for innovative manufacturing of products owing to its single-piece, low-volume, rapid manufacturing characteristics. AM technology can be classified into different types, such as selective laser melting, selective laser sintering, fused deposition modeling (FDM), etc. Among them, FDM technology is well known for its simplicity and low cost, which makes FDM unique among the many 3D printing processes. The conditions required during FDM are mild since the plastic has a low melting point and does not produce many harmful substances, allowing for desktop-level printing. The processes that need to be mastered by those operating them are also simple and the final products require little or no post-processing.

Environmental protection, green manufacturing, sustainable economic development, and the search for alternatives to nonrenewable resources have long been issues of global concern. Biodegradable materials derived from renewable sources are the most promising substitutes for petroleum-based resources [1, 2]. The AM of fully biodegradable materials combines the advantages of biodegradable materials and AM technology, therefore, products made of renewable biodegradable materials can be precisely designed and manufactured. FDM technology is the most suitable in AM to process biodegradable composites owing to its wide material compatibility, low cost, and structural design flexibility. This chapter mainly discusses the FDM printing of biodegradable polymer materials and attempts to reveal the relationship between material, equipment, and process parameters, and the performance of printed products, using filament and pellet extrusion FDM as two examples.

2. Filament extrusion 3D printing

2.1 Commonly used filament extrusion 3D printer

Since the commercialization of fused deposition modeling (FDM) 3D printing technology by Stratasys, it has become the most common and fastest-growing 3D printing method, accounting for over 50% of the 3D printing market. Currently, filament extrusion remains the mainstream commercial FDM technology. **Figure 1** shows the principle of FDM-based 3D printing. One of the most common processes today is to take a filament consumable, which must be of a specific diameter (1.75 mm or 3 mm) and is pushed through a pair of filament-feeding gears into a hot melt nozzle for melting. The material extruded from the nozzle is deposited on a three-dimensional

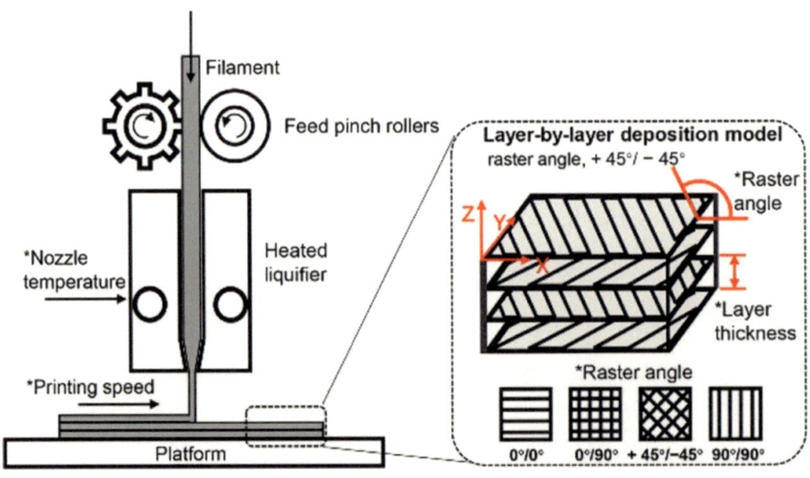

Figure 1.
Principle of FDM-based 3D printing.

platform. The motor drives the nozzle in the X-Y plane and the platform drops one layer in height after each print run, thus achieving layer-by-layer accumulation.

FDM technology is characterized by its simplicity and low cost, however, the quality of the molded part is low and the strength of the interlayer bond is the main factor affecting the mechanical properties. This is owing to the inherent printing principle of FDM technology and the performance of the printed part will vary significantly in different printing directions. This problem is also prevalent in other 3D printing technologies. For the same ABS material, the ultimate stress of an injection molded part was 26 MPa, whereas that of FDM printed bars could only reach a maximum of 22 MPa [3].

The performance of 3D-printed parts is often improved by optimizing the printing parameters. As shown in **Figure 1**, a large number of FDM parameters have different effects on product properties [4]. The effect of different printing parameters on the performance of printed parts has been studied extensively. The main influencing parameters of filament extrusion FDM are raster angle, layer thickness, nozzle diameter, feed rate, infill density, printing speed, etc. The significance of several process parameters in filament extrusion FDM has been reported in the literature: raster angle > layer thickness > printing speed > nozzle temperature [5–18].

2.2 Biodegradable materials commonly used for FDM 3D printing

The widely used biodegradable polymers include polylactic acid (PLA), polyhydroxyalkanoates (PHA), polybutylene succinate (PBS), polybutylene adipate-co-terephthalate (PBAT), and polycaprolactone (PCL), of which PLA, PHA, and some PBS (using fermentation-derived succinic acid) are both bio-based and biodegradable. The biodegradable materials commonly used for FDM are still mainly PLA. There is also a small amount of research on other biodegradable composites, such as PBS, PBAT, and PCL. **Table 1** lists the characteristics of biodegradable polymers commonly used for FDM 3D printing.

Polymer	Printing temperature (°C)	Biodegradation (industrial)*	Biodegradation (ocean)*	Printing characteristics and printed product performance
PLA	190–200	6–9 weeks	>1.5 years	Easier to print and no odor; less warpage; poor heat resistance; and impact resistance
PBS	130–200	2–5 months	>1 year	Heavy warpage; good heat resistance; and toughness
PBAT	150–200	2–3 months	>1 year	Difficult to feed filament; less warpage; good heat resistance, ductility, and impact resistance
PCL	80–100	4–6 weeks	6 weeks	Difficult to feed filament; poor heat resistant
PVA	/	1–2 weeks	4 months	Prone to thermal degradation; commonly used as a water-soluble support material

Biodegradation refers to industrial compostability under EN 13432 or ASTM D6400 conditions or degradation in ocean water according to the references [2].

Table 1.
Characteristics of biodegradable polymers commonly used for FDM 3D printing.

PLA is currently a widely used printing material for home 3D printers. PLA is biodegradable, has no irritating odor, and does not warp easily during the printing process. However, it is less heat resistant, with a glass transition temperature of around 60°C. It is also hard and brittle, with an elongation at break of less than 10% and poor impact resistance, which limits its use in applications requiring ductility and flexibility. Furthermore, PLA biodegrades slowly in the natural environment, mainly through hydrolysis and chain breaking. Overall, PLA biodegradability is dependent on numerous factors such as degradation environment, permeability to water, molecular weight, and crystallinity [19].

PBS is a thermally stable, processable, and biodegradable aliphatic polyester. However, the use of PBS for 3D printing is still relatively unexplored. Because of its low stiffness and high shrinkage, PBS can suffer from severe warpage during printing, leading to print failure, so it needs to be modified before it can be used for filament 3D printing.

To promote the development of FDM-based 3D printing, the design of new environmentally friendly polymeric materials that overcome the disadvantages of existing materials has attracted much research interest. By blending with PBS, which has better toughness and heat resistance, the disadvantages of PLA in terms of brittleness, low melt strength, and poor thermal stability can be avoided.

3. Blending modification of PLA to improve the performance of printed parts

The process of modification by blending PBS with PLA is shown in **Figure 2**. The dried PBS and PLA pellets were premixed and added to the twin-screw extruder for melt blending, and the specific formulations are listed in **Table 2**. The obtained blends were pelletized and extruded as 1.75 ± 0.02 mm filaments by a single-screw extruder for FDM 3D printing. This section provides a reference for determining

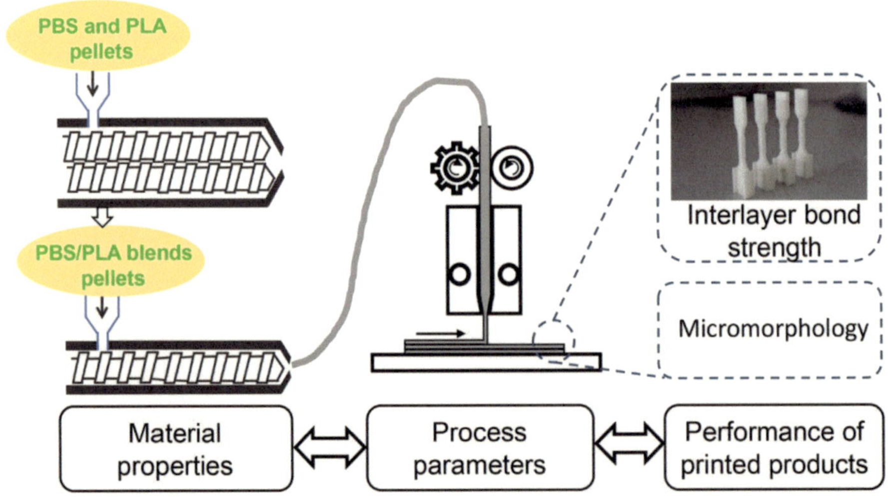

Figure 2.
Schematic diagram of the process of blending PBS and PLA, extrusion of filaments, and FDM 3D printing using the blend filaments.

Sample	Content (wt%)	
	PBS	PLA
PBS100	100	0
PBS80/PLA20	80	20
PBS60/PLA40	60	40
PBS40/PLA60	40	60
PBS20/PLA80	20	80

Table 2.
Sample information of PLA/PBS blends.

whether a thermoplastic polymer is suitable for FDM printing by studying the relationship between the rheological and crystallization properties of the PBS/PLA blends, process parameters, and the performance of the printed products [20].

3.1 Characterization of PBS/PLA blends

3.1.1 Rheological properties of the blends

The viscosity of the filament affects the FDM printing process. It has been reported that the modulus/viscosity ratio of the filament before entering the heating zone should be in the range of (3–5) $\times$ 10^5 s^{-1} to ensure continuous and stable conveying without bending near the feed pinch rollers [21]. If the filament melt viscosity is too high, it can lead to extrusion difficulties, affecting the performance of the printed part or even leading to print failure. Therefore, the proper melt viscosity is also critical for filament FDM printing. The modified blends were subjected to a rotational rheology test to investigate the rheological properties, and the results are shown in **Figure 3**. The zero-shear viscosity of the blends gradually decreased with the increase in PBS content. And the Newtonian plateau area widens and the shear thinning effect weakens at 60% PBS content and above. This is because the zero-shear viscosity of PLA is much higher than that of PBS at the test temperature. **Figure 3b** and **c** shows the storage modulus (G′) and loss modulus (G″) of the blends. Similarly, since the modulus of PLA is higher than that of PBS, the modulus of the blends gradually decreases with the increase in PBS content.

The FDM-3D printing process is accompanied by the cooling and solidification of the material, so it is necessary to study the thermal properties of the raw material. The crystallinity of the PBS/PLA blends increases with the PBS content, as shown in **Figure 4a**. Since the PLA used here is nearly amorphous, the apparent crystallinity in the blends decreases with the addition of PLA. The mechanical properties of the PBS/PLA blend bars produced by injection molding using a microinjection molding machine are shown in **Figure 4b** and **c**. The tensile strength and modulus of the blends had no significant changes when the PBS content was below 40%. However, when the PBS content continued to increase, the tensile modulus decreased sharply while the tensile strength decreased slightly. Elongation at break of blends increased with increasing PBS content, reaching a maximum of 356% at a PBS content of 80%. This was attributed to the effect of lower content of PLA as a rigid filler on the toughening of the blends.

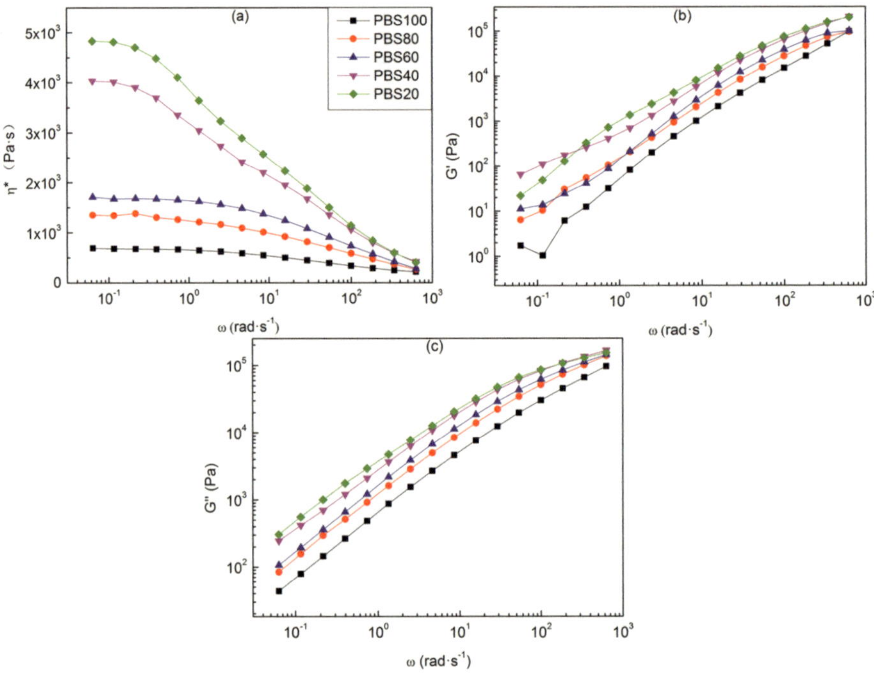

Figure 3.
(a) Plots of the complex viscosity (η), (b) storage modulus (G′), and (c) loss of crystalline and mechanical properties of the blends. (a–c) Reproduced under the terms of the CC-BY Creative Commons Attribution 4.0 International license (https://creativecommons.org/licenses/by/4.0/) [20]. Copyright 2018, The Authors, published by American Chemical Society.*

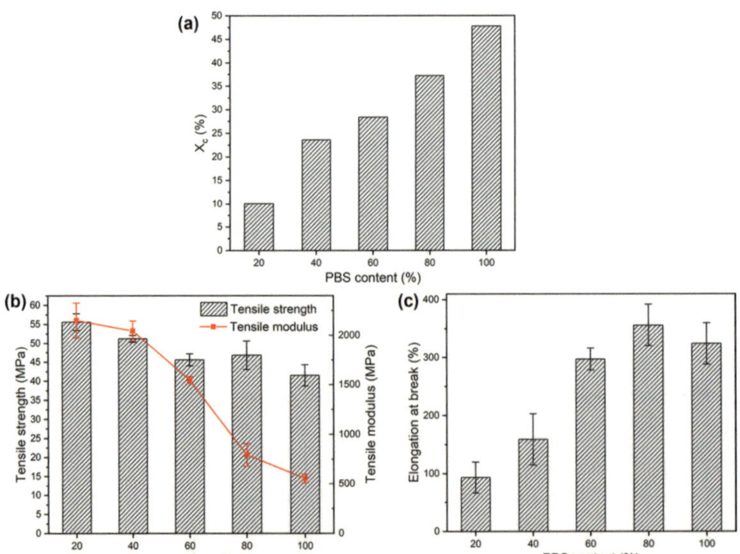

Figure 4.
(a) Crystallinity of PBS/PLA blends, (b) Tensile strength, modulus, and (c) elongation at break of the PBS/PLA blend bars produced by injection molding.

3.2 Performance characterization of PBS/PLA blends for printed parts

Filaments of different blends were used to print different specimens and models with an FDM printer (AOD Dreamer, Qingdao Autolay 3D printing Co., Ltd. China). The printing parameters are listed in **Table 3**. In particular, all tests were printed with the platform unheated.

Preprinting experiments were first performed, that is, the appearance of the printed bars as well as the cross-sectional morphology were characterized. As shown in **Figure 5**, the dimensional stability of the PBS/PLA blended bars was good at low PBS content, and a significant warpage deformation was observed at PBS content higher than 60%. Combined with **Figure 4a** and **b**, it can be seen that this is because the crystallinity of the blends increased and the modulus decreased as the PBS content increased. Therefore, the print bars generated a large thermal stress gradient during the layer-by-layer deposition process leading to an increase in volume shrinkage, which resulted in severe warpage. The warpage deformation is very detrimental to the FDM 3D printing process, especially for large-area products, which may lead to print failure owing to the detachment of the printed part from the platform. Therefore, the low crystallinity, as well as a high modulus of the raw material, is essential when choosing a printing material.

Process parameters	
Nozzle diameter	0.4 mm
Nozzle temperature	190°C
Printing speed	1.5 m/min
Layer height	0.1 mm
Raster angle	+ 45°/ – 45°
Infill ratio	100%

Table 3.
Printing parameters of PBS/PLA blend filament.

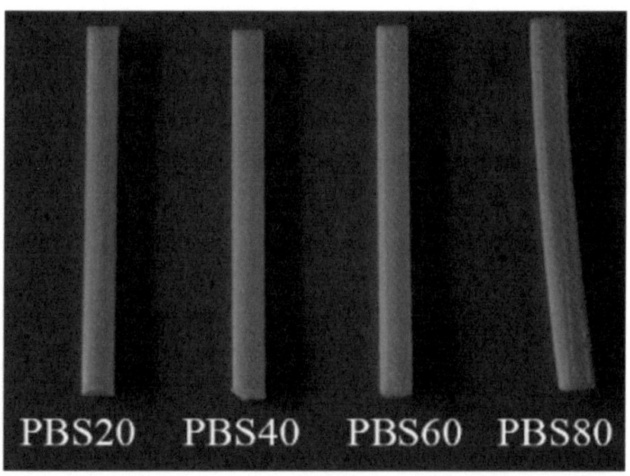

Figure 5.
Appearance of the PBS/PLA blend bars prepared by FDM printing. Reproduced under the terms of the CC-BY Creative Commons Attribution 4.0 International license (https://creativecommons.org/licenses/by/4.0/) [20]. Copyright 2018, The Authors, published by American Chemical Society.

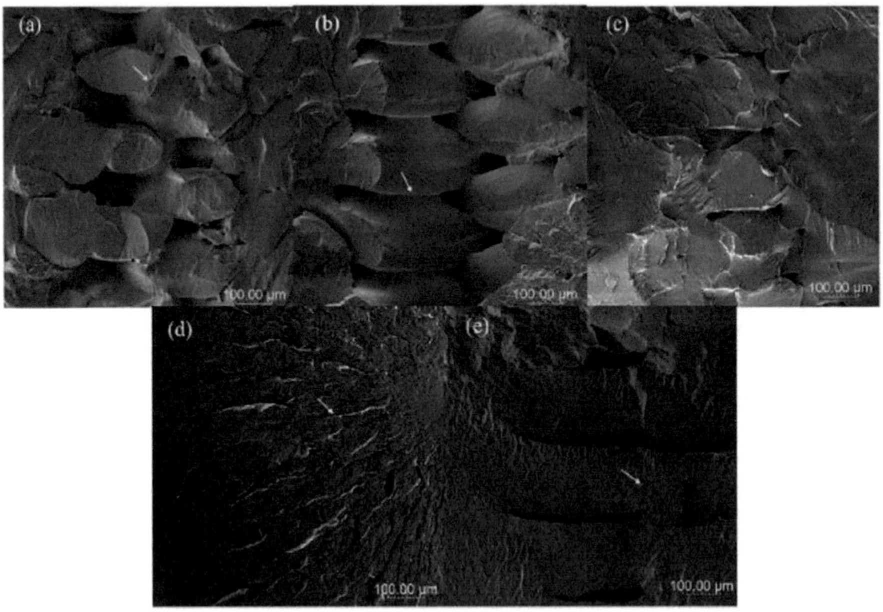

Figure 6.
SEM images of cross sections of the FDM-printed bars. (a) PBS20/PLA80, (b) PBS40/PLA60, (c) PBS60/PLA40, (d) PBS80/PLA20, and (e) PBS100. (a–e) Reproduced under the terms of the CC-BY Creative Commons Attribution 4.0 International license (https://creativecommons.org/licenses/by/4.0/) [20]. Copyright 2018, The Authors, published by American Chemical Society.

SEM images of cross sections of the FDM-printed bars are shown in **Figure 6**. The cross-sectional morphology after brittle fracture by liquid nitrogen shows that there are more obvious pores between the layers as well as the filaments at lower PBS content. The interlayer interface was gradually blurred when the PBS content was higher than 60%. This is owing to the lower viscosity and better flowability of the blends at higher PBS content (**Figure 2a**), which results in better interlayer interface bonding. PBS40/PBS60 was selected for the subsequent study considering the appearance, cross-section of the printed bars, the crystallization, and rheological properties of the blends.

From the above results, it can be inferred that the bonding of the interlayer interface is closely related to the viscosity of the melt. The main factor affecting the melt viscosity is the temperature. Therefore, the effect of temperature on the melt viscosity of the blends was further investigated. **Figure 7** shows the zero-shear viscosity of the PBS40/PLA60 blends at different temperatures. The zero-shear viscosity of the blend melt was below 1000 Pa·s and no longer decreased significantly above 190°C. The melt viscosity increased sharply when the temperature decreased from 190 to 160°C, which is detrimental to the printing process. Higher melt viscosity may clog the nozzle and cause underfilling, thus affecting the performance of the printed part. Therefore, the optimal printing temperature range is 190–230°C.

After determining the printing temperature window, the interlayer bond strength of the blends was further investigated at different printing temperatures. **Table 4** shows the tensile strength of the vertically printed samples at different printing temperatures and the melt viscosity of the corresponding blends. The interlayer bond strength decreased with PBS content at the same printing temperature. This is attributed to the increase in melt viscosity with increasing PLA content, which leads to the formation of larger interfilament gaps, resulting in a decrease in bond strength. For the PBS40/

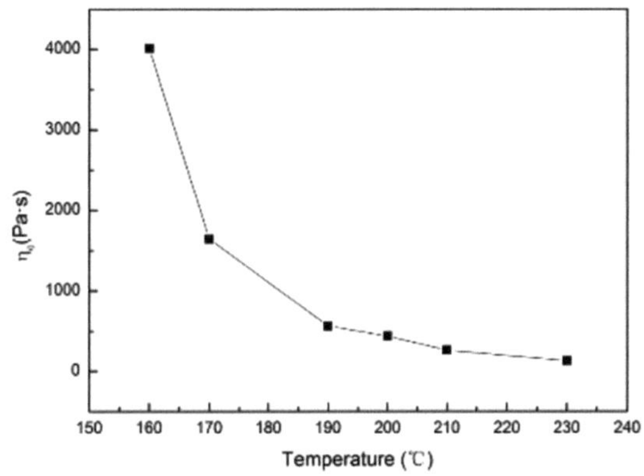

Figure 7.
Variation of the zero-shear viscosity of PBS40/PLA60 melt with temperature. Reproduced under the terms of the CC-BY Creative Commons Attribution 4.0 International license (https://creativecommons.org/licenses/by/4.0/) [20]. Copyright 2018, The Authors, published by American Chemical Society.

Blends	Printing temperature (°C)	Zero-shear viscosity (Pa·s)	Tensile strength (MPa)
PBS100	210	67	25.0 ± 1.2
PBS80/PLA20	210	96	20.5 ± 1.3
PBS60/PLA40	210	123	19.6 ± 1.1
PBS40/PLA60	190	564	21.4 ± 5.2
PBS40/PLA60	200	438	19.5 ± 2.7
PBS40/PLA60	210	262	18.4 ± 2.7
PBS40/PLA60	230	130	16.5 ± 2.7

Reproduced under the terms of the CC-BY Creative Commons Attribution 4.0 International license (https://creativecommons.org/licenses/by/4.0/) [20]. Copyright 2018, The Authors, published by American Chemical Society.

Table 4.
Interlayer bond strengths of the bars printed from different blend filaments or at different nozzle temperatures.

PLA60 blend, the melt viscosity gradually decreased as the printing temperature increased from 190 to 230°C, and the interlayer adhesion strength gradually decreased. Generally, the decrease in melt viscosity at an increasing temperature is beneficial to interlayer bonding. The PLA used here is amorphous and has low crystallinity. It is presumed that the reduction of interlayer bonding may be due to the thermal degradation of PLA at higher than 190°C. Therefore, for printing materials, it is important to understand the relationship between material properties, processing, and performance of printed products, which helps to determine the appropriate printing process and thus obtain the optimal print product performance.

FDM 3D printing usually requires polymer materials with moderate melt viscosity, a certain tensile strength and toughness, low coefficient of thermal expansion, and good adhesion properties, as shown in **Figure 8**. From the point of view of polymer properties, it is necessary to select polymer materials that are suitable for the printing conditions. In practice, it is the melt viscosity, volumetric shrinkage, warpage, modulus, strength, interlayer bonding, and other properties of raw materials and the printed product that determine whether the material can be used in 3D printing.

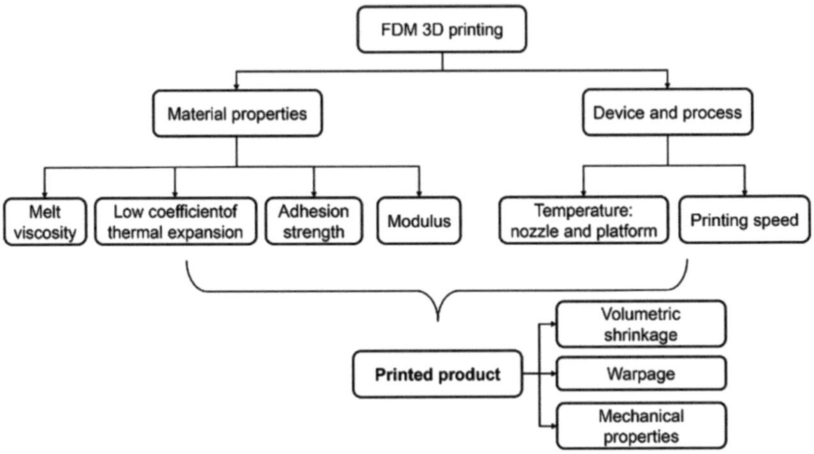

Figure 8.
Factors influencing the performance of FDM 3D printed products.

In addition, process parameters, such as nozzle temperature, platform temperature, printing speed, infill density, etc. are also important.

4. Pellet extrusion 3D printer based on micro-screw

Among biodegradable polymers, polyhydroxyalkanoates (PHAs) have attracted significant attention as a promising new family of sustainable polymers that are biobased and biodegradable in natural environments, with adjustable mechanical and physical properties [22]. PHAs are stable with respect to moisture and air exposure as they require an active microbial environment for degradation. Among various PHAs, the most common one, polyhydroxybutyrate (PHB), is brittle and hard. To solve the shortcomings, it can be copolymerized with an appropriate amount of 3-hydroxyvaleric acid (HV) monomers to improve the toughness. The melting point of the corresponding copolymer decreases with increasing HV fraction. Unlike polyolefins, some PHA melts have low thermal stability. Studies have shown that the molecular weight of PHAs may decrease during processing due to random chain scission [23].

PHA has a relatively high crystallinity like conventional polymers, such as polyethylene (PE) and polypropylene (PP). However, compared with commercial thermoplastic materials, it also has disadvantages such as higher cost and narrow processing window. Many studies have been conducted to modify product properties by blending PHA with PLA, improving its mechanical strength and customizable biodegradability [24, 25]. Wood is a natural organic material widely used in the form of sawdust, ground into smaller parts, or fine wood flour (WF) [19, 26]. Through the melt blending of WF and PHA, costs are reduced. In addition, the presence of wood filler can provide nucleation sites to promote rapid polymer crystallization from the melt, overcoming the slow crystallization of PHAs, which is a marked limitation for practical applications [23].

However, difficulties in preparing the filament and the feed section of biodegradable composites [27, 28] have limited the application of FDM for filament extrusion. Despite these difficulties, the inherent anisotropy of crystalline biodegradable composites can lead to a warpage of the printed part and thus prevent the print from

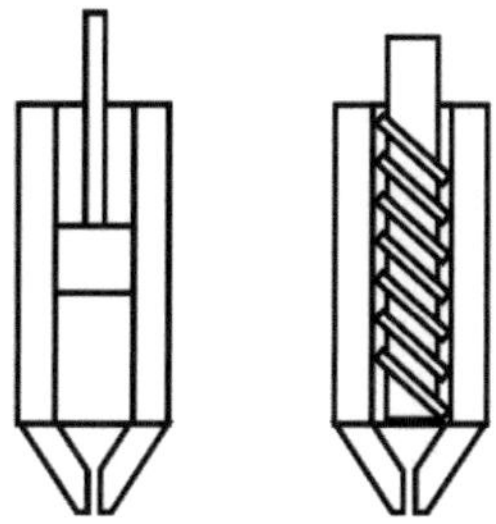

Figure 9.
Diagram of plunger and screw extrusion nozzles in FDM.

being completed. Therefore, FDM technology must be combined with traditional manufacturing techniques. FDM-based technology has developed two main types of structured print nozzles. As shown in **Figure 9**, one incorporates a screw nozzle design and the other a plunger nozzle. The screw nozzle is commonly used for industrial and biomedical printing applications, while the plunger is commonly used for applications such as food or clay printing.

In the case of plunger extrusion FDM technology, during the extrusion process, formulated "ink" are loaded into a cylinder (extruder) and the force generated by the hydraulic piston extrudes the "ink." Successive layers are deposited by guiding the cylinder to a specific point on a predetermined three-dimensional model. However, plunger extruders cannot print continuously due to the limited volume within the cylinder. Previous studies [26, 29, 30] have also shown that the deposited products suffer from underfilling and poor accuracy, so plunger extrusion cannot be used for fused deposition molding of biomass composites.

Therefore, this section describes a micro-screw extrusion AM system suitable for the printing of fully biodegradable composites [31]. Specifically, it covers the design of micro-screws for different types of materials (crystalline and noncrystalline materials), the preparation of wood flour-filled polyhydroxyalkanoate (PHA) composites, and the elaboration of the relationship between materials properties, device design, and printed part performance [32].

Figure 10 shows the process diagram for micro-screw-based FDM 3D printing of WF/PHA composites. The dried PHA pellets were premixed with WF in the proportions listed in **Table 5** and then extruded in a twin-screw extruder in melt blending. The obtained WF/PHA composite pellets are pulverized and used directly for printing on a micro-screw 3D printer. The relationship between material properties, equipment and process parameters, and printed product performance was investigated. Combining traditional processing techniques with 3D printing increased the variety of printable materials and the depth of processing, thus providing a reference for the development of new 3D printing materials and equipment.

4.1 Micro-screw extrusion-based FDM system

A schematic diagram of the proposed micro-screw extrusion FDM 3D printer is shown in **Figure 10**, which consists of a micro-screw extrusion system, motion system, and control system. The fully biodegradable composite was extruded into a 0.1–3 mm filament by a micro-screw extrusion device and then deposited on the printing platform by an AM system based on FDM technology to realize layer-by-layer stacking and shaping.

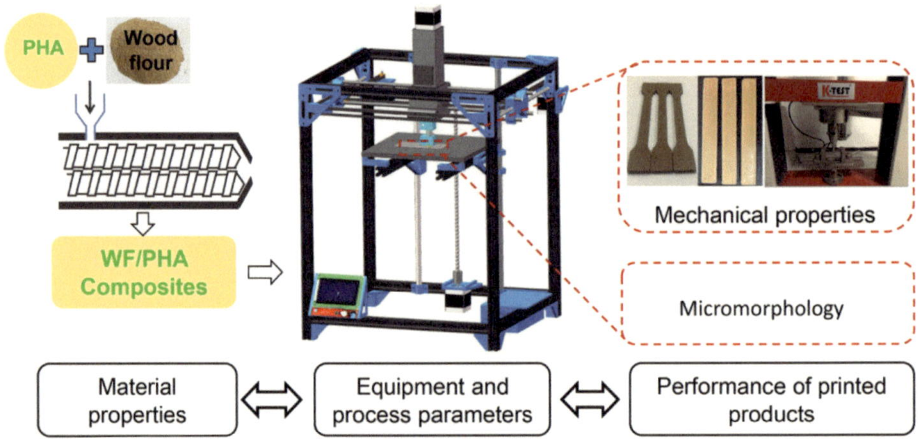

Figure 10.
Process diagram for micro-screw-based FDM 3D printing of WF/PHA composites.

Sample	Content (wt%)	
	PHA	WF
10WF/PHA	90	10
20WF/PHA	80	20
30WF/PHA	70	30

Table 5.
Parameters of WF/PHA composites.

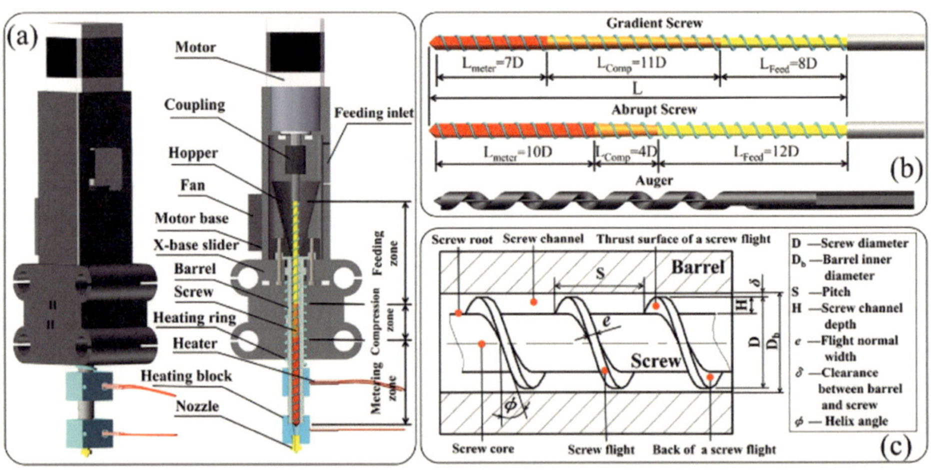

Figure 11.
Micro-screw extrusion system. (a) 3D and cross-sectional views of the micro-screw extrusion system, (b) a schematic diagram of two types of screws and an auger, and (c) notations for screw geometry. (a–c) Adapted with permission [31]. Copyright 2021, Elsevier.

Figure 11a shows a three-dimensional view and cross-sectional view of the micro-screw extrusion system. The micro-screw extrusion system consists of a planetary reduction stepper motor, motor base, hopper, X-base slider, barrel, micro-screw, heating block, and brass nozzle. The micro screw is driven by a stepper motor at one end and the other end penetrates the hopper into the barrel. The micro screw extrusion is, therefore, fed continuously and forcibly without feeding difficulties compared with filament extrusion. Three heating zones on the outside of the barrel allow the material to be completely molten. The micro-screw drives the material in the extrusion direction to obtain a stable extrusion pressure, which improves the melt plasticization quality and improves the performance of the 3D printed product. Overall, micro-screw extrusion 3D printing can achieve higher melting and printing efficiency than filament extrusion FDM 3D printing, resulting in continuous and stable printing. **Figure 11b** shows two types of screws designed according to the material properties and the 3D printing process, as well as the conveying-only auger used for comparison. For the conventional screw design, the gradient screw is suitable for amorphous polymer processing, while the abrupt screw is suitable for crystalline polymer. **Figure 11c** shows the specific parameters of the screw structure. In order to achieve excellent nozzle dynamics [33], a screw diameter of 6 mm, an L/D ratio of 26, and a compression ratio of 2.9 were designed. The ratios of the feeding, compression, and metering section lengths of the gradient screw and abrupt screw were 8D:11D:7D and 12D:4D:10D, respectively.

4.2 Rheological properties of WF/PHA composites

Investigating the rheological properties of raw materials prior to micro-screw extrusion 3D printing can help to optimize the printing process and thus improve the

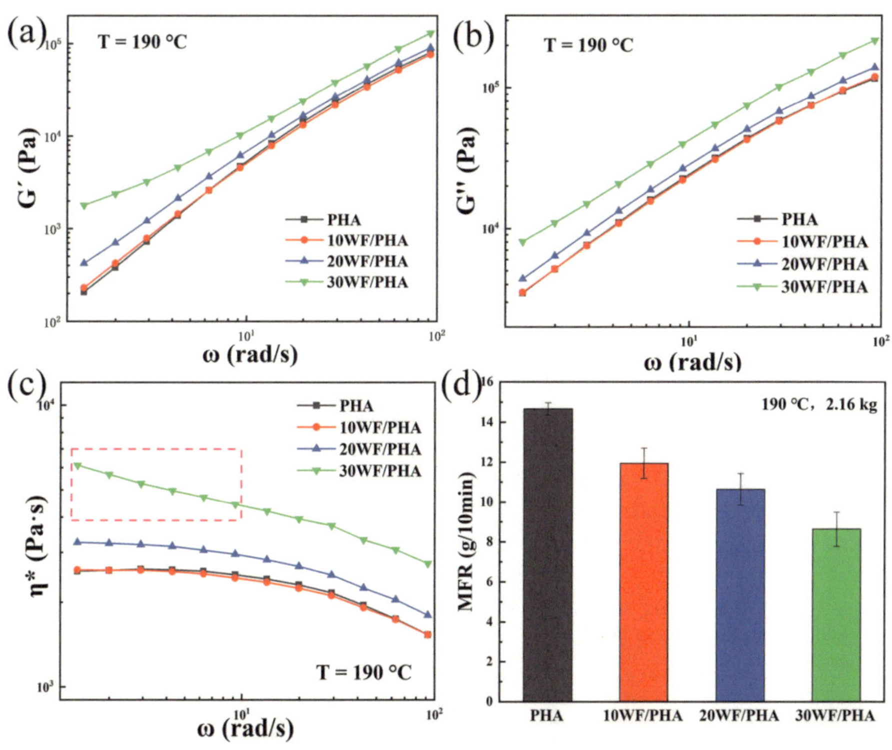

Figure 12.
Rheological properties of neat PHA and WF/PHA composites: (a) storage modulus, (b) loss modulus, (c) complex viscosity, and (d) MFR. (a–d) Reproduced with permission [31]. Copyright 2021, Elsevier.

performance of the printed product. The properties of 3D printed materials require, on one hand, sufficient fluidity to ensure adequate bonding to the previous layer after deposition and, on the other hand, sufficient melt strength to ensure shape stability. **Figure 12** shows the rheological properties of neat PHA and WF/PHA composites. WF is often used as a filler to improve the dimensional stability, rigidity, and heat resistance of the matrix resin and is widely used in the processing of wood-plastic composites. As a result, the modulus of WF/PHA composites increased compared with neat PHA. Furthermore, **Figure 12a** and **b** shows that the loss modulus (G″) of neat PHA and the composite is higher than the storage modulus (G′), exhibiting typical viscoelastic properties. The complex viscosity(η^*) of the composites increased with the WF content, as shown in **Figure 12c**. Furthermore, the shear thinning effect of the composites was stronger with the increase in WF content, indicating an improved processing range of the composites. **Figure 12d** shows the MFR of the neat PHA and WF/PHA composites. The fluidity of the composites decreased with the increase in WF content. This is consistent with the results of viscosity in the rotational rheology test, where the increased WF content in the composite impedes the movement of the polymer chain segments, thus reducing fluidity.

4.3 Stability of extrusion flow rate and analysis of pressure-building capacity

The extrusion stability and pressure-building capability of screw extrusion 3D printers are closely related to screw design and composite rheological properties. **Figure 13a** and **b** presents extrusion stability curves of neat PHA and WF/PHA composites extruded at various nozzle temperatures and different screw speeds. The effect of screw speed on the extrusion output (Q) was more significant than the effect of temperature. At the same temperature, the Q of the WF/PHA composites decreased as the WF content increased, which is consistent with the rheological test results. For micro-screw extrusion, the Q of neat PHA was not sensitive to temperature. This indicates that the screw working characteristic was not suitable for neat PHA. The Q of WF/PHA composites was stable at 190°C and did not increase significantly at higher temperatures. The discontinuity defect shown in **Figure 13a** appeared above 200°C, which was caused by partial degradation of the WF at higher temperatures. The Q of WF/PHA composites was linearly related to the screw speed at the same temperature. This is essential for controlling Q by screw speed and thus regulating the performance of the printed product.

Figure 13c shows the melt pressure of neat PHA and WF/PHA composites with various WF contents at different screw speeds. The SEN-PT series of the miniature pressure sensor (accuracy ±1%FS, mounting thread size M14*1.5) was adopted to measure the melt pressure at different screw speeds. The melt pressure signal can be output *via* the accompanying digital display instrument and monitored online in real-time and recorded when the melt pressure data is stable. Because Q increased with increasing screw speed, the melt pressure of neat PHA and WF/PHA composites increased with increasing screw speed. The melt pressure of WF/PHA composites decreased with increasing WF content at the same screw speed. The determining factor for melt pressure is actually the amount of melt built-up in the nozzle chamber and is, therefore, related to Q. As Q decreased with increasing WF content, the corresponding melt pressure also decreased.

It is worth noting that the rheological characteristic curves of neat PHA and 10WF/PHA in **Figure 12a–c** largely overlap, but neat PHA is unable to complete printing due to warpage and underfilling (**Figure 13d**). This is because the addition

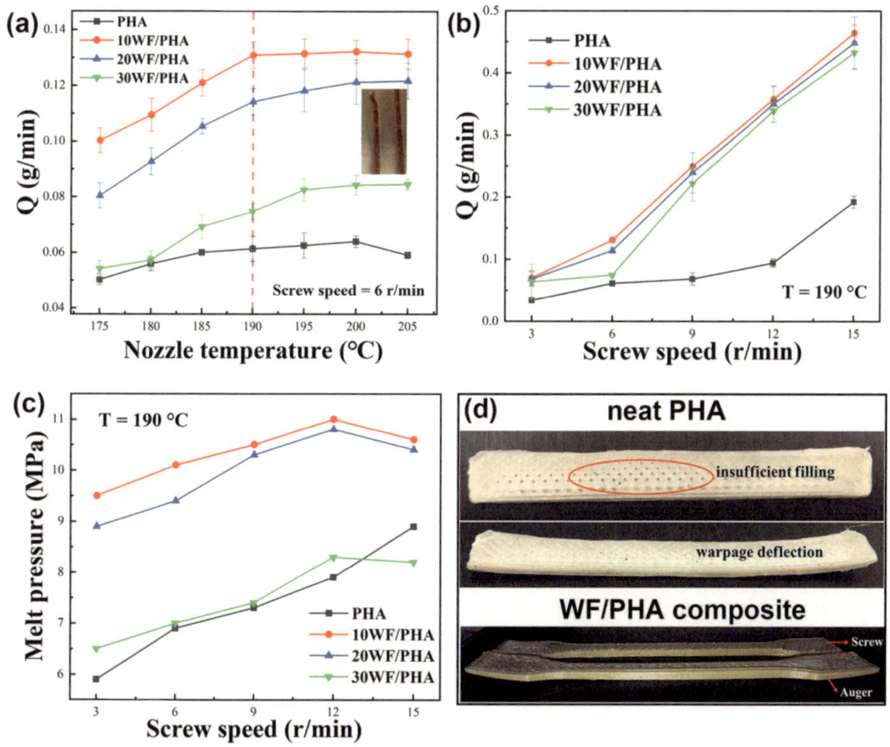

Figure 13.
Extrusion stability curves of WF/PHA composites extruded at (a) various nozzle temperatures and (b) different screw speeds. (c) Melt pressure of neat PHA and WF/PHA composites with various WF content at different screw speeds. (d) Printing defects of neat PHA and WF/PHA composite samples printed by micro-screw and auger. (a–d) Reproduced with permission [31]. Copyright 2021, Elsevier.

of WF filler increased the modulus of the PHA as well as the Q. Moreover, the melt pressure curve also shows that the melt pressure of 10WF/PHA is much higher than that of neat PHA. Combined with the picture of the WF/PHA composite samples printed by micro-screw and auger under the same printing parameters in **Figure 13d**, the results show that the pressure-building capacity of the micro screw has a positive effect on the resistance to warpage.

4.4 Performance of micro-screw extrusion 3D printed WF/PHA composites

4.4.1 Effect of WF content and different raster angles on the mechanical properties of WF/PHA composite

As PHA is a semicrystalline material, an abrupt screw was used to investigate the effect of WF content and different raster angles on the mechanical properties of the product. The specific printing parameters were: nozzle temperature of 190°C, layer thickness of 0.3 mm, printing speed of 15 mm/s, nozzle diameter of 1 mm, and 100% fill percentage. As shown in **Figure 14**, the impact, tensile, and flexural strengths of the composite decreased with increasing WF content, while the modulus increased significantly. This is due to the increased stiffness of the WF/PHA composites caused by the addition of WF. Furthermore, the stress-strain curves of the composites

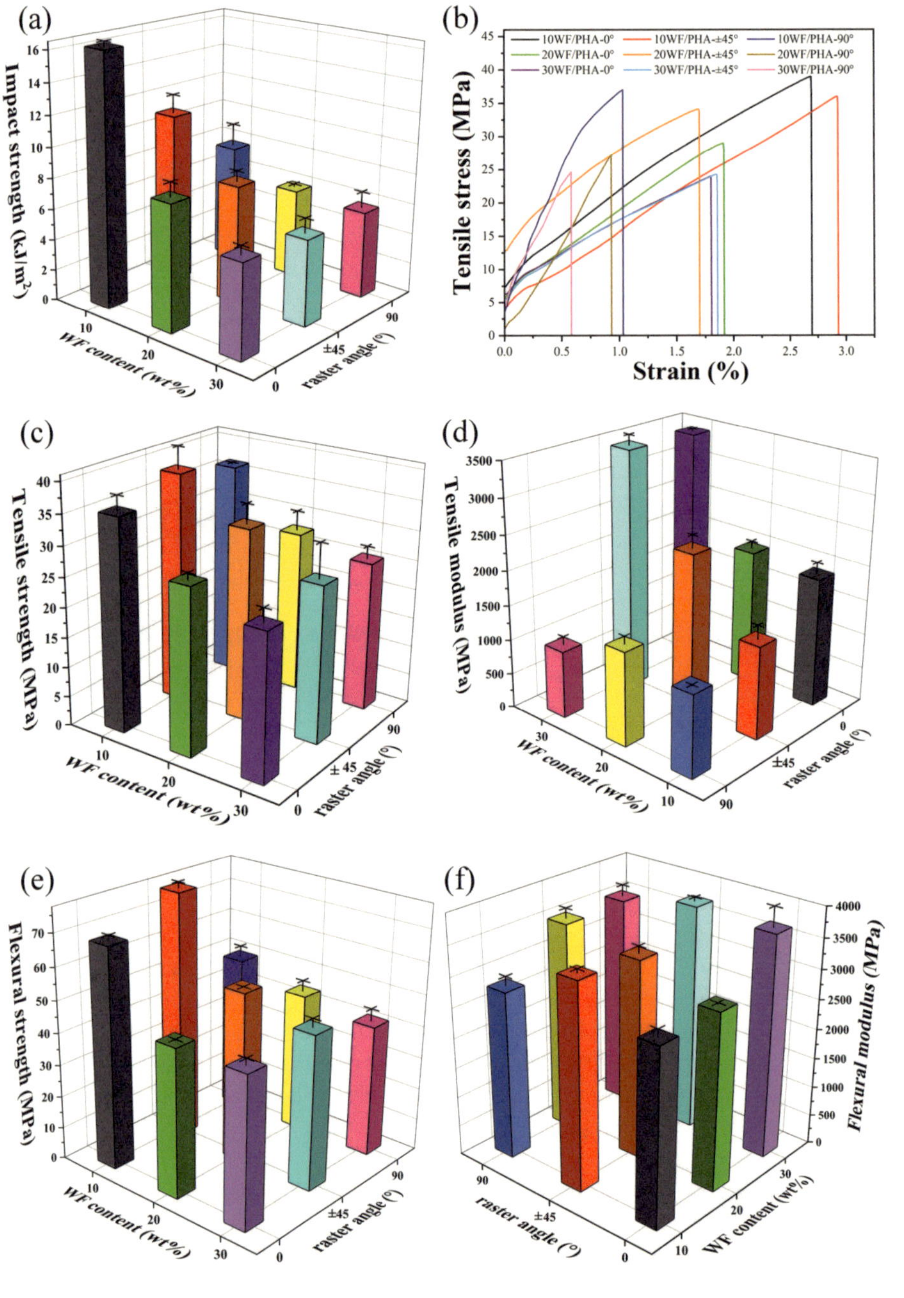

Figure 14.
Mechanical properties of printed WF/PHA composites with various WF contents and different raster angles: (a) impact strength, (b) tensile stress-strain curves, (c) tensile strength, (d) tensile modulus, (e) flexural strength, and (f) flexural modulus. (a–f) Adapted with permission [31]. Copyright 2021, Elsevier.

exhibited typical brittle fracture, which can also be verified by the cross-sectional morphology of the composites in **Figure 15**. In addition, raster angles have a significant effect on the mechanical properties of the composite, with the 45°/−45° print direction having the best mechanical properties. As can be seen in **Figure 15**, this is due to the denser deposition of the products printed at 45°/−45° compared with 0°

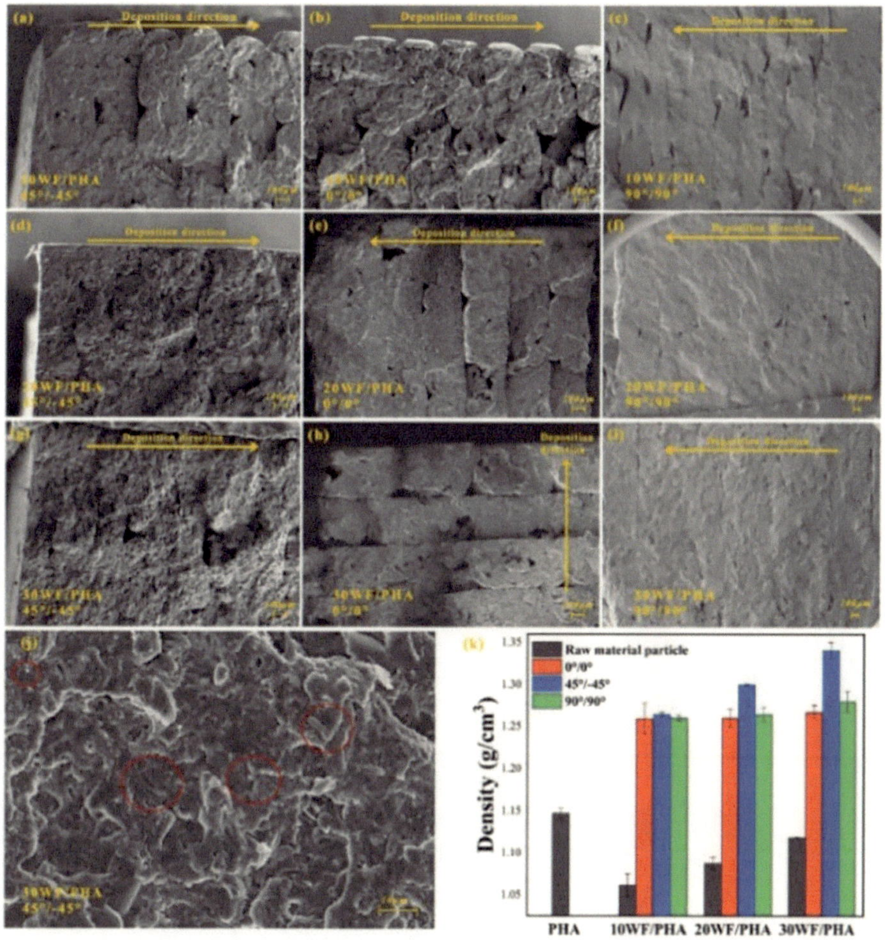

Figure 15.
(a–j) SEM images of the fractured cross-section of WF/PHA composites, and (k) density of neat PHA and WF/PHA composites. (a–k) Reproduced with permission [31]. Copyright 2021, Elsevier.

and 90° raster angles, as verified by the fact that the WF/PHA printed at 45°/−45° has the highest density in **Figure 15k**. However, the difference in mechanical properties of the composites with different raster angles decreased with the increasing WF content. This further supports the role of WF in improving the printability of PHA and reducing anisotropy.

4.4.2 Effect of screw structure on the mechanical properties of WF/PHA composite

Screw structure influences the performance of the printed product mainly by affecting the plasticizing quality and the extrusion output. Therefore, the difference in Q between the two screws at different screw speeds was first compared and the results are shown in **Figure 16a**. The Q of the gradient screw was higher than that of the abrupt screw at screw speeds below 13 r/min. The difference between the two types of screws is mainly in the length of the three sections of the screw (**Figure 11b**), while the Q of the single screw depends on the metering section. The metering section of the gradient screw is shorter than that of the abrupt screw and conveys the melt

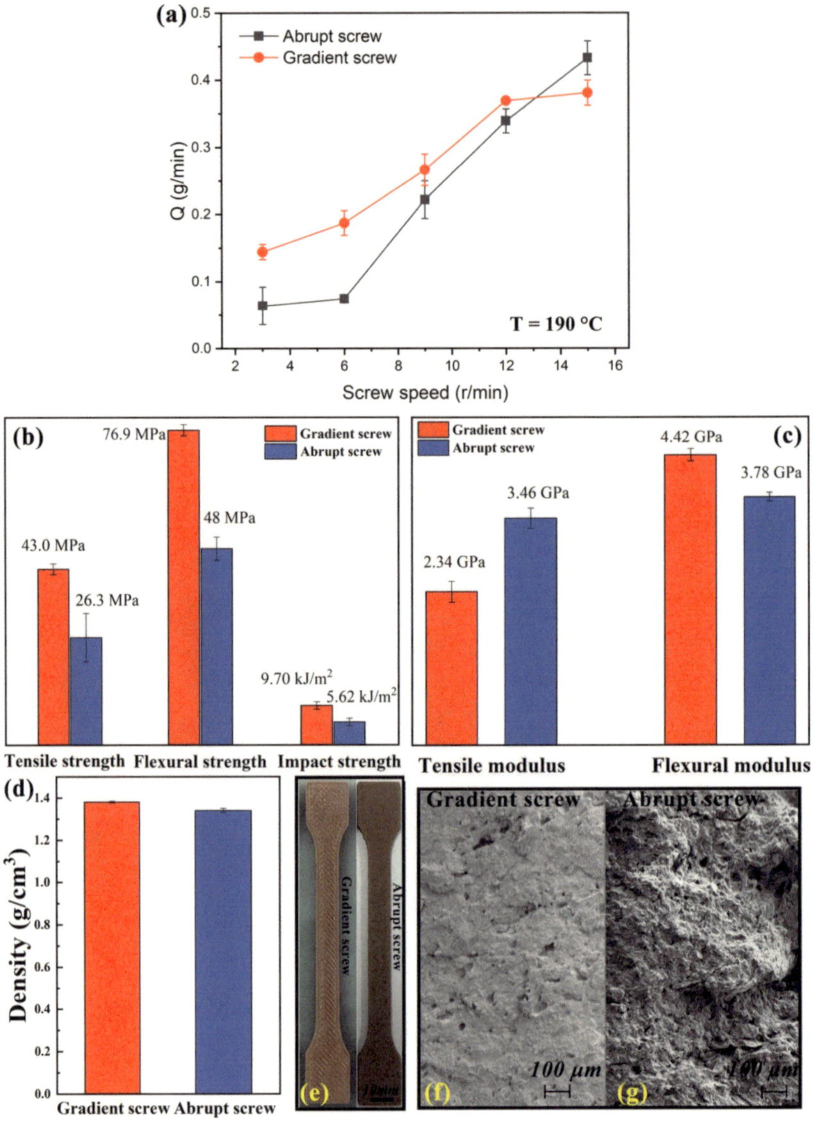

Figure 16.
(a) Extrusion output for two screws at different screw speeds, (b) Mechanical strength, (c) modulus, (d) density, (e) surface quality, and (f–g) SEM images of samples printed by two kinds of screws. (b–g) Adapted under the terms of the CC-BY Creative Commons Attribution 4.0 International license (https://creativecommons.org/licenses/by/4.0) [32]. Copyright 2021, The Authors, published by MDPI.

much more efficiently, resulting in a higher Q for the gradient screw. The higher Q results in a denser sample build-up, so the mechanical properties of the bars printed by the gradient screw were higher than those printed by the abrupt screw. This can be verified in **Figure 16d–g**, where the gradient screw-printed samples were denser than those of the abrupt screw, and the sample surface was uneven due to over-extrusion, while the abrupt screw printed samples had more porosity in the cross-section. This can be verified in **Figure 16d–g**: the densities of the samples printed by the gradient screw are higher than those of the abrupt screw; the surfaces of the samples printed by the gradient screw are uneven due to over-extrusion; and the cross-sections of the samples printed by the abrupt screw have more porosity.

The performance of micro-screw extrusion 3D-printed products is influenced in many ways by material properties, process conditions, and equipment parameters. Establishing a process-structure-performance relationship is, therefore, essential for choosing 3D printing material and design of equipment.

5. Conclusions

3D printing technology has shown rapid diversification in recent years and has unique advantages in the innovative manufacturing of customized products. To achieve the goal of green and sustainable economic development, combining 3D printing with biodegradable materials will not only broaden the range of methods for forming biodegradable products but also enrich the range of environment-friendly materials available for 3D printing.

In filament extrusion FDM, a promising biodegradable polymer blend PBS/PLA was formulated, which combined the advantages of high modulus and low warpage of PLA with the toughness of PBS. PBS60/PLA40 and PBS40/PLA60 blends have a better appearance, less warpage, and higher dimensional accuracy and can be used as an alternative to PLA for FDM 3D printing.

For pellet extrusion FDM, we have solved the problem that WF/PHA composites cannot be used in 3D printing owing to difficulties in filament extrusion by designing new 3D printing equipment. The micro-screw extrusion 3D printing device is characterized by wide material adaptability, high printing efficiency, and low cost. The printability of the composite material was evaluated by testing the extrusion flow rate stability and the pressure-building ability of the micro-screw extrusion device. The effects of material properties, equipment structure, and process parameters on the performance of printed products are explored. The process-structure-performance relationship of micro-screw extrusion 3D printing is further illustrated.

Searching for new biodegradable polymers that can be 3D printed is still in progress and will greatly widen the application of FDM-based 3D printing in both industry and our daily life.

Author details

Jing Tian[1], Yanyan Zheng[1], Qing Ouyang[1], Ping Xue[2], Baohua Guo[1] and Jun Xu[1*]

1 Department of Chemical Engineering, Tsinghua University, Beijing, China

2 Department of Mechanical and Electrical Engineering, Beijing University of Chemical Technology, Beijing, China

*Address all correspondence to: jun-xu@mail.tsinghua.edu.cn

References

[1] Kakadellis S, Rosetto G. Achieving a circular bioeconomy for plastics. Science. 2021;**373**(6550):49-50. DOI: 10.1126/science.abj3476

[2] Rosenboom JG, Langer R, Traverso G. Bioplastics for a circular economy. Nature Reviews Materials. 2022;**7**:117-137. DOI: 10.1038/s41578-021-00407-8

[3] Zhai Y, Lados DA, LaGoy JL. Additive manufacturing: Making imagination the major limitation. Journal of Metals. 2014;**66**(5):808-816. DOI: 10.1007/s11837-014-0886-2

[4] Casavola C, Cazzato A, Moramarco V, Pappalettere C. Orthotropic mechanical properties of fused deposition modelling parts described by classical laminate theory. Materials and Design. 2016;**90**:453-458. DOI: 10.1016/j.matdes.2015.11.009

[5] Mohamed OA, Masood SH, Bhowmik JL. Optimization of fused deposition modeling process parameters: A review of current research and future prospects. Advanced Manufacturing. 2015;**3**:42-53. DOI: 10.1007/s40436-014-0097-7

[6] Solomon IJ, Sevvel P, Gunasekaran J. A review on the various processing parameters in FDM. Materials Today Proceedings. 2020;**37**:509-514. DOI: 10.1016/j.matpr.2020.05.484

[7] Kumar NP, Kumar SA, Sankar MS. Effect of process parameters on the mechanical behavior of FDM and DMLS build parts. Materials Today Proceedings. 2019;**22**:1443-1451. DOI: 10.1016/j.matpr.2020.01.502

[8] Rajpurohit SR, Dave HK. Effect of process parameters on tensile strength of FDM printed PLA part. Rapid Prototyping Journal. 2018;**24**:1317-1324. DOI: 10.1108/RPJ-06-2017-0134

[9] Huang B, Meng S, He H, Jia Y, et al. Study of processing parameters in fused deposition modeling based on mechanical properties of acrylonitrile-butadiene-styrene filament. Polymer Engineering and Science. 2019;**59**:120-128. DOI: 10.1002/pen.24875

[10] Benali N, Hammami D, Khlif M, Bradai C. Optimization of FDM manufacturing parameters of a biodegradable thermoplastic (PLA). In: Proceedings of the Design and Modeling of Mechanical Systems - IV. CMSM 2019. Lecture Notes in Mechanical Engineering; 18-20 March 2019; Hammamet. Tunisia: Springer, Cham; 2020. pp. 355-362

[11] Chacón JM, Caminero MA, García-Plaza E, Núñez PJ. Additive manufacturing of PLA structures using fused deposition modelling: Effect of process parameters on mechanical properties and their optimal selection. Materials and Design. 2017;**124**:143-157. DOI: 10.1016/j.matdes.2017.03.065

[12] Wang P, Zou B, Xiao H, et al. Effects of printing parameters of fused deposition modeling on mechanical properties, surface quality, and microstructure of PEEK. Journal of Materials Processing Technology. 2019;**271**:62-74. DOI: 10.1016/j.jmatprotec.2019.03.016

[13] Dizon JRC, Espera AH, Chen Q, et al. Mechanical characterization of 3D-printed polymers. Additive Manufacturing. 2018;**20**:44-67. DOI: 10.1016/j.addma.2017.12.002

[14] Ahn SH, Montero M, Odell D, Roundy S, Wright PK. Anisotropic

material properties of fused deposition modeling ABS. Rapid Prototyping Journal. 2002;**8**:248-257. DOI: 10.1108/13552540210441166

[15] Chaudhry MS, Czekanski A. Evaluating FDM process parameter sensitive mechanical performance of elastomers at various strain rates of loading. Materials (Basel). 2020;**13**:1-10. DOI: 10.3390/ma13143202

[16] Zaldivar RJ, Witkin DB, McLouth T, et al. Influence of processing and orientation print effects on the mechanical and thermal behavior of 3D-Printed ULTEM ® 9085 material. Additive Manufacturing. 2017;**13**:71-80. DOI: 10.1016/j.addma.2016.11.007

[17] Hanon MM, Marczis R, Zsidai L. Influence of the 3D printing process settings on tensile strength of PLA and HT-PLA. Period Polytechnical and Mechanical Engineering. 2020;**65**:38-46. DOI: 10.3311/ppme.13683

[18] Guessasma S, Belhabib S, Nouri H. Microstructure and mechanical performance of 3D printed wood-PLA/ PHA using fused deposition modelling: Effect of printing temperature. Polymers (Basel). 2019;**11**:1778. DOI: 10.3390/ polym11111778

[19] Chan CM, Vandi LJ, Pratt S, et al. Composites of wood and biodegradable thermoplastics: A review. Polymer Reviews. 2018;**58**(3):444-494. DOI: 10.1080/15583724.2017.1380039

[20] Ou-Yang Q, Guo B, Xu J. Preparation and characterization of poly(butylene succinate)/Polylactide blends for fused deposition Modeling 3D printing. ACS. Omega. 2018;**3**(10):14309-14317. DOI: 10.1021/acsomega.8b02549

[21] Venkataraman N, Rangarajan S, Matthewson MJ, et al. Feedstock material property–process relationships in fused deposition of ceramics (FDC). Rapid Prototyping Journal. 2000;**6**:244-253. DOI: 10.1108/13552540010373344

[22] Vandi LJ, Chan CM, Werker A, et al. Wood-PHA composites: Mapping opportunities. Polymers (Basel). 2018;**10**:751. DOI: 10.3390/polym10070751

[23] Reinsch VE, Kelley SS. Crystallization of poly(hydroxybutrate-co-hydroxyvalerate) in wood fiber-reinforced composites. Journal of Applied Polymer Science. 1997;**64**(9):1785-1796. DOI: 10.1002/ (SICI)1097-4628(19970531)64:9<1785

[24] Gonzalez Ausejo J, Rydz J, Musioł M, Sikorska W, Janeczek H, Sobota M, et al. Three-dimensional printing of PLA and PLA/PHA dumbbell-shaped specimens of crisscross and transverse patterns as promising materials in emerging application areas: Prediction study. Polymer Degradation and Stability. 2018;**156**:100-110. DOI: 10.1016/j. polymdegradstab. 2018.08.008

[25] Gerard T, Budtova T. Morphology and molten-state rheology of polylactide and polyhydroxyalkanoate blends. European Polymer Journal. 2012;**48**(6):1110-1117. DOI: 10.1016/j. eurpolymj.2012.03.015

[26] Kariz M, Sernek M, Kuzman MK. Use of wood powder and adhesive as a mixture for 3D printing. European Journal of Wood Product. 2016;**74**(1):123-126. DOI: 10.1007/ s00107-015-0987-9

[27] Kariz M, Sernek M, Obućina M, Kuzman MK. Effect of wood content in FDM filament on properties of 3D printed parts. Materials Today Communication. 2018;**14**:135-140. DOI: 10.1016/j.mtcomm.2017.12.016

[28] Mazzanti V, Malagutti L, Mollica F. FDM 3D printing of polymers containing

natural fillers: A review of their mechanical properties. Polymers (Basel). 2019;**11**:1094. DOI: 10.3390/polym11071094

[29] Godoi FC, Prakash S, Bhandari BR. 3d printing technologies applied for food design: Status and prospects. Journal of Food Engineering. 2016;**179**:44-54. DOI: 10.1016/j.jfoodeng.2016.01.025

[30] Lanaro M, Forrestal DP, Scheurer S, et al. 3D printing complex chocolate objects: Platform design, optimization and evaluation. Journal of Food Engineering. 2017;**215**:13-22. DOI: 10.1016/j.jfoodeng.2017.06.029

[31] Tian J, Zhang R, Wu Y, et al. Additive manufacturing of wood flour/polyhydroxyalkanoates (PHA) fully bio-based composites based on micro-screw extrusion system. Materials and Design. 2021;**199**:109418. DOI: 10.1016/j.matdes.2020.109418

[32] Tian J, Zhang R, Yang J, et al. Additive manufacturing of wood flour/PHA composites using Micro-screw extrusion: Effect of device and process parameters on performance. Polymers (Basel). 2021;**13**(7):1107. DOI: 10.3390/polym13071107

[33] Yan Y, Li S, Zhang R, et al. Rapid prototyping and manufacturing technology: Principle, representative technics, applications, and development trends. Tsinghua Science and Technology. 2009;**14**:1-12. DOI: 10.1016/S1007-0214(09)70001-X

Section 4

3D Printing in Healthcare

Chapter 11

Application of Three-Dimensional Printing in Surgical Planning for Medical Application

Wei-Ling Chen, Tsung-Lung Yang, Jieh-Neng Wang and Chung-Dann Kan

Abstract

Three-dimensional printing (3DP) is an evolving technology with a wide range of medical applications. It complements the traditional methods of visualizing the cardiovascular anatomy and assists in clinical decision making, especially in the planning and simulation of percutaneous surgical procedures. The doctor–patient relationship has changed substantially, and patients have become increasingly aware of their rights and proactively make decisions regarding their treatment. We present our experience in using 3DP for aortic repair, preoperative surgical decision making for congenital heart disease, and simulation-based training for junior vascular surgeons. 3DP can revolutionize individualized treatment, especially for congenital heart disease, which involves unique anatomy that is difficult to examine using traditional computed tomography. As cardiovascular medicine and surgery require increasingly complex interventions, 3DP is becoming an essential technology for surgical instructors and trainees, who can learn to become responsible and humane medical doctors. 3DP will play an increasingly crucial role in the future training of surgeons.

Keywords: doctor–patient relationship, patient-specific geometry, three-dimensional printing (3DP), surgical decision making, personalized 3D-printed cast

1. Introduction

A thorough preoperative evaluation of anatomy and proper surgical planning are essential to successful operations. Clinicians must be able to assess the patient's anatomy to quickly make surgical plans and provide information to reduce surgical risk and operating time [1]. However, the availability of imaging modalities (computed tomography [CT] and magnetic resonance imaging [MRI]) and the cost of technology affect the acquisition of information and images for surgical plans. Clinicians can usually rely on their experience and two-dimensional (2D) medical imaging to guide their decision-making process. Although three-dimensional (3D) postprocessed images improve upon traditional 2D image sets, they usually do not provide sufficient information for surgical simulation. Medical 3D printing (3DP) is a rapidly advancing technology that can provide innovative solutions to problems in preoperative planning [2, 3]. The doctor–patient relationship may change substantially as patients become increasingly aware of their rights and proactively make decisions about their treatment; medical 3DP can help patients select their treatment from the available options [4]. Surgeons' decisions may sometimes be questioned. Therefore, surgeons must inform patients and their families about treatments, the process of the operation, and the pros and cons of the operation to help patients select their desired treatment method.

Because of considerable improvements in surgical intervention and the development of endovascular stent grafts since 2000, the use of 3DP for surgical planning may reduce surgical risk and improve outcomes and the doctor–patient relationship. Before any examination or surgery, clinicians should discuss the risks, benefits, and alternatives with patients. Although informed consent is integral to high-quality medical practice, doctors should advise their patients of all risks; negligence during this process can lead to a breakdown in the doctor–patient relationship [5–10]. However, this represents a challenge for many providers of healthcare services. Fortunately, 3DP has improved surgical processes and the quality of information on which patients may base their consent. However, the following limitations remain [11, 12]:

1. Accurate image capture by transthoracic echocardiography, MRI, and CT is essential for 3DP.

2. Risk scores are suboptimal for individuals with anatomical diversity or unique traits because they provide population estimates rather than patient-specific estimates, even though risk scores represent all potential variables.

3. Complex disorders make imaging, therapy, and intervention more challenging, which has an impact on the results.

The following chapters discuss clinical applications of 3DP in medicine, including a wearable orthopedic brace that can monitor vascular access stenosis dysfunction; the use of 3DP to enable zone zero thoracic aortic endovascular repair for ascending aorta disease; a case report of surgical planning for congenital heart disease; a simulation-based training program for junior vascular surgeons; and additional clinical cardiovascular applications.

2. History of 3D printing in clinical application

The term "3D printing" has come to refer to a group of related manufacturing processes that use digital data to produce physical models. In the broad spectrum of additive manufacturing techniques now used in the industry, 3D printing stands out. The term refers to the procedure through which a three-dimensional physical object is created from a digital model [13]. Nowadays, 3D printing technology represents an opportunity to help pharmaceutical and medical companies create more specific drugs, enabling rapid production of medical implants and changing how doctors and surgeons plan procedures [1]. In addition, 3D printing can significantly improve the research knowledge and skills of the new generation of surgeons. Improving patient and surgeon relationships [14] and medical applications was initially reported in the early 2000s [15]. Initially, these reports focused on custom prostheses [16], but as the technology improved, reports of using anatomic models for preoperative planning began appearing [17]. The recent rapid growth of 3D printing in medicine has been staggering. Specifically, the anatomic data radiologists receive and interpret daily can be used in 3D printing to provide tailored medication. Offering such a service could be a method for radiology to show its worth in in-patient treatment and a new avenue for interaction with referring clinicians. Moreover, radiologists have seen the development of medical imaging that enables 3D printing. CT and MRI's multiplanar imaging led to 3D reconstructions, which enhanced the assessment of intricate anatomy [18–20]. As described, 3D printing transfers digital image data from a flat-screen into the physical world's third dimension [20].

Yearly, the scope of applications for 3D printing in the medical field broadens, making it possible to save lives and improve quality of life in ways that were previously imagined. In point of fact, three-dimensional printing has been put to use in a wide variety of medical specialties, including cardiothoracic surgery, cardiology, gastroenterology, gastroenterology, neurosurgery, oral and maxillofacial surgery, orthopedic surgery, plastic surgery, podiatry, pulmonology, radiation oncology, transplant surgery, urology, and vascular surgery [21–34].

Below we list some of the most prominent direct uses of 3D printing in the clinical and medical arena [35–37]:

1. Integrating clinical and imaging data will be an integral part of the multi-step process that will determine the optimal treatment choice via personalized pre-surgical planning.

2. 3D-printed models educate medical students, residents, and patients.

3. Lowered expenditures associated with healthcare, shorter hospital stays after surgery, and fewer cases requiring further medical attention are all benefits.

4. With 3D printing, it is possible to choose the precise dimensions of prosthetic parts before implantation.

5. 3D printing may make personalized implants or surgical guidance and devices for specific procedures.

6. To follow a pharmacological treatment, 3D printing helps invalidate the results achieved by the patient.

7. The viability and efficacy of a cardiovascular system in preventing and treating peripheral and coronary artery disease.

8. Patient education: patients may not fully understand 2D images (CT or MRI) representation of a 3D anatomy. Therefore, 3D printing may improve doctor-patient communication by showing the anatomic model directly.

9. Boosting medical training using 3D-printed models of individual patients, which may improve outcomes and facilitate quick comprehension.

10. Bioprinted by pharmaceutical industries to replace animal models for analyzing the toxicity of new drugs.

3. 3D printing methods

3.1 Image acquisition modalities

Surgical outcomes depend on experience and technical skills. However, outcomes and operative times can be improved if anatomical information is acquired before surgery. MRI, CT, and echocardiography enable clinicians to obtain valuable information regarding the physiology and structural details of patients with cardiovascular disease. 3DP analysis software can help inform less-experienced surgeons. For example, the Mimics Care Suite (Materialize, Leuven, Belgium) is medical software that offers an effective and efficient method of evaluating stent appositions by using 3D imaging. Advances in 3DP technology have enabled clinicians to examine 3D structures in preoperative planning [38]. 3D Slicer is free, open-source, extensible operating software for medical image visualization and computation (www.slicer.org). 3D Slicer supports CT, MRI, positron emission tomography, X-ray, and ultrasound images [3, 38, 39].

3D models can be viewed from any angle. In preoperative planning, 3D models can be used to determine the proper location for an endograft and reduce surgical risk. Surgeons can apply the detailed anatomical information provided by 3D models to solve critical problems. Surgeons can also use 3D models to accurately determine the location of a pseudoaneurysm lumen and whether an aneurysm entrance may be covered by a stent during thoracic endovascular aortic repair [40–43]. 3DP software can also facilitate stent sizing. We used 3D Slicer to create an STL file for 3DP.

3.2 Methodology- creating an STL file for medical printing using 3D slicer

Cardiovascular images were obtained from contrast-enhanced CT scan data in accordance with standardized scanning protocols [44]. These CT images were used to build an STL package. 3D Slicer has function that exports reconstructed 3D images as STL files for 3DP. **Figure 1** shows the 3D printing framework: first, gather CT or MRI pictures of the patient; second, import DICOM images of the scanned component into the 3D Slicer; third, model generation; fourth, construct a printable 3D model; and fifth, print 3D objects.

3.3 CT with digital imaging and Communications in Medicine file format as the standard for storing images

Digital Imaging and Communications in Medicine (DICOM) is the most widely used standard for sharing information about medical imaging worldwide. Using

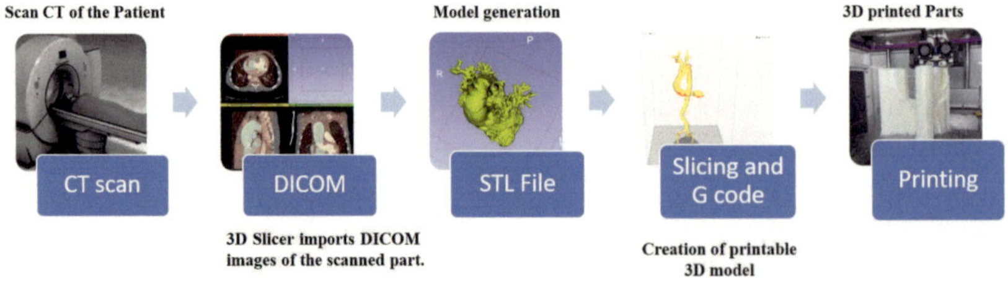

Figure 1.
3D printing framework. (A) Gather CT or MRI pictures of the patient (B) import DICOM images of the scanned component into the 3D slicer (C) model generation (D) construct a printable 3D model (E) print 3D objects.

DICOM images to print in 3D has become more popular. With the growth of technologies like medical and imaging engineering, as well as the improvement of hardware and software and their falling prices, life has become more accessible. For example, more and more patient-specific 3D models, such as teaching, planning, and simulating surgery, are used in cardiovascular areas. DICOM image 3D printing uses 2D pictures stacked on top of each other and then broken up into the data format that the 3D printer needs. DICOM pictures are being split into a 3D computer-aided design format for intermediate data, which can be used for first-stage processing, like setting up an ROI. Out of the roughly 100 file formats of 3D CAD data that are used as 3D native files and intermediate files, STL is the most common format for 3D printing. There are several commercials (paid) and open-source (free) software tools for converting DICOM images to STL data, and all of them can be run on a regular personal computer.

After creating a 3D picture, the following phase is to convert it into a mesh structure for verification before printing. The subsequent step of segmenting the pictures into regions of interest based on characteristics such as brightness or contrast is a crucial part of the process. The decision of how to segment a picture can be helped by virtual rendering. 3D Slicer is equipped with a virtual rendering module for rapid 3D modeling, enabling medical professionals to inspect lesions and see photographs quickly. On the other hand, virtual rendering models cannot be immediately transformed into the STL format necessary for 3D printing. The segmentation process can be carried out using a variety of approaches, the most common of which are completely automatic, semiautomatic, and entirely manual. The segmentation process is based on the built-in tools available in 3D Slicer for sketching and coloring tissue outlines and lesions. Layered structures can be helpful for segmentation because they enable medical professionals to distinguish tissues and highlight connections between lesions and neighboring healthy tissues using different colors. Segmentation may now be finished more rapidly as a result. 3D Slicer may also be used to do this.

The software that can be used to segment images and the software that creates STL files from DICOM data differ. Therefore, the characteristics of each must be understood (**Figure 2**).

Step A: The first step requires dragging and dropping the folder containing the DICOM images onto the 3D Slicer's welcome window.

Step B: To add layers, users must utilize the Segment Editor. Each Segment Editor module has a collection of default parameters matching a specific anatomical component.

Step C: Paint each layer using the paint tool (growing for seeds). The algorithm makes each seed grow at the same time. Since growth is faster in areas where everything is the same, the lines between segments are where there is a significant change in image intensity. Therefore, If the wrong segment is grown anywhere, then paint some more seeds there with the correct segment. Since this usually happens in low-contrast regions, typically, users need to paint with two colors on both sides of hardly visible boundaries.

Step D: A preview.

Step E: A new layer must be combined with an existing one using logical operators, and then the combined layer must be used before moving on to the next step.

Step F: Create 3D-printable vessel walls from contrast-CT volumes. "hollow" has been added to the Segment editor to create hollow objects from solid objects in a straightforward step.

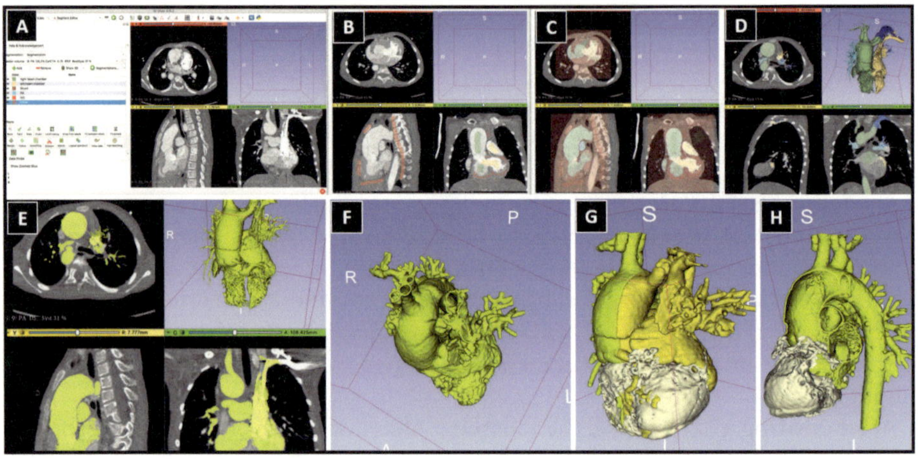

Figure 2.
Computer tomography uses DICOM as the standard format for storing patient images (A) drag and drop the folder containing your DICOM images onto the slicer welcome window (B) (C) (D) "segment editor" modules have a series of presets for different anatomy structures. (E) Using a new layer to join the layer just done logical operators. (F) Hollowing (G) ventricular wall (H) smooth tool.

Step G: Create a mask layer. Paint the ventricular wall by using the spherical brush tool. The primary goal of this step is to make a mask so that paint may be applied using a spherical brush to the ventricular compartment's wall.

Step H: Using the smoothing tool, make it even in this step (median, hole closing).

4. Using on-demand, multi-sensor wearable 3d-printed medical device for hemodialysis patient care

According to the 2020 Annual Data Report of the US Renal Data System, Taiwan has the highest end-stage renal disease (ESRD) prevalence. HD accounts for about 69% of all renal replacement therapies and 88% of all dialysis worldwide. According to Taiwan's National Health Insurance (NHI) Administration, chronic kidney disease is the most expensive to treat. The National Health Insurance (NHI) annually spends NT$53,3 billion ($1.8 billion) on hemodialysis (HD) for 87,000 patients, placing considerable pressure on the healthcare budget. Moreover, almost ninety percent of Taiwanese patients select hemodialysis as their primary dialysis treatment [45, 46].

Long-term dialysis effectiveness in HD patients depends on preserving Arteriovenous access (AVA) [47]. The high prevalence of AVF maturation failure is the first persistent issue in HD patients; AVA patency loss is also prevalent. Arteriovenous graft (AVG) has a much lower useful life than arteriovenous fistula (AVF) (AFV). With a combined incidence of 66–73% in AVF and 84% in AVG, stenosis, and thrombosis are the primary causes of access dysfunction. Juxta-anastomotic venous stenosis affects between fifty percent and seventy-one percent of AVAs [48, 49]. For hemodialysis therapy to be deemed efficient, AVA flow rates must reach between 600 and 1000 mL/min. After continuous use of these methods, pathogenic alterations occur (e.g., high venous pressure or insufficient blood flow). When the arterial lumen is decreased by 70%, the stenotic segment must be dilated, or the intraluminal thrombus must be removed.

Recent developments in wireless ultrasonography have made rapid scans possible. Although these devices are not yet suitable for independent professional use or comprehensive testing of AV operation, they may serve as an alternative, cost-effective portable device for stenosis identification. Clinicians have tried qualitative and quantitative investigation of bruits since 1970, resulting in the creation of AVA dysfunction monitoring devices. However, the sensitivity of present monitoring equipment varies from 35–80% due to the spectral characteristics of functional and dysfunctional vascular access [50–52]. The audible sound of turbulent blood flow is known as a bruit. Phonoangiography (PAG), the recording and statistical analysis of bruits, has been utilized to assess vascular access in HD patients for many decades. The features of turbulent blood flow induced by a restricted channel are objectively described by spectral analysis of bruits.

On the other hand, the acoustic approach correctly reflects the directions of hemodynamic changes related to the turbulence state. From the perspective of Spencer's and Reid's curve [53], if the cross-sectional area decreases by more than 96% or the diameter decreases by more than 85%, the velocimetry may underestimate the cross-sectional area by 90% or less or the lumen diameter reduction by 70%. To understand any blood flow velocity, one needs to consider whether it was measured on the upslope, downslope, or "opposite side" of Spencer's curve. Therefore, it is vital to ensure that flow volume and velocity changes correspond. The flow volume falling to or below 20–30% of the usual flow suggests a greater degree of stenosis than anticipated. Consequently, combining PAG and flow volume changes may boost the probability of early and accurate stenosis prediction.

Photoplethysmography (PPG) is a low-cost, non-invasive, and straightforward optical technology that detects changes in peripheral circulation volume. However, wearable PPG sensors can only be positioned at specific skin sites. The light source of a PPG device transmits light to a tissue, and the photodetector analyses the light reflected from the tissue. The light reflected is proportional to fluctuations in blood volume. PPG sensors that measure blood flow volume are often powered by an infrared light-emitting diode (IR-LED) or green LED.

In recent years, several medical institutions have used 3D printing technology to create personalized external prostheses, surgical guides for implant placement, simulated implants, and other devices for preoperative planning [45]. In 2018, we developed a prototype to capture AVA hemodynamic data using AVA-targeted sensors implanted in a customized 3D-printed cast. In addition, we will request on-duty nursing personnel to evaluate sensor readings using their stethoscopes qualitatively. In addition, statistically, we will employ Doppler and grayscale ultrasonography as the gold standard to determine the AVA stenosis's ultimate status. In situations of severe blockage (i.e., 95% occlusion), however, noises may not be heard due to limited blood flow, resulting in a significant chance of false-negative findings. For improved and more accurate stenosis diagnosis, we integrated PPG (for detecting volumetric changes in AV accesses) and PAG (for detecting changing pitch patterns in AV accesses).

In recent years, the use of 3D printing has expanded significantly. Certain medical institutions have used this technology to create bespoke external prostheses, surgical guides for implant placement, simulated implants, and other preoperative planning assistance. The device's position is an essential aspect that might influence the accuracy of PAG and PPG instruments. 3D-printed, multi-sensor, wearable medical devices are a feasible method for minimizing or eliminating the influence of location.

Chen et al. combined PPG and PAG sensors to create a wearable device with increased diagnostic accuracy for detecting AVA malfunction (**Figure 3**) [54, 55].

5. Surgical predictive plan to assistant the spatial relationship of patient-specific geometry

5.1 Case 1 effective thoracic endovascular repair in acute type a aortic dissection

Acute type A aortic dissection (ATAAD) is a catastrophic disease. The risk of death without surgery is approximately 65%; the mortality rate increases by 1–2% per hour in the first 24 hours. Surgical options are often limited by open interposition grafts through median sternotomy wounds [56]. The global prevalence of ATAAD ranges from 5 to 10 cases per 1 million people, and the estimated incidence is approximately 11.9 cases per 100,000 person-years. However, approximately 25% of cases with ATAAD are considered inoperable [57–60]. Open repair surgery for patients with critical illness can entail high surgical risks and operative and hospital mortality rates and long hospital stays. Thoracic endovascular stent graft implantation is an alternative for patients with certain types of ascending aortic disease who are not suitable for traditional open surgery.

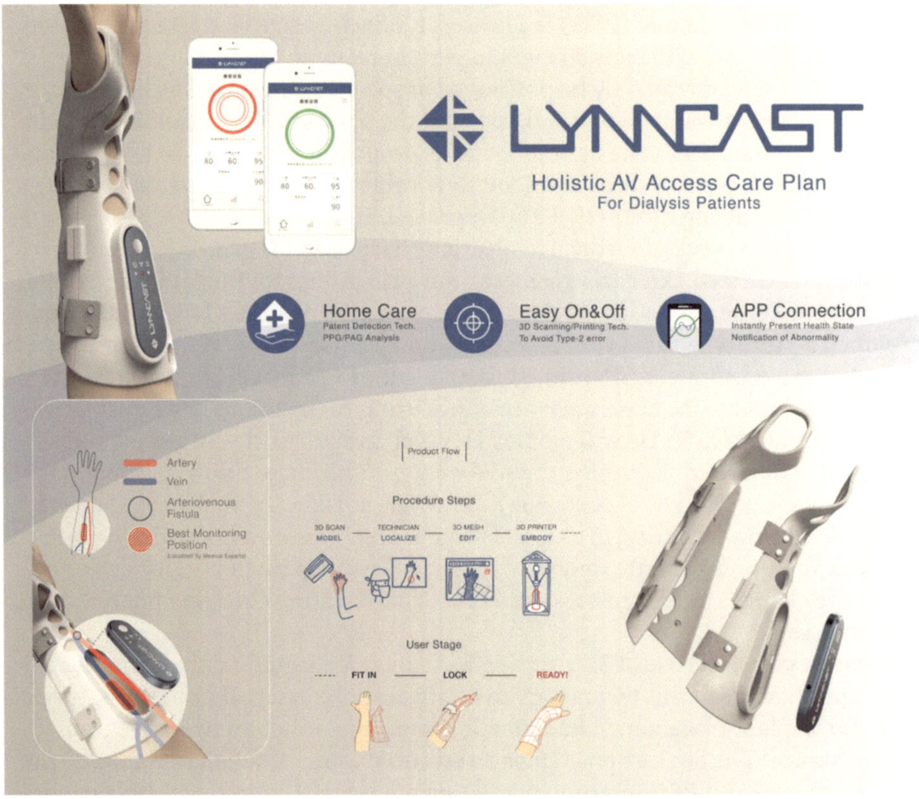

Figure 3.
Using on-demand, multi-sensor wearable 3D-printed medical device for hemodialysis patient care.

In the past two decades, endovascular aortic repair (EVAR) has been successfully proved and largely used to be a good alternative strategy for abdominal aortic and descending thoracic diseases. Due to the development of fenestrated and branched stent-grafts, diseases involving the visceral branches of the descending aorta and even diseases on the aortic arch have become diseases that can be treated with stent-grafts [61]. However, there is a fundamental difference between the ascending aorta and the descending aorta diseases. The ascending aorta is challenged by hemodynamic and anatomical limitations, especially under the condition of aortic dissection. The geometry of the ascending aorta is greatly affected by the aortic dissection, which leads to an increase in the diameter of the ascending aorta and loss of structural wall integrity through delamination. Therefore, it complicates the identification of a stable landing zone. At present, empirical reports on the endovascular repair of the ascending aorta are just limited to case series or case reports. Since there is no aortic stent-graft design specifically for the ascending zone, the FDA has not approved any commercial aortic stent-graft devices for the ascending aorta [62, 63].

We can try to use computed tomography 3D reconstruction and 3D Slicer software version 4.13.0 (3D Slicer contributor, http,//www.slicer.org) to help clarify the spatial relationship of the patient's specific geometry, and then successfully apply the aortic stent graft system to treat the type II acute aortic dissection diseased patient. In order to accurately understand the applicability of the TEVAR devices in patients, we can also use three-dimensional printing (3DP) technology to reconstruct the real condition of the specific landing zone (**Figures 4** and **5**). In this way, we may quickly gain the goals of early diagnosis and the tasks of meticulous planning, thereby achieving successful treatment results.

An 88-year-old man suddenly lost consciousness at the scene and was rushed to the emergency room by emergency medical services. He has a history of hypertension

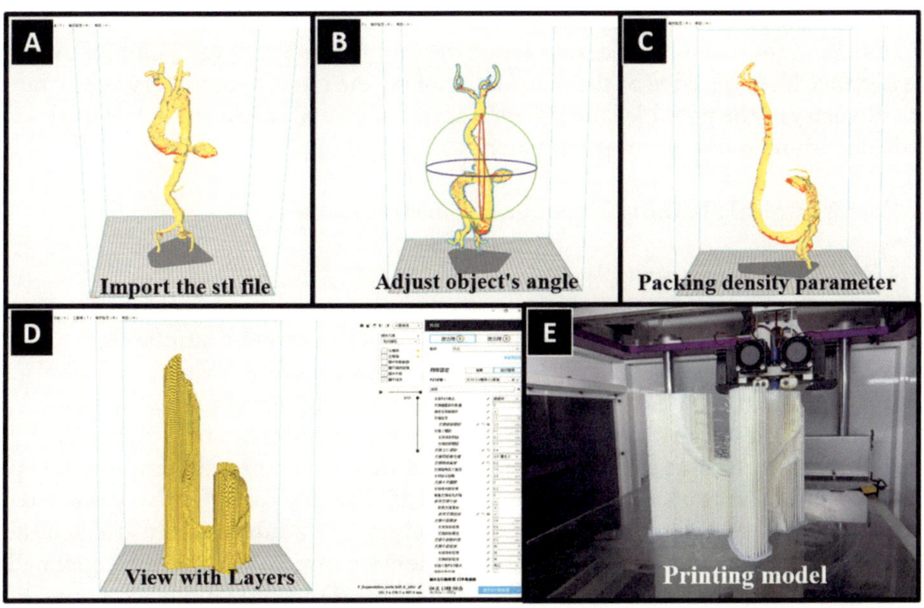

Figure 4.
(A) Import stereolithography of the patient (B) adjust the object's orientation (C) set the packing density parameter (D) view the model with layers (E) print the model.

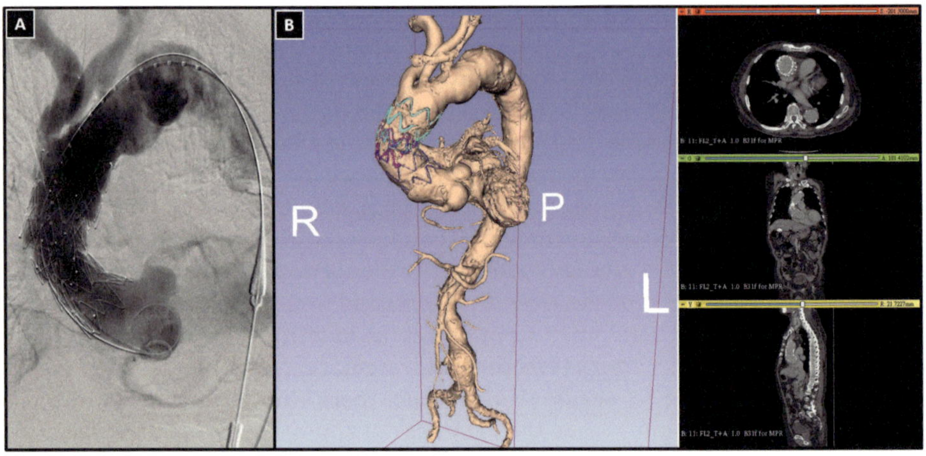

Figure 5.
3D reconstruction and 3D slicer software to reconstruct the 3D image of the aorta; (A) the patient's final aortic angiography showed complete coverage of the dissection sac orifice, satisfactory alignment of TEVAR stents, and no signs of endoleaks (B) with the assistance of computed tomography 3D reconstruction and 3D slicer software, the operation was completed as smoothly as our preoperative plan.

and infra-renal aortic dissection. An emergency computed tomography (CT) scan of the head showed no significant findings. However, his chest CT showed a small amount of pericardial effusion and acute DeBakey type II dissection. An extensive tear hole lesion originated in the middle of the ascending aorta, and the obvious giant pseudoaneurysm and hematoma extended down to the coronary ostium and up to the proximal orifice of the innominate artery. The maximum diameter of the pseudoaneurysm of the ascending aorta was measured to 55.6 mm. The sinotubular junction (STJ) measurement value is 34.6 mm × 31.9 mm. The distance from the edge of the tear hole at the proximal end of the greater curvature of the aorta to the STJ was 67.5 mm, and the distance from the side of the minor curvature was 38.0 mm. And, the distance from the edge of the distal tear hole to the innominate artery is 23.8 mm. The diameter of the possible landing zone at the distal end measured at the level of the innominate artery is approximately 40.8 × 40.3 mm.

5.2 Case 2 surgical planning for congenital heart disease

5.2.1 Double outlet right ventricle

Most cases of double outlet right ventricle (DORV) present a unique challenge to congenital cardiac surgeons. The relationships among the ventricles, ventricular septum, and great arteries can vary in DORV, and the condition presents a range of clinical manifestations secondary to changes in infundibular and intracardiac morphology. The unique spatial arrangement of these structures determines the optimal type of intervention. Although traditional 2D imaging can effectively represent these structures, the preoperative 3DP of models in case 2 substantially affected the time to extubation and length of stay in the intensive care unit. Zhao et al. operated on 25 patients with DORV and studied preoperative 3D-printed models for one-third of the patients [64]. Although whether these patients were selected randomly was unclear, cardiopulmonary bypass and cross-clamp times were lower in the 3DP group; however, these differences were nonsignificant. Significant differences in time to extubation and length of stay in the intensive care unit were observed, possibly suggesting improved surgical outcomes. In our case, a 7-year-old boy presented with complex congenital heart disease (DORV with a large subaortic ventricular septal defect [VSD], and d-looping transposition of the great arteries [D-TGA]), infundibular pulmonary stenosis, had received a modified Blalock–Taussig (BT) shunt at 6 months of age. His condition was stable after a palliative operation, but he was admitted for surgical evaluation and intervention after progressive exercise intolerance and cyanosis. Echocardiography revealed (1) situs solitus and levocardia; (2) a large VSD; (3) atrioventricular concordance and ventricular arterial discordance; (4) the aorta and pulmonary trunk arising from the right ventricle (RV) with the aorta right-anterior to the pulmonary trunk, compatible with D-TGA; (5) severe infundibular pulmonary stenosis; and (6) left-side aortic arch. Chest CTA revealed similar findings. Cardiac catheterization hemodynamic data revealed that aortic pressure was 116/75 mmHg with 90.7% oxygen saturation (SaO_2), superior vena cava (SVC) pressure was 8 mmHg (mean) with 78.5% SaO_2, inferior vena cava (IVC) pressure was 9 mmHg (mean) with 80.6% SaO_2, right atrium (RA) pressure was 8 mmHg (mean) with 80% SaO_2, RV pressure was 119/12 mmHg with 82.2% SaO_2, and pulmonary artery (PA) pressure was 12/2 mmHg (mean) with 84.1% SaO_2. Our surgical plans were total cavopulmonary connection (TCPC) or physiological biventricular repair (VSD patch repair with an intracardiac tunnel or right ventricular outflow tract reconstruction with RV-to-PA valved conduit [Rastelli operation]). We used CTA data for surgical simulation through 3DP (**Figure 6**). The 3DP analysis indicated that physiological biventricular repair with left BT shunt ligation was most feasible for the patient. The surgical procedure was conducted as planned, and the results were satisfactory. Currently, the patient's condition is stable. He has received regular follow-up in the outpatient clinic for 2 years and does not require further intervention.

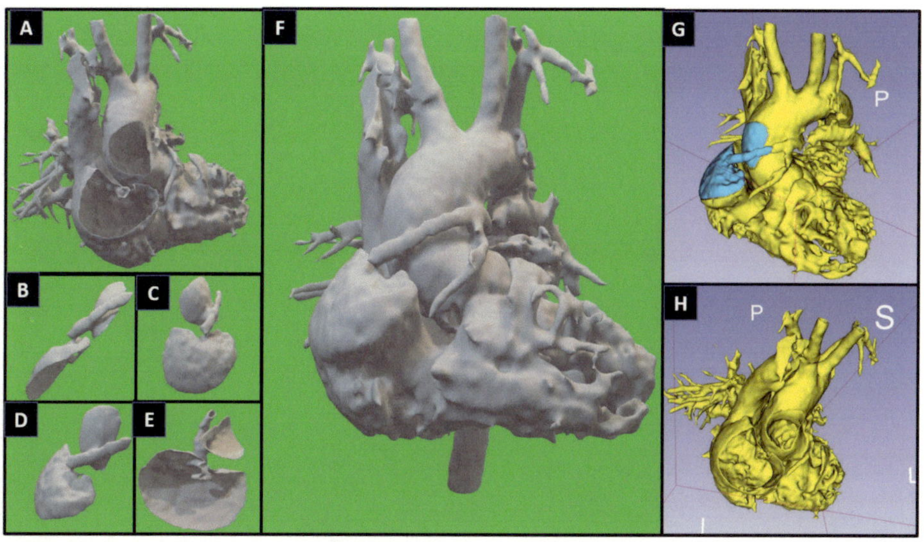

Figure 6.
Congenital heart for surgical planning; (A) 3D printing (B) cutting the 3D model for surgical planning (C) (D) (E) ventricular wall (F)3D dimensional printing congenital heart (G) (H) reconstruct the real condition of the congenital heart.

6. Transposition of the great arteries (TGA)

A 10-year-old girl presented with complicated congenital heart disease, situs inversus, levocardia, a large VSD, an atrial septal defect (ASD), and corrected transposition of the great arteries with pulmonary atresia. She had received a right modified BT shunt at 1 month of age and a left modified BT shunt at 3 years of age. Her condition stabilized after these palliative operations, but she was admitted for surgical evaluation and intervention because of progressive exercise intolerance and cyanosis. Echocardiography revealed (1) situs inversus and levocardia; (2) a large type II VSD with a bidirectional shunt; (3) atrioventricular discordance and ventricular arterial discordance, with the aorta right-anterior to the atretic PA; and (4) right aortic arch with patent bilateral BT shunts and bilateral SVCs. Chest CTA revealed similar findings. Cardiac catheterization hemodynamic data revealed that aortic pressure was 106/58 mmHg, SVC pressure was 9 mmHg (mean) with 75.1% SaO_2, IVC pressure was 9 mmHg (mean) with 79.1% SaO_2, RA pressure was 8 mmHg (mean) with 78.4% SaO_2, RV pressure was 87/14 mmHg with 88.2% SaO_2, left atrium pressure was 9 mmHg (mean) with 99.9% SaO_2, LV pressure was 93/14 mmHg with 85.5% SaO_2, and PA pressure was 16 mmHg (mean) with 93.4% SaO_2. Our surgical plan involved TCPC, physiological biventricular repair (morphological LV-to-PA valved conduit), and physiological biventricular repair with anatomical repair (double [arterial plus atrial] switch). We used the CTA data for surgical simulation through 3DP (**Figure 7**). We discovered that physiological biventricular repair with VSD and ASD repair was the most feasible option for this patient. The position for the bilateral BT shunt was identified using the 3D-printed anatomical structure, facilitating identification of the structure and adhesion tissue during surgery. The surgical procedure was conducted as planned, and the results were satisfactory. Currently, the patient's condition is stable. She has received regular follow-ups in the outpatient clinic for 2 years without requiring further intervention.

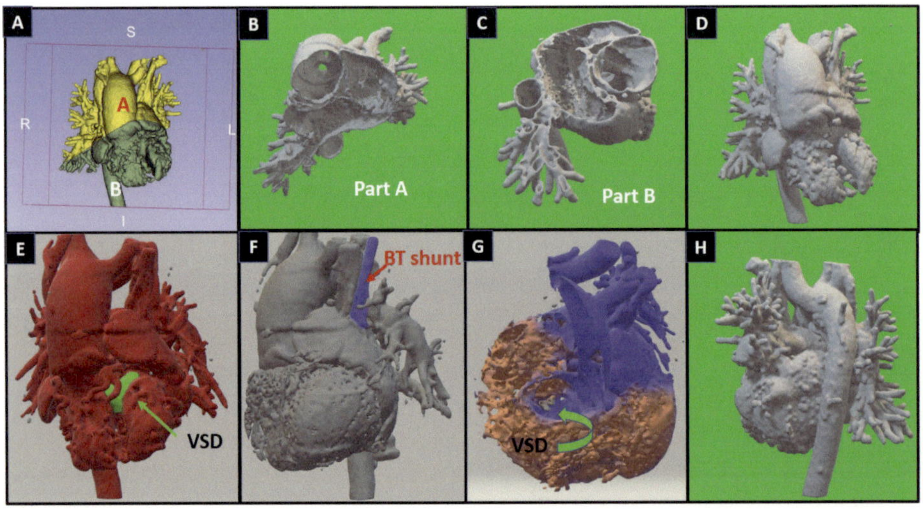

Figure 7.
Surgical planning in medical application; (A) (B) (C) (D) 3D printing for congenital heart, (E) ventricular septal defect (VSD), (F) BT shunt, (G) 3D printing model.

6.1 Training scenario: simulation-based training increases training efficiency for young surgeons

Simulation has been widely used in various surgical training programs for laparoscopic surgery, virtual aortic stent graft procedures, and ultrasound to allow surgeons to practice advanced surgery procedures and acquire new skills [6–8]. Various simulation platforms can be used. Anatomical simulations and presurgical planning may improve surgical results and patient safety, and 3DP can improve both. 3DP can provide anatomical information and a visualization of structural relationships. It can also be used to create a model of a patient's anatomy. We inserted silicone tubing into a printed leg structure to simulate arteries and veins and instructed participants to perform vascular exploration and anastomosis in deep and difficult-to-access areas [1–3]. We used a previous training plan to evaluate the results. A total of 28 students participated in the training course and completed a questionnaire. Most of the participants came from one of three job categories; 75% worked in medical centers, 58.3% were attending surgeons, and 66.6% had less than 5 years of working experience. The effect of the course was evaluated on the basis of the participants' perceptions of their disease familiarity, confidence, and competence, all of which improved significantly ($p < 0.001$). The effect on young surgeons was more substantial than that on attending surgeons. However, trainees from nonmedical centers progressed further than did trainees from medical centers. A total of 95.4% of the participants were satisfied with the course [6].

7. Discussion

3DP and fast prototyping have been used in the medical field since the early 2000s and gradually gained a foothold, possibly because of the widespread use of small, affordable printing units and advancements in imaging acquisition and postprocessing software [5, 8]. 3D-printed anatomical models created from imaging data are increasingly valuable in personalized precision medicine. The applications of 3DP in the medical field can be used for tissue and organ fabrication, prosthetics and implant production, and anatomical models.

In recent years, many medical institutes are already using 3D printing technology to fabricate customized external prosthetics, surgical guides for preoperative planning [45]. We worked out a prototype using AVA-targeted sensors embedded in a personalized 3D printed cast to collect AVA hemodynamic data. The wearable orthopedic brace can monitor vascular access stenosis dysfunction. The wearable medical device incorporated sensors are the PAG sensor for mainly examining vascular pitch patterns and the PPG sensor for calculating flow volume secondarily as a double-checking of the AV access status. By monitoring and detecting changes in the frequency spectrum domain, a new function of the autoregressive (AR) model was introduced to the PAG-based sensors to identify AVA stenosis and concurrently audit the state of its life cycle. It alerts hemodialysis patients about AVA malfunction early and urges them to return [5].

Although 3DP has not been applied as often to aortic surgery and congenital heart surgery as it has to orthopedic and spinal surgery because of the complexity and uniqueness of aortic surgery and congenital heart disease, research on 3DP to improve preoperative surgical planning and medical education for young surgeons is underway. 3DP can benefit almost all parties involved in patient care. Surgeons can use 3DP to improve their understanding of anatomy, preoperatively plan, and simulate procedures. 3DP materials may help patients and their families to understand diseases. Moreover, 3DP may help medical students, trainees at all levels, and nurses to improve their understanding of certain concepts [3].

We employed DICOM imaging data. DICOM is a standardized image format and network communications protocol that allows for the interoperable transfer of medical images and related information among systems. 3D Slicer and 3D modeling can be used to visualize the aorta before surgery, thereby allowing surgeons to quickly make surgical plans and offering reference information for determining the appropriate combination of stent grafts.

Section 5.1 Case 1 demonstrates that TEVAR stents can be used to treat severe ATAAD. However, because of the complexity of the related anatomical structures and the technological limitations of current devices, utilizing TEVAR stents for ATAAD treatment remains challenging. Dorros et al. (2000) demonstrated the feasibility of treating the ascending thoracic aorta with an endoprosthesis [59]. However, research on ascending TEVAR has been limited to several single-center clinical trials and case reports. In addition, the US FDA has not approved any endovascular stents specifically for the descending aorta [62]. This lack of development may be caused by anatomical and physiological challenges, such as those presented by the opposing ascending aorta, the descending anatomy, complex pathologies, and hemodynamic alterations [60–62].

The use of TEVAR devices designed to treat ascending aortic disease has gradually become feasible; however, whether ascending aortic disease can be predicted and how failure can be avoided remain unclear. In the anatomical structure of the ascending aorta, the greater curvature can be more than 30% longer than the lesser curvature. In addition, in cases of acute ascending aortic dissection, the diameter of the ascending aorta can suddenly increase by up to 32% [63]. The size and shape of commercial stent grafts with fixed diameters and lengths make them incompatible with ascending aortic diseases. The US FDA has yet to approve any commercial devices designed explicitly for ascending aortic diseases. We used 3DP to visualize the spatial relationships in patient-specific geometry and treated patients by using a combination of three stents. However, continued follow-up is required for the evaluation of long-term outcomes.

Case 2 demonstrates the application of 3DP to DORV and D-TGA in surgical planning for congenital heart disease (**Figures 6–8**). The heart has a complicated structure, and congenital heart defects can generate limitless variation in internal and external elements, often yielding unique and highly complex structures. Therefore, the precision with which 3DP can model rhythmic structure is essential. Although comparing the dimensions of a model with those of a digital source image is easy, it can only confirm that the printing process is accurate in relation to stereolithographic input data. Therefore, Hoashi et al. tested the accuracy of 3DP by using preserved cardiac autopsy specimens and discovered that the accuracy was ≥90% [38, 64].

Surgical residents have novice open surgical skills, limited work hours, short training programs, numerous endovascular procedures to perform, high costs, and pressure caused by public expectations. Therefore, basic lectures are insufficient to prepare students for surgical practice. Every operation requires precision and reliability. Simulation-based training may improve the learning process. The purpose of such training is to allow for new surgeons to practice operations and build confidence before surgical procedures and to allow experienced surgeons maintain and refine their skills.

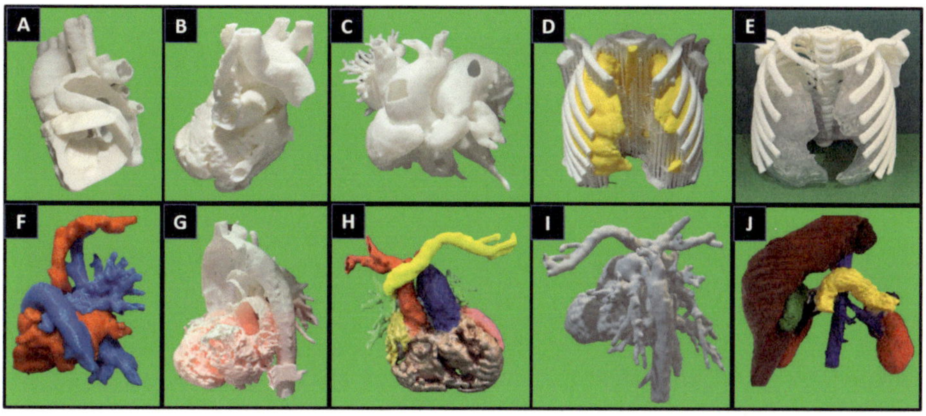

Figure 8.
Surgical planning in medical application; (A) (B) (C) 3D printing for congenital heart, (D) (E) 3D model for surgical planning, (F) (G) 3D printing of heart congenital heart, (H) front side of double outlet right ventricle (DORV), (I) Back side of double outlet right ventricle, (J) 3D printing model.

Employing 3DP for anatomical modeling based on scan data is increasingly valuable in personalized precision medicine. Implantable 3D-printed organs may become available in the future, reducing waiting times and increasing survival rates. The development of implantable living 3D-printed organs is also possible [65].

Acknowledgements

This work was supported by TAIWAN ASSOCIATION OF CARDIOVASCULAR SURGERY RESEARCH, under contract numbers A-190020; and also supported by the Ministry of Science and Technology, Taiwan, under contract numbers MOST109-2221-E-006-214 and MOST 110-2221-E-006-043.

Conflict of interest

The authors declare there are no conflict of interests.

Author details

Wei-Ling Chen[1,2], Tsung-Lung Yang[3], Jieh-Neng Wang[4] and Chung-Dann Kan[5*]

1 School of Biomedical Engineering, College of Biomedical Engineering, Taipei Medical University, Taipei, Taiwan

2 Institute Food and Drug Administration, Ministry of Health Welfare, Taipei County, Taiwan

3 KSVGH Originals and Enterprises, Kaohsiung Veterans General Hospital, Kaohsiung, Taiwan

4 Department of Pediatrics, National Cheng Kung University Hospital, College of Medicine, National Cheng Kung University, Tainan City, Taiwan

5 Division of Cardiovascular Surgery, Department of Surgery National Cheng Kung University Hospital, College of Medicine, National Cheng Kung University, Tainan, Taiwan

*Address all correspondence to: kcd56@mail.ncku.edu.tw

References

[1] Aimar A, Palermo A, Innocenti B. The role of 3D printing in medical applications: A state of the art. Journal of Healthcare Engineering. 21 Mar 2019;**2019**:5340616. DOI: 10.1155/2019/5340616

[2] Farooqi KM, Mahmood F. Innovations in preoperative planning: Insights into another dimension using 3D printing for cardiac disease. Journal of Cardiothoracic and Vascular Anesthesia. 2018;**32**:1937-1945. DOI: 10.1053/j.jvca.2017.11.037

[3] Min JK, Mosadegh B, Dunham S, Al'Aref SJ, editors. 3D Printing Applications in Cardiovascular Medicine. United States: Academic Press; 2018

[4] Dorr Goold S, Lipkin M Jr. The doctor-patient relationship: Challenges, opportunities, and strategies. Journal of General Internal Medicine. 1999;**14**:S26-S33. DOI: 10.1046/j.1525-1497.1999.00267.x

[5] https://trh.gase.most.ntnu.edu.tw/en/article/content/163

[6] Kan CD, Chen WL, Lin CH, Hsu CP, Chang JC, Wei J, et al. Simulation-based training program on the learning of open vascular surgery repair for junior vascular surgeons-2020 TSVS aorta cadaveric workshops. Academia Journal of Educational Research. 2020;**8**:500-508. DOI: 10.15413/ajer.2020.0709

[7] Qamar K, Osama M. Role of dissection in light of students' perceptions. The Journal of the Pakistan Medical Association. 2014;**64**:1021-1024

[8] Su W, Xiao Y, He S, Huang P, Deng X. Three-dimensional printing models in congenital heart disease education for medical students: A controlled comparative study. BMC Medical Education. 2018;**18**:178. DOI: 10.1186/s12909-018-1293-0

[9] Paige JT, Garbee DD, Kozmenko V, Yu Q, Kozmenko L, Yang T, et al. Getting a head start: High-fidelity, simulation-based operating room team training of interprofessional students. Journal of the American College of Surgeons. 2014;**218**:140-149. DOI: 10.1016/j.jamcollsurg.2013.09.006

[10] Ciobotaru V, Combes N, Martin CA, Marijon E, Maupas E, Bortone A, et al. Left atrial appendage occlusion simulation based on three-dimensional printing: New insights into outcome and technique. EuroIntervention. 2018;**14**:176-184. DOI: 10.4244/EIJ-D-17-00970

[11] Sherrah AG, Grieve SM, Jeremy RW, Bannon PG, Vallely MP, Puranik R. MRI in chronic aortic dissection: A systematic review and future directions. Frontiers in Cardiovascular Medicine. 2015;**2**:5. DOI: 10.3389/fcvm.2015.00005

[12] Bishawi M, Vemulapalli S. Chapter 12 - surgical predictive planning using 3D printing. In: Al'Aref SJ, Mosadegh B, Dunham S, Min JK, editors. 3D Printing Applications in Cardiovascular Medicine. Cambridge, Massachusetts, United States: Academic Press; 2018. pp. 227-241

[13] Ballard DH, Trace AP, Ali S, Hodgdon T, Zygmont ME, DeBenedectis CM, et al. Clinical applications of 3D printing: Primer for radiologists. Academic Radiology. 2018;**25**:52-65. DOI: 10.1016/j.acra.2017.08.004

[14] Hull CW, inventor; Uvp, Inc., assignee. Apparatus for production of three-dimensional objects by

stereolithography. United States patent US 4,575,330 1986. Available from: http://www.google.com/patents/us4575330. [Accessed: August 30, 2017]

[15] Curodeau A, Sachs E, Caldarise S. Design and fabrication of cast orthopedic implants with freeform surface textures from 3-D printed ceramic shell. Journal of Biomedical Materials Research. 2000;**53**:525-535

[16] Hong SB, Eliaz N, Leisk GG, Sach EM, Latanision RM, Allen SM. A new Ti-5Ag alloy for customized prostheses by three-dimensional printing (3DP). Journal of Dental Research. 2001;**80**:860-863

[17] Sodian R, Schmauss D, Marked M, et al. Three-dimensional printing creates models for surgical planning of aortic valve replacement after previous coronary bypass grafting. The Annals of Thoracic Surgery. 2008;**85**:2105-2108. DOI: 10.1016/j.athoracsur.2007.12.033

[18] Kijima S, Sasaki T, Nagata K, Utano K, Lefor AT, Sugimoto H. Preoperative evaluation of colorectal cancer using CT colonography, MRI, and PET/CT. World Journal of Gastroenterology. 2014;**20**:16964-16975. DOI: 10.3748/wjg.v20.i45.16964

[19] Sakamoto T. Roles of universal three-dimensional image analysis devices that assist surgical operations. Journal of Hepato-Biliary-Pancreatic Sciences. 2014;**21**:230-234. DOI: 10.1002/jhbp.88

[20] Fayad LM, Patra A, Fishman EK. Value of 3D CT in defining skeletal complications of orthopedic hardware in the postoperative patient. AJR. American Journal of Roentgenology. 2009;**193**:1155-1163. DOI: 10.2214/AJR.09.2610

[21] Kurenov SN, Ionita C, Sammons D, Demmy TL. Three-dimensional printing to facilitate anatomic study, device development, simulation, and planning in thoracic surgery. Journal of Thoracic and Cardiovascular Surgery. 2015;**149**(4):973-979. DOI: 10.1016/j.jcmg.2016.12.001

[22] Vukievic M, Mosadegh B, Little JK, Little SH. Cardiac 3D printing and its future directions. JACC: Cardiovascular Imaging. 2017;**10**(2):171-184. DOI: 10.1016/j.jcmg.2016.12.001

[23] Jeon H, Kang K, Park SA. Generation of multilayered 3D structures of HepG2 cells using a bio-printing technique. Gut and Liver. 2017;**11**(1):121-128. DOI: 10.5009/gnl16010

[24] Randazzo M, Pisapia JM, Singh N, Awani JP. 3D printing in neurosurgery: A systematic review. Surgical. Neurology International. 2016;7(34):801-809. DOI: 10.4103/2152-7806.194059

[25] Lino H, Igawa K, Kanno Y. Maxillofacial reconstruction using custom-made artificial bones fabricated by inkjet printing technology. Journal of Artificial Organ. 2009;**12**(3):200-205. DOI: 10.1007/s10047-009-0462-7

[26] Huang W, Zhang X. 3D printing: Print the future of ophthalmology. Investigative Opthalmology & Visual Science. 2014;**55**(8):5380-5381. DOI: 10.1167/iovs.14-15231

[27] Crafts TD, Ellsperman SE, Wannemuehler TJ, Bellicchi TD, Shipchandler TZ, Mantravadi AV. Three-dimensional printing and its applications in otorhinolaryngology-head and neck surgery. Otolaryngology-Head and Neck Surgery. 2017;**156**(6):999-1010. DOI: 10.1177/0194599816678372

[28] Auricchio F, Marconi S. 3D printing: Clinical applications in orthopaedics and traumatology. EFORT Open Reviews. 2016;**1**(5):121-127

[29] Chae MP, Rozen WM, McMenamin PG. Emerging applications of bedside 3D printing in plastic surgery. Frontiers in Surgery. 2015;**16**(2):25. DOI: 10.3389/fsurg.2015.00025

[30] Williams C, James A, Chae MP, Hunter-Smith DJ. 3D printing in clinical podiatry: A pilot study and review. Journal of Foot and Ankle Research. 2015;**8**(2):41. DOI: 10.1186/1757-1146-8-S2-O41

[31] Guilbert N, Mhanna L, Didier A. Integration of 3D printing and additive manufacturing in the interventional pulmonologist's toolbox. Respiratory Medicine. 2018;**134**:139-142. DOI: 10.1016/j.rmed.2017.11.019

[32] Zein NN, Hanouneh IA, Bishop PD. Three dimensional print of a liver for preoperative planning in living donor liver transplantation. Liver Transplantation. 2013;**19**(12):1304-1310. DOI: 10.1002/lt.23729

[33] Soliman Y, Feibus AH, Baum N. 3D printing and its urologic applications. Urology. 2017;**17**(1):20-24. DOI: 10.1007/s00345-019-02995-1

[34] Hangge P, Pershad Y, Witting AA, Albadawi H, Oklu R. Three-dimensional (3D) printing and its applications for aortic diseases. Cardiovascular Diagnosis & Therapy. 2018;**8**(1):19-25. DOI: 10.21037/cdt.2017.10.02

[35] Radenkovic D, Solouk A, Seifalian A. Personalized development of human organs using 3D printing technology. Medical Hypotheses. 2016;**87**:30-33. DOI: 10.1016/j.mehy.2015.12.017

[36] Ji S, Guvendiren M. Recent advances in bioink design for 3D bioprinting of tissues and organs. Frontiers in Bioengineering and Biotechnolgy. 2017;**5**(5):23

[37] Charbe N, McCarron PA, Tambuwala MM. Three dimensional bioprinting: A new frontier in oncology research. World Journal of Clinical Oncology. 2017;**8**(1):21-36

[38] Hermsen JL, Roldan-Alzate A, Anagnostopoulos PV. Three-dimensional printing in congenital heart disease. Journal of Thoracic Disease. 2020;**12**:1194-1203. DOI: 10.21037/jtd.2019.10.38

[39] Bortolotto C, Eshja E, Peroni C, Orlandi MA, Bizzotto N, Poggi P. 3D printing of CT dataset: Validation of an open source and consumer-available workflow. Journal of Digital Imaging. 2016;**29**:14-21. DOI: 10.1007/S10278-015-9810-8

[40] Bizzotto N, Sandri A, Regis D, Romani D, Tami I, Magnan B. Three-dimensional printing of bone fractures: A new tangible realistic way for preoperative planning and education. Surgical Innovation. 2015;**22**:548-551. DOI: 10.1177/1553350614547773

[41] Lei Y, Chen X, Li Z, Zhang L, Sun W, Li L, et al. A new process for customized patient-specific aortic stent graft using 3D printing technique. Medical Engineering & Physics. 2020;**77**:80-87. DOI: 10.1016/j.medengphy.2019.12.002

[42] Tong YH, Yu T, Zhou MJ, Liu C, Zhou M, Jiang Q, et al. Use of 3D printing to guide creation of fenestrations in physician-modified stent-grafts for treatment of thoracoabdominal aortic disease. Journal of Endovascular Therapy. 2020;**27**:385-393. DOI: 10.1177/1526602820917960

[43] Tang F, Hu C, Huang S, Long W, Wang Q, Xu G, et al. An innovative customized stent graft manufacture system assisted by three-dimensional printing technology. The Annals of

Thoracic Surgery. 2021;**112**:308-314. DOI: 10.1016/j.athoracsur.2020.07.013

[44] Paramasivam V, Singh G, Santhanakrishnan S. 3D printing of human anatomical models for preoperative surgical planning. Procedia Manufacturing. 2020;**48**:684-690. DOI: 10.1016/j.promfg.2020.05.100

[45] Chen WL, Yang TL, Lee PL, Kan CD. Feasibility of smart hemodialysis patient care using sensors embedded in personalized 3D printing cast. Journal of Mechchanics in Medicine Biology. 2019;**19**:1940017-1940025. DOI: 10.1142/S0219519419400177

[46] Lai TS, Hsu CC, Lin MH, Wu VC, Chen YM. Trends in the incidence and prevalence of end-stage kidney disease requiring dialysis in Taiwan: 2010-2018. Journal of the Formosan Medical Association. Feb 2022;**121**(Suppl 1):S5-S11. DOI: 10.1016/j.jfma.2021.12.013

[47] National Kidney Foundation. K/DOQI Clinical Practice Guidelines for Chronic Kidney Disease. New York, United States: The National Kidney Foundation, Inc.; 2002

[48] Nassar GM, Ayus JC. Infectious complications of the hemodialysis access. Kidney International. 2001;**60**:1-13

[49] Gulati S, Sahu KM, Avula S, Sharma RK, Ayyagiri A, Pandey CM. Role of vascular access as a risk factor for infections in hemodialysis. Renal Failure. 2003;**25**:967-973

[50] Chen WL, Lin CH, Chen T, Chen PJ, Kan CD. Stenosis detection using burg method with autoregressive model for hemodialysis patients. Journal of Medical and Biomedical Engineering. 2013;**33**:356-362

[51] Chen WL, Lin CH, Chen T, Chen PJ, Kan CD. Phonoangiography with a fractional order chaotic system – A new and easy algorithm in analyzing residual arteriovenous. Medical & Biological Engineering & Computing. 2013;**51**:1011-1019

[52] Chen WL, Kan CD, Lin CH, Chen T. A rule-based decision-making diagnosis system to evaluate arteriovenous shunt stenosis for hemodialysis treatment of patients using fuzzy petri nets. IEEE Journal of Biomedical and Health Informatics. 2014;**18**:703-713. DOI: 10.1109/JBHI.2013.2279595

[53] Alexandrov AV. The Spencer's curve: Clinical implications of a classic hemodynamic model. Journal of Neuroimaging. 2007;**17**:6-10. DOI: 10.1111/j.1552-6569.2006.00083.x

[54] Chen WL, Kan CD, Lin CH, Mai YC. Generalized regression estimator improved the accuracy rate of estimated dialysis accesses stenotic condition on in-vitro arteriovenous graft experimental model. IEEE Access. 2018;**6**:10381-10391. DOI: 10.1109/ACCESS.2018.2802479

[55] Chen WL, Lin YH, Kan CD, Yu FM, Lin CH. Assessment of flow instabilities in in-vitro stenotic arteriovenous grafts using an equivalent astable multivibrator. IET Science, Measurement and Technology. 2015;**9**:709-716

[56] Malaisrie SC, Szeto WY, Halas M, Girardi LN, Coselli JS, Sundt TM 3rd, et al. The American Association for Thoracic Surgery expert consensus document: Surgical treatment of acute type a aortic dissection. The Journal of Thoracic and Cardiovascular Surgery. 2021;**2021**(162):735-758.e2

[57] Matthews CR, Madison M, Timsina LR, Namburi N, Faiza Z, Lee LS. Impact of time between diagnosis to treatment in acute type a aortic

dissection. Scientific Reports. 2021;**11**:3519. DOI: 10.1038/s41598-021-83180-6

[58] Hsieh WC, Kan CD, Yu HC, Aboud A, Lindner J, Henry BM, et al. Ascending aorta replacement vs. total aortic arch replacement in the treatment of acute type a dissection: A meta-analysis. European Review for Medical and Pharmacological Sciences. 2019;**23**:9590-9611

[59] Petrov I, Stankov Z, Adam G. Endovascular treatment of type a aortic dissection. Journal of Cardiology and Cardiovascular Sciences. 2020;**4**:51-58

[60] Baikoussis NG, Antonopoulos CN, Papakonstantinou NA, Argiriou M, Geroulakos G. Endovascular stent grafting for ascending aorta diseases. Journal of Vascular Surgery. 2017;**66**:1587-1601

[61] Ronchey S, Serrao E, Alberti V, Fazzini S, Trimarchi S, Tolenaar JL, et al. Endovascular stenting of the ascending aorta for type a aortic dissections in patients at high risk for open surgery. European Journal of Vascular and Endovascular Surgery. 2013;**45**:475-480

[62] Rylski B, Blanke P, Beyersdorf F, Desai ND, Milewski RK, Siepe M, et al. How does the ascending aorta geometry change when it dissects? Journal of the American College of Cardiology. 2014;**63**:1311-1319

[63] Czerny M. Landing in zone 0: Is ascending thoracic endovascular aortic repair ready for takeoff? The Journal of Thoracic and Cardiovascular Surgery. 2018;**155**:1390

[64] Zhao L, Zhou S, Fan T, Li B, Liang W, Dong H. Three-dimensional printing enhances preparation for repair of double outlet right ventricular surgery. Journal of Cardiac Surgery. 2018;**33**:24-27. DOI: 10.1111/jocs.13523

[65] Badash I, Burtt K, Solorzano CA, Carey JN. Innovations in surgery simulation: A review of past, current and future techniques. Annals of Translational Medicine. 2016;**4**:453. DOI: 10.21037/atm.2016.12.24

Chapter 12

3D Printing and Airway Stents

Carlos Aravena and Thomas R. Gildea

Abstract

A central goal of an airway stent is to restore patency by preventing restenosis, holding the tracheobronchial wall, or occluding fistulas. Complications with stents, however, are frequent and can have grave repercussions. Stents are therefore viewed as a last resort in cases where other forms of treatment are ineffective. Furthermore, it is common for people with complex airways to have airway stents that do not fit them well, which can result in several complications. Three-dimensional printing technology was developed at the turn of the 20th century. It has been employed in a variety of applications and has transformed healthcare. This technology has mainly been employed in respiratory medicine to develop three-dimensional models of the airways and to make airway splints and prostheses to treat central airway diseases. In the past ten years, it has transformed and advanced personalized medicine, enabling the creation of patient-specific stents for people with complex airway diseases. Three-dimensional printing might be used to create a patient-specific stent that would lessen risks, enhance the quality of life, and eliminate the need for additional procedures. This chapter discusses the most recent developments in three-dimensional printing technology, how they are being used to create airway prostheses to treat complex airway illnesses and the current body of research that supports their use.

Keywords: three-dimensional printing, 3D-printing, bronchoscopy, airway stents, patient-specific airway stent, computer-aided design, three-dimensional airway mold, 3D airway mold

1. Introduction

A wide range of benign and malignant illnesses can impact the airway [1]. The diagnosis and treatment of these disorders rely significantly on bronchoscopy [2]. After non-invasive treatments are found to be ineffective, therapeutic bronchoscopy attempts to enhance the patient quality of life, reduce symptoms, and provide significant palliation [1, 3]. Flexible or rigid bronchoscopy (RB) is a procedure that can deliver a variety of therapeutic modalities, among them stent placement play an important role [1, 3–6]. Reestablishing patency, preventing restenosis, stabilizing the tracheobronchial wall, or occluding fistulas are the primary purposes of an airway prostheses [7, 8].

Stents could be made of metallic wire mesh, silicone, or a combination of these materials (hybrid), as well as various sizes and shapes [9]. It is typical to experience complications like migration, granulation, infection, and mucus clogging [7, 10–13].

IntechOpen

Therefore, stents should only be temporary when no other methods can achieve appropriate and long-lasting patency [5, 14–16].

Because implantation may last long and result in a high rate of complications, using stents in benign obstructions require caution [14]. Due to the possibility of excessive complication rates uncovered metallic stents are not recommended for use in most benign airway diseases [14]. Silicone stents are the most popular choice for benign conditions [6, 14, 17]. The synthetic substance used to create silicone stents has a low tissue reactivity and is simple to remove [14, 16].

Unfortunately, there are not many sizes and forms of commercially available stents (CAS). Sometimes it's crucial to modify them to enhance fit and performance, particularly in patients with complicated airway anatomy [18]. This Customization usually involves cutting and stitching stents together in the operating room and requires an expert and highly trained doctor [18].

Because of the intricate features of the tracheobronchial anatomy, Three-dimensional printing (3DP) technology is perfect for creating airway prostheses to treat difficult conditions. With the help of years of research and development, it is now possible to produce a patient-specific stent [19–22]. A patient-specific stent may help decrease risks, shorten healing time, and enhance patients' quality of life while alleviating symptoms and avoiding the need for additional bronchoscopies.

Recent research has examined the use of 3DP technology to produce silicone and hybrid airway prostheses and investigate biodegradable materials and drug-eluting stents (DES) [23].

This chapter discusses the most recent developments in three-dimensional printing technology, how they are being used to create airway prostheses to treat complex airway problems, and the body of research that supports their use.

2. Airway diseases and three-dimensional printing

Anatomical modeling of various bodily components for preoperative planning purposes by surgeons was one of the earliest uses of rapid prototyping techniques in medicine [24, 25].

In respiratory medicine, 3DP technology has been used, particularly for conditions of the central airways. In 2013, Tam et al. printed inspiratory and expiratory three-dimensional (3D) models of the tracheobronchial tree of a patient with airway disease due to relapsing polychondritis, and they explored the potential use for surgical or interventional planning and teaching [26].

In a case study published in 2013, Zopf and colleagues described how a bioresorbable airway splint was surgically inserted into a newborn's malacic left main bronchus [27].

Then, in 2015, George Z. Cheng et al. published the first case study of a 3D-modeled T-tube inserted into a patient's difficult upper airway. This prosthesis was made to allow for the three-dimensional reconstruction of the trachea from a computed tomography (CT) scan [28].

To treat a 56-year-old male patient with airway complications of granulomatosis with polyangiitis (GPA) who required numerous unsuccessful therapeutic bronchoscopies and multiple commercially and manually customized stents, Thomas R. Gildea created and implanted the first bronchial patient-specific airway stent (PSS) made of silicone under Food and Drug Administration clearance for compassionate use in 2016. He did this using CT imaging and 3DP technology [20, 22]. In 2017, Gildea and colleagues described the one-year experience of this patient and another with complex airway disease attributable to GPA. Both patients improved their average time between treatments and stent life following implantation after inserting PSS made utilizing 3DP technology [20, 22].

Guibert et al. presented a case of a patient with an airway problem following lung transplantation in 2016; the right airway had dehiscence, a stenotic bronchus intermedius, and complex morphology. After a 3D airway was built from a CT scan, and the difficulties were virtually removed, a planned 3D mold was made and used to create a unique stent. It was successfully implanted with RB [29].

Similar efforts have been made in treating tracheobronchomalacia in adult patients using 3DP technology [29]. Shan and colleagues recently published their experiences with hybrid stents that were assisted with 3DP technology to treat malignant airway obstruction and aerodigestive fistula [30, 31].

3. Airway stents and 3D printing procedures

To make airway stents, many 3D techniques and materials have been employed.

In order to create a 3D reconstruction of the trachea using a CT scan, Cheng and colleagues employed 3D slicer, a free, open-source, and multi-platform software application that is frequently used for medical, biomedical, and related imaging research. The virtual T-tube was created using Solidworks® (computer-aided design (CAD) software) then a 3D model was imported, matching his patient's virtual difficult upper airway. The silicone customized T-tube was created and put through the tracheostomy stoma under bronchoscopy supervision [19, 28].

Gildea and colleagues used specialized software created for orthopedic surgery to transfer the digital imaging from a CT scan (COS Inc., Cleveland, OH, USA). The airway is turned into a virtual 3D prototype. The ideal stent dimensions, including the area, diameter, angulation, branching, length, and wall thickness, were defined using this virtual model of each patient's anatomy. The doctor creates a virtual depiction based on clinical requirements by placing spheres in the 3D airway and adjusting their forms and sizes using the software tools. Using 3DP technology, a mold of the prescribed stent is created, and medical-grade silicone is then injected into this mold to create the stent (**Figure 1**). External studs are inserted after the stent has been finished, cleaned, and polished to produce a flat surface. The stent is sterilized using steam sterilization. The stent is then implanted utilizing RB and conventional procedures and instruments [20, 32].

Guibert employed a similar process, creating a virtual 3D mold (VGStudio MAX software). To create an Ertacetal POM mold, the 3DP (RolandDG MDX 40A) was fed with the 3D data. This mold was used to create a personalized silicone stent. The stent is placed during a therapeutic RB procedure [29, 33].

Using CAD software, Shan and associates could recreate 3D representations of the airway using information from a 64-slice multidetector spiral CT scan (Vitaworks, Shanghai, China). After different colors were given to the airway and tumor, the image was transformed into a 3D stereolithographic (STL) file. An airway mold composed of photosensitive polymers was produced using the 3D reconstruction data and a 3DP (RS600, Union Tech, Shanghai, China). The dimensions of the 3D-printed airway mold's area of interest were measured. The temperature-memory nickel-titanium alloy-covered self-expandable Y-shaped metallic airway stents (Micro-Tech, Nanjing, China) were then created utilizing the 3D printed airway model as a template.

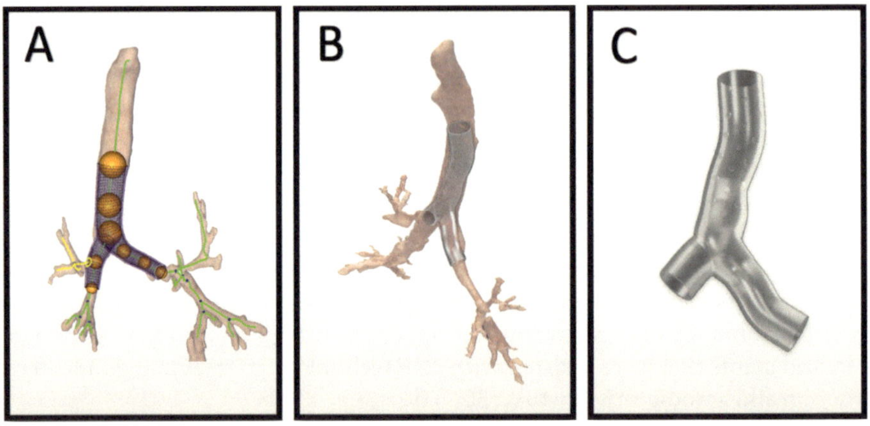

Figure 1.
Patient specific silicone stent (PSS) created with 3D printing technology. After a 3D virtual mold of the airway is created. The physician uses a software placing a series of spheres in the target airway to adjust the dimensions and make a virtual representation of the PSS. (A) Stent design with spheres. (B) Stent design. (C) Silicone stent. Picture authorized by Visionair, Cleveland, OH, USA.

Flexible bronchoscopy was used to evaluate the patient and install the guidewires before inserting the stents. A stent delivery system was advanced posteriorly out from the endotracheal tube, and the stent was deployed with fluoroscopy [30, 31].

4. Biodegradable stents

Research is being done on biodegradable stents (BDS) that might temporarily sustain patency in an airway [34]. They may be helpful when temporal stenting is desired in individuals with benign airway disorders [9]. The prosthetic material must be biocompatible, release harmless residues as it degrades, be strong enough to maintain the integrity of the airway, and be durable enough to allow the airway to reconstruct [34, 35]. Research has been done on several synthetic degradable polymers, including polyesters containing lactic acid, glycolic acid, dioxanone, caprolactone, polytrimethylene carbonates, polyanhydrides containing sebacic acid, and polyarylates generated from tyrosine [34, 36–38]. Studies in animal models showed that, depending on the polymer employed, stents had a high safety profile, were biocompatible, and degraded quickly over time [38–40].

According to studies conducted on patients, the BDS is safe and reduce symptoms. However, several patients might develop cough, mucosal hyperplasia, granulation tissue, biofilm, expectoration of stent parts after insertion, and other complications resulting from a lack of radiation force or re-stenosis [41–43].

Related to 3DP and BDS, research has been done on materials that can be directly printed to create an airway stent. In 2015, Nidah M. Hussain from the University of South Carolina used 3DP technology to design and print a bioresorbable tracheobronchial stent to investigate potential improvements on existing stents. He concluded that thermoplastic polyurethane is potentially viable as a biologically degradable silicone substitute and polycaprolactone is compatible with fused deposition modeling printing [44].

With the use of elastomeric polyurethane (EPU), Catherine Wood and colleagues created a platform for designing and manufacturing 3DP flexible airway EPU stents. They concluded that the 3D-printed EPU stent performs similarly to silicone stents after conducting comparison testing [45].

In an in-vivo examination of healthy rabbits, Paunovic et al. reported employing a digital light 3DP customized bioresorbable stent. The stents were made of biocompatible dual polymer that remained in situ for seven weeks [46].

Because of the potential for quick and direct manufacture, customization, biocompatibility, and degradability, these 3DP materials seem promising. However, more investigation should be done to enhance degradation time, radial force and reduce problems, particularly re-stenting before it is used in patients.

5. Drug eluting stents

DES have been widely used in cardiology to decrease coronary stent complications such as re-stenosis or stent thrombosis [47].

The use of stents in central airway obstruction has been related to numerous complications and looking to reduce adverse events (AEs) rate or to treat airway obstruction airway-DES research has been ongoing since the last decade [48].

In 2011, Zhu and colleagues randomized different types of stents in 20 rabbits, the bioabsorbable stents with mitomycin C had the best performance, with less mucus plugging and airway obstruction [48]. Other drugs have shown to decrease the granulation tissue or scar formation in animal models, such as airway DES with paclitaxel, sirolimus, methylprednisolone, or cisplatin [49–54].

The potential use of DES in airway diseases is not limited to preventing mucus plugging or granulation tissue formation. It could prevent stent-related infection, treat long-term malignant central airway, or manage benign central airway stenosis. Several chemotherapeutic, anti-proliferative, or antifibrotic agents have been proposed [55].

3DP technology can be used to create a personalized drug-eluting stent. According to the patient's need, it could have different polymers and drugs or a combination to produce a sustained drug release effect and prevent or treat different conditions [56].

6. 3D printing and clinical research

The number of studies on 3DP and airway stents has grown since 2015. Ten publications using 3DP and PSS in humans were identified (**Table 1**) [22, 28–33, 57–59]. Most studies have focused on benign airway conditions such as tracheobronchomalacia, post-radiotherapy airway complications, post-surgical airway complications, post-transplant airway illnesses, and GPA airways. Silicone was the most used stent material in benign airway disorders [22, 29, 32, 33, 57, 60, 61]. Malignant central airway obstruction and malignant aerodigestive fistula have been the subjects of more recent research. Except for one study, they utilized covered metal stents [30, 30, 58].

To produce the PSS, nine studies reported printing a 3D airway mold (there is no description in one study). Additionally, Y tracheobronchial or bronchial stents were the majority. In six studies, Y-bronchial stents or a bronchial branch to the right upper lobe were made using a 3D mold.

Author	Year	Type of study	Nº patients	Type of airway disease	Airway mold	Stent shape	Stent material2	Results	AEs
Cheng et al.	2015	Case report	1	Benign	Yes (virtual)	T-Tube	Silicone	Improvement in phonation, no granulation tissue after 4 months f/u	N/D
Guibert et al.	2017	Case report	1	Benign	Yes	Right bronchial stent branched to the RUL	Silicone	Improvement of symptoms, PEF	N/D
Gildea et al.	2018	Case report	2	Benign	Yes	Y-stent for LMB	Silicone	Increased time between procedures. Increase stent life. Decrease procedure time	N/D
Schweiger et al.	2018	Case report	2	Benign	Yes	Y Tracheobronchial	Silicone	Symptoms improvement	N/D
Guibert et al.	2019	Prospective	10	Benign	Yes	Y bronchial and tracheobronchial stent	Silicone	90% congruence, 80% improvement in dyspnea, quality of life and FEV1 or PEFR	Complication rate of 40% at 3 months f/u.
Aravena Leon et al.	2019	Retrospective	4	Benign	Yes	Y bronchial and tracheobronchial stent	Silicone	No difference compared to commercial stents in RB loading, placement, removal. Increase time between procedures. Increase stent life	PSS had a lower severity of migration. All other AEs were not statistically different between the two groups.

Author	Year	Type of study	N° patients	Type of airway disease	Airway mold	Stent shape	Stent material2	Results	AEs
Duong et al.	2020	Case report	1	Malignant	N/D	Right bronchial stent branched to the RUL	Polyurethane	Symptoms improvement	PSS replacement at 2 months and migration and removal of the 2nd PSS
Huang et al.	2020	Retrospective	6	Benign (Post-esophagectomy)	Yes	Y Tracheobronchial	Hybrid (Nitinol, Silicone and PTFE)	All fistulas sealed. Leaking control in 6 patients. KPS improvement	2 patients had mucus retention, 1 had excessive granulation tissue and stent removal
Shan et al	2021	Retrospective	12	Malignant	Yes	Y stent (1 bronchial, rest tracheobronchial)	Hybrid (Nitinol, Silicone and PTFE)	1 patient had 2 stents. 11 had significant symptoms improvement, KPS improvement	4 patients had mucus retention, 2 excessive granulation tissue, 0 migration, 0 removal
Shan et al	2021	Retrospective	26	Malignant TEF and post-esophagectomy	Yes	Y stent (1 bronchial, rest tracheobronchial)	Hybrid (Nitinol, Silicone and PTFE)	Clinical success rate 80% Resolution TEF post-esophagectomy (9/16). Resolution malignant TEF (0/10). KPS improvement	2 (7.69%) patients had granulation tissue treated with cryotherapy and stent removal. 5 (19.23%) sputum retention, treated w/ suction. 1 (3.84%) had stent migration underwent a second stent. 1 (3.84%) not tolerate the stent and was removed.

N/D=Not described, PTFE = polytetrafluoroethylene, KPS=Karnofsky Performance Scale, TEF: Tracheoesophageal fistula, RB: Rigid bronchoscopy, PEFR: peak expiratory flow rate, EV1: Forced expiratory volume in the first second.

Table 1.
Clinical studies description of PSS assisted by 3DP technology.

Improvement in symptoms was observed in all studies (**Table 1**). Guibert et al. used 3DP silicone PSS in 10 patients, the majority had post-transplant airway complications and reported high rates of congruence between the stent and the airway, 80% improvement in dyspnea (>1 New York Health Association score point gain), quality of life (>10% increase in VQ11 Chronic Obstructive Pulmonary Disease-specific quality of life score), and pulmonary function test (>10% Forced Expiratory Volume in 1 second (FEV1) or Peak Expiratory Flow (PEF) increase) [33].

A retrospective analysis of patients who got 3DP silicone PSS at the Cleveland Clinic was reported by Aravena et al. Interventional pulmonologists completed a survey and two physicians rated stent-related AEs using the Common Terminology Criteria for Adverse Events scoring system. Four patients received a total of 13 PSS. No difference was noted in the loading, positioning, or removal of the PSS in comparison to the CAS ($p > 0.05$). Following the placement of the PSS, bronchoscopists saw a substantial clinical improvement ($p = 0.03$). The PSS's average lifespan was considerably longer than the CAS's (300.2 days vs. 124.0 days, $p = 0.001$) by a large margin. With PSS, the median time between bronchoscopies was significantly longer than with CAS (65.6 days vs. 36.6 days, $p = 0.004$) [57, 61].

In a study by Shan et al., 12 individuals with malignant airway obstruction caused by lung or esophageal cancer had 13 covered metal PSS implants. Hugh-Jones dyspnea scale and Karnofsky performance status (KPS) improvements were reported ($P = 0.003$ and $P = 0.006$, respectively) [30].

Regarding the AEs that were portrayed in the different trials. At three months, Guibert and colleagues reported a 40% complication rate, one patient with a mucus plug, two stent migrations, and one untreatable cough. Three of them needed to have their stents removed. One patient experienced distal stenosis at a lobar level that needed balloon dilatation, and another mucus plug incident occurred at the end of the four-month follow-up period. There were no life-threatening problems found [33]. Aravena and associates found that silicone PSS had less severe migration than CAS ($p = 0.0225$). The statistical differences between the two groups did not exist for any other AEs [57, 61]. Shan et al. demonstrated a 50% (6/12) complication rate with coated metal PSS after 5.6 months of follow-up. Only two patients had significant granulation tissue development, four patients had mucus blocking, and no patients needed their stents removed or had migration [30].

7. Discussion

Using any stent is always a last resort in cases of complex benign or malignant airway diseases. There are several problems associated with stents. Migration, stent obstruction by either granulation tissue or mucus, and infection are the three most frequent adverse outcomes of silicone stenting. These disorders can also be interrelated [5, 6, 14–18]. Even with non-malignant diseases, many patients still benefit from long-term palliation despite the risk.

The technical success of the PSS congruence to the complicated airway is high. It can result in a decline in AEs because fit issues could be responsible for many stent-related consequences. The migration rate is greater for either too loose or too tight stents. Excessive pressure from a stent on the airway might cause tissue necrosis and perforation. An improperly fitted stent may promote granulation at the ends or result in poor secretion clearance. Additionally, the material used may have clinical implications. Even in silicone stents, there are prostheses with a different durometer and elastic modulus that can have a distinct impact on wall stress.

The benefits could be substantial. Through all the research mentioned, symptoms have consistently improved. In one study, the lung function test and quality of life improved [31]. The PSS and CAS were only contrasted in one research. This showed a longer stent life and longer intervals between treatments, which may be associated with an improvement in the quality of life for patients with complicated benign airway diseases who often need several consecutive bronchoscopies to try to achieve palliation [57, 61].

Improving performance status scale is an important additional finding related to malignant disorders [30, 31, 58]. It could be linked to decreased adverse events (AEs), better congruence, and airway obstruction relief, which would aid with symptoms and possibly enhance clinical performance, allowing for the potential of receiving oncologic therapy.

Additionally, the 3DP PSS is at least as secure as the CAS. The rate of complications is comparable, and no problems that threaten life have been reported. The development of 3DP PSS with Y-bronchial and Y-tracheobronchial stent shapes, which are compatible with and suit the intricate airways of those patients, likely contributed to a reduction in migration rate compared to CAS (**Table 1**) [57, 61].

The PSS may be loaded and inserted using traditional techniques, just like conventional stents [57, 61]. This is crucial since deploying some complex stent designs, such as the dynamic Y-stent, requires specialized equipment.

The research in 3DP PSS do not have a sufficient sample size to generalize the application of patient-specific stenting. Many studies are retrospective cohorts or cases, which are both highly biased. But the early experience has demonstrated that PSS are highly successful and safe in the palliative treatment of extraordinarily complicated airway diseases, above all currently available best practices, despite the small sample size.

8. Future directions

3DP technology is developing fast and will be an important tool for personalized medicine in patients with complex airway diseases. Research is still being done to make more advancements. New materials that might be directly printed and biodegradable may be preferred when temporal stents are needed [46]. Additionally, 3DP drug-eluting stents might be a feasible therapeutic strategy for avoiding excessive granulation tissue, precluding infections, and managing malignant or benign obstruction of the airway [23, 62].

We envision the future of 3DP of completely compatible or engineered biological tissue prosthesis that promotes improvement of the damaged tissue and replace part of the impaired airway.

9. Conclusion

Over the past few decades, 3DP technology has made significant progress. This technology has been applied to healthcare for preoperative planning, education, medical equipment, prostheses, implants, and medical models. Respiratory medicine is at the forefront of a revolution in personalized treatment by making individualized airway stents tailored to the unique requirements of patients with complicated malignant or benign airway disease. These patient-specific airway stents developed utilizing 3DP technology can potentially reduce the number of treatments needed and adverse events (AEs) and improve symptoms and quality of life.

To ascertain the real impact this technology will have on this group of patients, new research with markedly better methodology will be necessary yet challenging.

Conflict of interest

CA has no conflicts of interest to declare. Visionair is the manufacturer of one of the Stents created with 3DP technology presented in this review. TRG is the inventor and may be entitled to royalty payments from the company in accordance with Cleveland Clinic policy.

Author details

Carlos Aravena[1*] and Thomas R. Gildea[2]

1 Faculty of Medicine, Department of Respiratory Diseases, Pontificia Universidad Católica de Chile, Santiago, Chile

2 Department of Pulmonary, Allergy, and Critical Care Medicine, Respiratory Institute, Cleveland Clinic, Cleveland, OH, USA

*Address all correspondence to: caravenal@med.puc.cl

References

[1] Mehta AC, Jain P, Gildea TR, editors. Diseases of the central airways: a clinical guide. 1st ed. Cham: Springer International Publishing; 2016:1-383 DOI: 10.1007/978-3-319-29830-6

[2] Ernst A, Herth FJF, editors. Principles and Practice of Interventional Pulmonology. 1st ed. New York, NY: Springer; 2013. pp. 1-757. DOI: 10.1007/978-1-4614-4292-9

[3] Ost DE, Ernst A, Grosu HB, Lei X, Diaz-Mendoza J, Slade M, et al. Therapeutic bronchoscopy for malignant central airway obstruction: Success rates and impact on dyspnea and quality of life. Chest. 2015;**147**(5):1282-1298

[4] Chua AP, Santacruz JF, Gildea TR. Pulmonary complications of cancer therapy and central airway obstruction. In: Mellar P. Davis, Petra Ch. Feyer, Petra Ortner, Camilla Zimmermann, editors. Supportive Oncology. 1st ed. W.B. Saunders; 2011. P. 309-325. DOI: 10.1016/B978-1-4377-1015-1.00029-1

[5] Aravena C, Almeida FA, Mukhopadhyay S, Ghosh S, Lorenz RR, Murthy SC, et al. Idiopathic subglottic stenosis: A review. Journal of Thoracic Disease. 2020;**12**(3):1100-1111

[6] Dutau H, Dumon JF. Airway stenting revisited: 30 years, the age of reason? Journal of Bronchology and Interventional Pulmonology. 2017;**24**(4):257-259

[7] Martínez-Ballarín JI, Díaz-Jimétiez JP, Castro MJ, Moya JA. Silicone stents in the management of benign tracheobronchial stenoses: Tolerance and early results in 63 patients. Chest. 1996;**109**:626-629

[8] Dumon JF. A dedicated tracheobronchial stent. Chest. 1990;**97**(2):328-332

[9] Folch E, Keyes C. Airway stents. Annals of Cardiothoracic Surgery. 2018;7(2):273-283

[10] Ost DE, Ernst A, Grosu HB, Lei X, Diaz-Mendoza J, Slade M, et al. Complications following therapeutic bronchoscopy for malignant central airway obstruction: Results of the AQuIRE registry. Chest. 2015;**148**(2):450-471

[11] Dumon JF, Cavaliere S, Diaz-Jimenez JP, Vergnon JM, Venuta F, Dumon MC, et al. Seven-year experience with the dumon prosthesis. Journal of Bronchology. 1996;**3**(1):6-10

[12] Noppen M, Stratakos G, D'Haese J, Meysman M, Vinken W. Removal of covered self-expandable metallic airway stents in benign disorders: Indications, technique, and outcomes. Chest. 2005;**127**:482-487

[13] Cancer L, How DT, Photobiol P Jr, et al. Airway stents. Chest. 2003;**123**(5):1709-1710

[14] Flannery A, Daneshvar C, Dutau H, Breen D. The art of rigid bronchoscopy and airway stenting. Clinics in Chest Medicine. 2018;**39**(1):149-167. DOI: 10.1016/j.ccm.2017.11.011

[15] Shapshay SM, Valdez TA. Bronchoscopic management of benign stenosis. Chest Surgery Clinics of North America. 2001;**11**(4):749-768

[16] Mudambi L, Miller R, Eapen GA. Malignant central airway obstruction. Journal of Thoracic Disease. 2017;**9**(Suppl. 10):S1087-S1110

[17] Wood DE, Liu Y-H, Valliè E, Karmy-Jones R, Mulligan MS. Airway

stenting for malignant and benign tracheobronchial stenosis. The Annals of Thoracic Surgery. 2003;**76**:167-174

[18] Breen DP, Dutau H. On-site customization of silicone stents: Towards optimal palliation of complex airway conditions. Respiration. 2009;**77**:447-453

[19] Cheng GZ, Estepar RSJ, Folch E, Onieva J, Gangadharan S, Majid A. Three-dimensional printing and 3D slicer powerful tools in understanding and treating structural lung disease. Chest. 2016;**149**(5):1136-1142

[20] Young BP, Machuzak MS, Gildea TR. Initial clinical experience using 3d printing and patient-specific airway stents: Compassionate use of 3d printed patient-specific airway stents. American Journal of Respiratory and Critical Care Medicine. 2017;**195**:A1711

[21] Alraiyes AH, Avasarala SK, Machuzak MS, Gildea TR. 3D printing for airway disease. AME Medical Journal. 2019;**4**:14-14

[22] Gildea TR, Young BP, Machuzak MS. Application of 3D printing for patient-specific silicone stents: 1-year follow-up on 2 patients. Respiration. 2018;**96**(5):488-494

[23] Aravena C, Gildea TR. Patient-specific airway stent using three-dimensional printing: A review. Annals of Translational Medicine. 2022. DOI: 10.21037/atm-22-2878

[24] McGurk M, Amis AA, Potamianos P, Goodger NM. Rapid prototyping techniques for anatomical modelling in medicine. Annals of the Royal College of Surgeons of England. 1997;**79**:169-174

[25] Matsumoto JS, Morris JM, Foley TA, Williamson EE, Leng S, McGee KP, et al. Three-dimensional physical modeling: Applications and experience at mayo clinic. Radiographics. 2015;**35**(7):1965-1988

[26] Tam MD, Laycock SD, Jayne D, Babar J, Noble B. 3-D printouts of the tracheobronchial tree generated from CT images as an aid to management in a case of tracheobronchial chondromalacia caused by relapsing polychondritis. Journal of Radiology Case Reports. 2013;**7**(8):34-43

[27] Zopf DA, Hollister SJ, Nelson ME, Ohye RG, Green GE. Bioresorbable airway splint created with a three-dimensional printer. The New England Journal of Medicine. 2013;**368**(21):2043-2045

[28] Cheng GZ, Folch E, Brik R, Gangadharan S, Mallur P, Wilson JH, et al. Three-dimensional modeled T-tube design and insertion in a patient with tracheal dehiscence. Chest. 2015;**148**(4):e106-e108

[29] Guibert N, Didier A, Moreno B, Mhanna L, Brouchet L, Plat G, et al. Treatment of post-transplant complex airway stenosis with a three-dimensional, computer-assisted customized airway stent. American Journal of Respiratory and Critical Care Medicine. 2017;**195**(7):e31-e33

[30] Shan Q, Huang W, Shang M, Wang Z, Xia N, Xue Q, et al. Customization of stent design for treating malignant airway stenosis with the aid of three-dimensional printing. Quantitative Imaging in Medicine and Surgery. 2021;**11**(4):1437-1446

[31] Shan Q, Huang W, Shang M, Wang Z, Xia N, Xue Q, et al. Treatment of aerodigestive fistulas with a novel covered metallic Y-shaped segmented airway stent customized with the

assistance of 3D printing. Annals of Translational Medicine. 2021;**9**(13):1051

[32] Schweiger T, Gildea TR, Prosch H, Lang G, Klepetko W, Hoetzenecker K. Patient-specific, 3-dimensionally engineered silicone Y-stents in tracheobronchomalacia: Clinical experience with a novel type of airway stent. The Journal of Thoracic and Cardiovascular Surgery. 2018;**156**(5):2019-2021

[33] Guibert N, Didier A, Moreno B, Lepage B, Leyx P, Plat G, et al. Treatment of complex airway stenoses using patient-specific 3D-engineered stents: A proof-of-concept study. Thorax. 2019;**74**(8):810-813

[34] Dutau H, Musani AI, Laroumagne S, Darwiche K, Freitag L, Astoul P. Biodegradable airway stents-bench to bedside: A comprehensive review. Respiration. 2015;**90**:512-521

[35] Zopf DA, Flanagan CL, Wheeler M, Hollister SJ, Green GE. Treatment of severe porcine Tracheomalacia with a 3-dimensionally printed, Bioresorbable, external airway splint. JAMA Otolaryngology Head & Neck Surgery. 2014;**140**(1):66-71

[36] Korpela A, Aarnio P, Sariola H, Törmälä P, Harjula A. Comparison of tissue reactions in the tracheal mucosa surrounding a bioabsorbable and silicone airway stents. The Annals of Thoracic Surgery. 1998;**66**(5):1772-1776

[37] Korpela A, Aarnio P, Sariola H, Törmälä P, Harjula A. Bioabsorbable self-reinforced poly-L-lactide, metallic, and silicone stents in the management of experimental tracheal stenosis. Chest. 1999;**115**:490-495

[38] Robey TC, Välimaa T, Murphy HS, Tôrmâlâ P, Mooney DJ, Weatherly RA. Use of Internal Bioabsorbable PLGA "Finger-Type" Stents in a Rabbit Tracheal Reconstruction Model. Archives of Otolaryngology - Head and Neck Surgery; 2000;**126**(8):985-991

[39] Saito Y, Minami K, Kobayashi M, Nakao Y, Omiya H, Imamura H, et al. New tubular bioabsorbable knitted airway stent: Biocompatibility and mechanical strength. The Journal of Thoracic and Cardiovascular Surgery. 2002;**123**(1):161-167

[40] Liu KS, Liu YH, Peng YJ, Liu SJ. Experimental absorbable stent permits airway remodeling. The Journal of Thoracic and Cardiovascular Surgery. 2011;**141**(2):463-468

[41] Lischke R, Pozniak J, Vondrys D, Elliott MJ. Novel biodegradable stents in the treatment of bronchial stenosis after lung transplantation. European Journal of Cardio-Thoracic Surgery. 2011;**40**(3):619-624

[42] Stehlik L, Hytych V, Letackova J, Kubena P, Vasakova M. Biodegradable polydioxanone stents in the treatment of adult patients with tracheal narrowing. BMC Pulmonary Medicine. 2015;**15**:164. DOI: 10.4172/1747-0862.1000065

[43] Vondrys D, Elliott MJ, McLaren CA, Noctor C, Roebuck DJ. First experience with biodegradable airway stents in children. The Annals of Thoracic Surgery. 2011;**92**(5):1870-1874

[44] Hussain NM. Considerations for Development of 3D Printed Bronchial and Tracheal Stents and Tracheal Stents [Thesis]. Columbia, SC, USA: University of South Carolina; 2015

[45] Wood C, Cheng G, Miller A, Wahidi MM, Gall K. Mechanical characteristics of 3D printed airway stent. In: A45 Topics in Interventional Pulmonary. American Thoracic Society. San Diego. 2018;**197**. p. A1739

[46] Paunović N, Bao Y, Coulter FB, Masania K, Geks AK, Klein K, et al. Digital light 3D printing of customized bioresorbable airway stents with elastomeric properties. Science Advances. 2021;7(6):1-13

[47] Zarogoulidis P, Darwiche K, Tsakiridis K, Teschler H, Yarmus L, Zarogoulidis K, et al. Learning from the cardiologists and developing eluting stents targeting the Mtor pathway for pulmonary application; a future concept for tracheal stenosis. Journal of Molecular and Genetic Medicine. 2013;7:65. DOI: 10.4172/1747-0862.1000065

[48] Zhu GH, Ng AHC, Venkatraman SS, Boey FYC, Wee ALY, Trasti SL, et al. A novel bioabsorbable drug-eluting tracheal stent. The Laryngoscope. 2011;**121**(10):2234-2239

[49] Wang T, Zhang J, Wang J, Pei YH, Qiu XJ, Wang YL. Paclitaxel drug-eluting tracheal stent could reduce granulation tissue formation in a canine model. Chinese Medical Journal. 2016;**129**(22):2708-2713

[50] Duvvuri M, Motz K, Murphy M, Feeley M, Ding D, Lee A, et al. Engineering an immunomodulatory drug-eluting stent to treat laryngotracheal stenosis. Biomaterials Science. 2019;7(5):1863-1874

[51] Duvvuri M, Motz K, Tsai HW, Lina I, Ding D, Lee A, et al. Design of a biocompatible drug-eluting tracheal stent in mice with laryngotracheal stenosis. Journal of Visualized Experiments. 2020;(155). DOI: 10.3791/60483

[52] Sindeeva OA, Prikhozhdenko ES, Schurov I, Sedykh N, Goriainov S, Karamyan A, et al. Patterned drug-eluting coatings for tracheal stents based on pla, plga, and pcl for the granulation formation reduction: In vivo studies. Pharmaceutics. 2021;**13**(9):1-14

[53] Li Z, Tian C, Jiao D, Li J, Li Y, Zhou X, et al. Synergistic effects of silver nanoparticles and cisplatin in combating inflammation and hyperplasia of airway stents. Bioactive Materials. ;**2022, 9**:266-280

[54] Xu J, Ong HX, Traini D, Williamson J, Byrom M, Gomes Dos Reis L, et al. Paclitaxel-eluting silicone airway stent for preventing granulation tissue growth and lung cancer relapse in central airway pathologies. Expert Opinion on Drug Delivery. 2020;**17**(11):1631-1645

[55] Hohenforst-Schmidt W, Zarogoulidis P, Pitsiou G, Linsmeier B, Tsavlis D, Kioumis I, et al. Drug eluting stents for malignant airway obstruction: A critical review of the literature. Journal of Cancer. 2016;7(4):377-390

[56] Zarogoulidis P, Sardeli C, Konstantinou F, Sapalidis K. Conventional versus therapeutic stents for airway malignancies: Novel local therapies underway. eBioMedicine. 2018;**33**:10-11

[57] Aravena Leon C, Inaty H, Urbas A, Grafmeyer K, Machuzak M, Sethi S, et al. Early outcomes with 3D printing and airway stents. Chest. 2019;**156**(4):A199-A200

[58] Duong DK, Bedi HS, Guo HH, Cheng GZ, Sung AW, Free D. Custom 3D-printed airway stent for the Management of Complex Malignant Airway Obstruction. In: C36 Case Reports in Use of Interventional Pulmonology Stents. American Thoracic Society. Philadelphia. 2020;**201**. p. A4883

[59] Huang W, Shan Q, Wu Z, Li H, Zhou M, Ding X, et al. Retrievable covered metallic segmented Y airway stent for gastrorespiratory fistula of carina or main bronchi. The Journal of

Thoracic and Cardiovascular Surgery. 2021;**161**(5):1664-1671.e2

[60] Cheng GZ, Folch E, Wilson A, Brik R, Garcia N, Estepar RSJ, et al. 3D printing and personalized airway stents. Pulmonary Therapy. 2017;**3**(1):59-66

[61] Aravena C, Inaty H, Urbas A, Grafmeyer K, Machuzak M, Sethi S, et al. Clinical outcomes with 3D patient specific airway stents compared to commercially available airway stents. In: 21st World Congress of Bronchology and Interventional Pulmonology. November 2020:19-22

[62] Xu J. Development of 3D-Printed, Drug-Eluting Airway Stents for the Personalised and Local Treatment of Central Airway Pathologies Statement of Originality. Camperdown, Australia: The University of Sydney; 2021

Chapter 13

Making Use of Three-Dimensional Models of Teeth in Practical Teaching of Endodontics

Przemysław Kustra

Abstract

Making use of 3D-printed teeth models in teaching students offers an innovative approach. Empowering a highly efficient digital science to improve teaching. This gives opportunity to learn and enable intuitive dentist and student-patient communication. Clear and engaged satisfactory experience for teacher, student and patient. Thanks to the perfect representation of teeth anatomy, making use of 3D models in the teaching of endodontics may well be recommended as holding substantial potential in improving overall quality of training at the preclinical stage, with a view to appreciably reducing overall risk of encountering complications during the actual clinical work. The mistakes made by the students, for example, at the access cavity for root canal treatment stage were assessed with the help of 3D models, as well as their overall, hands-on learning progress was evaluated. Also in the clinical process, before the procedure with the participation of a patient, a student or a specialist may perform a treatment procedure on a tooth printed in 3D, based on tomography, under the supervision of an experienced specialist. 3D printing digital solutions and the popularization of these solutions around the globe are helping dental clinics and hospitals to effectively and efficiently achieve digital transformation.

Keywords: digital stomatology, digital oral medicine, three-dimensional printing, virtual endoscopy, three-dimensional teaching

1. Introduction

In the following chapter, I want to present the possibilities of using 3D printers and 3D printing in the practical clinical teaching of endodontics. In this respect, the 3D printing of tooth models, needed for root canal therapy practice, and the virtual endoscopy technique are the most relevant for teaching [1–5].

The adoption and adaptation of the latest advances in digital technology, such as three-dimensional (3D) printed dental objects, have influenced the teaching and treatment of cases that cover virtually the entire field of dentistry [6].

This technology offers a unique setting for the development of clinical and educational treatment, as demonstrated by publications in the field. Using a conjunction of three classic technologies, already used in medicine, but combined all at once, results in modern performance in education and clinical treatment. This contributes to creating opportunities for the development of dental medicine, which is constantly improving in clinical, educational and, of course, research contexts. In order to benefit from their treatment, dentists in the current era can already interact with the available multidisciplinary knowledge and 3D printing to understand the essence of the new technology and meet the challenges of the digital medicine era. It, therefore, becomes legitimate to introduce this as a subject in the teaching process of students [4, 7].

All over the world, the digital solutions of 3D printing and their popularisation are helping teaching units and hospitals to achieve digital transformation in an effective and efficient way, of which this book and the chapter dedicated to education is a good example. Furthermore, in dentistry and facial areas close to the oral cavity, this technique is widely used in head and neck surgery (craniofacial and orthognathic implants), personalised oral soft tissue regeneration, orthopaedics (fracture printing in orthopaedic trauma surgery, 3D imaging), 3D printing and virtual 3D planning in endodontics, ophthalmology, template printing in mandibular (and surgical) reconstruction, prosthodontics (replication techniques for making e.g., digital dentures and overdentures, also on implants), periodontal regeneration and repair (periodontal implants), orthognathic surgery, 3D physical models of teeth, printing of bone implants and virtual endoscopy, as well as in autotransplantation [8–15].

In the aforementioned areas, prior practice is recommended before starting treatment. The initial use of templates or 3D-printed models can facilitate treatment planning and reduce the risk of complications during clinical procedures performed as part of the teaching process [16].

3D printing consists of three stages, which ensure full integration of 3D digital dental solutions, namely, acquiring 3D data by means of cone-beam computed tomography (CBCT) (or other stationary and intraoral scanners, if necessary; in the study in which I participated, a 3D scanner ATOS Triple Scan III (GOM, Germany) was used), processing and designing the data using professional dental software, after transferring the obtained scans into an electronic version, for example, Model Creator (Exocad, Germany) or standard software provided by the manufacturer, and production of 3D dental objects from the obtained digital 3D dental models using, for example 3D resin printers, such as 3D Form 2 (Formlabs), employing materials with similar physical properties, for example hardness and brittleness, to the reproduced tissues (**Figures 1–4**) [1, 4, 16].

Dental applications of 3D printing adopt one or more of the following common technics: Stereolithography (SLA), fused deposition modelling (FDM), MultiJet printing (MJP), PolyJet printing, ColorJet printing (CJP), digital light processing (DLP) and selective laser sintering (SLS) also known as selective laser melting (SLM). The most popular technique in teaching root canal treatment is the first mentioned that is SLA printing, for example using the dental model resin [10, 17].

2. 3D printing teaching discussion

Endodontics is such a field of science that is based on approximately 90% of manual skills. 3D printing technologies significantly improve and accelerate the acquisition of root canal treatment qualifications by students, while the use of 3D replicas of teeth obtained by means of 3D printing has contributed to the enhancement of qualification of teaching centres and the correctness of performed root

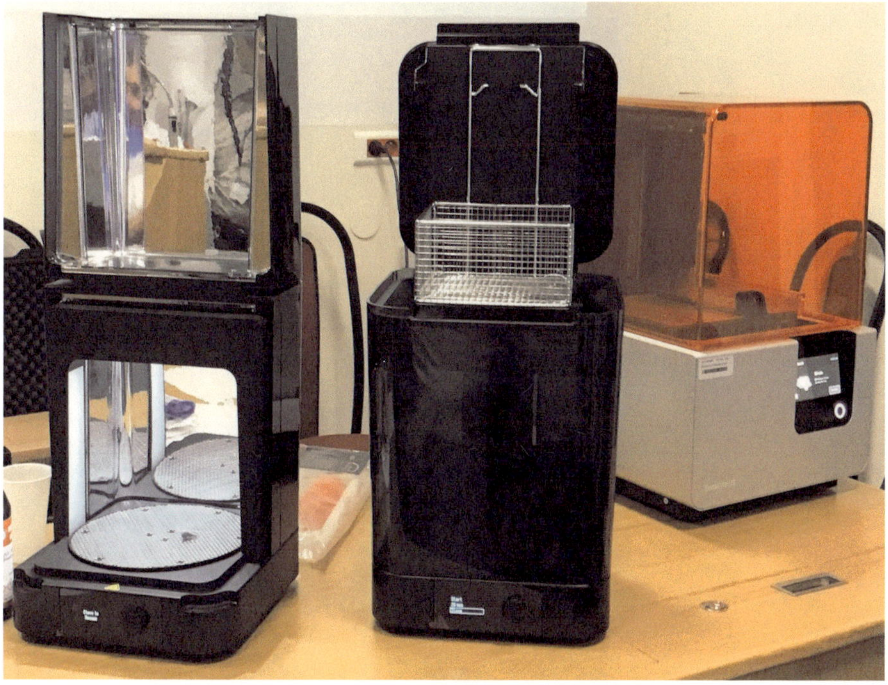

Figure 1.
3D printer kit (Formlabs), from left to right: Form cure, form wash and 3D printer.

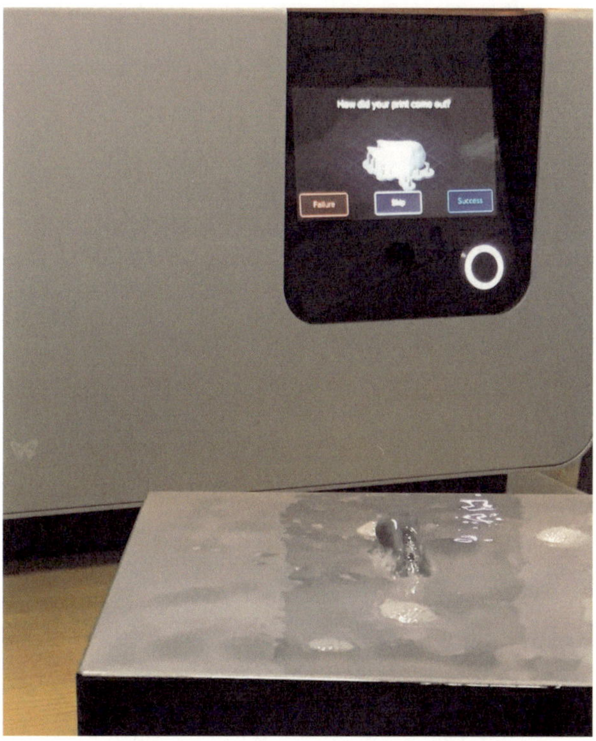

Figure 2.
Already printed tooth on the platform.

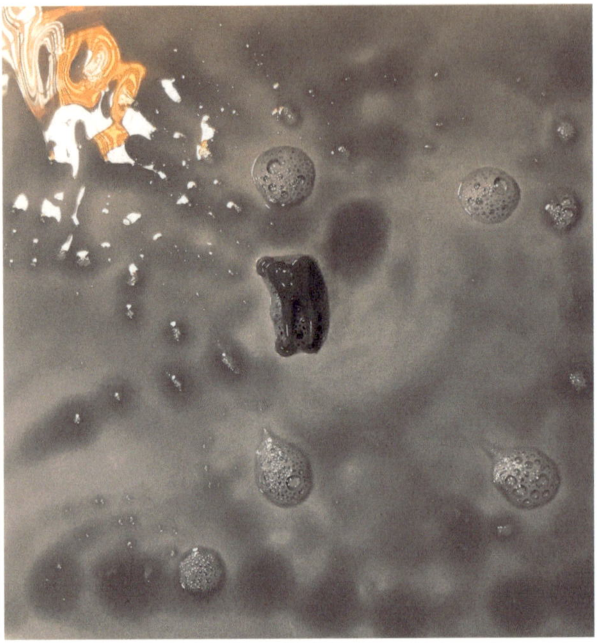

Figure 3.
Printed molar tooth on the platform, before inserting into the washer. Tooth length is 18,08 mm.

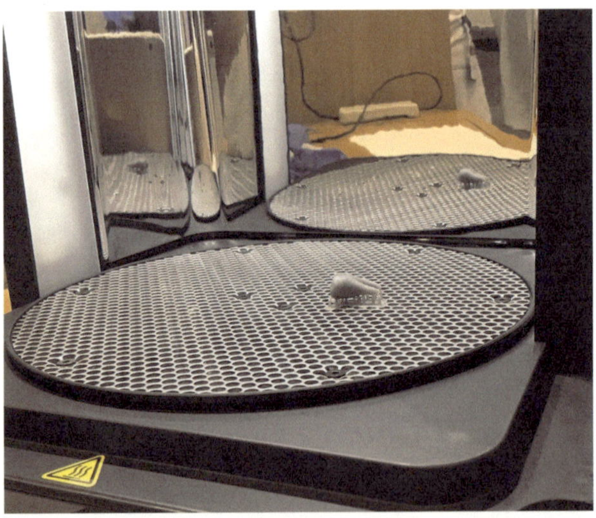

Figure 4.
Already printed molar tooth, after using form washer, on the platform of form cure.

canal treatments in undergraduate education. Printing 3D models is an innovative technique in the field of treatment and teaching (by reducing the likelihood of errors during treatments). 3D printing is the name used to describe the 'manufacturing approach' that creates a material by adding layer by layer. A student with little experience in root canal treatment will first have the opportunity to perform the procedure on a 3D replica of a real tooth. Thus, the aim

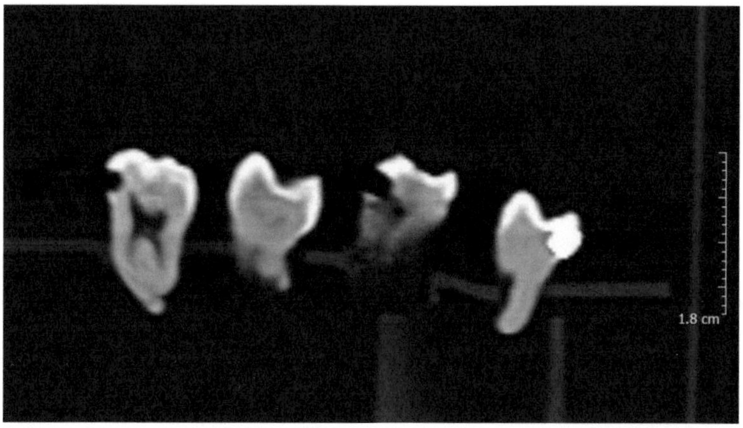

Figure 5.
Cone beam tomography (CBCT) image of molars during the assessment of cavity preparation and fillings.

of such a teaching process is to increase the effectiveness of correct root canal treatment in doctors during their speciality training in the area of restorative dentistry with endodontics [1, 4, 5, 17–19].

2.1 Teaching experience

On the basis of my experience and many years of teaching, I have identified the important elements needed in the development of teaching methodology in clinical classes. The most difficult part of the work for the student is to directly move from phantom classes to clinical classes, which significantly increases the demands placed on the student. A number of studies report that even the best-organised phantom classes are not able to translate the acquired skills, even if they are very good into the same tasks in clinical work. This is influenced by development of manual skills, which takes time. This phenomenon is related to 'brain plasticity', as well as the development of the right emotional attitude. From this one can conclude that the introduction of a didactic element of an intermediate nature, to play relatively smoothly a transitional role between phantom and clinical activities, is a very important element of the teaching process. It has turned out that 3D technology with 3D printing fits almost perfectly into these realities. This is possible because we use real-world models for pre-clinical and clinical learning [1, 2, 3, 4, 18].

Endodontics, or more precisely root canal treatment, cannot be taught strictly speaking; apart from the knowledge, which of course has to be learnt, manual skills need to be acquired. Clinical work involves performing tasks on obscure objects, which creates additional problems in gaining experience, which usually takes several years of continuous practice in this area of knowledge. In addition, the patient needs to be explained the course of treatment on the part of the tooth that cannot be seen and the need for necessary treatment. Many devices, instruments and materials are required for adequate endodontic treatment. Various types of diagnostic materials and devices have already been introduced over the years, including digital x-rays as well as digital cone tomography, intraoral cameras, fibre-optic endoscopes, 3D atlases (these are mainly physiology-based), operating microscopes and digital intraoral cameras. Fibre optic endoscopes (e.g., Ora scope) are particularly close to our field of application, as they are used in the tooth chamber to visualise the inside of the root canal (in order to obtain information for the doctor and patient and thus implement a favourable treatment plan). They can of course also be useful in the classroom to show a particular clinical issue on a monitor to a larger number of accompanying people, which without the use of a monitor would only be visible to the person performing the procedure. 3D printing and virtual endoscopy are better and more modern solutions. That is, what can be visualised better is of particular importance, so that the endodontist and the student are well-informed in the area of diagnostics, and modern technology has created new methods of image enhancement in the form of a model rather than the scans themselves. In this situation, in fact, any clinical problem can be visualised, printed and discussed in the wider community. In addition, the clinician relies on images, verbal information and numerical data, for example statistical or digital analyses (digital photography, digital x-rays and computerised visualisations), in communicating with the patient, in the process of assessing the patient and when selecting treatment options [2, 3, 4, 20].

2.2 Teaching objectives

Planning the technology acquisition strategy is the factor that determines the needs of an individual. Clinicians should focus on what is needed to provide the best solution for the patient and to establish clear needs. The adoption of new technologies, especially in teaching, must be based on clear objectives of the systems that are being implemented. Planning should be based on three areas: the main objective of the plan, the implementation plan and the measurement phase.

Over the years, I have relied on this scheme in the pattern of research I have done. The main objective of the plan is to convince us of the need to implement the technology, why it is needed and what the teaching benefits are? What goals must be achieved to have a problem-free implementation? All the information generated and processed in the practice and the way it is handled constitutes the strategic plan for the technology. Management of information received from the patient, students and other clinicians provides the basis for the transfer of materials before, during and after treatment. The implementation part involves planning and day-by-day practice until the implementation programme runs smoothly. Training and retraining of staff are essential in a fully functioning plan. The technology being implemented is advanced, so the fixed working hours of training and consultation with the available laboratory are essential. Courses and training are necessary to acquire basic skills. The implementation evaluation phase is important for verifying that the implementation plan is well-led and followed by improving performance, especially in terms of practical and manual skills. During the implementation of a technological system, a method of evaluating their progress towards the objectives implemented must be developed. Student and patient education is an important positive experience and part of any modern treatment, including endodontic treatment [1, 2].

In the first step, students were introduced to the analysis of 3D cone-beam tomography scans (**Figures 5** and **6**) to illustrate to students the results that would create in themselves the need to use this technology for their intended clinical purposes. To this end, a study (according to randomisation principles) was conducted to illustrate the quality of cavity preparation and the placement of fillings in extracted teeth in such a way that the student could learn about the quality and outcome of their work, with the realisation that this has clinical implications. That is, students were able to see the quality of their cavity preparation (e.g., the homogeneity of the prepared walls, as

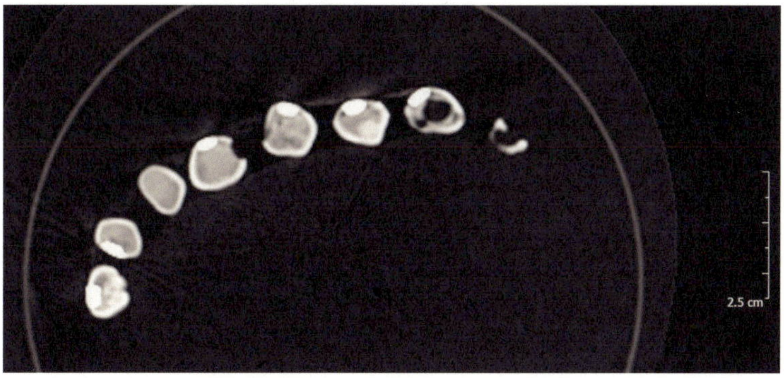

Figure 6.
Cone beam tomography (CBCT) image of molars during the assessment of cavity preparation and fillings, tooth cross-section view.

well as the seal of the placed fillings). Randomly allocated teeth were selected for the study and standardised cavity preparations were planned, followed by filling, using an adhesive technique with composite. The teeth prepared in this way were then scanned using CBCT and the scans were evaluated in IrfanView and Fiji Is Just ImageJ software. The teeth were then subjected to thermocycling and later subjected again to the same CBCT examination. The students were then able to evaluate the results of their work accurately on images not only from cone tomography scans but also after 3D visualisation (**Figure** 7). To make the work more tangible, of course, the examined teeth were also assessed under an optical microscope with a micrometre scale. The analysis results from the CBCT and virtual scan images were then compared with the images of the specimens themselves and similar results were obtained. This was proof that digitisation produces measurable comparable results and can be further implemented. The following pictures show the correctness of the preparation, that is the box-like shape of the cavity, as well as, in some situations, the loss of fillings, the protrusion beyond the standard curvature of the tooth crown, which was indicative of, for example, the care taken in finishing the filling or leaks in the adherence of the

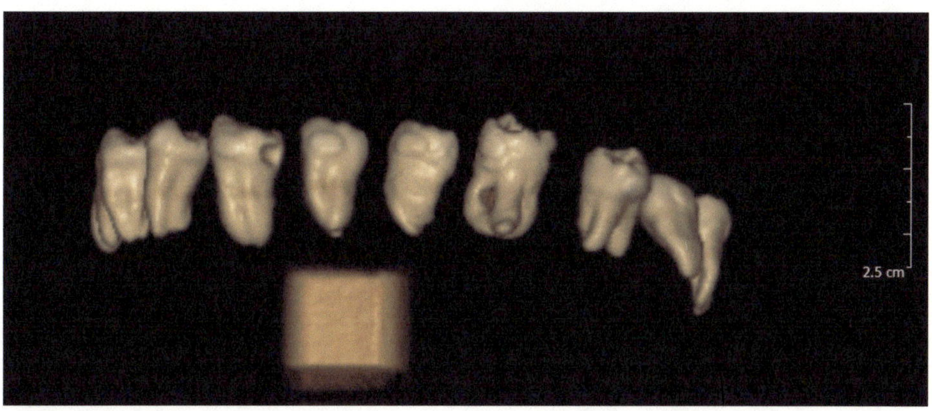

Figure 7.
Virtual 3D image of teeth obtained from cone-beam tomography scans. Visible preparation of cavities and applied fillings.

fillings. Only after these analyses did the students realise the need for quality in their work when they were able to make feedback visualisations of their work. They saw the possible consequences of developing pulpitis, for example in the case of a leakage, where it was not a phantom tooth but a real tooth with complications already present. The use of 3D model aids has significantly accelerated the progress of students' work and changes in thinking but has also created the need to introduce 3D techniques into the format of didactic teaching [1, 2].

Overall, when it comes to digitalisation, computers and software are invaluable in practice. Most clinicians are unable to consider practising with no digital solutions applied in their treatments. However, technology cannot replace the quality of care itself, especially in root canal treatment. This view may change with the implementation of 3D printing, as here the use of technology is an advancement in the acquisition of clinical skills, by working on real models of real clinical problems. This changes the student's thinking from a generic job, where a nameless phantom is treated, to the realisation that they are treating a specific tooth of a real person. In the study I proposed to the students, in my opinion, it should have been quite simple to open the chamber of the teeth, as this is the most visible part of the root canal treatment procedure. Obviously, this is the stage that determines proper access to the root canal and has an impact on the subsequent strength of the tooth crown as well as the ability to maintain patency of the root canal. We adopted a scale of 0–7 for the study (0—no mistakes, 1—one wall incorrectly prepared, and so on, 6—perforation and 7—damage eliminating the possibility of further work) (**Figure 8**). The entire study was

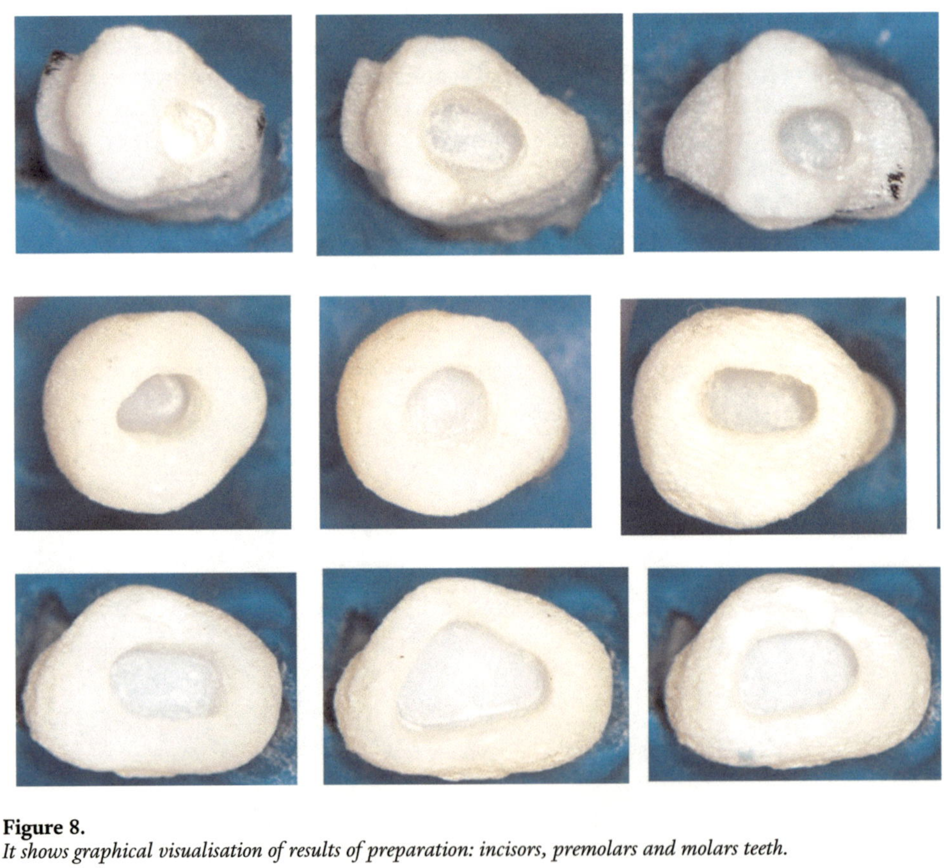

Figure 8.
It shows graphical visualisation of results of preparation: incisors, premolars and molars teeth.

Figure 9.
Example printout, 3D group of premolars.

performed according to randomisation rules. The principles of work and preparation were strictly defined in numerous lectures and training sessions in order to obviously standardise the study. The experience gained from this research work is described in this chapter [1–5].

A total of 9 students with little root canal experience took part in the study. The 3D prepared teeth included the lower incisors (there were a total of 30 3D teeth, 10 per student), the upper premolars (also 30 3D teeth, 10 per student) and finally the lower molars (also 30 3D teeth, 10 per student) (**Figure 9**). At the beginning of the study, there were errors in all central incisor preparations (**Figure 10**), in about 67% of the premolars and in about 72% of the molars. However, after ten consecutive preparations, this result was, respectively, only 19% of errors for incisors, around 14% of errors for premolars and around 33% of errors for molars. All these results between the start and end points are statistically different. The 90 3D tooth models used in the study were produced from 3D scans of extracted teeth [1].

The students gained once again a very large and interesting experience during their own assessment of their own specimens after performing only the root canal treatment procedure. The specimens were then visualised using a special scanner so that the preparation could be viewed in depth. In addition, during the practical classes, students were able to use the training microscope (only when assessing their own prepared earlier teeth as a feedback instrument) to evaluate their own work. They themselves witnessed their own growing experience with each model they prepared and concluded that the effectiveness of this method led to safer treatment for the patient. For example, all students preparing incisors only started to eliminate crown damage and perforations one by one by the fourth model, and by the seventh model, all three students were free of these two most serious complications. In molars, for example, one student in three, in this group, only avoided perforation and chamber damage in the fifth model. In general, it could be assumed that all nine students showed significant improvement in preparation from the seventh model. It can,

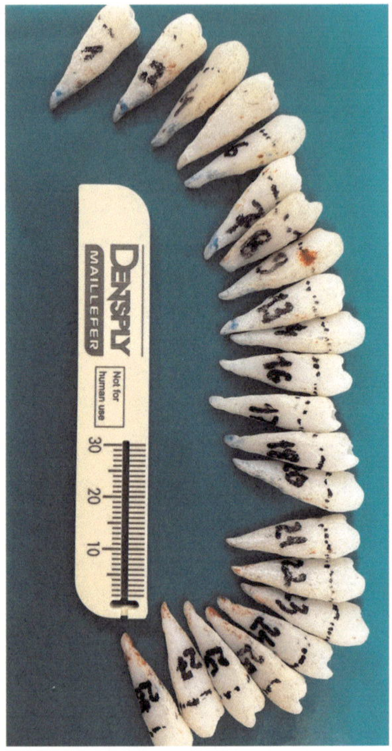

Figure 10.
Example of perforation of incisor tooth, number 20.

therefore, be assumed that practical coursework tends to make sense with ten preparation attempts. I would also like to mention that the students were assessed after the first preparation of the 3D model, just in case, to make sure they knew what a correct preparation was and to avoid copying mistakes. They also had in front of them constantly visible patterns of the correct preparation [1].

In our study with students, 3D models are produced based on real patient teeth. This provides the students with an almost realistic simulation of preparation in a real clinical situation.

I have noticed that even though it is possible to explain to students the whole procedure step by step, as well as to describe the complications that can occur after what we call an abnormality in the preparation, it is still only when they practise using real models from specific people that they realise the risks and responsibilities that accompany their work. Perforated tooth models are a fairly common complication, although the long axis of the tooth is visible. I would like to mention that in the clinical setting, except for examples of advanced periodontitis, the entire area of the long axis of the tooth and the curvatures of the root are not visible because, as we know, they are hidden in the alveolar bone. I would like to point out that the masses to be printed were very close to the hardness of the bone, so the preparation in the 3D model of the tooth was overall quite close to real conditions [1].

2.3 Historical background

In 1983, Charles Hull was the first to introduce a 3D printer. He applied the method of stereolithography in this technology. In order to characterise the 3D printing process, three basic steps can be identified: scanning of a selected tooth (e.g., using cone-beam tomography (CBCT)), digital reconstruction of the scanned image (e.g., using available dedicated software) and then 3D printing of a model of the tooth (e.g., using dedicated printers) [20].

The requirements for precision printing have enabled the use of 3D printing in medicine, since 1990, which, in the first instance, found its way into prosthetics. With regard to history, dental education was based on extracted teeth, resin blocks or commercially available resin teeth for pre-clinical training and the possibilities that the anatomy rooms had, according to accepted criteria [20].

2.4 Scope of knowledge

In the literature, there are isolated publications on the use of 3D printers in dentistry and in the education of students. Carrying out a study to improve the quality of didactic education will be another attempt to improve the skills in the field [4, 5, 18, 21].

The development of this technology makes it possible to improve treatment and teaching techniques by reducing the risk of errors during procedures primarily in clinical teaching, based predominantly on educational models and clinical simulations. From a didactic point of view in the teaching process, demonstrating the effectiveness of this teaching method is of particular importance. A student with little experience in root canal treatment will be given the opportunity to perform the procedure first on a 3D replica of a tooth that is in the root canal treatment plan in his or her class. This will enable them to avoid mistakes resulting from little clinical experience. The same is true for doctors—in difficult clinical cases.

2.5 Teaching pathway and controlled learning process

A trainee doctor with little experience in root canal treatment will have the opportunity to perform a complex procedure on a 3D replica of the tooth first before treating the patient's tooth in question. This will enable them to avoid mistakes resulting from little experience in different clinical situations. Take, for example, a situation where there are obliterations of the root canal, in which the still unobstructed canal lumen is present in the more apical parts, gradually narrowing the root as a result of, for example dentin deposition due to age, the presence of caries, orthodontic treatment, systemic diseases or the occurrence of trauma. If inflammation, such as pulpitis or periodontitis, is detected on clinical or radiographic examination, intervention is indicated. The risk of a complication when treating a root canal with obliteration can account for up to 75% of perforation incidents when attempting to localise and negotiate calcified canals. Students, as shown in the study, have difficulty even opening the chamber of a tooth that is not obliterated. In the classic procedure, the risk of perforation is reduced by methods that provide straightforward access to the canal orifice and the use of specialised instrumentation, such as the surgical microscope and ultrasonics. A trainee with little experience may choose unfavourable paths of access cavity preparation and under realistic conditions irreversibly damage the tooth. It is based on the fact that, when planning the opening of the tooth chamber or the angle of insertion of instruments for root canal negotiation or removal of obstruction during work, we may decide that a different insertion path may have been more advantageous, which then exposes the tooth to the loss of additional dental hard tissue [18].

The use of cone beam tomography in today's dentistry is widespread and frequent. Because of this, many patients now have a CBCT examination for various reasons (which, together with the favourable radiation dose, facilitates this examination), which is the basis for printing 3D models. The lack of 3D models may cause clinical errors, due, for example, to insufficient data on the treated tooth space, which, as I have already described, may contribute to further complications, for example strip perforation or unnecessary loss of hard tooth tissue.

Student training consists of learning how to operate the printer, resin selection, tooth preparation technique, virtual assembly of data obtained from cone beam tomography (simple single tooth models) to produce a virtual 3D tooth model, and the printing process. The plan of the digital model is already important, as already at this stage students draw a number of conclusions about the issues related to their planned treatment. In a didactic setting, the teaching assistant often helps with clinical work, but nevertheless cannot do the work for the student the whole time. In the end, the students have to perform the set procedure themselves. Due to the mere lack of manual dexterity, every clinical procedure is an emotional burden and is fraught with the risk of complications due to the lack of an imagined treatment endpoint. One of the first principles of medicine refers to the statement: 'before you start the treatment, imagine what the end will look like'. Only virtual planning and working on real models of the scanned tooth can help build up an idea of the intended positive expected treatment results and identify the cause in the case of unintended complications [18, 21, 22].

During the course of their studies, when operating a 3D printer, students will learn about various parameters, including XY resolution, which is the most common specification used to describe the quality or detail of a print and is analogous to size in pixels, it is the smallest movement the printer's laser can make in a horizontal layer. However, XY resolution does not take into account many variables that affect the quality of a component. In fact, professional 3D printers have more than 100 different settings that affect XY resolution. Layer thickness or Z-height usually describes the surface finish of the component, meaning that a lower layer height improves the surface finish. In addition, the thickness of the layer is influenced by the type of resin and the printer settings [18, 32].

In accordance with the growing interest in 3D printing technology during my consultations within the dental community (when presenting research at symposia and conferences), as a specialist in restorative dentistry with endodontics, when presenting the study described in this manuscript, I have received numerous communications regarding the provision of training in 3D printing of teeth in the context of considering the implementation of such courses. This demonstrates a particular interest in this technology. All those interviewed agreed on the potential of this technology in all its applications, particularly in decision support for complex root canal treatment cases.

3D printed tooth models and electronic images, used as a graphical guide to visualise the problem in question, can help operators plan and manage complex non-surgical and surgical endodontic treatment and develop skills, thus becoming an invaluable educational resource. Learning from one's mistakes without compromising the patient's health is a leading element of this technique. Three-dimensional (3D) volumetric images provide a three-dimensional view of the anatomy, facilitating treatment planning and teaching; when designing a 3D model, students must also model the area of their preparation, which forces them to familiarise themselves with the treatment.

2.6 Non-surgical endodontic treatment

Therefore, a study on non-surgical treatment using 3D printing of teeth, comparing a group of people who will be opening teeth for root canal treatment, having first opened 3D replicas of these teeth, with a group of people who will be opening teeth for root canal treatment without first being able to explore the anatomy of the teeth, could be used in endodontics as a means of teaching students. This contributes to the understanding of tooth morphology, for example the simulation of tooth chamber opening and root canal preparation [1].

In the studies I participated in, the study design was as follows. The extracted teeth (incisors, premolars and molars) were scanned using cone beam computed tomography (CBCT), and the resulting data was then analysed with the 3D model visualisation software, EXOCAD and Model Creator.

The 3D teeth were made from Dental Model compound and printed using a Form 2 3D printer (Form 2, CadXpert), using computer-based stereolithography (SLA) technology. The device is medically approved, with a special focus on dentistry, due to its high-resolution printing capabilities.

90 teeth were prepared. Thirty identical teeth, molars, premolars and incisors, were printed as replicas of the treated natural teeth, one from each tooth group, made on the basis of the corresponding natural tooth. They were divided into nine groups, randomly allocating 10 teeth of the same shape to each group. Within each group, the teeth were numbered in sequence. Dental students with little experience in root canal treatment were assigned to do the work, one to each group.

Differences between the start and end points of the tooth chamber opening procedure for root canal treatment according to the adopted criteria were studied for 90 teeth.

Dental students with little experience in root canal treatment were assigned to perform the preparation. Next, the correctness of performing these procedures was assessed by two independent researchers, according to the adopted criteria. The correct procedure of opening the tooth cavity and making the correct access to the root canal entrance was evaluated. Teeth were scanned using the cone beam computed tomography (CBCT) technique and the data was analysed in the three-dimensional visualization software.

The correctness of the execution of these procedures was then assessed by two independent investigators (experts and the students themselves under the microscope), according to the adopted criteria: 0—no errors, 1–4—wall correction, 5—chamber floor correction and 6—perforation (mesial and distal wall in incisors and lateral walls in molars). (The study was approved by the bioethics committee, no. 122.6120.235.2016) [1].

2.6.1 Results obtained

The data obtained were analysed using Statistica 12.0 software. The Kruskal-Wallis test and the multiple comparisons test were performed for the entire group, and significant statistical differences were found $P = 0.0001$. The results are also shown in Chart 1. Premolar teeth were the most favourably prepared, followed by the incisors and molars [1].

2.6.2 Presentation of the results

The study showed that each subsequent tooth cavity preparation procedure was more successful. This is best seen in the comparison between the first and last replicas. In the process, students gained experience.

Printing single objects in three dimensions, using special printers and software, is a very innovative and promising technology for pre-clinical teaching and treatment. The use of a 3D printer is justified in the teaching process [1].

2.7 Surgical endodontic treatment

In the surgical planning of endodontic treatment, we use 3D printing, with additive manufacturing and rapid prototyping techniques, which are used with satisfactory accuracy, mainly in diagnosis and surgical planning, and then in the direct production of implantable devices. The main limitation is the time and money spent on generating 3D objects, and the fact that the type, material and layer thickness of the printer affect the accuracy of the printed models [15, 28].

3. Virtual endoscopy

An interesting method based on the 3D printing process is a virtual simulation (**Figure 11**). It is used in medicine, and also in dentistry, to teach dental students,

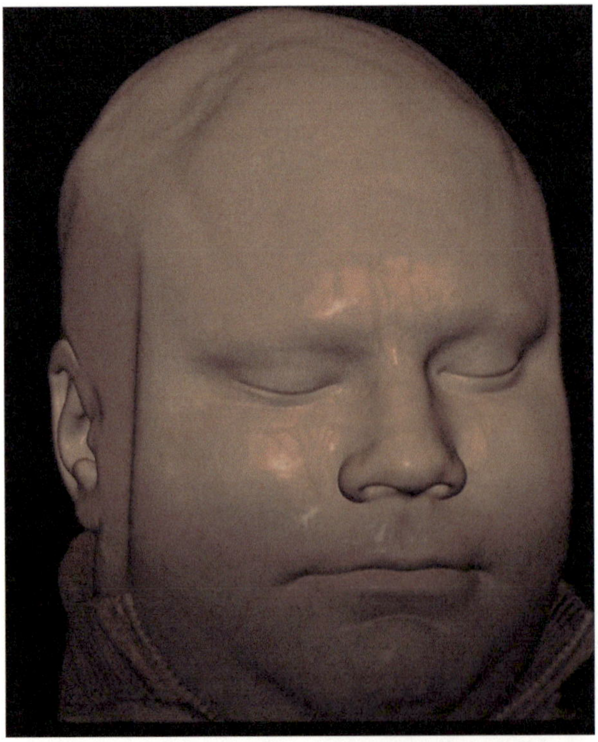

Figure 11.
3D reconstruction (of my head), frontal view and deeper structures of the skull visible under the skin surface.

including surgical techniques. On this basis, 3D prints of the areas of interest can be made, although digital 3D visualisation alone may already be enough. This means that we are referring to the clinical use of virtual endoscopy. Significant software developments in image processing, resulting in breakthroughs in image editing, as well as the development of the CT and MR scanners themselves, have made it possible to create virtual models. As a result, we now have a method for obtaining volumetric reconstructions, as well as virtual endoscopy, which is based on these reconstructions. This technique allows for the spatial presentation of the anatomical and morphological structures of the human being (**Figure 12**). Images obtained with this virtual technique are produced as a result of a CT or MR examination, without the need for classic clinical examinations such as endoscopy. This offers the possibility of non-invasive diagnosis and visualisation of small tissue structures (**Figures 13** and **14**). I have demonstrated the use of virtual endoscopy in a selection of images. I underwent a CT scan used in medicine. Based on the data acquired, soft and hard tissues, including teeth, were reconstructed (**Figure 15**). On the basis of the 3D images obtained, successive tissue layers can be inspected, for example by removing superficial structures to visualise the deeper ones. The following pictures show further examples (**Figures 16–20**). Particularly important are the images of the tooth after root canal treatment with the visualisation of the root canal filling and the canal lumen of the following tooth during root canal treatment, after preparation using a rotary system. In the visualisations, the texture of the scanned image is slightly overscaled, so a 3D programme is required to be able to accurately sharpen the object we want to print. This is why specialist knowledge and experience in the field, as well as training and courses, are needed. In practice, on the basis of the images reconstructed for the virtual endoscopy technique, once the images have been converted to the format used by the 3D printer, the given images can be printed. Virtual endoscopy is also a basis to learn, how to create a virtual 3D model for pre-bioprinting process based on computer

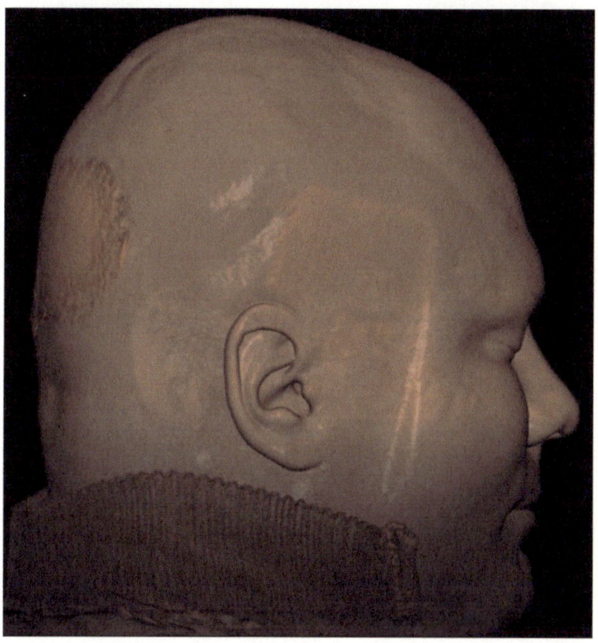

Figure 12.
3D reconstruction (of my head), side view and deeper cranial structures visible below the skin surface.

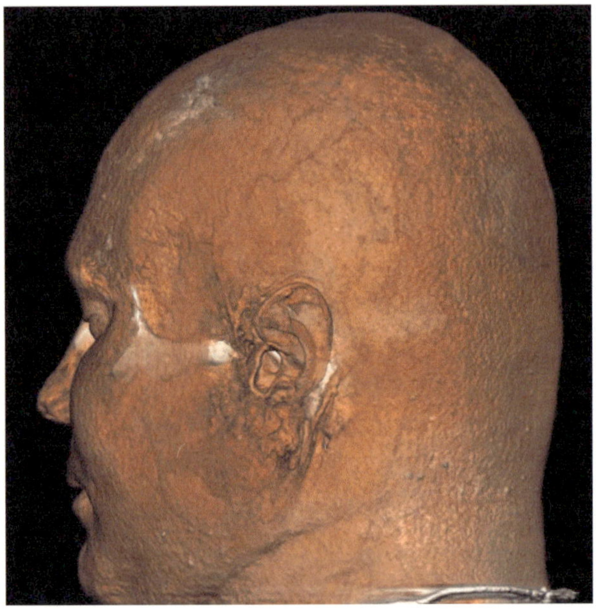

Figure 13.
3D reconstruction (of my head), lateral view and visible soft tissues including blood vessels and under the surface of the skin.

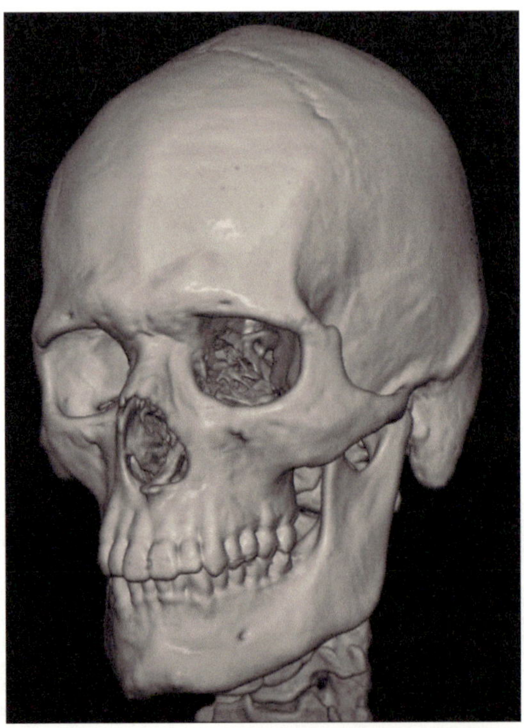

Figure 14.
3D reconstruction (of my head), oblique view and visible hard tissues including structures inside the skull.

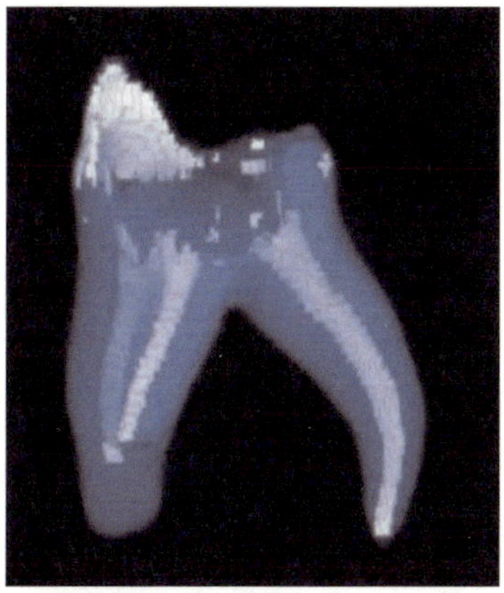

Figure 15.
3D reconstruction of a tooth after root canal treatment with filled root canals.

Figure 16.
3D virtual reconstruction of the inner part of the tooth root canal after preparation with the Flex Master rotation system.

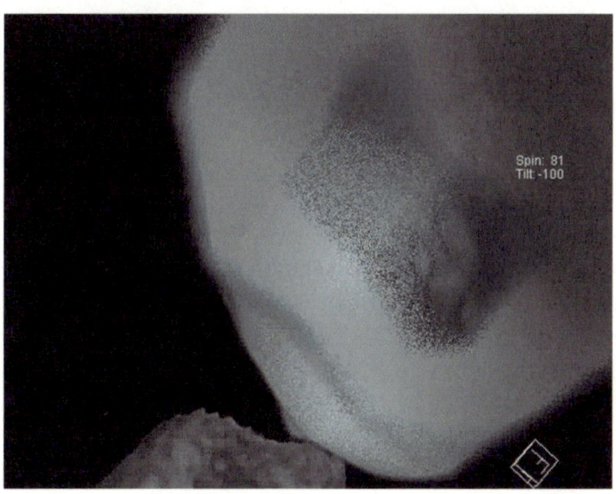

Figure 17.
3D virtual reconstruction of the outer part of the root canal, the tooth apex after preparation with the rotary system and Flex Master.

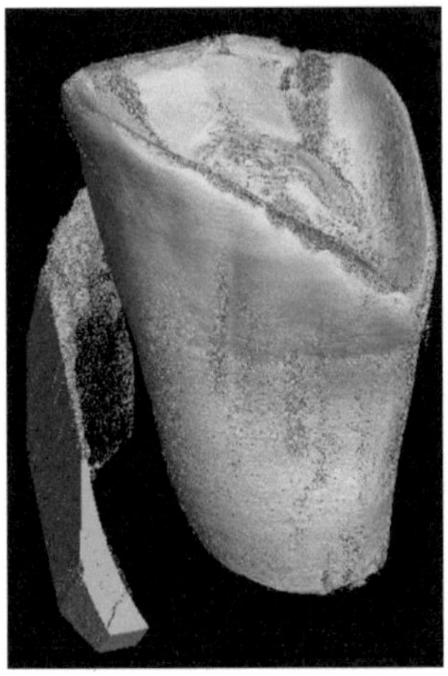

Figure 18.
3D virtual reconstruction of a molar tooth.

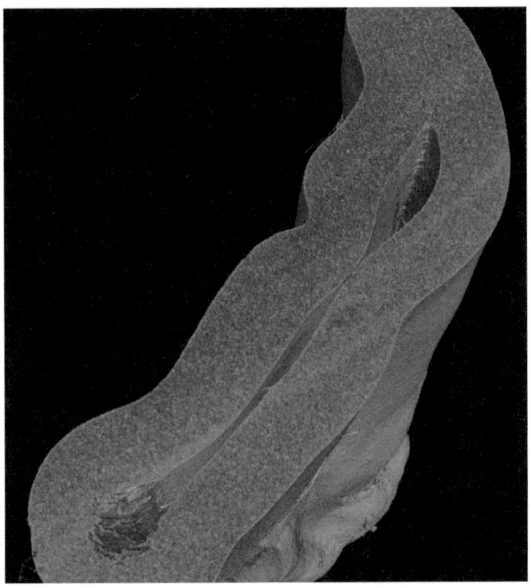

Figure 19.
3D virtual endoscopy, root canal and molar tooth.

numerical control machining processes. In this area, computed tomography is used and magnetic resonance imaging is used too. When a virtual model with endoscopy imaging of tomographic reconstruction is done, it is possible to print layer-by-layer, for example tissue-like structures, 3D models, using a special material known as bio-links [4, 22–29, 31].

4. Bioprinting

The 3D bioprinting needs three steps, as 3D printing,. Pre-bioprinting, bioprinting and post-bioprinting. Bio inks are used in bioprinting step using: a liquid mixture of

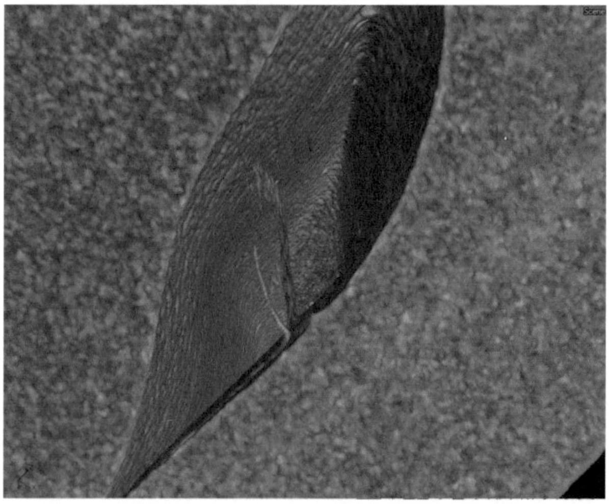

Figure 20.
3D virtual endoscopy, root canal, molar and view inside the orifice of the tooth.

cells, matrix and nutrients. Post-bioprinting is just a final process to create a stable structure from the biological material. 3D bioprinting is also used adapted stereolithography process, and also digital light processing is used too. Future directions of this study cover two main technologies: Bioprinting techniques and biomaterials.

Finally, it can be mentioned that students in the teaching process can already learn to use 3D printing to treat pulp tissue, in the sense of endodontic treatment, by using a bioprinted tissue scaffold. Bioprinting is a very innovative and promising technology for treatment and teaching. The technique has applications in all scientific fields and is also being implemented as a teaching component. Oral soft tissue engineering involves the reconstruction or reestablishment of oral and maxillofacial function and aesthetics. As an emerging technology of the early 21st century, three-dimensional

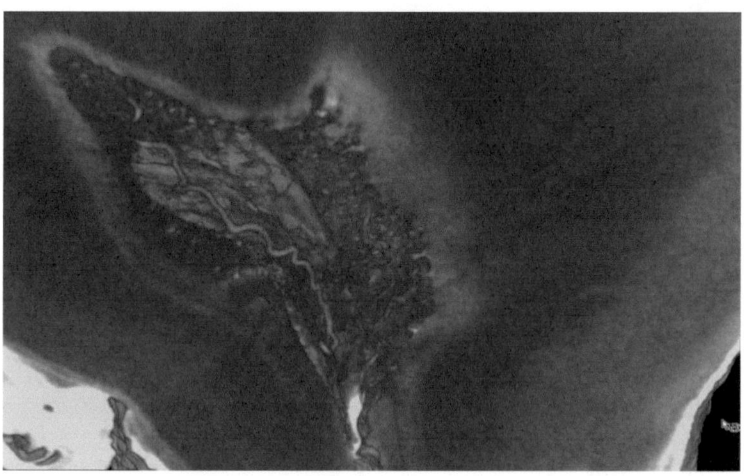

Figure 21.
3D virtual reconstruction of the spatial arrangement of the pulp tissues.

(3D) bioprinting offers great potential for application in scaffold development and tissue and organ engineering. Although oral soft tissues include the dental pulp, periodontium, gingiva, oral mucosa and salivary glands, as well as the associated skin in the maxillofacial area, vascular, muscular and neural tissue, the current use of 3D bioprinting in oral soft tissue reconstruction is mainly limited to dental pulp regeneration (**Figure 21**). A variety of bio-inks is used to introduce dental pulp cells into the dentin matrix to restore the dental pulp tissue. 3D bioprinting has only been described in a few in vitro studies on periodontal ligament reconstruction and salivary gland culture; 3D bioprinting used to regenerate gingival/oral mucosa tissue has not been demonstrated yet. In the 3D printing process, machines such as a robotic bioprinter are already used, and even compact microfluidic bioprinting platforms are in use. Also, it is important to make more easier and intuitive software for less advanced practitioners [14, 30–34].

5. Summary/conclusions

Enhancement of a highly efficient 3D digital solutions improves the overall clinical experience when printing 3D models. A set of available clinical tools helps teachers to assess and pre-design the scanned elements. Electronic 3D models help the dentist to have better, more intuitive communication with the student and the patient, which gives a clear message and the possibility to involve the student and the patient themselves in a satisfying experience. In the scan-to-print process, dental models can be edited and printed, enabling direct visualisation of the treatment plan (e.g., in the case of prosthetics, the restoration production cycle is shortened). Students could, by gaining experience, skilfully change the treatment plan. This was possible after using virtual endoscopy and 3D printing. I encourage you to broaden your knowledge in developing scientific topics such as biomaterials, biomedical engineering, additive manufacturing and tissue engineering, and to obtain insights into the future of biofabrication.

Acknowledgements

I would like to thank all of the students who participated in the study and colleagues from work.

Conflict of interest

'The authors declare no conflict of interest.'

Acronyms and abbreviations

SLA	Stereolithography
FDM	fused deposition modelling
MJP	MultiJet printing
PJP	PolyJet printing
CJP	ColorJet printing

DLP	digital light processing
SLS	selective laser sintering
SLM	selective laser melting
CBCT	Cone beam computed tomography
CT	computed tomography
MR	magnetic resnonance

Author details

Przemysław Kustra
Jagiellonian Univercity Medicall College, Institute of Dentistry, Kraków, Poland

*Address all correspondence to: przemyslaw.kustra@uj.edu.pl

References

[1] Kustra P, Dobroś K, Zarzecka J. Making use of three-dimensional models of teeth, manufactured by stereolithographic technology, in practical teaching of endodontics. European Journal of Dental Education. 2021;**25**:299-304

[2] Sasuła A, Kustra P, Wojas K, Jachymczyk A, Paw W, Stanisz-Wallis K, et al. Comparison of 3-D vizualization and dye penetration test in the evaluation of marginal seal of selected restorative composites. Journal of Stomatology. 2014;**67**:595-605

[3] Oberoi G, Nitsch S, Edelmayer M, Janjic K, Müller AS, Agis H. 3D printing-encompassing the facets of dentistry. Frontiers in Bioengineering and Biotechnology. 2018;**6**:1-13

[4] Shah P, Chong BS. 3D imaging, 3D printing and 3D virtual planning in endodontics. Clinical Oral Investigations. 2018;**22**:641-654

[5] Marty M, Broutin A, Vergnes JN, Vaysse F. Comparison of student's perceptions between 3D printed models versus series models in paediatric dentistry hands-on session. European Journal of Dental Education. 2019;**23**: 68-72

[6] Furlow B. Medical 3-D printing. Radiologic Technology. 2017;**88**:519-537

[7] Hu M. Three-dimensional printing and oral medicine. Chinese Journal of Stomatology. 2017;**52**:206-2011

[8] Bukhari S, Goodacre BJ, AlHelal A, Kattadiyil MT, Richardson PM. Three-dimensional printing in contemporary fixed prosthodontics: A technique article. The Journla of Prosthetic Dentistry. 2018;**119**:530-534

[9] Takeda Y, Lau J, Nouh H, Hirayama H. A 3D printing replication technique for fabricating digital dentures. Journla of Prosthetic Dentistry. 2020;**124**:251-256

[10] Khorsandi D, Fahimipour A, Abasian P, Saber SS, Seyedi M, Ghanavati S, et al. 3D and 4D printing in dentistry and maxillofacial surgery: Printing techniques, materials, and applications. Acta Biomaterialia. 2021; **122**:26-49

[11] Lin HH, Lonic D, Lo LJ. 3D printing in orthognathic surgery. Journal of the Formosan Medical Association. 2018;**117**: 547-558

[12] Lu T, Shao Z, Liu B, Wu T. Recent advance in patient-specific 3D printing templates in mandibular reconstruction. Journal of the Mechanical Behavior of Biomedical Materials. 2020;**106**:1-9

[13] Khalaja R, Tabriza AG. 3D printing advances in the development of stents. International Journal of Pharmaceutics. 2021;**609**:1-23

[14] Zhu W, Ma X, Gou M, Mei D, Zhang K, Chen S. 3D printing of functional biomaterials for tissue engineering. Current Opinion in Biotechnology. 2016;**40**:103-112

[15] Katkar RA, Taft RM, Grant GT. Three-dimensional volume rendering. Dental Clinics of North America. 2018; **62**:393-402

[16] Byun C, Kim C, Cho S, Baek SH, Kim G, Kim SG, et al. Endodontic treatment of an anomalous anterior tooth with the aid of a 3-dimensional printed physical tooth model. Joekull. 2015;**41**:961-965

[17] Gunpreet Oberoi G, Nitsch S, Edelmayer M, Janjic K, Müller AS, Agis H. 3D printing-encompassing the facets of dentistry. Frontiers in Bioengineering and Biotechnology. 2018; **6**:1-13

[18] Anderson J, Wealleans J, Ray J. Endodontic applications of 3D printing. International Endodontic Journal. 2018; **51**:1005-1018

[19] Gul M, Arif A, Ghafoor R. Role of three-dimensional printing in periodontal regeneration and repair: Literature review. Journal of Indian Society of Periodontology. 2019;**23**: 504-510

[20] Prasad TS, Sujatha G, Muruganandhan J, Patil S, Raj AT. Three-dimensional printing in reconstructive oral and maxillofacial surgery. The Journal of Contemporary Dental Practice. 2018;**19**:1-2

[21] Tian Y, Chen ChX XX, Wang J, Hou X, Li K, Lu X, et al. A review of 3D printing in dentistry: Technologies, affecting factors, and applications. Scanning Technology in Dental Science. 2021:1-19

[22] Dawood A, Marti BM, Sauret-Jackson V, Darwood A. 3D printing in dentistry. British Dental Journal. 2015;**11**: 521-529

[23] Tsui JKS, Bella S, Cruz L, Dick AD, Sagoo M. Applications of three-dimensional printing in ophthalmology. Survey of Ophthalmology. 2022;**67**: 1287-1310

[24] Colon RR, Nayak VV, Parente PEL, Leucht P, Nick Tovar N, Lin CC, et al. The presence of 3D printing in orthopedics: A clinical and material review. Journal of Orthopaedic Research. 2022;**2022**:1-13

[25] Crafts TD, Ellsperman SE, Wannemuehler TJ, Bellicchi TD, Shipchandler TZ, Mantravadi AV. Three-dimensional printing and its applications in otorhinolaryngology-head and neck surgery. Otolaryngology and Head and Neck Surgery. 2016;**156**: 999-1010

[26] Samaila EM, Negri S, Alessandro Z, Bizzotto N, Maluta T, Rossignoli C, et al. Value of three-dimensional printing of fractures in orthopaedic trauma surgery. The Journal of International Medical Research. 2019;**48**:1-9

[27] Bueno MR, Estrela C. A computational modeling method for root canal endoscopy using a specific CBCT filter: A new era in the metaverse of endodontics begins. Brazilian Dental Journal. 2022;**33**:21-30

[28] Khan R, Plahouras J, Johnston BC, Scaffidi MA, Grover SC, Walsh CM. Virtual reality simulation training in endoscopy: A Cochrane review and meta-analysis. Endoscopy. 2019;**51**: 653-664

[29] Tchorz JP, Brandl M, Ganter PA, Karygianni L, Polydorou O, Vach K, et al. Pre-clinical endodontic training with artificial instead of extracted human teeth: Does the type of exercise have an influence on clinical endodontic outcomes? International Endodonic Journal. 2015;**48**(9):888-893

[30] Bertassoni LE. Progress and challenges in microengineering the dental pulp vascular microenvironment. Journal of Endodontics. 2020;**46**: S90-S100

[31] Athirasala A, Tahayeri A, Thrivikraman G, França CM, Monteiro N, Tran V, et al. A dentin-derived hydrogel bioink for 3D bioprinting of cell laden scaffolds for

regenerative dentistry. Biofabrication. 2018;**10**:1-20

[32] Fahmy MD, Jazayeri HE, Razavi M, Masri R, Tayebi L. Three-dimensional bioprinting materials with potential application in preprosthetic surgery. Journal of Prosthodontics. 2016;**25**: 310-318

[33] Zhang Y, Zhang XM. Potential of three-dimensional bioprinting in oral soft tissue engineering applications. Journal of Clinical Medicine. 2021;**56**: 613-619

[34] Nesic D, Schaefer BM, Sun Y, Saulacic N, Sailer I. 3D printing approach in dentistry: The future for personalized oral soft tissue regeneration. Journal of Clinical Medicine. 2020;**9**:1-21

Chapter 14

3D-Printing Advances in Radiotherapy

Reza Shamsabadi

Abstract

As radiotherapy techniques have been becoming more applied in medicine, the success of radiotherapy treatment lies in an optimal radiation dosage distribution in tumor as well as dose limitation to the normal tissues. Accordingly, the application of three-dimensional (3D) printing technology, as an additive manufacturing (AM) process in radiotherapy technique, is proliferating rapidly due to the reduced manufacturing costs, improved printing precision, and the speed of 3D printers. The advent of 3D printers in medical fields, especially in radiotherapy, allows to produce any given specific design for patients from novel 3D printable materials. Generally, the applications of this modern industry in radiotherapy can be counted as the creation of traditional patient-specific bolus, brachytherapy applicators, personalized medical devices, physical phantoms for quality assurance (QA), compensator blocks, and patient-specific immobilization devices. Despite the technological advancements of 3D printing in radiotherapy practices, due to the high manufacturing cost, the printing speed, time-consuming workflows, poor conformability, and poor repeatability of applied materials, it is not currently well supported by most radiotherapy techniques. The applications of the 3D printing technology as well as its limitations in radiotherapy are discussed in following.

Keywords: radiotherapy, patient-specific bolus, brachytherapy applicators, personalized medical devices, physical phantoms

1. Introduction

1.1 Radiotherapy overview

Radiotherapy has been introduced as one of the most effective modalities for patients with various types of tumors, which can be given in several ways [1]. Totally, about 50% of all patients with cancer require radiotherapy during the course of their treatment process [2]. The success of a complex radiotherapy technique lies in an optimal radiation dosage distribution in tumor as well as dose limitation to the normal tissues [3]. Delivering an optimal radiation dosage highly depends on the precision in patient setup and immobilization [4]. Despite good therapeutic results, due to the damaging of the surrounding healthy tissues, radiotherapy can have negative side effects [5]. In this regard, combined modalities with radiotherapy have increasingly performed to improve local tumor control and negative effects reduction.

1.2 The 3D printing technology in radiotherapy field

Nowadays, the prospects of the medical technology have changed drastically due to the increasing rate of accessibility and versatility of 3D printing technology, especially when combined with medical imaging. Recently, owing to the advantages of 3D printing method such as versatility and commercial availability, it has been increasingly incorporated into medical practices. Totally, the 3D printing can be regarded as an ideal technology to optimize individual treatments [6], which provides practical and affordable ways for radiotherapy treatments. Due to the reduced manufacturing costs, improved printing precision, and the speed of the 3D printers, 3D printing technology has been exponentially becoming more commercial and accessible over the five decades.

The advent of commercially available 3D printers has been widely adopted in a diverse type of applications, ranging from training to therapeutic usages [7, 8]. Recently, due to the ease of creation and testing the novel designs, as well as the comparative affordability of 3D printers, 3D printing technology has been employed in radiotherapy for the construction of treatment accessories [9–13], and patient geometry or material reproduction [14–19]. Due to the personalization requirement, 3D printers can be used to create objects to accommodate a specific patient's treatment.

3D printing technology has been introduced as an efficient stage for production of custom-made devices to be applied in external beam radiotherapy. Preparation of organ models and rapid manufacturing of personalized medical devices can be obtained by 3D printing technology [20]. Successful printing of a 3D object highly depends on accurately segmentation of achieved patient's imaging data through the computed tomography (CT) or magnetic resonance imaging (MRI) [21].

The 3D printer's utilization in medical fields, especially in radiotherapy, allows to rapidly produce any given specific design for patients from anatomic images using a variety of stock materials [22]. Hence, 3D printing technology can truly enhance the patient care, especially in radiotherapy field [23]. Patient-specific models, which may not be readily obtainable with the traditional techniques, can be produced by 3D printing alternative method and ultimately increase the accuracy of the treatment [22].

The applications of the 3D printed objects in radiotherapy have been discussed in following.

2. The 3D printing applications in radiotherapy

In the last few years, 3D printing technology is being used in radiotherapy technique for a wide variety of applications, including creation of traditional patient-specific bolus and brachytherapy applicators, personalized medical devices, and reproducible and sophisticated physical phantoms production of various anatomical models for quality assurance. Further applications include compensator blocks, patient-specific immobilization devices, and beam modulators, which commonly have been less described [22]. The current applications of AM industry (3D printing technology) in radiotherapy are shown in **Figure 1**.

2.1 Traditional patient-specific bolus

To improve the radiation dose delivery during the high-energy radiotherapy treatment, patient-specific boluses are often applied, which on the one hand reduce

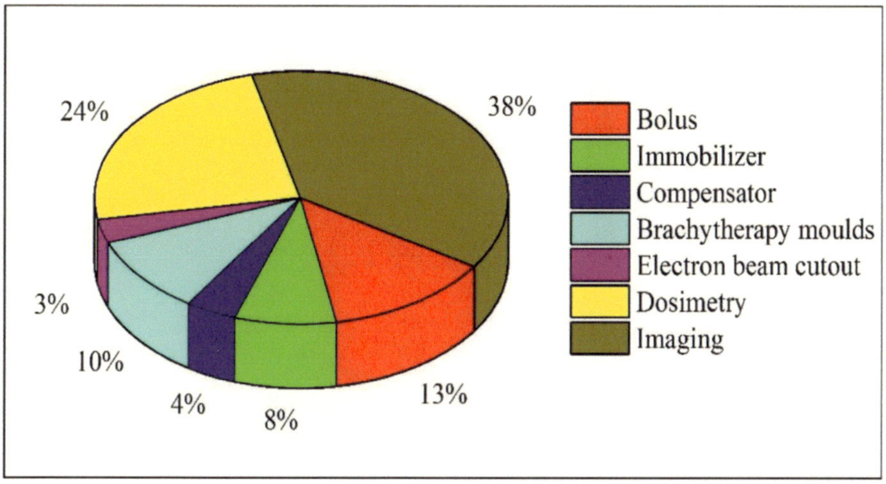

Figure 1.
The current applications of AM in radiotherapy as of 2018 [24].

the irradiation to healthy tissues and on the other hand increase the dose homogeneity for patients with complex surface contours [25]. Hence, more homogeneous dose can be delivered to the target by boluses through providing additional absorption and scattering of ionizing radiation. Besides, radiotherapy bolus is used to eliminate the skin-sparing effect [26].

In electron radiotherapy, customized bolus particularly acts as a layer of skin tissues, which provides a desired dose at superficial lesions as well as favorable curative radiotherapy outcomes [27]. The performance of the boluses in irradiation with photons is shifting the build-up region to ensure about the maximum absorbed dose at the tumor region [28].

The main problem of commercially available boluses is the poor accommodation with the irregular surfaces of patients such as the ear, nose, and scalp. This issue can lead to an air gap between the bolus and the irregular surfaces [29], and finally affect the prescribed dose. The 3D printing technology can afford these issues by creating anatomically matching boluses to achieve the individualized various complex structures [30].

Presently, 3D printed boluses are being increasingly applied in modern radiotherapy. Each 3D printed bolus must exhibit three attributes including uniformity and reproducible bulk density as well as closely match to the surface especially while using the rigid material [31]. Compared to the commercial flat bolus, the 3D printed bolus allows an appropriate fit to the patient's skin surface. The applications of the 3D printed boluses are shown in **Figure 2**.

Applied blouses are usually made of various water-equivalent materials (to control the radiation absorption and scattering in the bolus) such as wet gauze, paraffin, beeswax, and vaseline. Because of the unique physical properties (i.e., toughness, flexibility, and viscoelasticity) of soft polymers including plastics (resins), hydrogels, silicone elastomers, TangoPlus, and polyurethane soft polymers-made modern radiotherapy, bolus has begun to appear [27]. In spite of several advantages over the commercially available boluses, the ingredient materials of 3D printed boluses are often rigid thermo-plastics material, particularly acrylonitrile butadiene styrene (ABS). Besides, various tissue-equivalent printing materials such as polylactic acid (PLA), thermoplastic polyurethane (TPU), and polyvinyl acetate have been reported

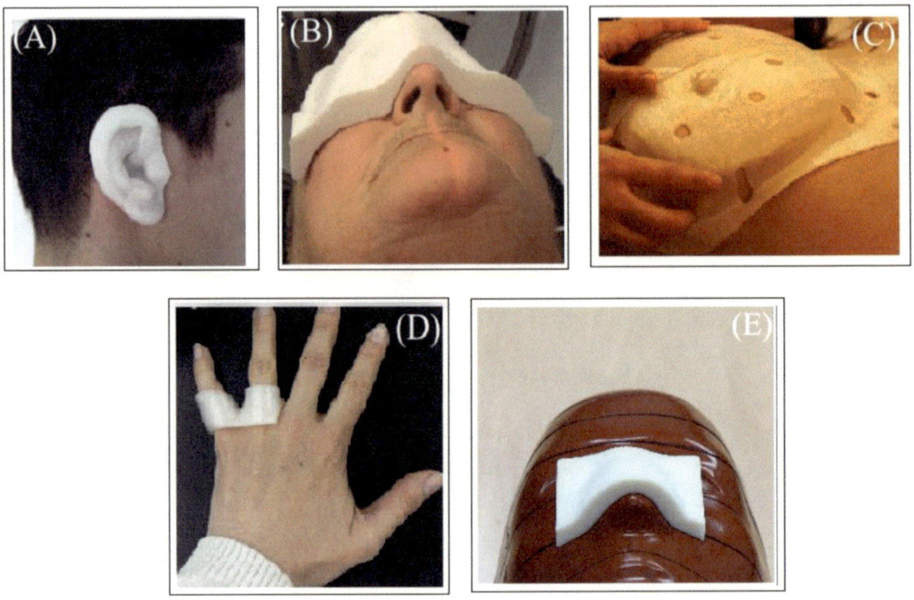

Figure 2.
Varying tissue-equivalent bolus materials for (A) ear [32], (B) mycosis fungoides of the face [13], (C) breast [32], (D) hand [32], (E) nose of the RANDO phantom [33].

and validated for radiotherapy applications [34]. Rigid-made boluses are uncomfortable for the patient that limits the proper contact with surface as well as anatomical changes over a treatment period [32].

The CT imaging data are a common way to design 3D patient-specific boluses, which can be obtained by two CT scans. The first scan acquires image data for reconstructing the shape designing. The second is conducted with the 3D bolus for dose calculation in the treatment planning. Hence, patient may receive an extra irradiation dosage. Moreover, some other designs can be produced by optical scanning, which are expensive and sometimes require complicated processing [31]. To achieve the similar characteristics such as irradiated biological tissues, 3D printed boluses are usually filled with water or paraffin wax [35].

Small gaps between the printed bolus and the head phantom surface are a problem, which are caused by the immobilization with the thermoplastic mask. This issue could be improved by directly putting the printed bolus onto the patient skin and under the thermoplastic mask [23].

Besides the method and the applied material for boluses, there is still a growing concern that radiations can affect physical properties of 3D blouses and even cause them to change or contract. The effect of ionizing radiations on the physical features of the printing materials in radiotherapy was investigated by Jezierska et al. [35]. The reported results demonstrated that owing to the radiotherapy process, the applications of therapeutic X-ray dosage do not significantly affect physical characteristics including hardness and dimensions of ABS-printed boluses [35].

2.2 Compensators

Compensator devices are manufactured with the molding and casting process for the radiotherapy of megavoltage X-ray [34]. During the total body irradiation,

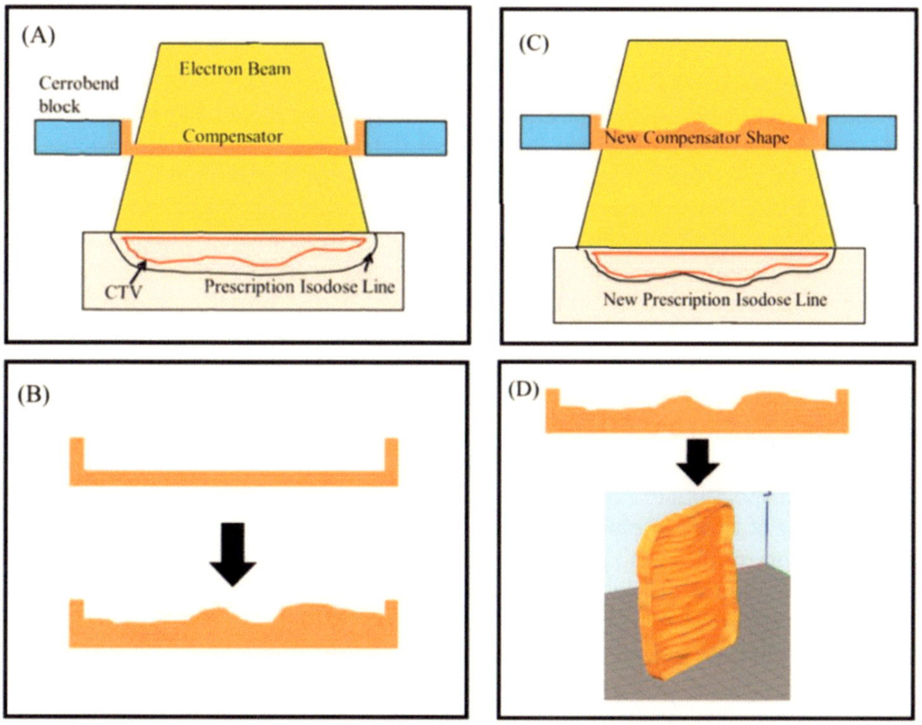

Figure 3.
The workflow of compensator devices. The four steps of the algorithm including (A) initial plan, (B) modulation and smoothing, (C) dose evaluation and recalculation, and (D) export and preparation for print [38].

patient-specific compensators are attached near the primary beam to attenuate the radiation dose and deliver a uniform dose distribution to the whole body [36].

During fabrication of the patient-specific compensators, the treatment field size, beam energy, and depth of the interested point should be considered. Due to the fabrication process and involved materials, patient-specific compensators are often expensive [34]. Hence, additive manufacturing of 3D printed patient-specific compensators not only reduces the costs and manufacturing time, but also has the ability for developing the complex geometries as the beam attenuation [37]. The workflow of compensator devices is shown in **Figure 3**.

2.3 Brachytherapy molds

Besides the radiotherapy boluses, further novel applications of the 3D printing technology in radiotherapy have primarily been performed to produce brachytherapy applicators to improve radiation dose delivery. Due to the well-unconformity to the uneven surfaces (such as the eyes, lips, and nose), it seems essential to use the surface applicators directly to the patient's thermoplastic mask or develop patient-specific molds in accordance with various anatomic sites.

Owing to the 3D printer advances, it can be a proper solution to create the patient-specific molds [39]. 3D printed applicators eliminate the need for a mask, which generally provide more convenient approach for the patient. The uniform dose distribution to the irregular curved surfaces by performing the 3D printed applicator results in decreasing the radiation toxicity and normal tissue complications [40].

Brachytherapy (as the short distance radiotherapy) is a form of radiotherapy technique, which provides a fast dose fall-off by using small radioisotopes, which are connected to applicators or catheters placed to treat small superficial lesions [41]. Due to the steep dose fall-off around the source, performance of the brachytherapy applicator is crucial, where inappropriate placement of the source results in inexpedient dose distribution [12]. To obtain a safe and effective brachytherapy treatment, the applied applicators should have a proper accordance with the treatment surface in the stable position, the source of which uniformly covers the treatment volume [12]. 3D printers may offer a more convenient and affordable method to produce surface mold applicators for a wide variety of irregular sites including the hands, breasts, and the face [42].

It has been affirmed that brachytherapy is an efficient and well-tolerated modality for skin cancer, low toxicity of which can be achieved. Due to the irregular surface of the skin lesions in the radiotherapy of melanoma cancer, brachytherapy mold is often used to fit along the surface [43].

Water equivalent-made surface brachytherapy applicators may be used to accurately guide the position of radioactive sources and fit closely to surface of patient's skin. This type of brachytherapy applicator is typically manufactured by a wax layer, built upon a thermoplastic shell. Some small hollows guide the radioactive seed implantation as well as helping the flexible drive cable [12].

Currently, different types of commercial standard applicators such as vaginal brachytherapy applicators and flap-style flexible silicon rubber applicators such as Harrison-Anderson-Mick (HAM) or Freiburg Flap applicators (for extended superficial malignancies) have been used [12].

3D printing offers cost-effective and more convenient method to produce surface applicators for various sites including the hands, breasts, and the face. Typically, for non-melanoma skin cancer brachytherapy, reductions about 34% in time and 49.5% in financial cost have been reported. Furthermore, 3D printing technology has been proposed to produce the desired applicator shapes for gynecological tumors (especially with narrow vaginal vaults) and surface molds for skin applicators. Besides, the application of 3D-printed templates has been reported for the treatment of recurrent head and neck malignancies [29].

Due to the requirement of biocompatible material for printing brachytherapy applicator, selected materials for such a purpose depend on the type of tissue and the length of the contact [44]. It should be noted that employed materials for rapid prototyping of customized brachytherapy applicators must be biocompatible, sterilizable, and free of CT scanning artifacts with the similar dose attenuation properties such as water [45]. The dosimetric properties of employed materials for brachytherapy applicators have been assessed previously. For example, the dosimetric properties of PC-ISO material as the water-equivalency material were evaluated by Cunha et al. [44], in 2015 for gynecological brachytherapy applicators. In 2016, a personalized 3D printed vaginal template was developed by Lindegaard et al. [9], to guide commercially available applicators.

2.4 Anthropomorphic phantoms production

Since the radiation dose cannot be directly measured in patients during the radiotherapy, to ensure about the accuracy of the dose delivery and conformal dose distributions, the treatment planning system algorithms, and the quality assurance, it is common to build phantoms, which mimic radiation attributes of humans. To achieve ideal or optimal radiotherapy outcomes, the results of patient-specific

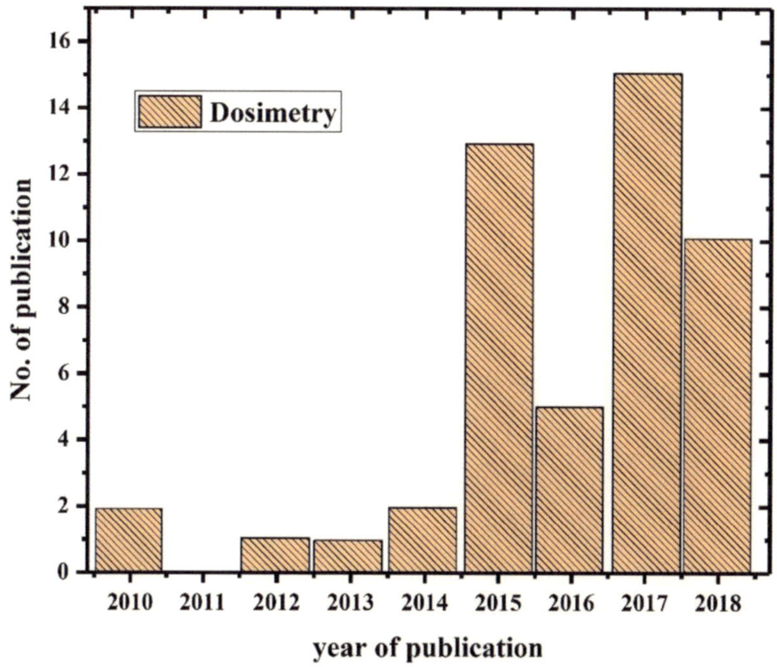

Figure 4.
The number of publications for AM dosimetry radiotherapy phantoms until 2018 [24].

phantoms should be combined with treatment planning systems. Using 3D printed phantoms, all parts of preclinical radiation examinations including planning, image guidance, and treatment delivery can be assessed by clinicians to verify the treatment planning and the calibration of the applied treatment techniques. An observed linear ascending trend for dosimetry applications of manufactured radiotherapy phantoms through the 3D printing technology is indicated in **Figure 4**.

To assess the QA of new modalities or validate the patient treatment plans using the calculated absorbed dose by means of thermoluminescent dosimeters (TLD), film dosimetry, and ionization chambers, commercially available phantoms are often utilized [46]. The QA process can be affected by the phantom limitations including not fully pretending the patient-specific anatomical and pathological characteristics, which eventually lead to errors in absorbed dose measurement. Owing to the advances of 3D printed radiotherapy phantoms including fast and high-resolution workflows as well as multi-material printing, the need for QA procedure is becoming more achievable [47].

The four main steps of 3D printed phantoms fabrication can be observed in **Figure 5**. These steps include 1) imaging, 2) segmentation for region of interests (ROI), 3) slicing the segmented 3D models, numerically controlled process of desktop 3D printers (called gcode), as well as printing parameter modifications, and finally, 4) printing models and final implementation for imaging or dosimetry applications.

To create the 3D printed phantoms, utilized printing technologies include fused deposition modeling (FDM) and polymer material jetting (PJT/MJT) [48].

The applications of the 3D printed phantoms have been demonstrated in various literatures [34, 49, 50]. The simulation of realistic lung movements has been

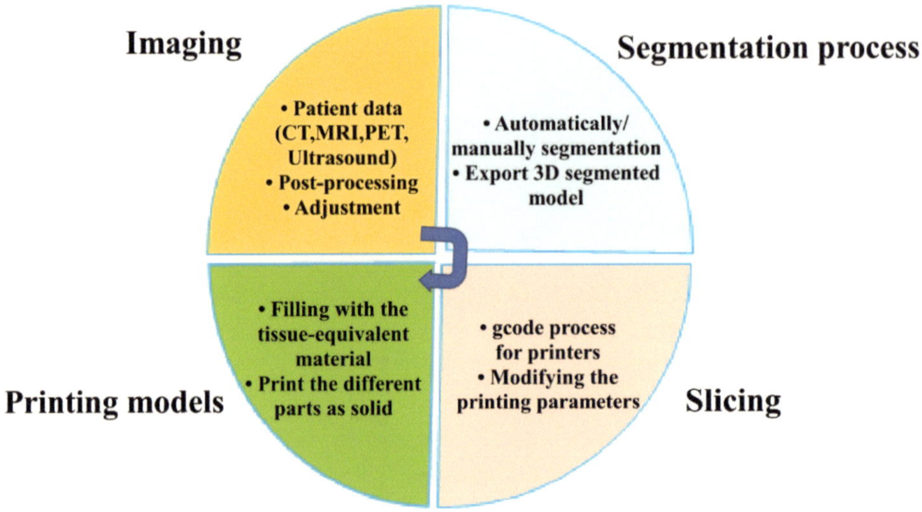

Figure 5.
The patient-specific phantoms workflow through the 3D printing technology [24].

demonstrated by Yoon et al. [51], through the 3D-printed lung phantom with printed lung lesions. Furthermore, Jahnke et al. [52] have evaluated the feasibility of manufacturing the realistic head phantom through i) paper-based 3D printing method and ii) customized laser object manufacturing printer. However, a promising result has been achieved in this study, and the CT tissue-equivalence and dosimetry assessments need further quantifiable assessments.

However, the performance of traditionally available phantoms may be limited by the geometry and physical properties, and 3D printed phantoms enable to develop a viable, low-cost alternative. Recently, 3D printed phantoms with inhomogeneous materials have been increasingly taken into consideration. Ehler et al. [16] employed the 3D printing of a head phantom using ABS material, which was filled with a modified M3 mix (with the density of about 1.05 g/cc).

2.5 3D printed organs

With the developments of 3D printing technology, production of organ models has also experienced massive progress [20]. 3D printed organ models still have some difficulties such as high cost, time-consuming, low production accuracy, and limited simulation characteristic, which limit their further applications [53]. Sometimes, complicated image segmentation and 3D model creating processes may require several days of labor, expensive materials and software as well as a long-learning period. There is still limitation for currently employed materials to perfectly mimic the characteristics of soft tissues [53]. Furthermore, errors would be inevitable while dividing large models into several small parts for printing and assembly process.

2.6 Immobilization devices

Before the radiotherapy treatment, immobilizing devices are utilized to accurately set the position and rigidly immobilize patients. Beaded bags, polyurethane foam castings, orthopedic plastics, and head masks are commonly used as the immobilization

Figure 6.
The workflow of 3D printed immobilization devices.

devices [54, 55]. To minimize patient movements during the radiotherapy process, external immobilization of the equipment (such as headrests, vac-bags, thermoplastic masks, or shells) is frequently applied, which can result in localized dose distribution on the tumor site [56]. Besides, healthy tissues can be protected from the radiation exposure and the effectiveness of radiotherapy treatments can be improved.

Besides the head and neck, 3D printed immobilization devices can be applied for immobilization of the abdominal area, mouth, and breasts [57]. The workflow of 3D printed immobilization devices is shown in **Figure 6**.

Besides the CT data, surface laser scanner (generation of a reference model for immobilizing shell construction) is the other procedure to develop printable models. 3D printed immobilizers can be created with a range of printing techniques including FDM, stereolithography (SLA), selective laser sintering (SLS), and PJT/MJT. Laycock et al. [58] have assessed the feasibility of VisiJet SL Clear, for immobilization of devices through the FDM printing technology. To achieve an accurate position in radiation therapy, thermoplastic material is routinely performed. For example, the head immobilization during the radiotherapy is most practically accomplished by thermoplastic masks. Since the skill of the human fabricator for manufacturing the immobilizers is crucial, these mentioned devices have a variable quality [59].

Today, immobilization masks are commonly utilized for head and abdominal fixation. Thermoplastic and PLA materials are commonly accomplished for head and abdominal immobilization mask generation, respectively [34]. However, 3D printed fixation masks improve patient comfort and eliminate the thermoforming process for mask manufacturing, and immobilization masks modeling may be a time-consuming process, which strongly depends on the healthcare professional and potentially results in inconvenience for the patient [60].

2.7 Other applications

For more applications of 3D printed objects in radiotherapy, the fabrication of Cerrobend® grids for spatially modulated therapy can be mentioned. Grid therapy is part of a novel treatment method, which provides passive-scattering proton beams with a fixed range for stereotactic radiosurgery as well as the manufacturing of oral stents from CT images used during radiotherapy of head and neck [29].

3. Applied material for 3D objects in radiotherapy

The most commonly applied materials for manufacturing the 3D objects are ABS and PLA, which may differ from metallic alloys with other thermoplastics [61]. PLA as the most widely used biodegradable polyester has the strong potential to be applied in industrial applications or in medicine as a biomaterial. PLA production has numerous advantages as well as ecology, biocompatibility, and better thermal resistance than other biopolymers. Since PLA products are nontoxic, they are potentially able to be applied in biomedical applications [62].

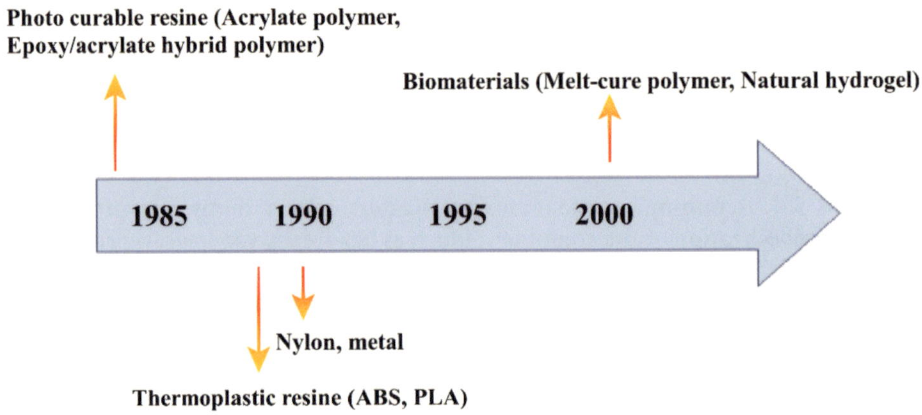

Figure 7.
The timeline of discovered important materials for 3D printing applications [66].

Due to the bolusing effect of thermoplastic materials on the patient surface, the dose buildup can be affected [60]. To reduce this bolusing effect, it has been reported that multi jet fusion (MJF) method, which is uncommonly flexible in various additives, may be an opportunity to develop low physical/electron density materials [60].

The dosimetric, biological, and physical characteristics of applied materials for 3D printed objects have been investigated in several literatures [63–65]. Suitable materials for 3D printing models in radiotherapy applications should have similar characteristics to water such as radiation attenuation, scatter properties, and biocompatibility [27]. For example, ABS and PLA have the advantages of electron density similar to water and low failure rate, respectively. The ABS is prone to layer separation. Besides, the PLA has the potential of high electron density and easy wrapping [12].

The timeline of discovered materials for 3D printing applications is summarized in **Figure 7**.

4. The 3D printing barriers in radiotherapy

In spite of the advantages of 3D printing technology in current radiotherapy practices, for the clinical implementations, this modern technology appears to be low. Despite the recent significant developments, the size, capability, and print speed are still considered as the limitation factors for affordable 3D printers [63]. Hence, for successful clinical implementation of 3D printing in radiotherapy, besides the mentioned factors, QA of 3D-printed objects as well as the material characteristics and efficient workflow should be considered [29].

In radiotherapy technique, the manufacturing cost of printed objects is probably the major determining factor for 3D printing technology in radiotherapy clinical practice, especially for creation of the small targets [59]. Despite the manufacturing cost decrements in recent years, high resolution of commercial printers is still expensive. Sometimes, the workflows of 3D printing objects may require several days, precise information about the target, and expensive materials. Besides, some applied materials have disadvantage such as rough craftsmanship, poor conformability, and poor repeatability [64]. Utilized raw materials to perfectly mimic the characteristics of soft tissues may be the other limitation and determined factor for 3D model creation in radiotherapy. Low resolution, the necessity of post-processing stage, and poor mechanical properties are the other disadvantages of 3D printing technology in radiotherapy. Hence, there are still some limitations to be overcome to achieve the requirements for such clinical practices [65, 67].

5. Conclusion

The 3D printing technology is rapidly evolving, where at present the AM-radiotherapy field can be regarded as an exciting phase of exploration and innovation. This technology, like any new technology, has introduced many advantages and possibilities in medical field. It holds the potential to be applied in radiotherapy for a wide variety of objects including patient-specific bolus, brachytherapy applicators, personalized medical devices, physical phantoms for radiotherapy QA, compensator blocks, and patient-specific immobilization devices. Low-cost 3D printed objects can result in the accurate and comfortable custom-made devices, which will ultimately improve the treatment outcomes. Due to the rapid developments and increased clinical applications, the additional use of 3D printing within radiotherapy will likely emerge.

Due to the importance of personalized radiotherapy devices, 3D printing technology can be applied to improve the quality of life, reduce the setup times, improve patient comfort, and enhance reproducibility and reliability. The modern 3D printers have facilitated the design and fabrication of patient-specific radiotherapy phantoms and ultimately result in precise, cost-effective, high-resolution phantoms with the multi-material printing workflows.

In spite of the impressive growth of 3D printing technology in medicine, the advantages in radiotherapy field are not still fully realized. However, it requires an updated and current legislation to guarantee its correct performance in radiotherapy.

Author details

Reza Shamsabadi
Hakim Sabzevari University, Sabzavar, Iran

*Address all correspondence to: rshams220@gmail.com

References

[1] Mortezaee K, Najafi M. Immune system in cancer radiotherapy: Resistance mechanisms and therapy perspectives. Critical Reviews in Oncology/ Hematology. 2021;**157**:103180

[2] Atun R, Jaffray DA, Barton MB, Bray F, Baumann M, Vikram B, et al. Expanding global access to radiotherapy. The Lancet Oncology. 2015;**16**:1153-1186

[3] Elith C, Dempsey SE, Findlay N, Warren-Forward HM. An introduction to the intensity-modulated radiation therapy (IMRT) techniques, Tomotherapy, and VMAT. Journal of Medical Imaging and Radiation Sciences. 2011;**42**(1):37-43

[4] Van der Merwe D, Van Dyk J, Healy B, Zubizarreta E, Izewska J, Mijnheer B, et al. Accuracy requirements and uncertainties in radiotherapy: A report of the International Atomic Energy Agency. Acta Oncologica. 2017;**56**(1):1-6

[5] De Ruysscher D, Niedermann G, Burnet NG, Siva S, Lee AW, Hegi-Johnson F. Radiotherapy toxicity. Nature Reviews: Disease Primers. 2019;**5**:13

[6] Wong KC. 3D-printed patient-specific applications in orthopedics. Orthopedic Research and Reviews. 2016;**14**(8):57-66

[7] Jamróz W, Szafraniec J, Kurek M, Jachowicz R. 3D printing in pharmaceutical and medical applications–recent achievements and challenges. Pharmaceutical Research. 2018;**35**(9):1-22

[8] Beg S, Almalki WH, Malik A, Farhan M, Aatif M, Rahman Z, et al. 3D printing for drug delivery and biomedical applications. Drug Discovery Today. 2020;**25**(9):1668-1681

[9] Lindegaard JC, Madsen ML, Traberg A, Meisner B, Nielsen SK, Tanderup K. Individualised 3D printed vaginal template for MRI guided brachytherapy in locally advanced cervical cancer. Radiotherapy and Oncology. 2016;**118**:173-175

[10] Jones EL, Baldion AT, Thomas C, Burrows T, Byrne N, Newton V, et al. Introduction of novel 3D-printed superficial applicators for high-dose-rate skin brachytherapy. Brachytherapy. 2017;**16**:409-414

[11] Canters RA, Lips IM, Wendling M, Kusters M, van Zeeland M, Gerritsen RM, et al. Clinical implementation of 3D printing in the construction of patient specific bolus for electron beam radiotherapy for non-melanoma skin cancer. Radiotherapy and Oncology. 2016;**121**:148-153

[12] Ricotti R, Vavassori A, Bazani A, Ciardo D, Pansini F, Spoto R, et al. 3D-printed applicators for high dose rate brachytherapy: Dosimetric assessment at different infill percentage. Physica Medica. 2016;**32**(12):1698-1706

[13] Zhao Y, Moran K, Yewondwossen M, Allan J, Clarke S, Rajaraman M, et al. Clinical applications of 3-dimensional printing in radiation therapy. Medical Dosimetry. 2017;**42**(2):150-155

[14] Kim MJ, Lee SR, Lee MY, Sohn JW, Yun HG, Choi JY, et al. Characterization of 3D printing techniques: Toward patient specific quality assurance spine-shaped phantom for stereotactic body radiation therapy. PLoS One. 2017;**12**:e0176227

[15] Oh D, Hong CS, Ju SG, Kim M, Koo BY, Choi S, et al. Development of

patient-specific phantoms for verification of stereotactic body radiation therapy planning in patients with metallic screw fixation. Scientific Reports. 2017;**7**:40922

[16] Ehler ED, Barney BM, Higgins PD, Dusenbery KE. Patient specific 3D printed phantom for IMRT quality assurance. Physics in Medicine and Biology. 2014;**59**:5763-5773

[17] Kamomae T, Shimizu H, Nakaya T, Okudaira K, Aoyama T, Oguchi H, et al. Three-dimensional printer-generated patient-specific phantom for artificial in vivo dosimetry in radiotherapy quality assurance. Physica Medica. 2017;**4**:205-211

[18] Madamesila J, McGeachy P, Villarreal Barajas JE, Khan R. Characterizing 3D printing in the fabrication of variable density phantoms for quality assurance of radiotherapy. Physica Medica. 2016;**32**:242-247

[19] Yea JW, Park JW, Kim SK, Kim DY, Kim JG, Seo CY, et al. Feasibility of a 3D-printed anthropomorphic patient-specific head phantom for patient-specific quality assurance of intensity-modulated radiotherapy. PLoS One. 2017;**12**:1-10

[20] Yan Q, Dong H, Su J, Han J, Song B, Wei Q, et al. A review of 3D printing technology for medical applications. Engineering. 2018;**4**(5):729-742

[21] Young PG, Beresford-West TB, Coward SR, Notarberardino B, Walker B, Abdul-Aziz A. An efficient approach to converting three-dimensional image data into highly accurate computational models. Philosophical Transactions of the Royal Society A: Mathematical, Physical and Engineering Sciences. 1878;**2008**(366):3155-3173

[22] Rooney MK, Rosenberg DM, Braunstein S, Cunha A, Damato AL, Ehler E, et al. Three-dimensional printing in radiation oncology: A systematic review of the literature. Journal of Applied Clinical Medical Physics. 2020;**21**(8):15-26

[23] Amor-Coarasa A, Goddard L, DuPré P, Wake N. 3D printing in nuclear medicine and radiation therapy. In: Wake N, editor. 3D Printing for the Radiologist. Health Sciences. Amsterdam: Elsevier; 2022. pp. 143-156

[24] Tino R, Yeo A, Leary M, Brandt M, Kron T. A systematic review on 3D-printed imaging and dosimetry phantoms in radiation therapy. Technology in Cancer Research & Treatment. 2019;**18**:1533033819870208

[25] Gentry JR, Steeves R, Paliwal BA. Inverse planning of energy-modulated electron beams in radiotherapy. Medical Dosimetry. 2006;**31**(4):259-268

[26] Supratman AS, Sutanto H, Hidayanto E, Jaya GW, Astuti SY, Budiono T, et al. Characteristic of natural rubber as bolus material for radiotherapy. Materials Research Express. 2018;**5**(9):095302

[27] Lu Y, Song J, Yao X, An M, Shi Q, Huang X. 3D printing polymer-based bolus used for radiotherapy. International Journal of Bioprinting. 2021;**7**(4):414

[28] Islam S, Mahmoud KA, Sayyed MI, Alim B, Rahman MM, Mollah AS. Study on the radiation attenuation properties of locally available bees-wax as a tissue equivalent bolus material in radiotherapy. Radiation Physics and Chemistry. 2020;**172**:108559

[29] Burnley JC. 3D-Printing for Radiotherapy Using Flexible Filament Materials [thesis]. United Kingdom: University of Manchester; 2021

[30] Meyer-Szary J, Luis MS, Mikulski S, Patel A, Schulz F, Tretiakow D, et al. The role of 3D printing in planning complex medical procedures and training of medical professionals—Cross-sectional multispecialty review. International Journal of Environmental Research and Public Health. 2022;**19**(6):3331

[31] Sasaki DK, McGeachy P, Alpuche Aviles JE, McCurdy B, Koul R, Dubey A. A modern mold room: Meshing 3D surface scanning, digital design, and 3D printing with bolus fabrication. Journal of Applied Clinical Medical Physics. 2019;**20**(9):78-85

[32] Dipasquale G, Poirier A, Sprunger Y, Uiterwijk JWE, Miralbell R. Improving 3D -printing of megavoltage X-rays radiotherapy bolus with surface-scanner. Radiation Oncology. 2018;**13**(1):203

[33] Kim S-W, Shin H-J, Kay CS, Son SH. A customized bolus produced using a 3-dimensional printer for radiotherapy. PLoS One. 2014;**9**:e110746

[34] Yeo A, Brandt M, Leary M, Kron T. The interlace deposition method of bone equivalent material extrusion 3D printing for imaging in radiotherapy. Materials & Design. 2021;**199**:109439

[35] Jezierska K, Sękowska A, Podraza W, Gronwald H, Łukowiak M. The effect of ionising radiation on the physical properties of 3D-printed polymer boluses. Radiation and Environmental Biophysics. 2021;**60**(2):377-381

[36] Chen HH, Wu J, Chuang KS, Lin JF, Lee JC, Lin JC. Total body irradiation with step translation and dynamic field matching. BioMed Research International. 2013;**2013**:216034

[37] Redler G, Pearson E, Liu X, Gertsenshteyn I, Epel B, Pelizzari C, et al. Small animal IMRT using 3D-printed compensators. International Journal of Radiation Oncology Biology Physics. 2021;**110**(2):551-565

[38] Craft DF, Balter P, Woodward W, Kry SF, Salehpour M, Ger R, et al. Design, fabrication, and validation of patient-specific electron tissue compensators for postmastectomy radiation therapy. Physics and Imaging in Radiation Oncology. 2018;**8**:38-43

[39] Robinson SS, Alaie S, Sidoti H, Auge J, Baskaran L, Avilés-Fernández K, et al. Patient-specific design of a soft occluder for the left atrial appendage. Nature Biomedical Engineering. 2018;**2**(1):8-16

[40] Bridger CA, Douglass MJ, Reich PD, Santos AMC. Evaluation of camera settings for photogrammetric reconstruction of humanoid phantoms for EBRT bolus and HDR surface brachytherapy applications. Physical and Engineering Sciences in Medicine. 2021;**44**(2):457-471

[41] Carrara M, Ziglio F. Brachytherapy Introduction to Medical Physics. Boca Raton: CRC Press; 2022. pp. 323-355

[42] Arenas M, Sabater S, Sintas A, Arguís M, Hernández V, Árquez M, et al. Individualized 3D scanning and printing for non-melanoma skin cancer brachytherapy: A financial study for its integration into clinical workflow. Journal of Contemporary Brachytherapy. 2017;**9**:270-276

[43] Guinot JL, Rembielak A, Perez-Calatayud J, Rodríguez-Villalba S, Skowronek J, Tagliaferri L, et al. GEC-ESTRO ACROP recommendations in skin brachytherapy. Radiotherapy and Oncology. 2018;**126**(3):377-385

[44] JAM C, Flynn R, Bélanger C, Callaghan C, Kim Y, Jia X, et al.

Brachytherapy future directions. Seminars in Radiation Oncology. 2020;**30**:94-106

[45] Sethi R, Cunha A, Mellis K, Siauw T, Diederich C, Pouliot J, et al. Clinical applications of custom-made vaginal cylinders constructed using three-dimensional printing technology. Journal of Contemporary Brachytherapy. 2016;**8**(3):208-214

[46] Rivera-Montalvo T. Radiation therapy dosimetry system. Applied Radiation and Isotopes. 2014;**83**:204-209

[47] Tino RB, Yeo AU, Brandt M, Leary M, Kron T. A customizable anthropomorphic phantom for dosimetric verification of 3D-printed lung, tissue, and bone density materials. Medical Physics. 2022;**49**(1):52-69

[48] Tajik M, Akhlaqi MM, Gholami S. Advances in anthropomorphic thorax phantoms for radiotherapy: A review. Biomedical Physics & Engineering Express. 2021;**8**(5)

[49] Rai R, Wang YF, Manton D, Dong B, Deshpande S, Liney GP. Development of multi-purpose 3D printed phantoms for MRI. Physics in Medicine & Biology. 2019;**64**(7):075010

[50] Sommer K, Izzo RL, Shepard L, Podgorsak AR, Rudin S, Siddiqui AH, et al. Design optimization for accurate flow simulations in 3D printed vascular phantoms derived from computed tomography angiography. Medical Imaging 2017: Imaging Informatics for Healthcare, Research, and Applications. 2017;**10138**:180-191

[51] Yoon K, Jeong C, Kim SW, Cho B, Kwak J, Kim SS, et al. Dosimetric evaluation of respiratory gated volumetric modulated arc therapy for lung stereotactic body radiation therapy using 3D printing technology. PLoS One. 2018;**13**(12):e0208685

[52] Jahnke P, Schwarz S, Ziegert M, Schwarz FB, Hamm B, Scheel M. Based 3D printing of anthropomorphic CT phantoms: Feasibility of two construction techniques. European Radiology. 2019;**29**(3):1384-1390

[53] Jin Z, Li Y, Yu K, Liu L, Fu J, Yao X, et al. 3D printing of physical organ models: Recent developments and challenges. Advanced Science. 2021;**8**(17):2101394

[54] Jakobsen A, Iversen P, Gadeberg C, Hansen JL, Hjelm-Hansen M. A new system for patient fixation in radiotherapy. Radiotherapy and Oncology. 1987;**8**:145-151

[55] Dickens C. Personalized fixation using a vacuum consolidation technique. The British Journal of Radiology. 1981;**54**:257-258

[56] Cleland S, Chan P, Chua B, Crowe SB, Dawes J, Kenny L, et al. Dosimetric evaluation of a patient-specific 3D-printed oral positioning stent for head-and-neck radiotherapy. Physical and Engineering Sciences in Medicine. 2021;**44**(3):887-899

[57] Enke C, Saw CB, Yakoob R, Enke CA, Lau TP, Ayyangar KM. Immobilization devices for intensity-modulated radiation therapy (IMRT). Medical Dosimetry. 2001;**26**(1):71-77

[58] Laycock SD, Hulse M, Scrase CD, Tam MD, Isherwood S, Mortimore DB, et al. Towards the production of radiotherapy treatment shells on 3D printers using data derived from DICOM CT and MRI: Preclinical feasibility studies. Journal of Radiotherapy in Practice. 2015;**14**(01):92-98

[59] Asfia A, Novak JI, Mohammed MI, Rolfe B, Kron T. A review of 3D printed patient specific immobilisation devices

in radiotherapy. Physics and Imaging in Radiation Oncology. 2020;**13**:30-35

[60] Robar JL, Moran K, Allan J, Clancey J, Joseph T, Chytyk-Praznik K, et al. Intrapatient study comparing 3D printed bolus versus standard vinyl gel sheet bolus for postmastectomy chest wall radiation therapy. Practical Radiation Oncology. 2018;**8**(4):221-229

[61] Ujfalusi Z, Pentek A, Told R, Schiffer A, Nyitrai M, Maroti P. Detailed thermal characterization of acrylonitrile butadiene styrene and polylactic acid-based carbon composites used in additive manufacturing. Polymers. 2020;**12**(12):2960

[62] Lopera AA, Bezzon VD, Ospina V, Higuita-Castro JL, Ramirez FJ, Ferraz HG, et al. Obtaining a fused PLA-calcium phosphate-tobramycin-based filament for 3D printing with potential antimicrobial application. Journal of the Korean Ceramic Society. 2020;**1-14**

[63] McGarry CK, Grattan LJ, Ivory AM, Leek F, Liney GP, Liu Y, et al. Tissue mimicking materials for imaging and therapy phantoms: A review. Physics in Medicine & Biology. 2020;**65**(23):23TR01

[64] Alssabbagh M, Abdulmanap M, Zainon R. Evaluation of 3D printing materials for fabrication of a novel multi-functional 3D thyroid phantom for medical dosimetry and image quality. Radiation Physics and Chemistry. 2017;**135**:106-112

[65] Mostafaei A, Elliott AM, Barnes JE, Li F, Tan W, Cramer CL, et al. Binder jet 3D printing—Process parameters, materials, properties, modeling, and challenges. Progress in Materials Science. 2021;**119**:100707

[66] Tetsuka H, Shin SR. Materials and technical innovations in 3D printing in biomedical applications. Journal of Materials Chemistry B. 2020;**8**(15):2930-2950

[67] DeSimone E, Schacht K, Jungst T, Groll J, Scheibel T. Biofabrication of 3D constructs: Fabrication technologies and spider silk proteins as bioinks. Pure and Applied Chemistry. 2015;**87**(8):737-749

Chapter 15

Applications of Three-Dimensional Printing Technology in Radiotherapy

Seyed Hamid Zoljalali Moghaddam

Abstract

Nowadays, three-dimensional (3D) printing technology has been used for rapid prototyping of high quality printed objects. This technology has taken a special place in the field of medicine, and today this technology plays an important role, especially in the field of radiotherapy. Radiotherapy is a main option for treating and management of various types of cancers. Personalized radiotherapy requires precise details. For this reason, it is very important to carry out the exact treatment design at the clinical. 3D printing technology is considered a promising method that can be effective in the treatment of each person in a specific way and as a complementary and promising method to help in integrated treatment and special equipment for each patient. In this chapter, various applications of this technology in radiation therapy have been discussed. This narrative review summarizes the applications of 3D printing technology to develop patient-specific bolus, brachytherapy applicators, phantoms, filters, immobilization and grid therapy devices for more personalized radiation treatment.

Keywords: 3D printing technology, radiotherapy, cancer, bolus, brachytherapy applicators

1. Introduction

Surgery, radiotherapy and chemotherapy are common methods in cancer treatment [1]. Almost more than two thirds of patients with cancer, are treated via radiotherapy [2]. The main goal in radiotherapy is maximum dose delivery to the tumor while minimizing the side effects caused by the treatment. In the past decades, new radiotherapy technologies such as Intensity Modulated Radiation Therapy (IMRT), Image Guided Radiation Therapy (IGRT), Stereotactic Radiosurgery (SRS), Stereotactic Body Radiation Therapy (SBRT), and three dimensional treatment planning in Brachytherapy have been introduced. Such advances in radiotherapy have led to an increment in the absorbed dose in the tumor area and a decrement in the received dose by healthy tissues in the treatment area [3–5]. Therefore, it is very important to carefully implement the treatment design. Although the advent of new radiotherapy technologies has led to the reduction of errors relevant to the reconstruction of the patient's geometry, there are some errors in the accessory's productions. So, the use of medical devices is not suitable for each patient.

Nowadays, 3D printing technology is considered as a promising method that can be effective in the treatment of each person individually. Hence, there is a great interest in application of the 3D printing technology in the radiotherapy process. Today, this technology has attracted a lot of attention for making patient-specific accessories, dose modifiers, phantoms and several other devices in radiotherapy [6–9].

Recently, several studies have been conducted in the field of 3D printing technology in radiotherapy, in terms of making patient-specific boluses, brachytherapy applicators, fixing devices, etc. Therefore, in this chapter, the basic applications of 3D printing technology in radiotherapy have been summarized.

2. Three-dimensional printing

The concept of 3D printing was conceived in the 1970s, but the first experiments date back to 1981. One of the main applications of 3D printing technology is the manufacturing of medical equipment. The first use of 3D printing technology in medicine appeared shortly after its invention. In 1990, Palser et al., first transferred Computed Tomography images of the human skull and knee joint in the solid three-dimensional polymer model making system and printed the 3D models using the stereolithography technique [10]. The process of creating a physical object from a digital model is considered as a simple definition of 3D printing technology. In comparison with the common printers, 3D printers create a 3D physical model from the desired target. Creating an object with a 3D printer requires a 3D-digital model. A 3D-digital model can be created by scanning a set of 3D images or drawing it using CAD design software, as well as using data from Computed Tomography (CT) or Magnetic Resonance Imaging (MRI), obtained [11]. Then, this 3D digital model is sent to the printer in STereoLithography (STL) format. Finally, a 3D model is created layer by layer. The whole above-mentioned process is called rapid prototyping or 3D printing [12, 13]. **Figure 1** shows the formation process of a three-dimensional physical model.

There are various examples of the methods for producing a 3D sample by 3D printing technologies, such as Binder Jetting (BJ), Photopolymer jetting (PJ), Selective Laser Sintering (SLS), Fused Deposition Modeling (FDM). It is among these technologies. Among the different methods of making 3D samples, FDM and SLS are two common types of 3D printing technology which many researchers and manufacturers

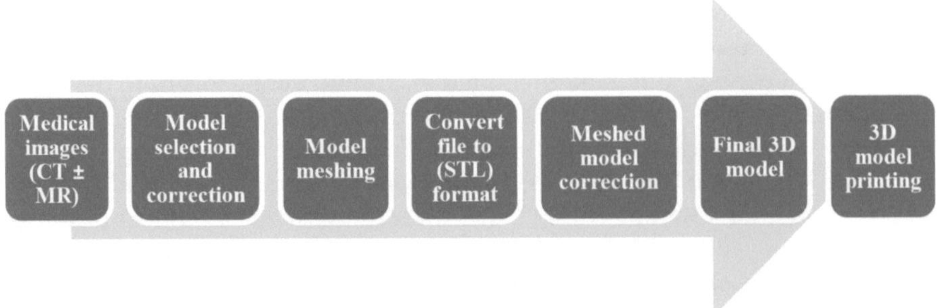

Figure 1.
The process of creating a 3D physical model using 3D printing technology based on medical imaging data.

have taken advantage of this method. One of the most famous 3D printers is the RepRap device [14].

Rapid advances in 3D printing technology have caused this technology to be increasingly used in the fields of medicine and health. In the field of medicine, this technology is used in the production of personal medical devices, implants, models for medical education, simulations, medical research, and also models for designing pre-operative treatment [11, 14].

3. Performed materials in 3D printers

So far, several materials have been used to produce the desired 3D object by the 3D printer, including polylactic acid (PLA), acrylonitrile butadiene styrene (ABS), polyethylene terephthalate glycol (PETG), thermoplastic elastomers (TPE), polyamide (PA also known as nylon), thermoplastic polyurethane (TPU), and polyvinyl acetate have been evaluated for use in radiation therapy [15, 16]. The success of 3D printing technology depends on the applied materials to create the desired 3D object [15]. The physical properties of the most important 3D printing materials are shown in **Table 1**.

PLA and ABS are two examples of the most common materials used in 3D printing technology [15]. PLA is a kind of odorless plastic polymers. This material is used in many industries, including degradable implants and food packaging. ABS has more resistance than PLA, and it is also resistant to high temperatures. As a plastic polymer, PA material is very resistant. In addition, it is flexible and very consistent. The most common application of TPE is in the construction of flexible objects, so that by using TPE, it is possible to create an object in a short time. PETG material is a combination of PET and glycols with different concentrations. Similar to PLA, PETG is used as a safe plaster for food containers and has been approved by the Food and Drug Administration (FDA). Compared to PLA, PETG is strong and hard. All the materials mentioned above are available in the form of filaments with diameters of 1.75 mm and 3 mm [15].

4. Applications of 3D printing technology in radiotherapy

Today, in spite of considering the tremendous advances in the radiotherapy, some radiotherapy steps are performed completely manually by the operator, therefore, a level of uncertainty is introduced in the clinical use of radiotherapy. Practical use of

Filament	Tension	Density	Flexibility	Durability	Printing problem	Printing temperature (°C)
ABS	Medium	1.01	Medium	High	Medium	210–250
PLA	Medium	1.24	Low	Medium	Low	180–230
PETG	Medium	1.27	High	High	Medium	220–235
TPE	Low	—	High	Medium	High	225–235
Nylon	High	—	High	High	Medium	220–260

ABS: acrylonitrile butadiene styrene; PLA: polylactic acid; PETG: polyethylene terephthalate glycol; TPE: thermoplastic elastomers.

Table 1.
Physical properties of 3D printed materials.

3D printing technology in radiotherapy can improve treatment results and reduce treatment errors. Recently, in the field of radiotherapy, the creation of objects from CT and MRI imaging data for various applications using professional and conventional 3D printers has been investigated by several groups of researchers. The main applications of this technology in radiotherapy, as shown in **Figure 2**, include making patient-specific blouses, fixation devices, brachytherapy applicators, compensation blocks, quality assurance phantoms, filters and grade blocks are therapeutic.

4.1 Radiotherapy bolus

In radiotherapy, one of the most common applications of 3D printing is the production of patient-specific bolus. So far, many studies have been done in the field of 3D printing to produce special blouses (**Figure 2E**). Bolus is a synthetic substance that is placed on the surface of the treatment area to correct the dose at the surface and depth of the skin, which can be used in both photon and electron beam radiotherapy [16]. Radiotherapy boluses can be created from commercial materials such as synthetic gel sheets, wet gauze, wax and thermoplastic sheets [22]. The preparing boluses methods are based on wax and thermoplastic materials is completely manual, which involves the level of uncertainty in its creation and use. In addition, the mismatch between the level of the bolus and the patient's level can lead to the creation of air cavities, which itself leads to a dose difference between the treatment plan and the dose delivered to the patient [16, 23–26].

Today, 3D printing technology provides the manufacturing of patient-specific bolus, so that it can lead to improved uniformity and dose distribution on complex and irregular surfaces [16]. It has been established that the 3D printed boluses used at the bedside can optimize the treatment preparation time and reduces costs of manufacturing process [27]. Besides, it has been shown that the use of this technology to create a bolus is safe and practical which also increases the work efficiency while increasing the quality of radiation dose delivery [27].

Park et al., have been used 3D printing technology to make a blouse for a 45-year-old man with Kimura disease who had received several sessions of radiation. Placing the 3D printed bolus in the treatment area placed the target position at an acceptable level and the air gap between the patient's body and the bolus was less than 5 mm. Most of the desired area was covered by a 3D printed blouse with a 95% uniform dose

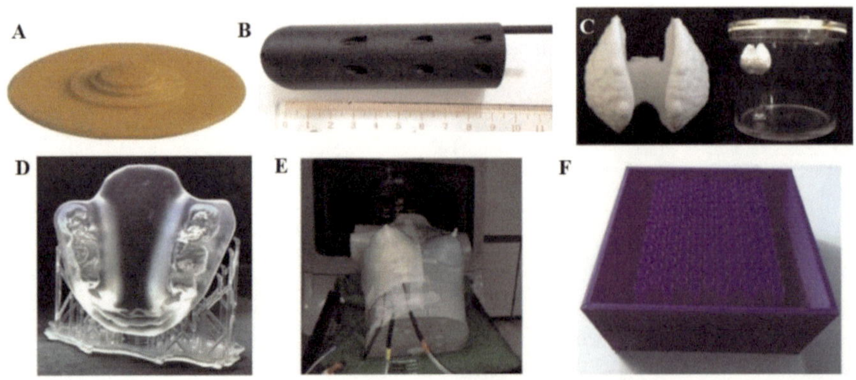

Figure 2.
3D printer filter (A), brachytherapy applicators (B), phantom (C), oral stent (D), blues (E) and block grade therapy (F). Adapted with permission from references [15, 17–21].

curve. Finally, the results of 3D printed boluses compared to the non-bolus state and even with paraffin wax blouse led to the improvement of the treatment target coverage [28]. The clinical effectiveness of bolus in head and neck cancers, head angiosarcoma, and after mastectomy has been investigated in many studies [29–31].

The presence of an air gap between the bolus and the patient's body has a significant effect on surface doses which can lead to a skin dose reduction [32, 33]. Fujimoto et al. designed a 3D bolus for the patient using CT images data. After placing the 3D printed bolus on the phantom, the results of dosimetry showed that the use of this special bolus can reduce the air gap and facilitate the dose coverage of the studied volume [34]. Besides, Ricotti et al., determined the dosimetric properties of common applied materials of ABS and PLA as printing materials that are usually. In this study it has been showed that the different filling percentages of these two substances lead to the creation of different densities, which can lead to differences between the calculated and measured dose distributions [35].

4.2 Brachytherapy applicators

Another application of 3D printing technology is the creation of applicators in brachytherapy. Several studies have investigated the 3D printing patterns to produce specific applicators that accommodate to the anatomy of each patient during the brachytherapy. The feasibility of 3D printing mold in brachytherapy was determined by Harris et al. [36]. The results of their study showed that a low-cost 3D printer with ABS plastic can accurately produce specific mold structures and catheter channels [36].

Recently, the production of templates for skin cancer brachytherapy using a 3D printer has been reported, so that the 3D printed template has led to a reduction in financial costs and clinical workflow [37]. In head and neck brachytherapy, applications of 3D printed applicators for implanting radioactive iodine-125 can accurately transfer a CT-based treatment plan to the brachytherapy needle insertion process [38]. In addition, using a separate 3D printed template (to guide the brachytherapy needle) the time required to insert the needle can be reduced and minimized the complications associated with incorrect implants [38].

In Seki et al. study, two cases of personalized 3D printed templates which were designed inversely from CT or MRI data were investigated for interstitial brachytherapy for cervical cancers (**Figure 2B**) [18]. As investigated in other studies, the accuracy of 3D printed model was accepted at a high level [39, 40]. There are two main weaknesses in conventional brachytherapy applicators. First, inability of the adaptation to the anatomy of the patient's body which leads to a change in the operator's position in a treatment department. Second, daily repeatability based on the requester's situation is considered a challenging issue [12]. However, the use of inexpensive 3D printers is a promising solution for creating high-dose brachytherapy applicators [12]. Jones et al., have shown that 3D printed applicators can improve high-dose surface brachytherapy using pre-programmed catheter orientation [41].

4.3 Phantom construction

Phantoms are widely used for quantitative and qualitative evaluations in medical imaging and radiotherapy (**Figure 2C**) [12]. Recently, several studies of 3D printing technology for making phantoms have been reported [11, 42]. In Tino et al. study, a significant increase in the publication of articles during the last 10 years in the field of imaging and dosimetry phantoms produced by 3D technology (number of 52 articles) has been reported. One of the main features of 3D printing technology was the production of customized additives and the possibility of density variations [43].

In medical imaging, researchers investigated the feasibility of an anthropomorphic chest phantom produced with a 3D printer for medical imaging purposes. Using 3D printing technology, they created a phantom based on the CT image taken from the chest of a patient with lung cancer. The phantom printed by 3D technology was examined in terms of size, shape, and structure by PET scan and CT scan. The results of the investigation showed that the phantom is able to withstand radiation doses of more than 24,000 Gy [7]. Today, many 3D printers produce 3D images of human organs [15]. 3D printing technology enables researchers to produce commercial phantoms and organs in a cost-effective manner compared to commercially available phantoms. Recently, several studies have reported that the use of 3D printing technology can create phantoms with variable density to ensure the quality of radiation therapy [6, 44]. 3D printing provides an inexpensive way to design and manufacture phantoms [45]. Also, the feasibility of 3D printed phantom for in-body artificial dosimetry in ensuring the quality of IMRT before treatment was examined. The results showed that the dose difference between the anthropomorphic head phantom and the 3D printed phantom was generally less than 2% [46]. In another study, to verify the accuracy of CyberKnife Xsight Lung Tracking System (XLTS) compared to Fiducial-based target tracking system (FTTS), a lung phantom made by 3D technology was used. XLTS and FTTS are Synchrony's two real-time respiratory tracking systems in CyberKnife robotic surgery. 3D printing technology has a very good ability to create lung phantoms [47]. It has been suggested that in the future multi-material printing using polymer jet technology will be used as an important printing process with the ability to create heterogeneous phantoms for dosimetry in radiotherapy [43].

4.4 Stabilizer devices

Recently, several advanced radiotherapy methods such as SBRT have been clinically used. In these techniques, accurate adjustment and immobilization of the patient is very crucial to ensure about the optimal dose coverage of the target volume and preservation of healthy tissues [44]. For this reason, immobilization of the patient is essential, especially in the head and neck area due to the tumor is usually located in the vicinity of endangered organs such as the brain stem or the spinal cord [48].

Nowadays, thermoplastic masks are usually used to immobilize the patient body [49]. Other immobilization devices include stereotactic frames, Scotchcast masks and bit blocks [44]. The most important concern about fixation devices can be attributed to the accuracy of fixation during the treatment session as well as different treatment sessions. In addition, the head mask can lead to physical discomfort and claustrophobia in some patients [50, 51]. Generally, it is necessary to improve the comfort of fixed devices in order to provide the best coverage and the least discomfort for the patient. Recently, a few studies have focused on the printing of fixation devices based on the anatomy of each patient by 3D printing technology.

In Haefner et al. study, a new method for making head fixation devices according to individual anatomy with 3D printing technology based on MRI data has been introduced. In this study, eight volunteers were studied and the 3D MRI data obtained from the head was processed using software and a meshed model of the surface of the fixing mask was obtained. Then, a fixing mask for the head was produced by a 3D printer and ABS material. The results showed that there is a high level of adjustment precision and also the patients had the least discomfort during use [44].

Besides, an oral stent was used during head and neck radiation therapy to reduce the side effects of radiation to healthy tissues. When an oral stent is used, the normal tissue is moved away from the high dose areas. Normally, the dentist makes an oral stent based on the image of the patient's teeth and a model of the relationship between the upper and lower jaw. The main purpose of using an oral stent is to increase the level of confidence and place the jaw in a reproducible position during radiation therapy. Nevertheless, there are various problems in using this device, among which it can be mentioned that it is difficult and time-consuming [20]. Recently, a new method for creating a patient-specific oral stent from common CT imaging data and 3D printing technology has been described by Wilke et al. (**Figure 2D**). Oral stent production using 3D printing method does not require the physical presence of the patient and will lead to a reduction in treatment time [20].

In Asfia et al. study, the result of published articles from 2000 to 2019 in the field of 3D printer fabrication of stabilizers was reviewed. The results of this study showed that with the advent of 3D printing technology, the manufacture of stabilizers by 3D printer is more affordable and accessible, so that functional parts They are able to produce more. Also, in this study a favorable agreement of the fixators in terms of matching the unique body geometry of the patient and also the possibility of repeatable adjustment for treatment have been reported [52].

4.5 Other applications of 3D printing technology in radiotherapy

In addition to the applications of 3D printing technology in above mentioned radiotherapy advances over the past few years, few studies have used 3D printing technology to make individual-oriented, compensating, and block grid radiation therapy filters [16, 17, 53, 54]. Filter production based on the individual characteristics and anatomy of the patient is considered an essential factor. Creating a filter is very time-consuming and its widespread use has not been reported [55, 56]. In this regard, a new method for creating special filters for the formation of electron beam fields using 3D printing technology has been described (**Figure 2A**) [17].

A very useful treatment method for large tumors is spatially fractionated radiotherapy (SFRT), which uses a special block to create a grid-like pattern. Grid therapy is an effective method for treating large tumors. Although radiotherapy with grade is effective, the clinical application of this method is limited due to inadequate understanding of radiobiological mechanisms [57, 58]. In addition, it is time-consuming and difficult to make specialized blocks that create radiation beams in the form of a network.

Today, the use of 3D printing technology can solve the limitations of access to specialized blocks in grid therapy. In 2015, Zhu et al. investigated cerrobend networks and used a three-dimensional technique for radiation modulation to create them (**Figure 2**). The block grid template was designed with tubes that lead to beam divergence. The mold was printed by a 3D printer using resin at a temperature of less than 230°C. Cerrobend liquid was melted at 120°C and poured into a resin mold and prepared for a block with a thickness of 7.4 cm. By using a small field dosimeter including a pinpoint ionization chamber and a stereotactic diode, the dosimetric characteristics of the grid block were investigated. For the 6 MW photon beam, the valley-to-peak ratio was 20% at dmax and 30% at a depth of 10 cm. The output factor was 84.9% at dmax and 65.1% at a depth of 10 cm. Their study showed that the 3D printing method can be used in grid therapy [21].

Another application of 3D printing technology in radiotherapy is the creation of compensatory blocks in IMRT [52]. Multileaf collimators and compensating blocks are two basic techniques used to modulate the intensity of photon beams. The conventional method of making IMRT compensating blocks is the use of milling machines, one of the disadvantages of which is the high operating and production cost compared to the lead multi-leaf collimator method. To eliminate the need for milling machines, 3D printing technology has been used to make IMRT compensatory blocks. The main advantage of this approach is reducing costs and production time [53].

5. Conclusion

The use of 3D printing technology will reduce the cost of radiotherapy, and as a promising method, it can lead to the treatment of cancer in a specific and person-centered way. Studies conducted on 3D printing showed that this method is a fast, practical and inexpensive method to deliver a uniform dose to the target volume and at the same time protect healthy tissues in the radiation field. In addition, this technology reduces the patient's discomfort and provides special radiation therapy devices suitable for each patient.

In this chapter, the main applications of 3D printing technology in radiotherapy in the manufacture of special treatment devices such as bolus, phantom, brachytherapy applicators, filters, patient fixation devices, compensatory blocks and grid blocks. There are various materials for 3D printing that can be used for better delivery of radiotherapy. The use of 3D printed devices based on the anatomical features of each patient in radiotherapy, such as bolus and fixing devices, can reduce daily uncertainty and also increase the accuracy of treatment.

The conducted investigations showed that patient-specific devices can be produced by 3D printing from volumetric CT images or MRI data. In practice, 3D printing technology has a great potential to improve accuracy and efficiency in the field of personal radiotherapy. 3D printing technology offers a relatively low-cost and effective way to produce devices based on individual anatomy in radiotherapy. When a new technique is introduced at the bedside, it is necessary to develop appropriate quality assurance programs to protect patients and healthcare professionals. Further developments in the field of 3D printing technology can create more flexibility in design, so that this method can be applied to the bedside as well. Besides, new 3D printing materials and methods that lead to better results in the treatment of patients will be introduced in the not-so-distant future.

Abbreviations

IMRT	Intensity Modulated Radiation Therapy
IGRT	Image Guided Radiation Therapy
SRS	Stereotactic Radiosurgery
SBRT	Stereotactic Body Radiation Therapy
CT	Computed Tomography
MRI	Magnetic Resonance Imaging
STL	STereoLithography

BJ	Binder Jetting
PJ	Photopolymer jetting
SLS	Selective Laser Sintering
FDM	Fused Deposition Modeling
PLA	polylactic acid
ABS	acrylonitrile butadiene styrene
PETG	polyethylene terephthalate glycol
TPE	thermoplastic elastomers
PA	polyamide
TPU	thermoplastic polyurethane
FDA	Food and Drug Administration
SFRT	spatially fractionated radiotherapy

Author details

Seyed Hamid Zoljalali Moghaddam
Iran University of Medical Science, Tehran, Iran

*Address all correspondence to: zoljalali.h@iums.ac.ir

References

[1] Ghaffari H, Beik J, Talebi A, Mahdavi SR, Abdollahi H. New physical approaches to treat cancer stem cells: A review. Clinical & Translational Oncology. 2018;**20**(12):1502-1521. DOI: 10.1007/s12094-018-1896-2

[2] Jermann M. Particle therapy statistics in 2014. International Journal of Particle Therapy. 2015;**2**(1):50-54. DOI: 10.14338/IJPT-15-00013

[3] Garibaldi C, Jereczek-Fossa BA, Marvaso G, Dicuonzo S, Rojas DP, Gattani F, et al. Recent advances in radiation oncology. Ecancermedicalscience. 2017;**11**:785. DOI: 10.3332/ecancer.2017.785

[4] Ghaffari H, Navaser M, Mofid B, Mahdavi SR, Mohammadi R, Tavakol A. Fiducial markers in prostate cancer image-guided radiotherapy. Medical Journal of the Islamic Republic of Iran. 2019;**33**:15. DOI: 10.34171/mjiri.33.15

[5] Mahdavi SR, Ghaffari H, Mofid B, Rostami A, Reiazi R, Janani L. Rectal retractor application during image-guided dose-escalated prostate radiotherapy. Strahlentherapie und Onkologie. 2019;**195**(10):923-933. DOI: 10.1007/s00066-019-01445-6

[6] Cerviño L, Soultan D, Cornell M, Yock A, Pettersson N, Song WY, et al. A novel 3D-printed phantom insert for 4D PET/CT imaging and simultaneous integrated boost radiotherapy. Medical Physics. 2017;**44**(10):5467-5474. DOI: 10.1002/mp.12495

[7] Hazelaar C, Eijnatten M, Dahele M, Wolff J, Forouzanfar T, Slotman B, et al. Using 3D printing techniques to create an anthropomorphic thorax phantom for medical imaging purposes. Medical Physics. 2018;**45**(1):92-100. DOI: 10.1002/mp.12644

[8] Jones E-L, Baldion AT, Thomas C, Burrows T, Byrne N, Newton V, et al. Introduction of novel 3D-printed superficial applicators for high-dose-rate skin brachytherapy. Brachytherapy. 2017;**16**(2):409-414. DOI: 10.1016/j.brachy.2016.11.003

[9] Kong Y, Yan T, Sun Y, Qian J, Zhou G, Cai S, et al. A dosimetric study on the use of 3D-printed customized boluses in photon therapy: A hydrogel and silica gel study. Journal of Applied Clinical Medical Physics. 2019;**20**(1):348-355. DOI: 10.1002/acm2.12489

[10] Palser R, Jamieson R, Sutherland J, Skibo L. Three-dimensional lithographic model building from volume data sets. Canadian Association of Radiologists journal= Journal l'Association canadienne des radiologistes. 1990;**41**(6):339-341

[11] Abdullah KA, McEntee MF, Reed W, Kench PL. Development of an organ-specific insert phantom generated using a 3D printer for investigations of cardiac computed tomography protocols. Journal of Medical Radiation Sciences. 2018;**65**(3):175-183. DOI: 10.1002/jmrs.279

[12] Ricotti R, Vavassori A, Bazani A, Ciardo D, Pansini F, Spoto R, et al. 3D-printed applicators for high dose rate brachytherapy: Dosimetric assessment at different infill percentage. Physica Medica. 2016;**32**(12):1698-1706. DOI: 10.1016/j.ejmp.2016.08.016

[13] Squelch A. 3D printing and medical imaging. Journal of Medical Radiation Sciences. 2018;**65**(3):171-172. DOI: 10.1002/jmrs.300

[14] Bowyer A. 3D printing and humanity's first imperfect replicator. 3D printing and additive manufacturing. 2014;**1**(1):4-5. DOI: 10.1089/3dp.2013.0003

[15] Alssabbagh M, Tajuddin AA, Abdulmanap M, Zainon R. Evaluation of 3D printing materials for fabrication of a novel multi-functional 3D thyroid phantom for medical dosimetry and image quality. Radiation Physics and Chemistry. 2017;**135**:106-112. DOI: 10.1016/j.radphyschem.2017.02.009

[16] Zhao Y, Moran K, Yewondwossen M, Allan J, Clarke S, Rajaraman M, et al. Clinical applications of 3-dimensional printing in radiation therapy. Medical Dosimetry. 2017;**42**(2):150-155. DOI: 10.1016/j.meddos.2017.03.001

[17] Miloichikova I, Krasnykh A, Danilova I, Stuchebrov S, Kudrina V, editors. Formation of electron beam fields with 3D printed filters. In: AIP Conference Proceedings. AIP Publishing LLC; 2016. DOI: 10.1063/1.4964598

[18] Park S-Y, Choi CH, Park JM, Chun M, Han JH, Kim J-i. A patient-specific polylactic acid bolus made by a 3D printer for breast cancer radiation therapy. PLoS One. 2016;**11**(12):e0168063. DOI: 10.1371/journal.pone.0168063

[19] Sekii S, Tsujino K, Kosaka K, Yamaguchi S, Kubota H, Matsumoto Y, et al. Inversely designed, 3D-printed personalized template-guided interstitial brachytherapy for vaginal tumors. Journal of Contemporary Brachytherapy. 2018;**10**(5):470. DOI: 10.5114/jcb.2018.78832

[20] Wilke CT, Zaid M, Chung C, Fuller CD, Mohamed AS, Skinner H, et al. Design and fabrication of a 3D–printed oral stent for head and neck radiotherapy from routine diagnostic imaging. 3D Printing in Medicine. 2017;**3**(1):12. DOI: 10.1186/s41205-017-0021-4

[21] Zhu X, Driewer J, Li S, Verma V, Lei Y, Zhang M, et al. Fabricating Cerrobend grids with 3D printing for spatially modulated radiation therapy: A feasibility study. Medical Physics. 2015;**42**(11):6269-6273. DOI: 10.1118/1.4932223

[22] Benoit J, Pruitt AF, Thrall DE. Effect of wetness level on the suitability of wet gauze as a substitute for superflab® as a bolus material for use with 6 MV photons. Veterinary Radiology & Ultrasound. 2009;**50**(5):555-559. DOI: 10.1111/j.1740-8261.2009.01573.x

[23] Behrens C. Dose build-up behind air cavities for Co-60, 4, 6 and 8 MV. Measurements and Monte Carlo simulations. Physics in Medicine & Biology. 2006;**51**(22):5937. DOI: 10.1088/0031-9155/51/22/015

[24] Kong M, Holloway L. An investigation of central axis depth dose distribution perturbation due to an air gap between patient and bolus for electron beams. Australasian Physics & Engineering Sciences in Medicine. 2007;**30**(2):111. DOI: 10.1007/BF03178415

[25] Li XA, Yu C, Holmes T. A systematic evaluation of air cavity dose perturbation in megavoltage x-ray beams. Medical Physics. 2000;**27**(5):1011-1017. DOI: 10.1118/1.598966

[26] Sharma S, Johnson M. Surface dose perturbation due to air gap between patient and bolus for electron beams. Medical Physics. 1993;**20**(2):377-378. DOI: 10.1118/1.597079

[27] Canters RA, Lips IM, Wendling M, Kusters M, van Zeeland M, Gerritsen RM, et al. Clinical implementation of 3D printing in the construction of patient specific bolus for electron beam radiotherapy for non-melanoma skin cancer. Radiotherapy and Oncology. 2016;**121**(1):148-153. DOI: 10.1016/j.radonc.2016.07.011

[28] Park J, Yea J. Three-dimensional customized bolus for intensity-modulated radiotherapy in a patient with Kimura's disease involving the auricle. Cancer/Radiothérapie. 2016;**20**(3): 205-209. DOI: 10.1016/j.canrad.2015.11.003

[29] Guadagnolo BA, Zagars GK, Araujo D, Ravi V, Shellenberger TD, Sturgis EM. Outcomes after definitive treatment for cutaneous angiosarcoma of the face and scalp. Head & Neck. 2011;**33**(5):661-667. DOI: 10.1002/hed.21513

[30] Kai M, Kanaya N, Wu SV, Mendez C, Nguyen D, Luu T, et al. Targeting breast cancer stem cells in triple-negative breast cancer using a combination of LBH589 and salinomycin. Breast Cancer Research and Treatment. 2015;**151**(2):281-294. DOI: 10.1007/s10549-015-3376-5

[31] Tieu MT, Graham P, Browne L, Chin YS. The effect of adjuvant postmastectomy radiotherapy bolus technique on local recurrence. International Journal of Radiation Oncology Biology Physics. 2011;**81**(3): e165-ee71. DOI: 10.1016/j.ijrobp.2011.01.002

[32] Khan Y, Villarreal-Barajas JE, Udowicz M, Sinha R, Muhammad W, Abbasi AN, et al. Clinical and dosimetric implications of air gaps between bolus and skin surface during radiation therapy. Journal of Cancer Therapy. 2013;**4**(7):1251. DOI: 10.4236/jct.2013.47147

[33] Su S, Moran K, Robar JL. Design and production of 3D printed bolus for electron radiation therapy. Journal of Applied Clinical Medical Physics. 2014; **15**(4):194-211. DOI: 10.1120/jacmp.v15i4.4831

[34] Fujimoto K, Shiinoki T, Yuasa Y, Hanazawa H, Shibuya K. Efficacy of patient-specific bolus created using three-dimensional printing technique in photon radiotherapy. Physica Medica. 2017;**38**:1-9. DOI: 10.1016/j.ejmp.2017.04.023

[35] Ricotti R, Ciardo D, Pansini F, Bazani A, Comi S, Spoto R, et al. Dosimetric characterization of 3D printed bolus at different infill percentage for external photon beam radiotherapy. Physica Medica. 2017;**39**: 25-32. DOI: 10.1016/j.ejmp.2017.06.004

[36] Harris BD, Nilsson S, Poole CM. A feasibility study for using ABS plastic and a low-cost 3D printer for patient-specific brachytherapy mould design. Australasian Physical & Engineering Sciences in Medicine. 2015;**38**:399-412. DOI: 10.1007/s13246-015-0356-3

[37] Arenas M, Sabater S, Sintas A, Arguís M, Hernández V, Árquez M, et al. Individualized 3D scanning and printing for non-melanoma skin cancer brachytherapy: A financial study for its integration into clinical workflow. Journal of Contemporary Brachytherapy. 2017;**9**(3):270. DOI: 10.5114/jcb.2017.68134

[38] Huang M-W, Zhang J-G, Zheng L, Liu S-M, Yu G-Y. Accuracy evaluation of a 3D-printed individual template for needle guidance in head and neck brachytherapy. Journal of Radiation

Research. 2016;**57**(6):662-667. DOI: 10.1093/jrr/rrw033

[39] Ji Z, Jiang Y, Guo F, Sun H, Fan J, Zhang L, et al. Dosimetry verification of radioactive seed implantation for malignant tumors assisted by 3D printing individual templates and CT guidance. Applied Radiation and Isotopes. 2017;**124**:68-74. DOI: 10.1016/j.apradiso.2016.12.009

[40] Jiang Y, Ji Z, Guo F, Peng R, Sun H, Fan J, et al. Side effects of CT-guided implantation of 125 I seeds for recurrent malignant tumors of the head and neck assisted by 3D printing non co-planar template. Radiation Oncology. 2018; **13**(1):18. DOI: 10.1186/s13014-018-0959-4

[41] Sethi R, Cunha A, Mellis K, Siauw T, Diederich C, Pouliot J, et al. Clinical applications of custom-made vaginal cylinders constructed using three-dimensional printing technology. Journal of Contemporary Brachytherapy. 2016; **8**(3):208. DOI: 10.5114/jcb.2016.60679

[42] Kim S-W, Shin H-J, Kay CS, Son SH. A customized bolus produced using a 3-dimensional printer for radiotherapy. PLoS One. 2014;**9**(10):e110746. DOI: 10.1371/journal.pone.0110746

[43] Tino R, Yeo A, Leary M, Brandt M, Kron T. A systematic review on 3D-printed imaging and dosimetry phantoms in radiation therapy. Technology in Cancer Research & Treatment. 2019;**18**:1533033819870208. DOI: 10.1177/1533033819870208

[44] Haefner MF, Giesel FL, Mattke M, Rath D, Wade M, Kuypers J, et al. 3D-printed masks as a new approach for immobilization in radiotherapy—A study of positioning accuracy. Oncotarget. 2018;**9**(5):6490. DOI: 10.18632/oncotarget.24032

[45] Craft DF, Howell RM. Preparation and fabrication of a full-scale, sagittal-sliced, 3D-printed, patient-specific radiotherapy phantom. Journal of Applied Clinical Medical Physics. 2017; **18**(5):285-292. DOI: 10.1002/acm2.12162

[46] Kamomae T, Shimizu H, Nakaya T, Okudaira K, Aoyama T, Oguchi H, et al. Three-dimensional printer-generated patient-specific phantom for artificial in vivo dosimetry in radiotherapy quality assurance. Physica Medica. 2017;**44**: 205-211. DOI: 10.1016/j.ejmp.2017.10.005

[47] Jung J, Song SY, Yoon SM, Kwak J, Yoon K, Choi W, et al. Verification of accuracy of CyberKnife tumor-tracking radiation therapy using patient-specific lung phantoms. International Journal of Radiation Oncology Biology Physics. 2015;**92**(4):745-753. DOI: 10.1016/j.ijrobp.2015.02.055

[48] Cacicedo J, Perez J, de Zarate RO, Del Hoyo O, Casquero F, Gómez-Iturriaga A, et al. A prospective analysis of inter-and intrafractional errors to calculate CTV to PTV margins in head and neck patients. Clinical and Translational Oncology. 2015;**17**(2): 113-120. DOI: 10.1007/s12094-014-1200-z

[49] Dimitriadis A, Kirkby KJ, Nisbet A, Clark CH. Current status of cranial stereotactic radiosurgery in the UK. The British Journal of Radiology. 2016; **89**(1058):20150452. DOI: 10.1259/bjr.20150452

[50] Goldsworthy SD, Tuke K, Latour JM. A focus group consultation round exploring patient experiences of comfort during radiotherapy for head and neck cancer. Journal of Radiotherapy in Practice. 2016;**15**(2):143-149. DOI: 10.1017/S1460396916000066

[51] Oultram S, Findlay N, Clover K, Cross L, Ponman L, Adams C. A comparison between patient self-report and radiation therapists' ability to identify anxiety and distress in head and neck cancer patients requiring immobilization for radiation therapy. Journal of Radiotherapy in Practice. 2012;**11**(2):74-82. DOI: 10.1017/S1460396911000136

[52] Asfia A, Novak JI, Mohammed MI, Rolfe B, Kron T. A review of 3D printed patient specific immobilisation devices in radiotherapy. Physics and Imaging in Radiation Oncology. 2020;**13**:30-35. DOI: 10.1016/j.phro.2020.03.003

[53] Avelino SR, Silva LFO, Miosso CJ, editors. Use of 3D-printers to create intensity-modulated radiotherapy compensator blocks. In: 2012 Annual International Conference of the IEEE Engineering in Medicine and Biology Society. IEEE Publishing; 2012. DOI: 10.1109/EMBC.2012.6347293

[54] Ju SG, Kim MK, Hong C-S, Kim JS, Han Y, Choi DH, et al. New technique for developing a proton range compensator with use of a 3-dimensional printer. International Journal of Radiation Oncology Biology Physics. 2014;**88**(2):453-458. DOI: 10.1016/j.ijrobp.2013.10.024

[55] Kudchadker R, Antolak J, Morrison W, Wong P, Hogstrom K. Utilization of custom electron bolus in head and neck radiotherapy. Journal of Applied Clinical Medical Physics. 2003; **4**(4):321-333. DOI: 10.1120/jacmp.v4i4.2503

[56] Stuchebrov SG, Miloichikova IA, Melnikov A, Pereverzeva M. Numerical simulation of the microtron electron beam absorption by the modified ABS-plastic. In: Journal of Physics: Conference Series. IOP Publishing; 2016. p. 012036. DOI: 10.1088/1742-6596/671/461/012036

[57] Asur R, Butterworth KT, Penagaricano JA, Prise KM, Griffin RJ. High dose bystander effects in spatially fractionated radiation therapy. Cancer Letters. 2015;**356**(1):52-57. DOI: 10.1016/j.canlet.2013.10.032

[58] Mohiuddin M, Fujita M, Regine WF, Megooni AS, Ibbott GS, Ahmed MM. High-dose spatially-fractionated radiation (GRID): A new paradigm in the management of advanced cancers. International Journal of Radiation Oncology Biology Physics. 1999;**45**(3): 721-727. DOI: 10.1016/S0360-3016(99)00170-4

Chapter 16

Health and Safety in 3D Printing

Hector Garcia Gonzalez and Mª Teresa Lopez Pola

Abstract

Nowadays, it is possible to find 3D printers everywhere, at homes, schools, work offices, etcetera. 3D printing is an additive manufacturing process that is increasingly gaining popularity, and it can create functional parts with a wide variety of shapes and sizes. But on the other hand, there are health risks associated with 3D printers, like nanoparticles and volatile organic compounds (VOCs), which are important to know to improve health and safety and avoid diseases such as asthma, allergic rhinitis and chronic obstructive pulmonary disease, among others. This chapter analyses techniques for sampling the nanoparticles and VOCs exposure during 3D printing and a health effects review, giving tools to evaluate the risks and recommendations to avoid or minimise these risks using engineering controls like extraction systems or good ventilation.

Keywords: nanoparticles, VOCs, health, air quality, 3D printing

1. Introduction

Every day 3D printing is more present in our lives. In 2018, 1.42 million units were sold; it is expected that by 2027 more than eight million units will be sold. Their relatively low price (from around 140 euros) means that they are present in many companies, homes and even nursery schools; with them, it is possible to produce prototypes of unique designs at a meagre cost; in addition, there are already models of games, figures or pieces that can be download for free from the Internet and print them with 3D printers.

However, they are not all advantages; with 3D printing, a series of emerging risks are associated, such as exposure to nanoparticles and volatile organic compounds (VOCs) [1–3]. The most investigated materials are poly lactic acid (PLA) and acrylonitrile butadiene styrene (ABS) since they were the most used; however, many compounds and other printing technologies are entering the market strongly, such as 3D resin printers.

In addition, it should be noted that there is no occupational exposure limit value (OELV) for nanoparticles in most countries. However, there are some reference values to help evaluate the exposures of nanoparticles from 3D printers. Regards the VOCs, not all of them are regulated with an OELV.

2. Nanomaterials and nanoparticles

According to the ISO/TS 27687, the nanoscale size ranges from approximately 1 to 100 nm, and the materials in these sizes usually show new or unusual properties. The European Union [4] defined "nanomaterial" as a natural, incidental or manufactured material consisting of solid particles that are present, either on their own or as identifiable constituent particles in aggregates or agglomerates, and where 50% or more of these particles in the number-based size distribution fulfil at least one of the following conditions:

a. One or more external dimensions of the particle are in size range of 1 to 100 nm;

b. The particle has an elongated shape, such as a rod, fibre or tube, where two external dimensions are smaller than 1 nm, and the other dimension is larger than 100 nm;

c. The particle has a plate-like shape, where one external dimension is smaller than 1 nm, and the other dimensions are larger than 100 nm.

No air quality standards are in place to regulate exposure to airborne nanoparticles because there is still debate over the number of nanoparticles that are acceptable to be exposed. There are no consensus measurement methods or tools [5]. However, it is evident that the harmful effects mainly depend on the nanoparticle's surface. Because nanoparticles have a large surface area to volume ratio compared to the same substance in bulk, they are a better metric for assessing risks than the conventionally (and officially) employed mass-based approach [6, 7].

To explain the importance of measuring nanoparticles, it is better to show an example; if there is a mass of 0.1 g in the air with a density of 2.65 g/cm^3, it can be calculated how many particles could be in the air from different sizes of particles. **Table 1** shows the number of particles and the superficial area of a mass as a function of the particle size.

If all the particles in the mass of 0.1 g had a size of 1 μm, the particle area would be 0.22 m^2, but on the other hand, if all particles had 1 nm, the total area would be 226 m^2. If these particles could irritate or damage the lungs, a person with the exact mass exposure could be exposed to a surface of 0.22 m^2 or 226 m^2; potentially, the health effects will be very different. It is estimated that the lung area is around 70 to 100 m^2 (**Figure 1**), but the nanoparticles are not deposited uniformly in the lungs; it is a complex issue, and there are different deposition probabilities according to the particle size [8].

Size	Particles number	Area (cm^2)	Area (m^2)
1 μm	72,071,806,726.62	2264.14	0.2264
10 μm	72,071,806.73	226.41	0.0226
5 μm	576,574,453.81	452.83	0.0453
100 nm	72,071,806,726,621.50	22641.36	2.2641
10 nm	72,071,806,726,621,500.00	226413.58	22.6414
1 nm	72,071,806,726,621,500,000.00	2264135.81	226.4136

Table 1.
Simulation of particle numbers and superficial area in a mass of 0.1 g with different sizes of particles.

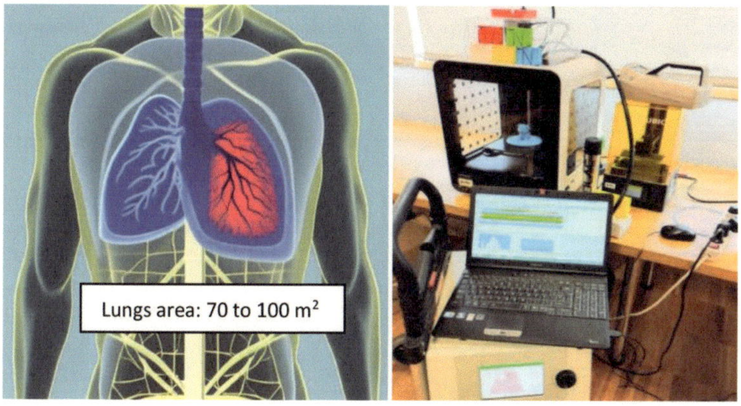

Figure 1.
Lungs area (left) and 3D printer (right).

The circle graph in **Figure 2** shows **Table 1**, which represents the area of mass of 0.1 g (density of 2.65 g/cm^3) in relation to the particle size. It is clear that the area increases with the particle size reduction.

Once the nanoparticle size's importance is known, it is critical to know reference levels; as previously exposed, it is still debatable because various types of nanoparticles have different or unknown health effects. When sampling nanoparticles, it is common to find different sizes distributions and heterogeneous compositions. In 2011 the Institut für Arbeitsschutz (IFA) proposed a benchmark concentration level based on particle numbers concentrations, **Table 2**.

According to **Table 2**, it is possible to carry out a nanoparticle health and safety assessment considering the material density and the particle number concentration. In materials with low density (most of the 3D printing filaments, poly lactic acid (PLA),

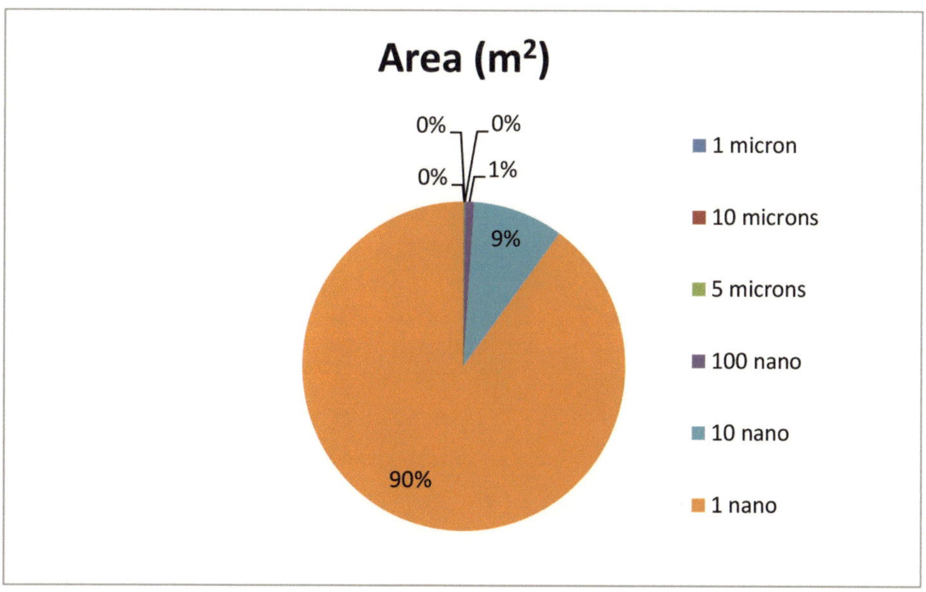

Figure 2.
Representation, surface in relation to particle size for the same mass.

Description	Density	Benchmark level (8-h time-weighted average (TWA))
Biopersistent granular nanomaterial in the range 1–100 nm	>6000 kg/m^3	20,000 particles/cm^3
Biopersistent granular nanomaterial in the range 1–100 nm	<6000 kg/m^3	40,000 particles/cm^3
Non-bio-persistent nanomaterial in the range of 1–100 nm		Applicable OELV

Table 2.
IFA nanoparticles benchmark level (IFA, 2009) [9].

acrylonitrile butadiene styrene (ABS), acrylonitrile styrene acrylate (ASA), ...), the reference level is 40,000 particles/cm^3, and for heavy materials 20,000 particles/cm^3. It is essential to know the sampling device characteristics because most of them also count water particles, and when sampling, it is crucial to take a sample for the background levels. In some cases, if the intention is to evaluate a process, the nanoparticles could be from the background (nanoparticles from diesel engines or pollen).

There are multiple nanoparticle sampling devices in **Table 3**, **Figure 3** shows two examples of sampling devices, and **Figure 4** shows filaments and a 3D printer.

There are many studies about particles emission during 3d printing. All the international literature agrees that respirable dust measurements are well below the OELV

Instrument	Sampling principle	Main metric	Range
CPC 3007 (TSI Inc., Shoreview, MN, USA)	Condensation nuclei counter	Particle number concentration	0.010 to 1 μm
EEPS 3090 (TSI Inc., Shoreview, MN, USA)	Engine exhaust Particle sizer Spectrometer	Ultrafine particle size distribution (nanoparticles)	0.0056 to 0.56 μm 32 Channels

Table 3.
Examples of devices for sampling nanoparticles.

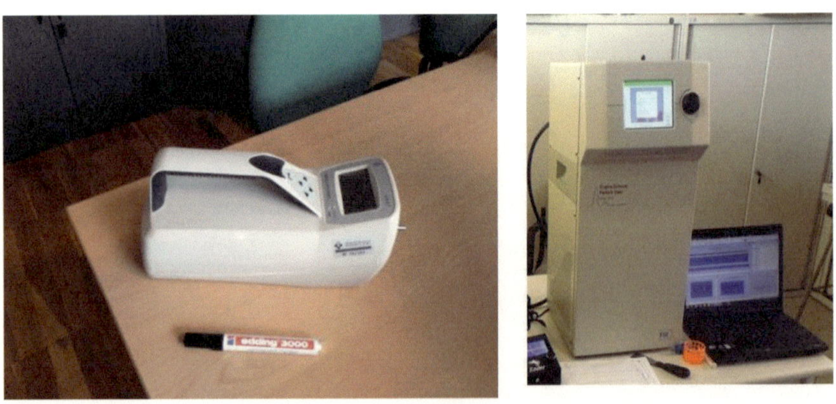

Figure 3.
Nanoparticles sampling devices: CPC 3007 (left) and EEPS3090 (right).

Figure 4.
Different 3D printing filaments (left) and 3D printer (right).

(3 mg/m^3), so 3d printing is not a risk if only the traditional and official assessment of dust mass (concentration in mg/m^3) is considered.

Chan et al. obtained concentrations of 700 μg/m^3 of total dust and 400 μg/m^3 of respirable dust well below the occupational limits, with the predominant sizes being between 27 and 115 nanometers [10]. Runström et al. obtained total dust and respirable dust values below the laboratory detection limit (negligible) and a maximum concentration of nanoparticles of 25,000 particles/cm^3 [11]. Jensen et al. obtained concentrations of 50.4 μg/m^3 in respirable dust [12].

As for nanoparticles, for different filaments in studies carried out in a chamber by Floyd et al. [13], values of up to 1,000,000 particles/cm^3 were obtained, which coincides with the studies by Zhang et al. [14] reporting this last author differences between various ABS manufacturers.

In the same way, Kwon et al. [1] detailed a maximum concentration of 54,000 nanoparticles/cm^3 in ABS and 1326 nanoparticles/cm^3 in PLA filament printing at the temperature recommended by the manufacturer, increasing the emission of nanoparticles at higher temperatures, very similar results than García-González and Lopez-Pola [7].

Azimi studies in a chamber showed concentrations between 108-1011 particles/cm^3 [15].

In metal printers, Jensen et al. measured maximum values of 10,000 particles/cm^3 [12], while studies by Yi et al. observed differences in the emission of nanoparticles and their size depending on the material filament colour, both in PLA and ABS filaments, obtaining concentrations in a chamber up to 2.18 x 10^{11} particles/cm^3 [16].

3. Volatile organic compounds (VOCs)

According to the United States Environmental Protection Agency (EPA), VOCs are gases containing various chemicals released from liquids or solids with a high vapour pressure at room temperature [17].

VOCs are classified into four groups:

- Very volatile organic compounds (VVOCs) with Tb: < to 50–100°C (e g. propane, butane and methyl chloride).

- Volatile organic compounds (VOCs) with 50–100°C < Tb < 240–260°C (Formaldehyde, d-Limonene, toluene, acetone, ethanol (ethyl alcohol) 2-propanol (isopropyl alcohol), and hexanal).
- Semi-volatile organic compounds (SVOCs) with 240–260°C < Tb < 380–400°C.
- Particulate organic matter (POM) with Tb > 380°C; e.g., pesticides (DDT, chlordane, plasticisers (phthalates)), fire retardants (Polychlorinated Biphenyl (PCBs) and Polybrominated Biphenyl (PBB)) [18].

VOCs are usually sampled with direct reading devices to obtain the total volatile organic compounds (TVOC), defined as the sum of all VOCs that elute between and include n-hexane and n-hexadecane on a non-polar capillary column. Also, the VOCs can be sampled with a personal pump, a tygon tubing air sampling and a Tenax® TA sorbent tube (**Figure 5**) and analysed in the laboratory with a gas chromatograph; this procedure requires a low flow of 50 ml/min or less, and the sample volume must be around 6 litres, with this technique, it is possible to quantify multiple individual VOCs, this method is based on NIOSH 2549.

There are multiple VOCs, and not of them are well studied. **Table 4** shows the OELV recommended by the Instituto Nacional de Seguridad y Salud en el Trabajo (INSST) for some VOCs.

The International Agency for Research on Cancer (IARC) has classified some volatile organic compounds as carcinogenic, such as acrylonitrile, 1–3 butadiene, benzene and formaldehyde.

In the same way, the legislation says that exposure to any carcinogen substance must be as low as technically possible. In non-occupational uses, for example homes, the reference levels are usually 200 μg/m**3** of TVOC [20], and the benchmark levels are proposed in the public health England guide for indoor air quality (**Table 5**).

Volatile organic compounds are emitted mainly in FDM (fused deposition modelling) printing processes [26], although concentrations are generally below occupational limits. In this type of printing, styrene is emitted with ABS material, and methyl methacrylate is more frequent when printing with PLA material [27].

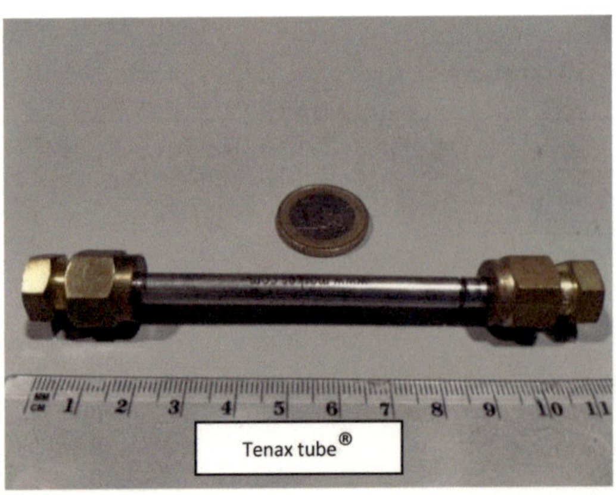

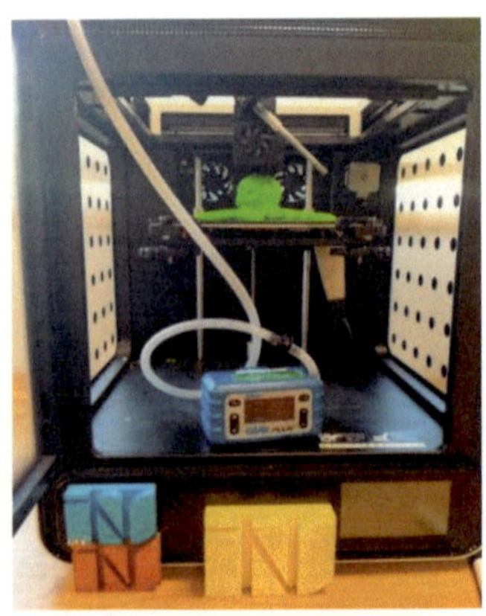

Figure 5.
Tenax sorbent tube (left), sampling VOCs during 3D printing (right).

VOCs	OELV (mg/m³)	VOCs	OELV (mg/m³)
Hexane	72	b-Pinene	113
n-Heptane	2085	Ethylbenzene	441
n-Octane	1420	p-Xylene	221
Carbon tetrachloride	6.4	m-Xylene	221
1.1.1-Trichloroethane	555	n-Butanol	61
n-Nonane	1065	o-Xylene	221
Ethyl acetate	734	d-Limonene	168
Benzene	3.25	1,3,5-Trimethylbenzene	100
Trichlorethylene	54.7	Styrene	86
α-Pinene	113	1,2,4- Trimethylbenzene	100
Toluene	192	1,2,3- Trimethylbenzene	100
1,2-dichloropropane	47	Formaldehyde	0.37
1.2-Dicloroethane	8.2	Naphthalene	53
N-butyl acetate	241		

Table 4.
The occupational limit for VOCs (INSST) [19].

VOCs	Limit values in μg/m³		Source
	Short term	Long term	
Acetaldehyde (75–07-0)	1420 (1 h)	280 (1 day)	Health Canada (2018) [22]
α-Pinene (80–56-8)	45,000 (30 min)	4500 (1 day)	EPHECT (Trantallidi et al., 2015) [23]
Benzene (71–43-2)	No recommended level of exposure		World Health Organisation (2010) [24]
D-Limonene (5989-27-5)	90,000 (30 min)	9000 (1 day)	EPHECT (Trantallidi et al., 2015) [23]
Formaldehyde (50–00-0)	100 (30 min)	10 (1 year)	World Health Organisation (2010) [24]
Naphthalene (91–20-3)	—	3 (1 year)	Agency for Toxic Substances and Disease Registry (2005), USA [25]
Styrene (100–42-5)	—	850 (1 year)	Health Canada (2018) [22]
Tetrachloroethylene (127–18-4)	—	40 (1 day)	Health Canada (2018) [22]
Toluene (108–88-3)	15,000 (8 h)	2300 (1 day)	Health Canada (2018) [22]
Trichloroethylene (71–01-06)	No recommended level of exposure		World Health Organisation (2010) [24]
Xylenes-mixture (1330-20-7)	—	100 (1 year)	Health Canada (2018) [22]

Table 5.
Indoor air quality guidelines for selected VOCs [21].

Runström et al. observed maximum peaks of TVOC (total volatile organic compounds) of 3200 μg/m³, indicating in their study that levels potentially harmful to health were not detected, which also coincides with the studies carried out on camera by Floyd [13] and García-González [7].

Zhang et al. state that the levels of VOCs in 3D printing are lower than in laser printers [14], coinciding with Stefaniak et al., who assesses the TVOC levels printing ABS at values of 3500 μg/h, 131 μg/h for PLA and around 6000 μg/h for toner printers [28]. Azimi concludes that the levels of VOCs emitted vary between 2 and 180 μg/min [15]. Stefaniak also states that the VOC levels in both PLA and ABS filaments vary according to colour [28].

4. Other pollutants in 3D printers

It is normal to find monomers of the individual materials used in filaments of resins like, for example, styrene [29]. Other contaminants detected in 3D printing are bisphenol A (BPA), bisphenol S (BPS) (endocrine disruptors), aldehydes, CO2, CO, SO2, H2S and CS2 [30].

Ozone levels of 9 $\mu g/m^3$ were detected in closed 3D printers. The dyes for coloured filament can contain chromium, nickel, aluminium and other chemicals that could be emitted during printing, generating reactive oxygen species.

Ozone's interaction with unsaturated VOCs can change the chemistry of indoor air by creating reactive products such as carbonyl compounds and secondary organic aerosols (e.g., aldehydes and ketones) [28].

Printing with nylon material, concentrations of CO_2 have been determined, higher than the established limit, in addition to other compounds such as carbon monoxide, hydrocarbon, ammonia, caprolactam and hydrogen cyanide [27].

In powder metal printers, metals such as aluminium, chromium, nickel, cobalt, etc., have been detected [27] in concentrations that can become harmful.

Another critical issue is observed in the resin printer when the cleaning alcohol is mixed with the resin, generating nano plastics with server impact on the environment if they are not disposed of adequately [31].

Other ancillary chemicals are used in 3D printing; for example, a thin layer of hair-spray usually is applied to the bed surface to provide a tacky structure and increase the filament adhesion in the bed. Add the hairspray pollutants (VOCs, resins, bisphenol, etc.) for a correct risk assessment.

Filaments are evolving very quickly; new additives such as inorganic colourants, metal particles, nanomaterials, metal-containing flame retardants, antioxidants, heat stabilisers and catalysts, among others, are putting in the environment metals and other associated pollutants [32].

5. Health effects

The health effects of 3D printing are still not well studied because it is a quite recent technology, some of the effects could appear with decades of exposure, and there is not enough information about the 3D printer emissions and the worker's exposure; nevertheless, even with few years or days of exposure, there is some bibliography with health effects [33].

Chronic obstructive pulmonary disease (COPD) is a progressive disease that may start after exposure to vapours, gases, fumes and dust, produced by 3D printers, which, with continued exposure, can lead to decreased lung function [27].

A case of asthma was detected in a 28-year-old patient after only 10 days of exposure to pollutants emitted by the ABS filament with 10 printers in operation, observing an improvement after the change to PLA filament, reduction in the number of printers and the use of air purifiers. However, it still required the use of an inhaler [34]. Aldehydes and ketones, among other carbonyl substances, are linked to the emergence of asthma [28].

Caprolactam emitted during nylon printing can irritate the eyes and mucous membranes and cause nervous system alterations [27].

The nanoparticles can penetrate the bronchial tubes producing ROS (reactive oxygen species), being able to cause inflammation and DNA damage and even the ability to pass directly into the bloodstream, penetrating biological barriers. Aluminium nanoparticles can cause fibrosis and microcytic hypochromic anaemia. A study of 46 workers in 17 3D printing companies showed that 59% reported respiratory symptoms, 20% reported skin symptoms and 17% reported headaches at least once a week in the past year [35].

Clinical studies by Ljunggren et al. compared welders with metal 3D printing workers, obtaining very similar parameters [36].

Studies in rats exposed to 3D printing nanoparticles for 3 hours report increased blood pressure [37]; the pollutants emitted can cause cardiovascular disease and stroke [38].

Some metals emitted in 3D printing are irritants (aluminium, arsenic, copper, tin and zinc), asthmagens (chromium, manganese, nickel and vanadium) and potentially carcinogenic to humans (arsenic, nickel and vanadium) [30].

Lastly, cases of sarcomas have been reported in teachers using 3D printers for at least 2 years in poorly ventilated rooms, even though the relationship has not been proven yet [39].

6. Conclusions

There are several studies on pollutants emitted during 3D printing; however, different criteria and strategies are adopted, so an international standardised method must be defined so that the results of further studies will be comparable [40, 41].

3D printing can emit contaminants that are potentially dangerous to health, such as volatile organic compounds, nanoparticles and metals, so precautions must be taken before working with this technology.

The current legislation on the particulate matter is based on mass concentration ($\mu g/m^3$), establishing limit values for PM10, PM4 and PM2.5, the levels emitted by 3D printing are generally much lower than these values. However, despite not being legislated, the nanoparticle concentration levels can harm health.

Runström et al. did not find dangerous concentrations of contaminants during 3D printing; however, the 3D printing centres studied had preventive measures such as extractions and confined printers [11].

In order to minimise the health effects of 3D printing, it is recommended (among other measures) that it be confined with an extraction system, using high-efficiency HEPA filters in the printing room and a good ventilation system in place [42], and whenever possible, select the lowest possible print temperature.

Because some 3D printer models have a relatively low cost (less than 200 euros), it is recommended that before buying this equipment, consider places carefully to instal it. The selected locations should have good ventilation and engineering controls like localised extraction, avoiding, as far as possible, placing them in closed rooms with people present.

Use materials from trusted manufacturers. In recent years, PLA and other filaments with unique colours have been reported made (elsewhere) with chemicals prohibited in the European Union.

Lastly, it is strongly recommended to carry out VOCs and nanoparticle measurements and air quality surveys.

Acknowledgements

The Fundación PREVENT financed this research in the Fundacion Prevent 2021 R&D awards.

We thank the Instituto Nacional de Silicosis (INS) and Fundación Prevent for supporting the research and the HUCA library for all the reference material and information provided.

Conflict of interest

The authors declare no conflict of interest.

Author details

Hector Garcia Gonzalez* and Mª Teresa Lopez Pola
Instituto Nacional de Silicosis, Oviedo, Spain

*Address all correspondence to: hectorg@ins.es

References

[1] Kwon O, Yoon C, Ham S, Park J, Lee J, Yoo D, et al. Characterization and control of nanoparticle emission during 3D printing. Environmental Science & Technology. 2017;**51**(18):10357-10368

[2] Gu J, Wensing M, Uhde E, Salthammer T. Characterization of particulate and gaseous pollutants emitted during operation of a desktop 3D printer. Environment International. 2019;**123**:476-485

[3] Byrley P, George BJ, Boyes WK, Rogers K. Particle emissions from fused deposition modeling 3D printers: Evaluation and meta-analysis. Science of the Total Environment. 10 Mar 2019;**655**:395-407. DOI: 10.1016/j.scitotenv.2018.11.070. Epub 2018 Nov 12. PMID: 30471608; PMCID: PMC8350970

[4] European Commission. Joint Research Centre. The NanoDefine methods manual: 2020 [Internet]. LU: Publications Office; 2020 [cited 2022 Nov 10]. Available from: https://data.europa.eu/doi/10.2760/79490

[5] Garcia-Gonzalez H, Domat M, Lopez-Pola T, Fernandez-Rubio P, Fernandez-Rodriguez P. Particulate matter characterization in a hospital's underground car park. Powders. 2022;**1**(4):194-206

[6] van Broekhuizen P, van Veelen W, Streekstra WH, Schulte P, Reijnders L. Exposure limits for nanoparticles: Report of an international workshop on nano reference values. The Annals of Occupational Hygiene. 2012;**56**(5):515-524

[7] García H, Pola TL. Health and safety in 3D printing: Article. International Journal of Occupational and Environmental Safety. 2022;**6**(1):14-25

[8] Instituto Nacional de Seguridad e Higiene en el Trabajo (INSHT). Seguridad y salud en el trabajo con nanomateriales [Internet]. Madrid: Servicio de Ediciones y Publicaciones del INSHT; 2015 [cited 2020 Sep 2]. Available from: https://www.insst.es/documentacion/catalogo-de-publicaciones/seguridad-y-salud-en-el-trabajo-con-nanomateriales?redirect=https%253A%252F%252Fwww.insst.es%252Fresultados-de-busqueda%253Fp_p_id%253D3%2526p_p_lifecycle%253D0%2526p_p_state%253Dmaximized%2526p_p_mod%2520e%253Dview%2526_3_keywords%253D%252522seguridad%252By%252Bsalud%252Ben%252Bel%252Btrabajo%252Bcon%252Bnanomateriales%252522%2526_3_struts_action%253D%25252Fsearch%25252Fsearch&inheritRedirect=true

[9] Deutsche Gesetzliche Unfallversicherung DGU. IFA - Technical information: Ultrafine aerosols and nanoparticles at the workplace [Internet]. 2009 [cited 2022 Jun 22]. Available from: https://www.dguv.de/ifa/fachinfos/nanopartikel-am-arbeitsplatz/beurteilung-von-schutzmassnahmen/index-2.jsp

[10] Chan FL, Hon CY, Tarlo SM, Rajaram N, House R. Emissions and health risks from the use of 3D printers in an occupational setting. Journal of Toxicology and Environmental Health. Part A. 2020;**83**(7):279-287

[11] Runström Eden G, Tinnerberg H, Rosell L, Möller R, Almstrand AC, Bredberg A. Exploring methods for surveillance of occupational exposure from additive manufacturing in four different industrial facilities. Annals of Work Exposures and Health. 2022;**66**(2):163-177

[12] Jensen ACØ, Harboe H, Brostrøm A, Jensen KA, Fonseca AS. Nanoparticle exposure and workplace measurements during processes related to 3d printing of a metal object. Frontiers in Public Health [Internet]. 2020;**8**. [cited 2022 Feb 21]. Available from: https://www.researchgate.net/publication/345725356_Nanoparticle_Exposure_and_Workplace_Measurements_During_Processes_Related_to_3D_Printing_of_a_Metal_Object

[13] Floyd EL, Wang J, Regens JL. Fume emissions from a low-cost 3-D printer with various filaments. Journal of Occupational and Environmental Hygiene. 2017;**14**(7):523-533

[14] Zhang Q, Sharma G, Wong JPS, Davis AY, Black MS, Biswas P, et al. Investigating particle emissions and aerosol dynamics from a consumer fused deposition modeling 3D printer with a lognormal moment aerosol model. Aerosol Science and Technology. 2018;**52**(10):1099-1111

[15] Azimi P, Zhao D, Pouzet C, Crain NE, Stephens B. Emissions of ultrafine particles and volatile organic compounds from commercially available desktop three-dimensional printers with multiple filaments. Environmental Science & Technology. 2016;**50**(3):1260-1268

[16] Yi J, LeBouf RF, Duling MG, Nurkiewicz T, Chen BT, Schwegler-Berry D, et al. Emission of particulate matter from a desktop three-dimensional (3D) printer. Journal of Toxicology and Environmental Health. Part A. 2016;**79**(11):453-465

[17] US EPA O. Volatile Organic Compounds' Impact on Indoor Air Quality [Internet]. US EPA. 2014 [cited 2021 Apr 27]. Available from: https://www.epa.gov/indoor-air-quality-iaq/volatile-organic-compounds-impact-indoor-air-quality

[18] US EPA O. Technical Overview of Volatile Organic Compounds [Internet]. US EPA. 2014 [cited 2021 Apr 27]. Available from: https://www.epa.gov/indoor-air-quality-iaq/technical-overview-volatile-organic-compounds

[19] INSST. Límites de exposición profesional para agentes químicos 2022 - Portal INSST - INSST [Internet]. 2022 [cited 2022 Apr 26]. Available from: https://www.insst.es/documentacion/catalogo-de-publicaciones/limites-de-exposicion-profesional-para-agentes-quimicos-2022

[20] Mečiarová Ľ, Vilčeková S, Krídlová Burdová E, Kiselák J. Factors effecting the total volatile organic compound (TVOC) concentrations in slovak households. International Journal of Environmental Research and Public Health. 2017;**14**(12):1443

[21] Public Health England. Indoor Air Quality Guidelines for selected Volatile Organic Compounds (VOCs) in the UK. London: Public Health England; 2019. p. 9

[22] Health Canada - Canada.ca [Internet]. 2018 [cited 2022 Nov 28]. Available from: https://www.canada.ca/en/health-canada.html

[23] Trantallidi M, Dimitroulopoulou C, Wolkoff P, Kephalopoulos S, Carrer P. EPHECT III: Health risk assessment of exposure to household consumer products. Science of the Total Environment. 2015;**536**:903-913

[24] WHO World Health Organisation. WHO Guidelines for Indoor Air Quality [Internet]. Copenhagen: WHO Regional Office for Europe; 2010. Available from: https://www.euro.who.int/__data/assets/pdf_file/0009/128169/e94535.pdf

[25] ATSDR. Agency for Toxic Substances and Disease Registry [Internet]. 2005

[cited 2022 Nov 28]. Available from: https://www.atsdr.cdc.gov/

[26] Karwasz A, Osiński F, Łukaszewski K. Pollutants Emitted from 3D Printers onto Operators. Sustainability. 2022;**14**(3):1400

[27] Mohammadian Y, Nasirzadeh N. Toxicity risks of occupational exposure in 3D printing and bioprinting industries: A systematic review. Toxicology and Industrial Health. 2021;**37**(9):573-584

[28] Stefaniak AB, LeBouf RF, Yi J, Ham J, Nurkewicz T, Schwegler-Berry DE, et al. Characterization of chemical contaminants generated by a desktop fused deposition modeling 3-dimensional Printer. Journal of Occupational and Environmental Hygiene. 2017;**14**(7):540-550

[29] Dobrzyńska E, Kondej D, Kowalska J, Szewczyńska M. Exposure to chemical substances and particles emitted during additive manufacturing. Environmental Science and Pollution Research. 2022;**29**(26):40273-40278

[30] Stefaniak AB, Bowers LN, Martin SB Jr, Hammond DR, Ham JE, Wells JR, et al. Large-format additive manufacturing and machining using high-melt-temperature polymers. Part II: Characterization of particles and gases. ACS Chemical Health & Safety. 2021;**28**(4):268-278

[31] Rodríguez-Hernández AG, Chiodoni A, Bocchini S, Vazquez-Duhalt R. 3D printer waste, a new source of nanoplastic pollutants. Environmental Pollution. 2020;**267**:115609

[32] Tedla G, Jarabek AM, Byrley P, Boyes W, Rogers K. Human exposure to metals in consumer-focused fused filament fabrication (FFF)/3D printing processes. Science of the Total Environment. 2022;**814**:152622

[33] García González H, López Pola T. Contaminantes y efectos en la salud de la impresión 3d. In: Acercamiento Multidisciplinar a la salud: Implicaciones prácticas hacia el bienestar [Internet]. ASUNIVEP. Almería: ASUNIVEP; 2022. pp. 13-20. Available from: https://ciise.es/9/contenido/textos/descargar_libro/24

[34] House R, Rajaram N, Tarlo SM. Case report of asthma associated with 3D printing. Occupational Medicine (London). 2017;**67**(8):652-654

[35] Chan FL, House R, Kudla I, Lipszyc JC, Rajaram N, Tarlo SM. Health survey of employees regularly using 3D printers. Occupational Medicine. 2018;**68**(3):211-214

[36] Ljunggren SA, Ward LJ, Graff P, Persson A, Lind ML, Karlsson H. Metal additive manufacturing and possible clinical markers for the monitoring of exposure-related health effects. PLoS One. 2021;**16**(3):e0248601

[37] Stefaniak AB, LeBouf RF, Duling MG, Yi J, Abukabda AB, McBride CR, et al. Inhalation exposure to three-dimensional printer emissions stimulates acute hypertension and microvascular dysfunction. Toxicology and Applied Pharmacology. 2017;**335**:1-5

[38] Karwasz A, Osiński F. Analysis of emission solid particles from the 3D printing process. In: Gapiński B, Ciszak O, Ivanov V, editors. Advances in Manufacturing III. Cham: Springer International Publishing; 2022. pp. 216-226 (Lecture Notes in Mechanical Engineering)

[39] Joo MW, Lee YS, Chung YG, Lee HK. Sarcomas in teachers using three-dimensional printers: A report of three patients and literature review. Clinics in Orthopedic Surgery. 2022;**14**(2):310-317

[40] Leso V, Ercolano ML, Mazzotta I, Romano M, Cannavacciuolo F, Iavicoli I. Three-dimensional (3D) printing: Implications for risk assessment and management in occupational settings. Annals of Work Exposures and Health. 2021;**65**(6):617-634

[41] Tang CL, Seeger S. Systematic ranking of filaments regarding their particulate emissions during fused filament fabrication 3D printing by means of a proposed standard test method. Indoor Air. 2022;**32**(3):e13010

[42] Yeom S, Kim H, Hong T, Jeong K. Analysis of ways to reduce potential health risk from ultrafine and fine particles emitted from 3D printers in the makerspace. Indoor Air. 2022;**32**(5):e13053

Section 5

3D Printing in Industrial and Digital Manufacturing

Chapter 17

Plasma Nitriding-Assisted 3D Printing for Die Technology in Digital Micro-Manufacturing

Tatsuhiko Aizawa, Tomomi Shiratori and Yohei Suzuki

Abstract

A plasma nitriding-assisted 3D printing method was developed to build up the micro-punch and micro-die systems. Two dimensional punch head and core-die cavity geometries were ink-jet printed or screen-printed onto the AISI316 and SKD11 tool substrate surfaces in following their two-dimensional computer-aided design (CAD) data. The low-temperature plasma nitriding process was utilized to make nitrogen supersaturation only into the unprinted substrates. The sand-blasting and chemical etching were utilized to mechanically or chemically remove the printed parts from punch and die substrate. As sand-blasted and chemically etched AISI316 and SKD11 punches and core-dies were simply finished and used as a die set for micro-embossing, micro-piercing and micro-punching processes. In particular, a micro-pump was selected as a miniature mechanical element. Its 3D CAD geometry was sliced to 2D CAD data for each functional AISI304 stainless steel sheet. A pair of punch and die for each 2D CAD geometry for constituent sheet was prepared by the plasma nitriding-assisted 3D printing. Each sheet was punched out by using this set of punch and die to functionalize each sheet unit in correspondence to the sliced CAD data. These constituent sheets were assembled and joined to a structural unit of micro-pump.

Keywords: micro-parts, micro-tools, 3D plasma printing, 2D screen printing, micro-embossing, micro-piercing, surface activation, low-temperature joining, micro-pump

1. Introduction

The additive manufacturing with the use of three-dimensional (3D) printing method has grown up as a fundamental scheme to fabricate the polymer, the ceramic, and the metal products with complex geometry [1]. Its feedstock is usually a powder or a particle with the specially designed alloying elements [2]. These powder materials are melt by application of high power supply via the lasers [3], the electron beams [4], and the plasma torch [5]. Since these materials are melt, solidified, and sintered pointwise within the power hot spot, the resolution in geometry is determined by the scanning in power supply and the hot spot size [6]. The total configuration of product in geometry is roughly built up by solidification and sintering processes in these 3D printing; the finishing step is still necessary to satisfy the design requirement [7].

Without the high-resolution techniques and material selection, a miniature product with strict requirement in dimension is difficult to be yielded by 3D printing, since its resolution is much lower than the product dimensional requirement. It is also unsuitable to mass production of parts and devices with complex geometry, e. g. micro-element for micro-electric mechanical system (MEMS), micro-parts and device for medical applications, mm-sized punch and die, and miniature mechanical parts with high-quality proof.

Let us change the normal feedstock for additive manufacturing to the metal and plastic foils, films, sheets, and plates. Most of them have been utilized as a feedstock of metal forming and plastic product processing with sufficient proof of strength and ductility. In addition, relatively wider material selection is available in this new additive manufacturing with the use of them. **Figure 1** illustrates the candidate scheme of additive micro-manufacturing to build up a micro-pump [8]. The three-dimensional CAD model of the micro-pump is sliced into a constituent two-dimensional model with each mechanical function. Except for the PZT unit to drive the micro-pump, its main unit is divided into an assembly of a grid, a shield, the reservoir plate, two diaphragm plates and valve plates, and an end plate. The metallic sheets and plates are prepared as a starting material with the suitable thickness to each constituent part of micro-pump unit. If the punch and die sets are simply prepared with sufficient accuracy in dimensions, each part is yielded even in mass production by blanking, punching, embossing, and piercing these plates as a feedstock.

In the normal die technology [9], the grinding, polishing, and machining processes are utilized to fabricate the punches and dies. In micro-manufacturing of punches and dies, the sub-mm and μm-ordered milling tools must be prepared to shape the miniature die substrates. In addition, lots of duration for preparation of Computer-Aided Machining (CAD) data and actual operations is needed only to fabricate a single punch even with the microtextures in 10 μm to 100 μm orders. When using the austenitic stainless steels or tool steels as a die substrate, the surface treatment is also necessary to strengthen and harden the punch under the severe dry stamping and

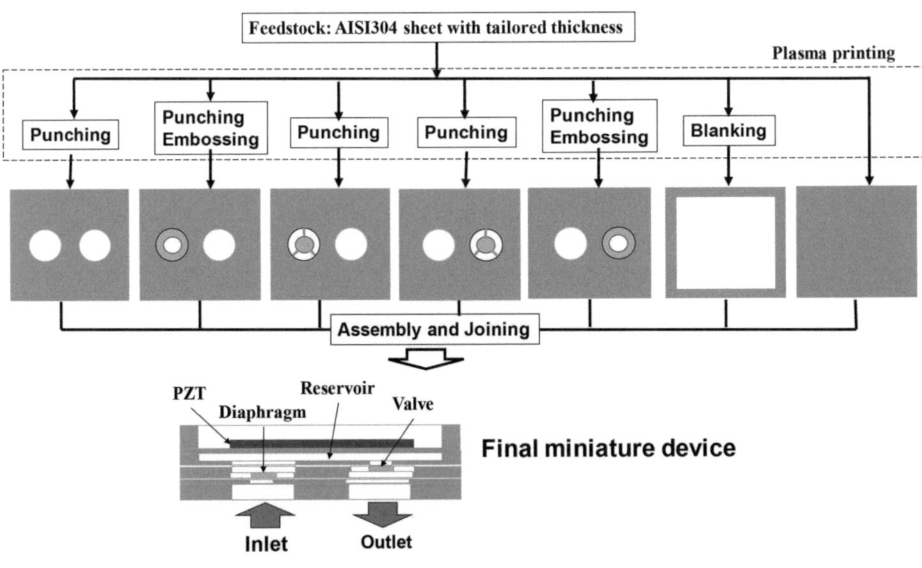

Figure 1.
Advanced additive micro-manufacturing from the foils, films, sheets, and plates as a feed stock to fabricate a metallic miniature product.

forging conditions. Hence, it is difficult or nearly impossible to prepare a set of punch and die for blanking, punching, and embossing processes to yield the constituent sheets in correspondence to the sliced data of product CAD data.

The plasma nitriding-assisted 3D printing [10] provides a solution to the above issue of difficulty in the die technology. The positive micro-pattern to the sliced 2D product model is printed onto the die surface. Its unprinted parts are selectively hardened by plasma processing to transform the unprinted parts to the significantly hardened ones in the die substrate. The mechanical and chemical processes are utilized to selectively remove the printed and unhardened parts from the substrate. The punch and die are directly built up to have the designed punch head and die cavity for embossing, piercing, blanking, and punching the feedstock sheets and plates to each constituent element in **Figure 1**.

In the present chapter, this new additive micro-manufacturing is proposed to fabricate the mm−/sub-mm-ordered mechanical sheet parts with the sliced 2D CAD data of 3D product model by embossing, piercing, and punching processes and to assemble and join them to a miniature mechanical unit and device. The plasma nitriding-assisted 3D printing method is utilized as the first step to fabricate the punch and die pair with its complex head and core geometry from the sliced 2D CAD of product model. This approach has three merits as a die technology: 1) high surface hardness for long tool life in stamping and forging, 2) flexible and short-time response to 2D CAD, and 3) dimensional accuracy in geometry enough to be used in fabrication of products. Since the austenitic stainless steel and titanium sheets are selected as a feedstock, this high hardness leads to prolongation of punch and die lives in practical stamping and forging operations. The tailored complex geometries directly reflect on the punch head and core-die cavity shapes in much shorter duration than needed in the conventional die technology. Owing to the autonomous dimensional accuracy in the plasma printing, the 3D-printed punch and die is used after a bit duration in the finishing step.

First, the plasma nitriding-assisted 3D printing procedure is explained in the procedure from the CAD of cross-meshing pattern to build up the hardened punch with meshing-textured head. In second, the copper substrate for the plastic packaging of semiconductors is precisely stamped to have microgroove loops by using the micro-textured punch after optimum geometry design to minimize the thermal stresses under the thermal loading in usage of the mold packages. Owing to this optimized microgrooves, the leak proof is certified in 100%. AISI304 sheets are pierced to have a micro-valve unit by using the plasma-printed punch and die. A constituent sheet is yielded when starting from the sliced 2D geometry of the micro-pump model. An electrical steel sheet is also blanked to have a T-shaped motor core unit by using the simultaneously plasma-printed punch and die. High burnished surface area ratio is attained even in the sheared surface by using the as-blasted punch and die. Finally, a micro-pump unit is produced in trial after the scheme in **Figure 1**. The pierced AISI304 constituent sheets by the plasma-printed punch and die are assembled and joined to a micro-pump unit.

2. Plasma nitriding-assisted 3D printing

2.1 Two-dimensional patterning and texturing

This 3D printing starts from drawing the two-dimensional model after CAD data onto the tool surfaces as shown in **Figure 2a**. This printed pattern works as a mask to

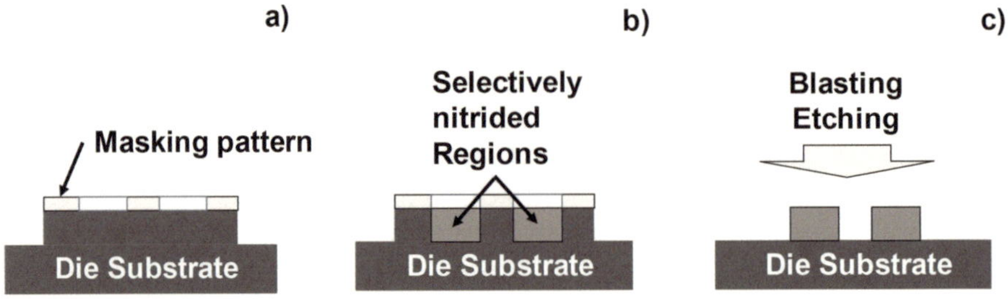

Figure 2.
A normal plasma nitriding-assisted 3D printing procedure. a) Patterning the 2D model of CAD data onto the die surfaces, b) plasma nitriding the unmasked die, and c) removing mechanically, or, chemically etching the printed parts to leave the 3D-shaped tools.

prevent the printed tool surface from plasma nitriding as shown in **Figure 2b**. The unprinted parts of tool are selectively nitrogen-supersaturated to their specified depth. Due to their high nitrogen contents, they have selectively much higher hardness and corrosion toughness than the printed tool parts. Through mechanical removal and chemical etching in **Figure 2c**, the printed parts are cut off to leave the designed three-dimensional tool and product configurations in adaptive to the original 2D surface by CAD.

2.2 Two-dimensional printing of CAD data onto tools

Several methods are available to draw 2D CAD data onto the tool surfaces, e.g. masking [11], inkjet printing [12], screen printing [13], lithography [14], and short-pulse laser machining [15]. As shown in **Figure 3a**, the masking plate is first prepared and fixed onto the tool surface. No actual printing process is necessary; its spatial resolution is predetermined by the dimensional accuracy of masking plate. The inkjet printing is feasible to draw 2D CAD data onto the flat and curved tool surfaces as depicted in **Figure 3b**. A wider tool surface can be printed. Its resolution is also limited

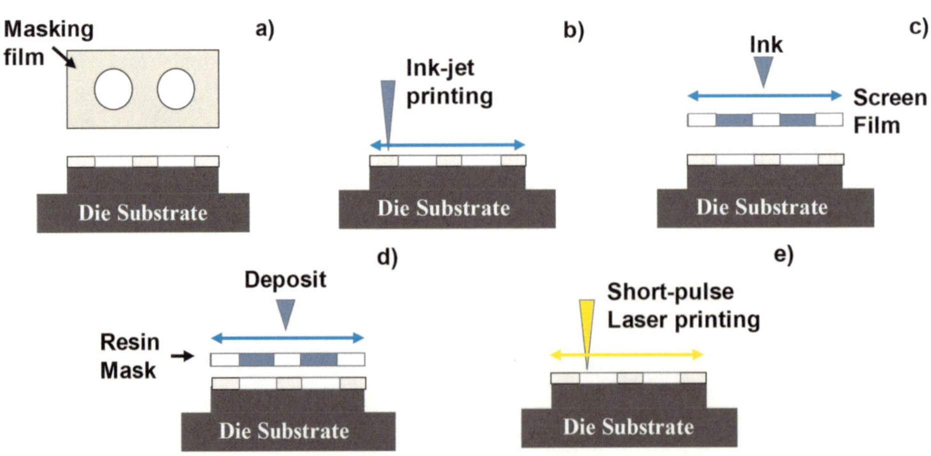

Figure 3.
Five methods to pattern the 2D model of tools onto the die substrate surface. a) Masking, b) inkjet printing, c) screen printing, d) lithography, and e) laser printing.

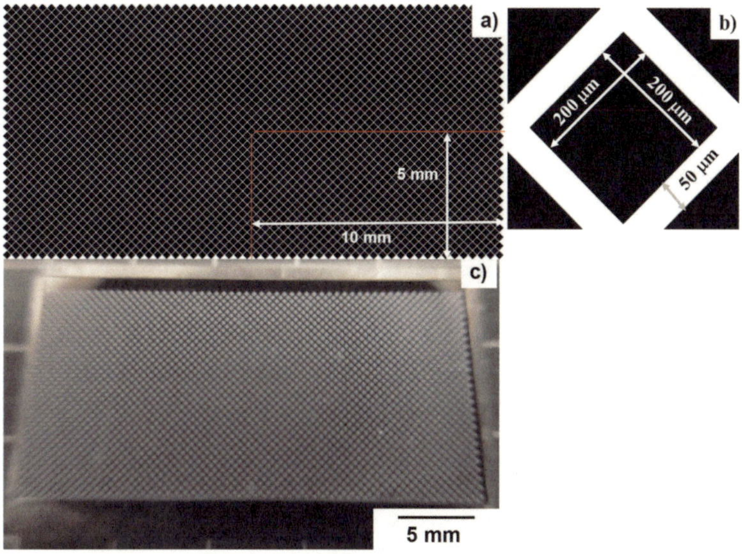

Figure 4.
Screen printing of the meshing line pattern onto AISI316 die. a) Screen film with meshing pattern, b) its unit cell, and c) printed micro-pattern onto the die surface.

by the inner diameter of nozzle. The viscosity of ink influences on the accuracy in dispensing.

The positive pattern to original CAD data is formed onto the screen as shown in **Figure 3c**. In this screen printing, its resolution is determined by the meshing technique in the preparation of screen films. The lithography is a powerful tool to draw any patterns with higher resolution in dimension as depicted in **Figure 3d**. The trade-balancing must be taken into account between the dimensional accuracy in CAD and the cost in usage of lithography system. The short-pulse laser machining also provides a method to make fine drawing onto the thin sacrifice film on the tool surface as depicted in **Figure 3e**. Its spatial resolution is functional in the laser frequency, the laser pulse control, and the film thickness. The original CAD data are directly transformed into CAM data for laser machining to print the complex geometry in short duration.

Let us use the screen printing method to draw a micro-pattern onto AISI316 die surface with the area of 10 mm x 20 mm [16, 17]. A screen film was prepared in corresponding to the meshing pattern in 2D CAD, as depicted in **Figure 4a**. The ink penetrates through the black squares to leave the meshing lines unprinted. Its unit cell is depicted in **Figure 4b**; the line width is 50 µm and its pitch is 250 µm. **Figure 4c** shows the square pattern, printed onto the AISI316 die surface.

2.3 Nitrogen supersaturation into unprinted parts

The plasma nitriding has been widely utilized as a typical surface treatment of various steels and high chromium alloys as surveyed in [18]. Most of the commercial nitriding processes, so-called by the ion nitriding (DC nitriding) and the radical nitriding (DC pulse nitriding), were utilized to harden the steel die materials by fine nitride precipitate formation in the nitrided layer [19]. After [20], the surface hardness increased above 1000 HV in the DC-plasma-nitrided Fe-19Cr alloy at 773 K for

57.6 ks. This inner nitriding process at high holding temperature is governed by the nitrogen diffusion process as analyzed in [20–22]. Hence, high holding temperature is necessary to sustain the nitrogen diffusion and synthesis of chromium nitride (CrN) by Cr + N ➔ CrN during the plasma nitriding. On the other hand, the stainless steels as well as tool steels were nitrogen-supersaturated by the plasma nitriding at 673 K without nitride precipitation reactions [23–25]. The most different features from the high-temperature plasma nitriding processes in the above are as follows: 1) surface hardness increases over 2000 HV, 2) microstructure in the nitrided layer is refined to have very fine two-phase granular structure, 3) nitrogen solute distributes uniformly and homogeneously in the nitrided layer with the average content of 4 mass%, and 4) nitrogen supersaturation accompanies with the significant lattice expansion. In addition, the nitrided layer thickness turns to be more than 50 μm after plasma nitriding at 673 K for 14.4 ks. This inner nitriding process at low holding temperature is not only governed by the nitrogen diffusion process but also controlled by the plastic straining. After [26, 27], the mismatched strains between the un-nitrided and nitrided zones are induced by the lattice expansion in the nitrogen supersaturation process. Most of nitrogen solute diffuse through the plastically strained zone boundaries so that the high nitrogen content is preserved in all the nitrided layer.

When using the printed pattern as a mask in this low-temperature plasma nitriding, the unprinted parts are only nitrogen-supersaturated to have higher hardness than 1400 HV and more nitrogen solute content than 4 mass% in average. Let us prove this selective nitrogen supersaturation process by plasma nitriding the printed AISI316 die in **Figure 4** at 673 K for 14.4 ks. As shown in **Figure 5a**, the square unit cell printed on the die remains the same as shown in **Figure 4b**, although the meshing line edges and corners became dull. **Figure 5b** depicts the nitrogen solute distribution by the SEM (scanning electron microscopy)–EDX (electron-dispersive X-ray spectroscopy) analysis. The nitrogen solute is only present in the unprinted regions but not in the printed regions. This proves that unprinted regions are selectively nitrogen-supersaturated.

2.4 Mechanical removal and chemical etching

The nitrided or unmasked regions have much higher hardness than the hardness of shooting media in the dry sandblasting. In addition, their corrosion toughness is much improved never to be in pitting corrosion by using the normal etchant. This difference in hardness and corrosion toughness between the nitrided and masked regions drives the selective removal of masked regions from the die substrate by sandblasting or by chemical etching processes [28, 29].

The nitrided AISI316 die in **Figure 5** was subjected to sandblasting. The silica particles with the diameter of 30 μm and the hardness of 700 HV were used as a shooting medium for sandblasting. **Figure 6** depicts SEM image on the AISI316 die surface after slightly sandblasting for 300 s. The masked square regions in **Figures 4c** and 5 were selectively dug; the nitrided regions became a mesh-textured punch head. To be noticed, the original 2D boundaries between the nitrided and masked regions changed to the side surfaces of mesh-textured punch. This proves that nitrogen supersaturation and diffusion processes advances in straight way from the surface to the depth of die substrate to form the selectively hardened zones in correspondence to the mesh textures.

Figure 7 shows an overview on the sandblasted AISI316 die. The whole die surface is modified to a mesh-textured punch head. This proves that mechanical removal of masked region advanced homogeneously so that the mesh-textured punch head has a uniform height with lower maximum surface roughness than 0.6 μm [30].

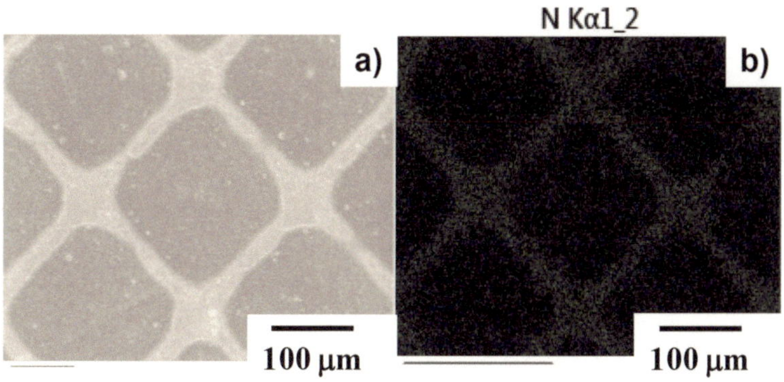

Figure 5.
Meshing pattern-printed on the AISI316 die surface. a) SEM image on the meshing pattern, and b) nitrogen solute distribution on the AISI316 die surface.

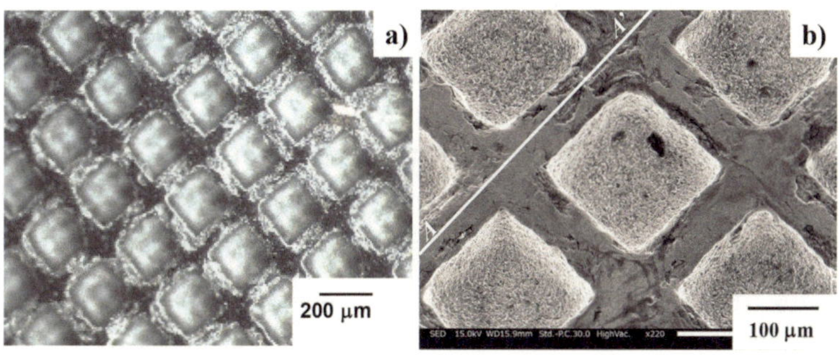

Figure 6.
AISI316 die surface after sandblasting. a) SEM image in low magnification, and b) SEM image on high magnification.

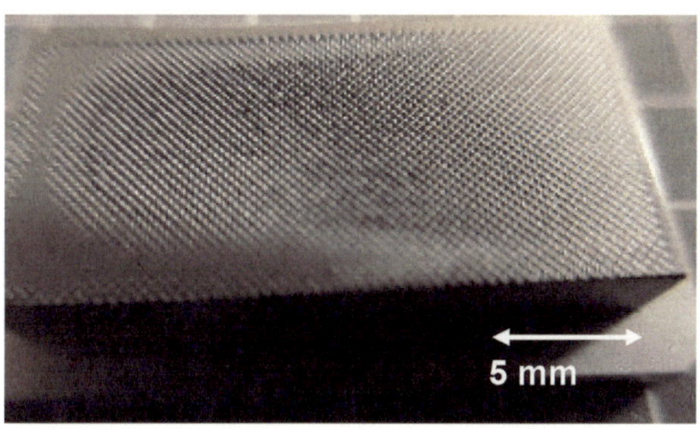

Figure 7.
AISI316 die with the mesh-textured heads.

2.5 Dimensional measurement and characterization

Three-dimensional surface profiling system was utilized to measure the cross-sectional profile of mesh-textured die surface along the path in **Figure 8a**. As shown in **Figure 8b**, the clearance between adjacent nitrided pillars was uniformly formed to have constant depth and width in 80 μm and 250 μm, respectively. Each pillar has the same height and width of 80 μm and 40 μm, respectively.

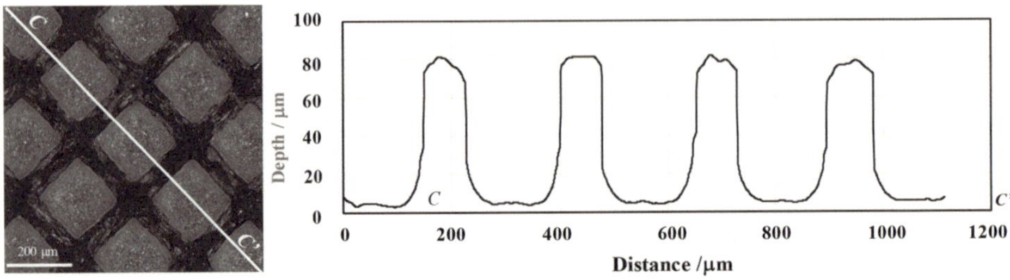

Figure 8.
Three-dimensional die surface profile. a) SEM image on the measurement path, and b) cross-sectional view of mesh-textured die surface.

3. 3D printing of complex-shaped embossing punch for electronic packaging

The normal 3D printing has little means to deal with the multi-material product except for alloyed powders. In the semiconductor packaging, especially in the high-frequency GaN multi-chip packaging, the copper substrate must be strictly joined to a packaging plastic mold with sufficient robustness and tightness against the thermal transient loading [31–33]. The microgroove array to be dug into the copper substrate is first designed to reduce the thermal stresses in the substrate under the joined state to plastic mold. The tailored microgroove design is transformed into the embossing punch by using the 3D printing procedure in second. Fine coining system is utilized to form the microgroove array by embossing the 3D-printed punch. This copper substrate is mechanically joined with the plastic mold for packaging. These packages are subjected to thermal transient loading test for tightness proof in the actual thermal transients.

3.1 Topological design on the microgroove array in the mold packaging

The high-power and high-frequency semiconductor unit with alignment of several gallium nitride (GaN) chips is packaged by joining the copper substrate with the plastic mold. As illustrated in **Figure 9**, both are mechanically joined through the microgrooves in the inside of copper substrate. Considering that thermal stresses are induced into the substrate during the thermal transient loading in operation, the topology of microgrooves must be optimized to reduce the thermal stress level in the inside and outside of joined area.

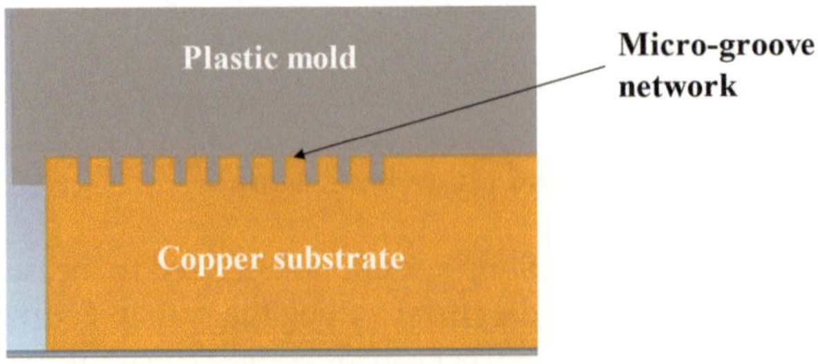

Figure 9.
A schematic view on the cross section of copper-plastic mold package.

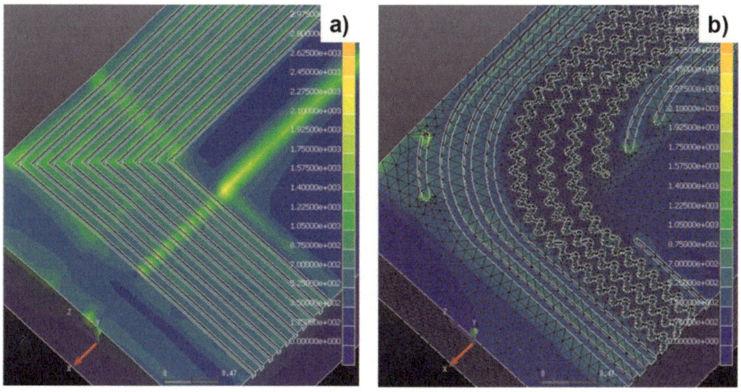

Figure 10.
Comparison of the thermal stress distribution between the normal microgrooved package and the optimally tailored package.

When using the simple linear loops of microgrooves, the thermal stress is enhanced to be high even in the inside and outside of joined interface to make distortion of chips and gold pins. This simple topology in **Figure 10a** is redesigned to a new configuration with continuous and discontinuous loops. As shown in **Figure 10b**, no significant stresses are observed both in the outside and inside of jointed interface. The continuous loops prevent the inside area from high thermal stress state; the discontinuous loops at the four corners of substrate preserve the low stress state even at the edges of substrate.

3.2 3D printing procedure

This tailored topology of microgrooves is cut in by using the plasma nitriding-assisted 3D printing procedure. After the general scheme in **Figure 2**, the screen film was prepared to print the two-dimensional CAD data of microgrooves onto the AISI316 die. **Figure 11a** depicts the 1/4 corner of screen-printed CAD data on the die. Two discontinuous loops (A and B) are placed at each corner of die in addition to three families of continuous loops {(C, D), (E, F, G), (H, K)} from the outside to the inside of die. As explained in the session 2, the black area in **Figure 11a** corresponds to a screen-printed mask and the white lines are left as an unprinted area. **Figure 11b** shows the SEM image on the die surface just after nitriding. As already demonstrated in **Figure 5**, the nitrogen solutes are supersaturated only into these discontinous lines

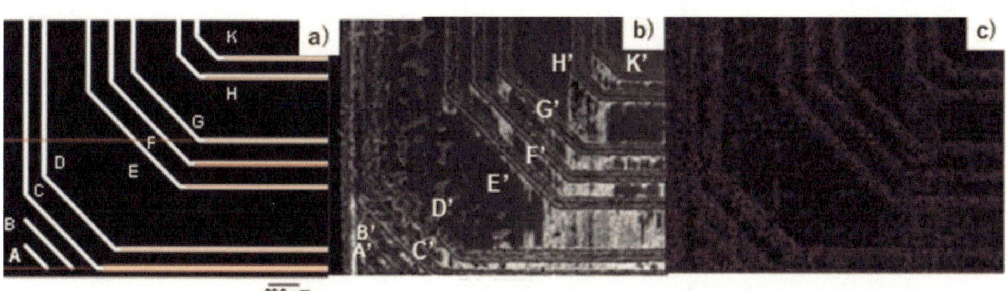

Figure 11.
Plasma nitriding-assisted 3D printing from CAD data on the optimum microgroove topology to the 3D-printed embossing punch. a) Original topology design on the screen film, b) SEM image on the AISI316 die after nitriding, and c) nitrogen element mapping on the die surface.

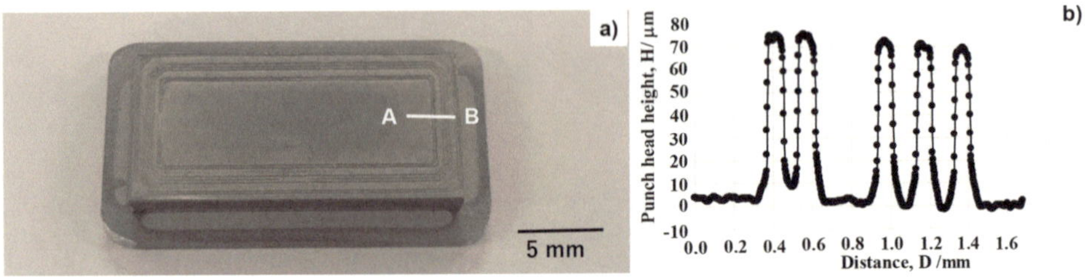

Figure 12.
A plasma-printed AISI316 punch for embossing. a) Overview of the embossing punch, and b) its cross-sectional profile along A–B.

(A', B′) and three continuous loops {(C′, D′), (E', F′, G'), (H′, K′)} in **Figure 11c**. This also proves that the original tailored topology design is transformed to the nitrogen-supersaurated lines.

3.3 Fabrication of embossing punch with the tailored topology

The sandblasting was utilized to mechanically remove the printed regions from the die substrate in correspondence to **Figures 6–8** in the Section 2. **Figure 12a** depicts the AISI316 die after sandblasting for 300 s. The discontinuous loop heads and three families of continuous loop heads are automatically formed by the 3D printing as depicted in **Figure 12a**. Let us investigate the height profile of continuous loop families along A–B in **Figure 12b**. The bottom line of punch heads is common among them; the printed regions of die is uniformly removed by the sandblasting. The punch heads have homogeneously the height of 70 μm and the width of 60 μm, respectively.

3.4 Fine coining into copper substrate using the 3D-printed punch

The multi-head punch in **Figure 12a** was utilized for micro-embossing of the copper substrate plate with the size of 22 mm x 12 mm x 1 t mm. The CNC (computer numerical control) stamping system (ZEN; Hoden-Seimitsu, Yokohama, Japan) was used for this precise embossing. **Figure 13a** depicts the embossed copper plate with microgrooves. As compared between **Figures 12b** and **13b**, this microgroove texture is just corresponding to the multi-head pattern in the punch. The average microgroove depth is 50 μm and its width is 60 μm. Since its profile is transcribed by embossing the multi-head configuration in **Figure 12b**, its side walls are steeply formed against the substrate surface. This steep microgrooving configuration has influence on the robustness in mechanical joining between the plastic mold and the copper substrate.

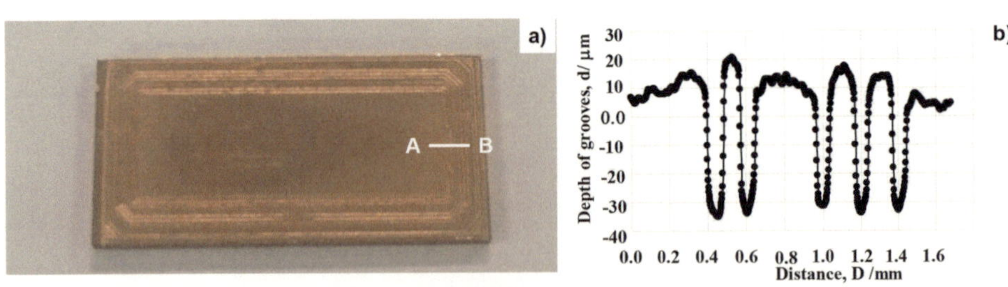

Figure 13.
A pure copper plate embossed by using the plasma-printed punch. a) Overview of the embossed copper plate, and b) its cross-sectional profile along A–B.

Figure 14.
A plastic mold package with joining the copper substrate plate through the microgrooves.

3.5 Evaluation on the mechanical integrity of joined package

Twelve package specimens were prepared to evaluate their mechanical integrity by thermal transient loading and reflow testing. This testing procedure consists of three steps: 1) baking test at 398 K (or 125°C) for 24 h, 2) humidity test at 273 K (or 0°C) at 60% RH for 192 h, and 3) reflow testing at 533 K (or 260°C) for three cycles. A typical package specimen was depicted in **Figure 14**. The terminals were also housed into the specimen.

No leaks were noticed among 12 specimens; this demonstrated that perfect tightness can be accommodated to this packaging by using the 3D-printed microgrooves onto the copper substrate. Let us evaluate on the cost-competitiveness to the present manufacturing. In the normal laser processing, the CAM (computer-aided manufacturing) data must be built up before laser machining; this time reaches 36 ks for calculation of paths, editing the data and checking them. Excluding the setup time for laser path control, the actual processing time is accounted to be 150 s per each machining step. In the present procedure, the screen printing time to die surface is 300 s, the plasma nitriding processing time including the heating and cooling durations is 18.0 ks, and the sandblasting time is 300 s. The coining process by stamping requires 10 s per a copper substrate. The total setup time for laser machining is reduced in half by the present 3D printing. The actual laser machining time is shortened by 1/15. Consider that N copper substrates are processed to compare the real processing duration between two. In the normal laser machining, the duration is estimated by 150 x N s. On the other hand, the punch in **Figure 12a** is prepared to have N modules without excessive increase in the same manner as shown in the above. The total duration is nearly the same as 10 s. That is, the cost-competitiveness to the present procedure for multi-substrate fabrication is enhanced by 15 x N.

4. 3D printing of complex-shaped punch for piercing

The micro-parts and micro-tools must have functional geometries in their inside [34]. **Figure 15** depicts the CAD data of micro-valve to push up the liquid matter by the circular pendulum with three U-shaped springs. In particular, this U-shaped spring has accurate dimensions to satisfy the tailored elastic stiffness. The plasma nitriding-assisted 3D printing was utilized to build up the complex-shaped punch to

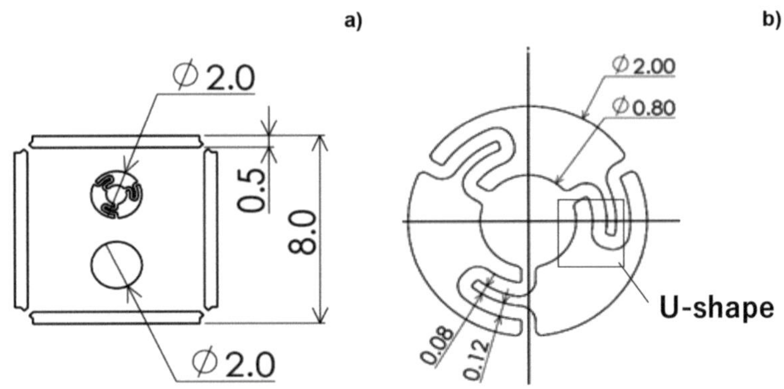

Figure 15.
A micro-valve design as a unit functional sheet of micro-pump. a) AISI304 sheet with a through-hole and a valve section, and b) detail structure of micro-valve unit.

yield this valve sheet [35, 36]. This piercing punch was shaped after the general procedure in the Section 2. The core-die was directly shaved by using the hardened punch head into the die substrate. The mechanical finishing step was utilized to build up the narrow clearance between the 3D-printed punch and its shaved core-die. This pair of 3D-printed punch and shaved core-die was utilized for piercing the AISI304 sheet with the thickness of 0.05 mm to accommodate the micro-valve into the sheet in corresponding to CAD data as shown in **Figure 15** [37].

4.1 Plasma printing of piercing punch

The plasma nitriding-assisted 3D printing procedure to fabricate the piercing punch was illustrated in **Figure 16**. The inkjet printing was utilized to draw the CAD data of micro-valve geometry onto the AISI316 punch surface as shown in **Figure 16a**. This punch was plasma-nitrided to selectively nitride the unprinted punch surfaces as shown in **Figure 16b**. The nitrided punch was sandblasted to remove the printed parts mechanically and to make near-net shaping of punch head as shown in **Figure 16c**. After the schematic steps in **Figure 16**, the inkjet printer (Mimaki, Tokyo, Japan) was utilized to draw the two-dimensional pattern onto the punch surface as depicted in **Figure 17a** in correspondence to **Figure 16a**. This printed punch was subjected to the plasma nitriding process to harden selectively the unprinted surfaces as shown in **Figure 17b**, corresponding to **Figure 16b**. The sandblasting was further employed to remove mechanically the soft parts. **Figure 17c** reveals that the complex-shaped punch heads are formed in fairly good reproduction of CAD data as shown in **Figure 16c**.

4.2 Fine piercing of AISI304 sheets

The plasma-printed punch and the shaved core-die were utilized to punch out the AISI304 stainless steel sheet with the thickness of 50 μm. CNC stamping system (Komatsu-Seiki-Kosakusho, Nagano, Japan) was employed in this punching process. **Figure 18** depicts the SEM image on the pierced micro-valve unit into the AISI304 stainless steel sheet. Comparing this with the CAD data in **Figure 16a**, the micro-valve unit is fabricated in exclusively negative to the CAD data for punch head in

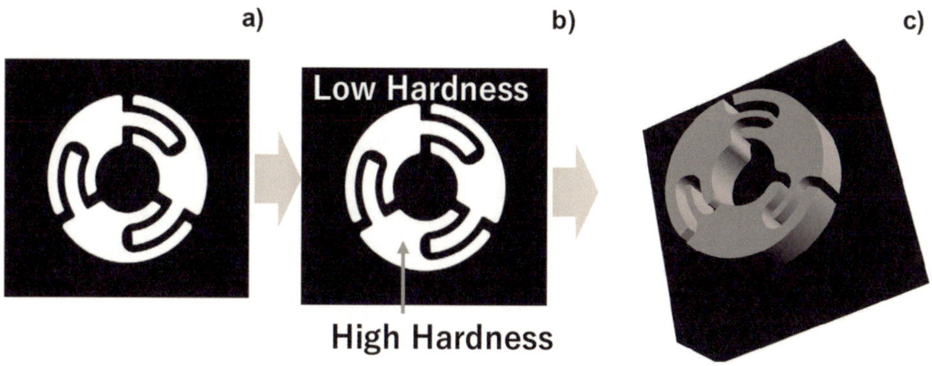

Figure 16.
Plasma nitriding-assisted 3D printing scheme to fabricate the piercing punch. a) Inkjet printing step onto the punch top surface, b) low-temperature plasma nitriding step, and c) sandblasting step.

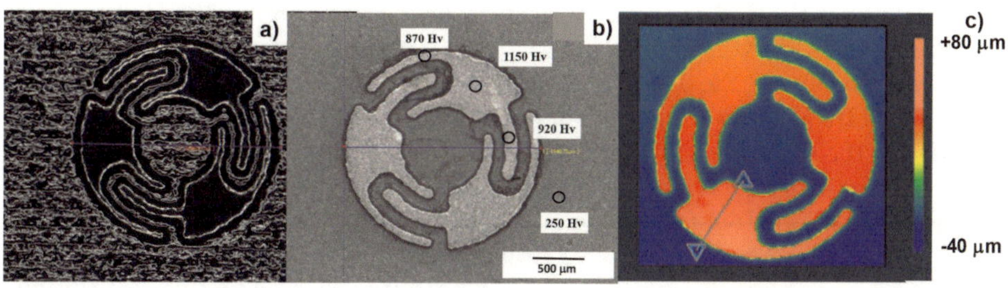

Figure 17.
Plasma nitriding-assisted 3D printing procedure to fabricate the piercing punch. a) Inkjet printing, b) plasma nitriding, and c) sandblasting.

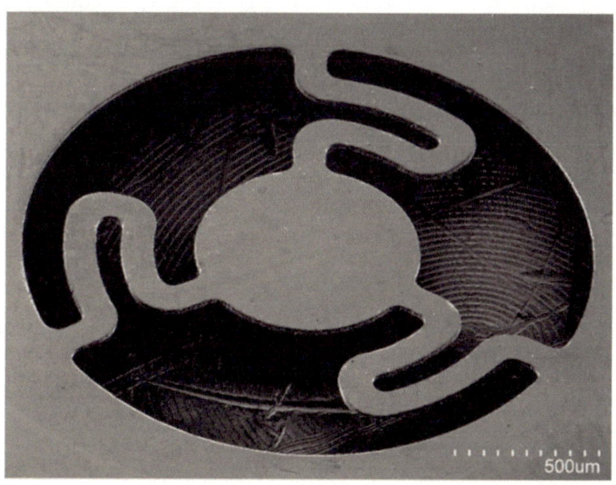

Figure 18.
AISI304 stainless steel sheet with a micro-valve.

Figure 16a. This assures the dimensional accuracy of this plasma nitriding-assisted 3D printing to accommodate the functional units into the metallic sheets or films. To be noticed, a set of functional units can be punched out by using this 3D-printed punch with the shaved die.

5. 3D printing of complex-shaped dies for punching

Let us develop the other way to simultaneously fabricate a pair of punch and core-die for precise punching. Various mechanical parts have two-dimensional product shape such as the spur gears and wheels, the metallic belts, and the motor cores. In particular, a motor core consists of complex-shaped electrical steel sheets as a two-dimensional product. Then, each sheet has to be accurately punched out by stamping or forging [38]. The final motor core shape is controllable by joining or laminating the same constituent blanked sheets [39]. The plasma nitriding-assisted 3D printing is advanced to simultaneously fabricate the punch and core-die to punch out the fine blanked electrical steel sheets with sufficient accuracy in dimension [40].

5.1 Simultaneous plasma printing of punch and die

Every step starts from the same CAD data of product shape. In the present section, the CAD data for the T-shaped electrical steel sheet unit in **Figure 19a** is common to the whole steps in the plasma nitriding-assisted 3D printing process. In the procedure to fabricate the punch, the negative pattern to these original data was screen-printed onto the AISI316 substrate. On the other hand, the positive pattern was also screen-printed onto the substrate for fabrication of a core-die, as depicted in **Figure 19b**. In the subsequent plasma nitriding and sandblasting steps, both printed substrates were subjected to the same treatment. The T-unprinted part is selectively nitrided, and other parts are automatically removed from die substrate as depicted in **Figure 19c** and **d**. The piercing punch with the T-shaped head is fabricated by this 3D printing. While, other die substrates than T-printed part is selectively nitrided so that T-shaped part is automatically removed from die substrate also in **Figure 19c** and **d**. Then, the piercing core-die with the T-shaped cavity is fabricated by this 3D printing. To be discussed later, the clearance between the 3D-printed punch and core-die is determined by the dimensional tolerance in CAD data.

5.2 Plasma nitriding-assisted 3D printing of AISI316 punch and core-die

After the schematic procedure in **Figure 19**, AISI316 punch and core-die are simultaneously fabricated to demonstrate that a pair of special tools for fine stamping and forging is automatically yielded from the original CAD data.

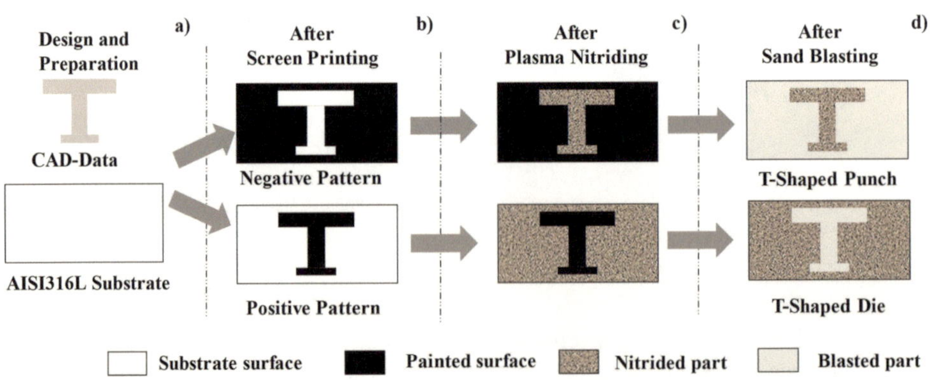

Figure 19.
Simultaneous plasma nitriding-assisted 3D printing procedure to fabricate the piercing punch and core-die.

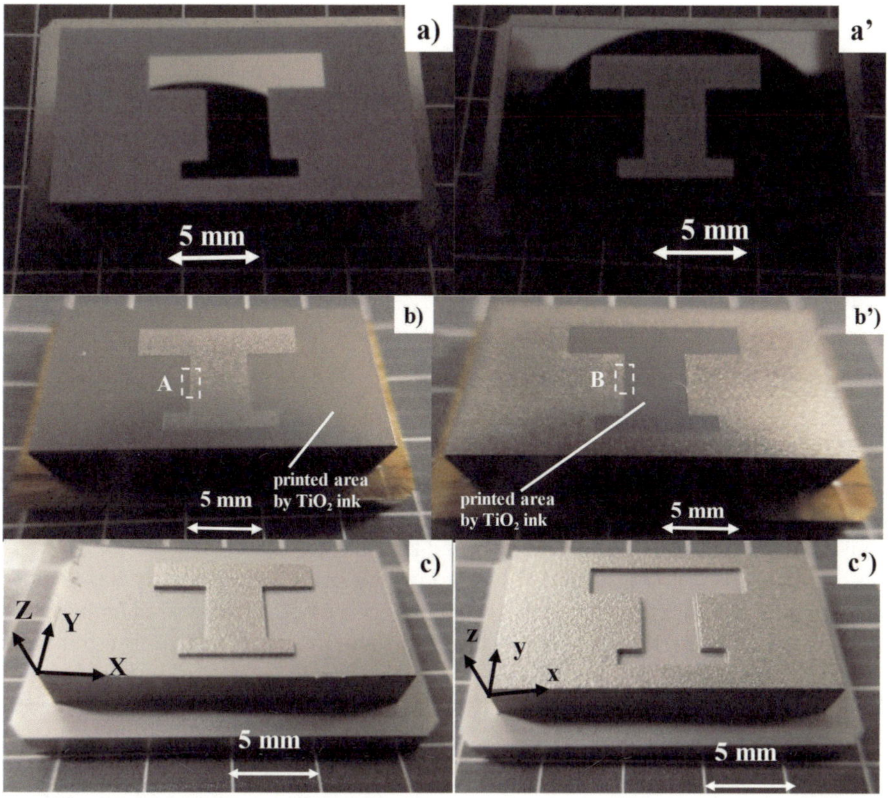

Figure 20.
Comparison of the plasma nitriding-assisted 3D printing procedures to simultaneously fabricate the T-shaped AISI316 punch and the T-shaped AISI316 core-die.

In corresponding to the screen-printing step in **Figure 19b**, the negative region to T-shaped model is all printed in **Figure 20a**, and the T-shaped region is only printed onto the substrate surface in **Figure 20a'**. Both AISI316 substrates are plasma-nitrided together at 673 K for 14.4 ks. The unprinted T-shaped surface is selectively nitrided in **Figure 20b**, while the printed T-shaped surface is selectively un-nitrided in **Figure 20b'**. Both nitrided AISI316 substrates are sandblasted to remove the softer parts of substrates than the hardness of shooting media. As shown in **Figure 20c**, other parts of substrate than T-shaped region are removed to leave the T-shaped punch head. On the other hand, the T-shaped region is exclusively dug and removed from the substrate to leave the T-shaped die cavity as depicted in **Figure 20c'**.

Three-dimensional profilometer was utilized to measure the T-shaped head of punch in **Figure 20c** and the T-shaped cavity of core-die in **Figure 20c'**. As shown in **Figure 21a**, the T-shaped head has uniform height of 350 μm and other surface is homogeneously ground by the present 3D printing. That is, the AISI316 punch is fabricated after the printing scheme in **Figure 19** to have a T-shaped head with uniform height. On the other hand, the T-shaped cavity is uniformly dug by 350 μm in depth; the cavity walls have steep gradient against the die surface in **Figure 21b**. AISI316 core-die is fabricated to have a T-shaped cavity with the uniform depth in correspondence to the T-shaped punch.

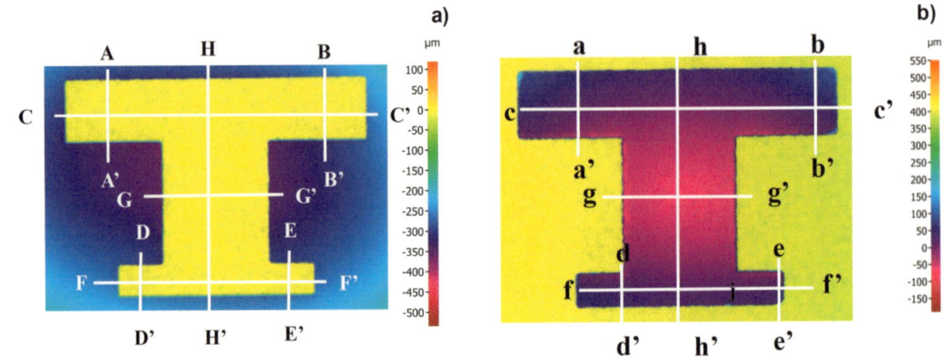

Figure 21.
Comparison of three-dimensional profile between the T-shaped punch head and the T-shaped core-die cavity.

Among several cross sections from A-A' to F-F′ in **Figure 21a** and from a-a' to f-f' in **Figure 21b**, the cross sections of F-F′ and f-f' were selected to describe the three-dimensional profile of T-shaped punch head and die cavity.

As compared in **Figure 22**, the T-shaped punch head and die cavity has uniform height of 350 μm and uniform depth of 350 μm with small roughness. The punch head and die cavity side surfaces are steeply formed to be perpendicular to the punch head and cavity bottom surfaces, respectively. The punch edge width and die cavity edge width are 15 μm and 20 μm, respectively. These dull edges influence on the shearing process in piercing process.

The clearance between the punch head and die cavity is an essential parameter with significant influence on the product quality through stamping and forging and on the shearing behavior [41]. In the present 3D printing, no tolerance for clearance control was performed in CAD; the effective clearance is autonomously determined by the accuracy in the plasma printing steps. Let us describe the allowable clearance by three-dimensional profiling on the cross sections along A-A' to F-F′ and a-a' to f-f'.

Table 1 compares the original CAD data on the edge lengths with the measured T-shaped punch and core-die dimensions along A-A' to F-F′ in **Figure 20a** and a-a' to f-f' in **Figure 20b**, respectively. In the plasma printing step of T-shaped punch, the dimensional deviation per the punch edge length ranges from - 3% to +0.2%. This printing of punch accompanies with negative tolerance. On the other hand, the edge length of T-shaped core-die deviates from +0.5% to +5% per the core edge length. The printing of core-die accompanies with positive tolerance. Hence, the clearance is

	A-A' a-a'	B-B′ b-b'	C-C′ c-c'	D-D′ d-d'	E–E' e–e'	F-F′ f-f'	G-G' g-g'	H-H′ h-h'
Screen (mm)	2.0	2.0	10.0	1.0	1.0	6.5	3.5	7.0
T-letter punch (mm)	2.001	1.991	10.001	0.978	0.971	6.512	3.476	6.990
T-letter core-die (mm)	2.023	2.039	10.045	1.047	1.039	6.531	3.561	7.073
Clearance (mm)	0.011	0.024	0.022	0.034	0.035	0.0095	0.043	0.0415

Table 1.
Comparison of the geometric accuracy among the CAD data, the measured T-shaped punch, and core-die dimensions and the calculated clearances.

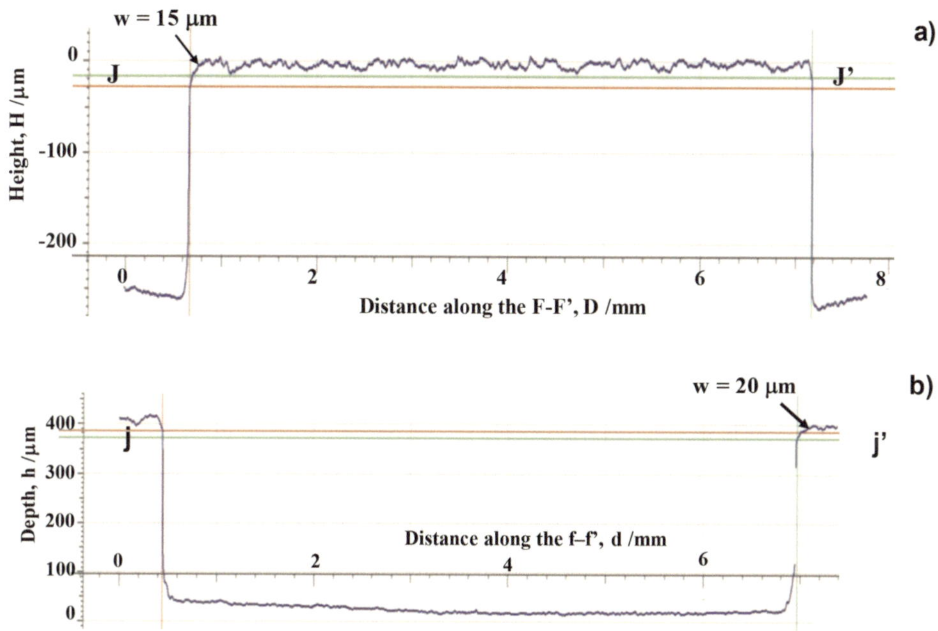

Figure 22.
Comparison of the three-dimensional profiles between the T-shaped punch head and die cavity at the cross section along F-F′ in ***Figure 21a*** *and f-f′ in* ***Figure 21b****.*

determined by the sum of two tolerances; the maximum clearance reaches +3.5% in **Table 1**.

5.3 Fine punching of electrical steel core unit sheets

The CNC stamping system (Precise Metal Forming Laboratory; Hachi-Oji, Japan) was utilized for punching out the electrical steel sheet with the thickness of 200 μm by using this T-shaped punch and die pair. **Figure 23** depicts the pierced electrical steel core sheet and its skeleton. No distortion was observed on both sheets; this pair is useful in continuous punching operations for mass production. As shown in **Figure 24**, the burnished cross-sectional surface area reaches 70% just in correspondence to highly qualified sheared surface with much lower clearance in [38].

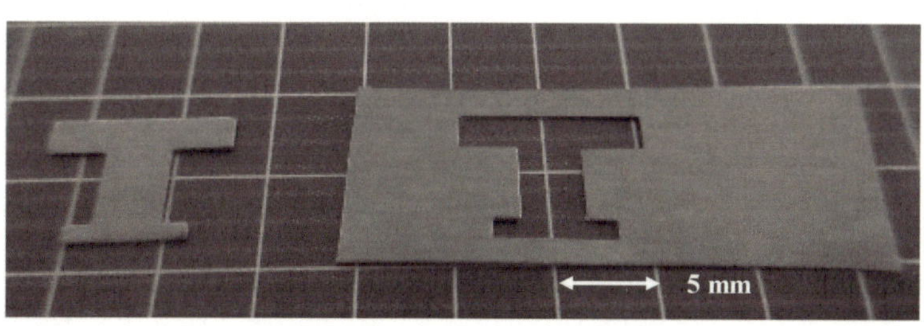

Figure 23.
The pierced electrical steel core unit and its skeleton.

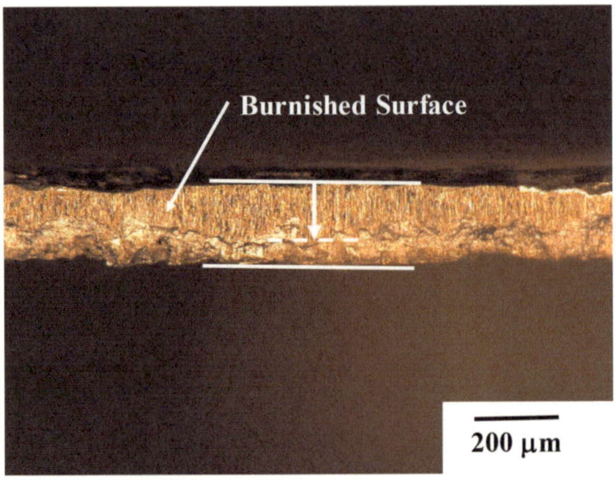

Figure 24.
SEM image on the sheared surface of pierced electrical steel sheets along F-F′ in ***Table 1****.*

6. Production line for fabrication of micro-pumps with aid of 3D printing

Most of medical tools and devices must have sufficient durability to be free from damage and risk of failure [42]. A micro-pump has to equip the sufficient leak proof against the pressurized blood, the drug solution, the liquid food, and so forth, never to splash them even into the human body [43, 44]. In addition to the original strength and toughness of its constituent materials, it has sufficiently high joining strength among these constituent units and parts.

In this section, the mother AISI304 sheet with the thickness of 50 μm is punched out by using each plasma-printed punch and die pair to fabricate each constituent sheet unit with the accommodated function [45]. Several to ten sheet units are assembled and joined to a micro-pump. The punching process of constituent sheet units and their assembling and joining processes work in parallel as a production line [46] as illustrated in **Figure 1**.

6.1 Production line design for fabrication of micro-pump

A production line was designed and developed after the scheme in **Figure 1** to make 3D printing of the micro-pump from the AISI304 austenitic stainless steel sheets as a feedstock. A whole procedure is shown in **Figure 25**. This system mainly consists of two subsystems. The sliced 2D CAD data from the original product 3D model of micro-pump unit is transformed into the punch head and die cavity by the plasma nitriding-assisted 3D printing. Each plasma-printed punch and die is utilized in blanking, embossing, piercing, and punching the AISI304 sheet with the tailored thickness. The assembly of the blanked, embossed, and punched-out sheets, is polished and cleaned by the argon and hydrogen plasmas. Through the bombardment by argon ions and the chemical reduction by activated hydrogen atoms, the thickness of passive oxide film on the work surfaces is reduced or removed to lower the joining temperature. Under the inert atmosphere, the cleaned and polished assembly is stacked into a preform with laminated work sheets and joined by hot stamping with

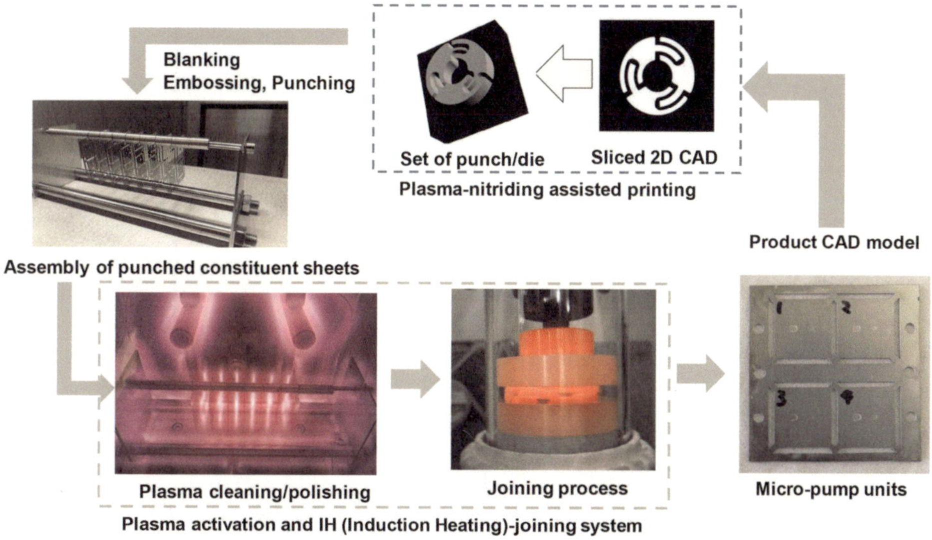

Figure 25.
Production line for micro-manufacturing of micro-pump from the feedstock of AISI304 sheets.

the use of high-frequency induction heating (HFIH) units. Four micro-pump units are produced by this procedure in a single shot.

In the same manner as explained in the sections 4 and 5, the finished set of punch and die is utilized to blank, emboss, pierce, and punch out the AISI304 sheet with the selected thickness from 200 μm to 10 μm, respectively. The shaped sheet is set as an assembly and transferred to the plasma activation and HFIH joining system.

6.2 Plasma polishing and cleansing of AISI304 sheet unit assembly

The normal hot-pressing in the nitrogen or argon atmosphere was effective to join the stacked sheets. However, the joining performance is completely governed by the separation of oxide layer through the oxygen atom diffusion into the matrix [47]. Since the onset temperature of oxide layer separation is around 1223 K (or 950°C), the joining temperature is estimated to be 1250 K. This holding temperature in hot pressing is possible to induce the thermal distortion to disturb the dimensional accuracy and the damage to stainless steels. Hence, the pretreatment before joining is needed to reduce the oxide layer thickness and to lower the holding temperature for joining.

In the present study, the argon and hydrogen plasma activation process was utilized to clean and polish the top and bottom surfaces of constituent sheets in the assembly. As shown in **Figure 25**, the oxide layer thickness was reduced by the bombardment of argon ions and the chemical reduction reaction via $MO_x + 2H \rightarrow M + xH_2O$ (M: Fe or Cr) in the plasma sheath. **Figure 26** compares the variation of peeling strength with increasing the holding temperature for joined specimens with and without the plasma surface activation.

At T_H = 1043 K (or 770°C), the peeling strength of joined specimens with plasma activation becomes nine times higher than that without plasma activation. This proves that the passive oxide layer thickness of AISI304 stainless steel sheets is reduced enough to significantly improve the joining strength.

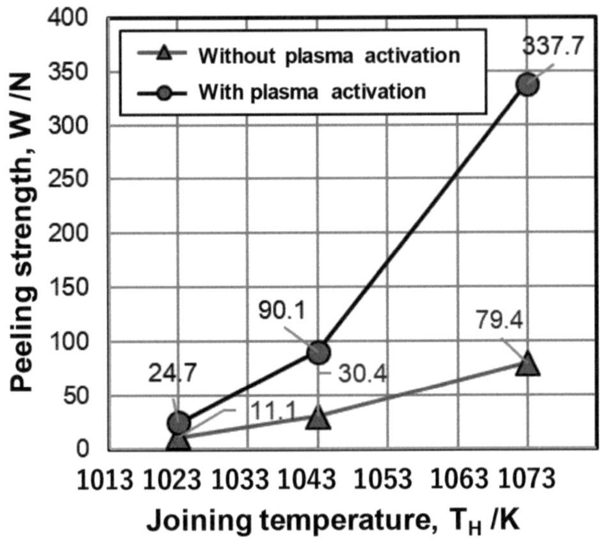

Figure 26.
Variation of the peeling strength (W) with increasing the joining temperature (T_H) for the joined specimens at 30 MPa for 1.8 ks with and without plasma activation in the plasma polishing and cleansing process.

6.3 Low-temperature joining of assembly to micro-pump part

The sufficient joinability is essential to certify the proof of mechanical integrity for a micro-pump in practical operation [48]. The higher peeling strength than 100 N is a necessary condition for this integrity proof as an essential mechanical engineering item. In addition, the fractography [49] on the peeled interfaces between the stainless steel sheets provides a material science tool to validate the sufficient joinability. In general, the brittle fractured surface is characterized by the smooth surfaces with little micro-voids [50]. On the other hand, the ductile fractured surface consists of the micro-dimples and micro-voids [51]. Hence, the integrity of joined surface is investigated by the fractography on the peeled surface. **Figure 27** compares the SEM image on the cross section of joined specimens at 1073 K for 1.8 ks by 30 MPa with and without plasma activation as well as their fractography. In the joined specimen without plasma activation, the boundary between the joined sheets is distinguished by each granular micro-structure. The peeled-out interface is flat including little micro-voids. While, in the joined specimen with plasma activation, the recrystallized grains are seen on the sheet boundary. Two adjacent stainless steels are joined to form new grains across the previous interface boundaries. The whole peeled-out surface consists of the micro-dimples and micro-voids. They were formed by the ductile fracture during the peeling test. This difference assures the high joinability on the interfaces of stainless steel assembly.

7. Discussion

The present plasma nitriding-assisted 3D printing has three merits superior to the traditional die technologies by mechanical milling [52] and laser machining [53]. As seen in comparison between the CAD data and the punch head and die cavity in **Figures 11, 17**, and **20**, their geometric configuration is just corresponding to the CAD data. This assures that any punch and die can be fabricated through the duplication of

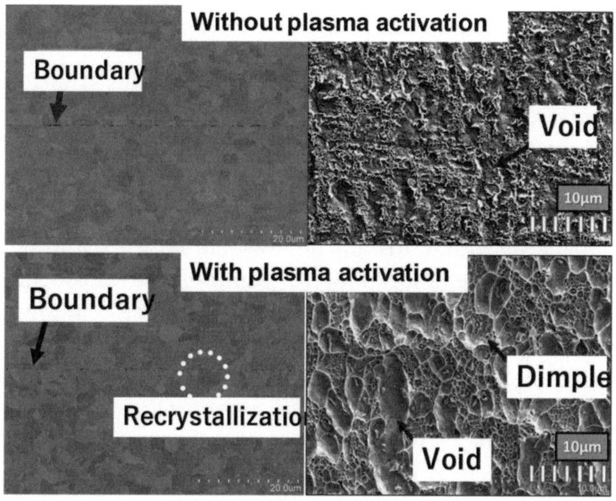

Figure 27.
Comparison of the SEM image on the microstructure of joined interface in low and high magnifications between the joined specimens at 1073 K by 30 MPa for 1.8 ks with and without plasma activation.

two-dimensional CAD in the present approach. In particular, the clearance between the punch head and die cavity is autonomously determined by the present plasma printing. This first merit is effective to build up the complex-shaped punch and die as well as the multi-head punch and multi-cavity die for mass production of mechanical parts [54].

Higher hardness than 800 HV in the plasma-printed punch and die substrates becomes the second merit for die technology. Since the lower hardness parts are removed by sandblasting, the punch head and die surface are proved to have much higher hardness than their mother substrate materials by plasma nitriding. Various stainless steels and tool steels are available as an original substrate material for this plasma printing [55].

The autonomous formation of steep side surfaces in the punch and die turns to be the third merit to prepare a die pair for embossing, piercing, and punching processes as shown in **Figures 12, 17, 21**, and **22**. In particular, this merit works well in fabrication of sub-mm and sub-100 μm-sized products by stamping with the use of plasma-printed punch and die pair [56]. As noticed in **Figures 22** and **25**, the punch and die edges must be sharped to improve the shearing behavior in embossing, piercing, and punching processes. At first, the plasma-printed punch and die might well be ground to reduce their edge widths of 15–20 μm down to a few μm. Then, an ion milling is employed to finish the edge width down to 1 μm or less than. As reported in [57, 58], the edge-sharpened punch and die with higher hardness than 1000 HV by the plasma nitriding has a capacity to improve the quality of sheared surfaces without loss of die life.

In the present 3D printing method, its dimensional accuracy of products is first determined by the spatial resolution (D) in drawing the 2D micro-pattern onto the die substrates. As depicted in **Figure 2**, the metal masking, the inkjet, and screen printing are only available to micro-patterning with $D > 10$ μm. When aiming at fine resolution for $D < 10$ μm, the lithography and the laser masking techniques are only available in this method. As reported in [59, 60], the normal maskless lithography was used to prepare the complex-shaped micro-patterns by $D = 0.5$ μm for plasma-printing the multi-head DLC (Diamond-Like Carbon) punches and the micro-nozzles. In order to improve the spatial resolution toward $D \ll 1$ μm, these two methods must be advanced together with suitable selection of inks, resins, and metallic deposits.

Owing to the principle of plasma nitriding assistance in this 3D printing, the whole surfaces of punch head and die cavity are hardened and enriched by nitrogen solutes in high content. After [61, 62], they have sufficient wear resistivity and corrosion roughness to be working in long-term usage of mass production. In addition to the titanium and titanium alloy sheets with high chemical affinity to die substrate materials [63], the thermoplastic plastic sheets are embossed, pieced, and punched to the constituent element of micro-parts by using the plasma-printed dies [64].

In the additive sheet manufacturing, the low-temperature joining process plays an essential role to reproduce the tailored micro-pump unit after the 3D CAD model with sufficient integrity proof. High interfacial strength by ductile joining among the sheet interfaces assures the leak proof to be working at the specified operation conditions for the designed product life. When this proof standard is relaxed, other joining methods are also available in the present additive sheet manufacturing. Among them, the mechanical clinching method [65] provides the elasto-plastic interlocking between adjacent sheets with sufficient interfacial strength. This pointwise joining has no rooms of leak proof against the fluids and solutions. The mechanical anchoring method [66] also becomes a suitable joining of dissimilar plastic sheets into a product with sufficient strength and water proof in addition to the joining of plastic mold with the metallic substrate in the Section 3.

A micro-pump was produced in two-step procedure. At first, its constituent stainless steel sheets are punched out by using the plasma-printed punch and die in correspondence to the 2D sliced data of its 3D model. In second, these punched-out sheets are assembled, plasma-cleaned and polished, and HFIH-joined to yield the four micro-pump units. Most of MEMS units are produced in the same manner. In particular, the pendulum in the micro-valve in **Figure 18** has no risk of local distortion and no hot adhesion to other sheet parts in wrong. This allows us to design various inner spaces for device movement in the MEMS.

The micro-inspection units [67] are a typical micro-part, working in vitro. They consist of several functional subunits including the micro-fluidic unit to control the chemical reaction rates for inspection, the micro-reservoir unit to control the pressure in the micro-channel flow, and so forth. Each subunit consists of two or three AIS304 stainless sheets shaped by the present printing process. The micro-injectors [68] embedded in vivo into deceased parts of body are also fabricated by the present 3D printing for local drug delivery and for curing the local tissue.

8. Conclusion

The plasma nitriding-assisted 3D printing was proposed to fabricate the complex-shaped punch and die for fine embossing, piercing, and punching processes and to yield the functionalized constituent sheets by using the CNC stamping. Through the embossing process, the tailored microgroove array is accommodated to the copper substrate for mechanical anchoring of the plastic molds in the leak-proof packaging. Through the piercing process, the micro-valve unit is housed into the stainless steel sheet for functionalization of sliced 2D CAD data from 3D solid model of micro-pump. The electrical steel sheets are punched out to yield the T-shaped motor core units for assembly to a motor core by mechanical joining or mechanical anchoring. These sheet elements with functional units for sliced 2D model are plasma-cleaned and polished and joined at low temperature to build up the micro-pump unit in correspondence to the original product model.

The design flexibility is preserved by transformation of the 2D CAD data to the constituent metal or polymer sheet through the embossing, piercing, and punching processes with the use of plasma-printed punch and die. The hot stamping, as well as the mechanical joining, is useful to integrate the discrete functionalized sheets into a continuous micro-element and part. This 3D printing with the use of metal and polymer films, sheets and plates as a feedstock, provides a new way of additive micro-manufacturing to yield the MEMS units with high mechanical integrity and the medical tools to be working in vitro and in vivo.

Acknowledgements

The authors would like to express their gratitude to Kurozumi S-I. and Morita H. (Nano-Film Coat, llc.), Yoshino T. (Komatsu-Seiki Kosakusho, Co., Ltd.), and graduate students both in the Shibaura Institute of Technology and the University of Toyama for their help in experiments. This study was financially supported in part by the Grand-In-Aid from METI, Japan.

Conflict of interest

The authors declare no conflict of interest.

Abbreviations

CAD	Computer-Aided Design
CAM	Computer-Aided Machining
CNC	Computer Numerical Control
DC	Direct Current
EDX	Energy-Dispersive X-ray spectroscopy
GaN	Gallium Nitride
PZT	Lead (Pb) Zirconate Titanate
RF	Radio Frequency
SEM	Scanning Electron Microscopy
2D	Two Dimension
3D	Three Dimension

Author details

Tatsuhiko Aizawa[1*], Tomomi Shiratori[2] and Yohei Suzuki[3]

1 Surface Engineering Design Laboratory, Shibaura Institute of Technology, Tokyo, Japan

2 Faculty of Engineering, University of Toyama, Toyama, Japan

3 Komatsu-Seiki Kosakusho, Co., Ltd., Suwa, Japan

*Address all correspondence to: taizawa@sic.shibaura-it.ac.jp

References

[1] Kumar SA, Prasad RVS, Basic principles of additive manufacturing: different additive manufacturing technologies. In: Additive Manufacturing Woodhead Pub. Ch. 2. 2021. pp. 17-35

[2] ISO/TC 261 additive manufacturing, ISO.ASTM 52907:2019 Additive manufacturing feedstock materials – methods to characterize the metal powders. 2019. pp. 1-19

[3] Munsch M, Laser additive manufacturing of customized prosthetics and implant- feedstock amts for biomedical applications. In: Laser Additive Manufacturing – Materials, Design, Technologies and Applications – Woodhead Pub. Ch. 15. 2017. pp. 399-420

[4] Zhong Y, Raenner L-E, Lie L, Koptyung A, Wikman S, Olsen J, et al. Additive manufacturing of 316L stainless steel by electron beam melting for nuclear fusion applications. Journal of Nuclear Materials. 2017;**486**:234-245

[5] Alberti EA, Bueno BMP, D'Oliveira ASCM. Additive manufacturing using plasma transferred arc. The International Journal of Advanced Manufacturing Technology. 2016;**83**:1861-1871

[6] Loterie D, Derrot P, Moser C. High-resolution tomographic volumetric additive manufacturing. Nature Communications. 2020;**11**:852-861

[7] Zhang J, Jung YG, editors. Additive Manufacturing –Materials, Processes, Qualifications and Applications. London, UK: Elsevier; 2018

[8] Shiratori T, Nakano S, Suzuki Y, Katoh M, Sato N, Komatsu T, et al. Development of metal MEMS manufacturing technologies using pierced metal foil and diffusion bonding process at low temperature. In: Proc. 4M/IWMF2016 Conference. Copenhagen, Denmark: 4M Conf. Committee; 2016. pp. 209-212

[9] Arnold J. Die Makers Handbook. 1st ed. Cumberland, USA: Industry Press; 2020

[10] Katoh T, Aizawa T, Yamaguchi T. Plasma assisted nitriding for micro-texturing onto martensitic stainless steels. Manufacturing Review. 2015;**2**(2): 1-7

[11] Aizawa T, Redationo NT, Mizushima K. Precise micro-texturing onto DLC coating via high density oxygen plasma etching. Proceedings of 4M-Conference. 2013;**1**:129-132

[12] Aizawa T, Yamaguchi T. High-density plasma nitriding assisted micro-texturing onto martensitic stainless steel mold-die. Proceedings of 9th IWFM. 2014;**2A**:11-f7

[13] Aizawa T, Wasa K. Fine micro-fabrication of stainless steel nozzle array by plasma printing. Proceedings of WCMNM-2018. 2018;**1**:65-68

[14] Aizawa T, Tamaki M, Fukuda T. Motion-controlled CNC stamping for micro-texturing onto metallic sheet. Proceedings of 9th ICOMM. 2014;**14**:1-8

[15] Aizawa T, Inohara T. Micro-texturing onto glassy carbon substrates by multi-axially controlled pico-second laser machining. Proceedings of 7th ICOMM. 2012;**1**:66-73

[16] Aizawa T, Shiratori T, Saito Y. Fabrication of micro-punch array by

plasma printing for micro-embossing into copper substrates. Proceedings of 12th AWMFT. 2019;**1**:93-98

[17] Shiratori T, Aizawa T, Saito Y, Wasa K. Plasma printing of an AISI316 micro-meshing punch array for micro-embossing onto copper plates. Journal of Metals. 2029;**9**(396):1-11

[18] Aizawa T, Sugita Y. High density RF-DC plasma nitriding of steels for die and mold technologies. Research Reports SIT. 2013;**57**(1):1-10

[19] Aizawa T, Sugita Y. High density plasma nitriding of tool and die steels. Proceedings of 6th SEATUC Symposium. 2012;**1**:134-137

[20] Kuwahara H. Fundamentals and Application of DC-Plasma Nitriding and Carburizing. PhD Thesis. Kyoto, Japan: Kyoto University; 1992

[21] Hiraoka Y, Inoue K. Prediction of nitrogen distribution in steels after plasma nitriding. Denki-Seiko. 2010;**86**: 15-24

[22] Granito N, Kuwahara H, Aizawa T. Normal and anormal microstructure of plasma nitrided Fe-Cr alloys. Journal of Materials Science. 2002;**37**(4):835-844

[23] Farghali A, Aizawa T. Phase transformation induced by high nitrogen content solid solution in the martensitic stainless steels. Materials Transactions. 2017;**58**:697-700

[24] Farghali A, Aizawa T. Nitrogen supersaturation process in the AISI420 martensitic stainless steels by low temperature plasma nitriding. ISIJ International. 2018;**58**(3):401-407

[25] Aizawa T. Low temperature plasma nitriding of austenitic stainless steels. In: Stainless Steels. London, UK: IntechOpen; 2018. pp. 31-50

[26] Aizawa T, Yoshihara S-I. Inner nitriding behavior and mechanism in stainless steels at 753 K and 623 K. SEATUC Journal of Science and Engineering (SJSE). 2019;**1**:13-20

[27] Aizawa T, Shiratori T, Yoshino T, Suzuki Y, Komatsu T. Nitrogen supersaturation of AISI316 base stainless steels at 673 K and 623 K for hardening and microstructure control. In: Stainless Steels. Ch. 1 ed. London, UK: IntechOpen; 2021

[28] Aizawa T, Shiratori T, Saito Y. Fabrication of micro-punch array by plasma printing for micro-embossing into copper substrates. In: Proc. 12th Asia Workshop on Metal Forming Technology. Tokyo, Japan: JSTP; 2019. pp. 93-98

[29] Aizawa T., Shiratori T., Wasa K., Plasma-printed AISI316L multi-punch array for fabrication of aluminum heatsink with micro-pillar fins. Proceedings of 3rd WCMNM (Raleigh, USA) 2019;**1**: 220-223

[30] Aizawa T, Yoshihara S-I. Microtexturing into AISI420 dies for fine piercing of micropatterns into metallic sheets. Journal of JSTP. 2019;**60**(698): 53-57

[31] Saito Y, Aizawa T, Wasa K, Nogami Y. Leak-proof packaging for GaN chip with controlled thermal spreading and transients. Proceedings of 41st BCICTS2018. 2018;**1**:243-246

[32] Aizawa T, Saito Y, Hasegawa H, Wasa K. Fabrication of optimally micro-textured copper substrates by plasma printing for plastic mold packaging. International Journal of Automation Technology. 2020;**14**(2):200-207

[33] Aizawa T, Nakata H, Nasu T, Nogami Y. Micro-textured graphitic substrate – Copper packaging for robustness. In: Proc. 5th WCMNM. Leuven, Belgium: Res. Pub.; 2022. pp. 291-295

[34] Geiger M, Vollertsen F, Kals R. Fundamentals on the manufacturing of sheet metal microparts. CIRP Annals. 1996;**45**(19):277-282

[35] Aizawa T, Takashima T, Shiratori T. Plasma printing to fabricate the micro-piercing dies for miniature metal products. In: Proc. 8th AWMFT. Suwa, Japan: CD-ROM; 2015

[36] Aizawa T, Takashima T, Shiratori T, Suzuki Y. Fabrication of micro-piercing dies and punches via plasma printing. In: Proc. 11th ICOMM. Vol. 2. California, USA: I2M2; 2016. pp. 1-5

[37] Shiratori T, Nakano S, Suzuki Y, Aihara T, Yang M. Development of multiple shape one process piercing method by laser processing punch. In: Proc. 2016 JSTP. Kyoto, Japan: JSTP; 2016. pp. 313-314

[38] Katsuta E, Aizawa T, Morita H, Dohda K, Anzai M. Fine piercing of electromagnetic steel sheets by micro-punches under nearly zero clearance. Procedia Manufacturing. 2018;**15**: 1459-1466

[39] https://www.emobility-engineering.com/motor-laminations/ [Accessed: October 15, 2022]

[40] Aizawa T, Suzuki Y, Yoshino Y, Shiratori T. Fabrication of punch and die using plasma-assisted 3D printing technology for piercing sheet metals. Journal of Manufacturing and Materials Processing. 2022;**6**(49):1-15

[41] Schmidt K. Manufacturing Processes for Engineering Materials. 5th ed. Hoboken, NJ, USA: Prentice Hall; 2008

[42] Aizawa T, Satoh T, Shiratori T, Proc. 22nd Int. ESAFORM Conf. Mater. Form. AIP Conf. Proc. 2019;**2113**(050007):1-6

[43] Satoh T, Aizawa T, Shiratori T, Sugita Y, Anzai M. Micro-joining of multi stainless steel sheets into mechanical element by low temperature diffusion process. Procedia Manufacturing. 2018;**15**:1475-1480

[44] Satoh T, Aizawa T, Shiratori T, Yoshihara S-I. Micro-joining of multiple stainless-steel sheets into a mechanical element by low-temperature diffusion process. J. JSTP. 2019;**60**:80-84

[45] Aizawa T, Shiratori T. Microforming of mechanical elements, devices and systems by additive sheet-manufacturing. Proceedings of 12th AWMFT-2019. 2019;**1**:1-6

[46] Aizawa T, Shiratori T. Microforming of stainless steel miniature pump by additive sheet-manufacturing. Materials Transactions. 2020;**60**(2):266-271

[47] Sugimoto K. Passive films on stainless steels – Present state of analysis and understanding. Ziryo-to-Kankyo. 2008;**57**:375-384

[48] Kelly T, Childers RW, Busby D, Rodolfo Roger R, Sayyid WME, Din S. Medical fluid pump valve integrity test methods and systems. US-Patent. US9072831B2. 2004

[49] Hayes MD, Edwards DB, Shah AR. Fractography in Failure Analysis of Polymers. London, UK: Elsevier; 2015

[50] Quinn GD. Fractography of brittle materials: Analysis of fractures in ceramics and glasses. Microscopy and Analysis. 2008;**5**:21-24

[51] Li H, Fu M. Deformation-Based Processing of Materials. London UK: Elsevier; 2019

[52] Rahman M, Asad ANMA, Wang YS, Shinno H, Li X. Advanced maching technologies. In: Comprehensive Materials Processing. Vol. 11, London, UK: Elsevier; 2014

[53] Aizawa T, Inohara T. Pico- and femtosecond laser micromachining for surface texturing. In: Stanimirovi'c Z, Stanimirovi'c I, editors. Chapter 1 in: Micromachining. London, UK: IntechOpen; 2019. pp. 1-23

[54] Aizawa T, Morita H. Dry progressive stamping of copper-alloy snaps by the plasma nitrided punches. Materials Science Forum. 2018;**920**:28-33

[55] Aizawa T, Shiratori T, Komatsu T. Micro−/nano-structuring in stainless steels by metal forming and materials processing. Chapter 5 In: Electron Crystallography. London, UK: IntechOpen; 2020. pp. 101-122

[56] Pratap A, Patra K, Dyakonov M. Manufacturing miniature products by micro-grinding; a review. Procedia Engineering. London, UK: Elsevier; 2016;**150**:969-974

[57] Aizawa T, Yoshino T, Shiratori T, Dohda K. Material characterization on the affected zones by fine piercing into single crystal Fe-6Si steel sheet. In: Proc. 4th WCMNM. Mumbai, India; 2021

[58] Aizawa T, Shiratori T, Yoshino T, Suzuki Y, Dohda K. Quantitative characterization of the affected zones in a single crystal Fe-6Si steel sheet by fine piercing. Journal of Micromachines. 2022;**13**(562):1-15

[59] Aizawa T. Micro-texturing onto amorphous carbon materials as a mold-die for micro-forming. Applied Mechanics and Materials. 2013;**289**:23-37

[60] Aizawa T, Wasa K. Plasma printing of micro-nozzles with complex shaped outlets into stainless steel sheets. Journal of Micro-Nano-Manufacturing, ASME. 2019;7:034502/1-034502/6

[61] Aizawa T, Fukuda T, Morita H. Low temperature high density plasma nitriding of stainless steel molds for stamping of oxide glasses. Manufacturing Review. 2016;**3**(5):1-6

[62] Borgioli F, Galvanetto E, Bacco T. Low temperature nitriding of AISI300 and 200 series austenitic stainless steels. Vacuum. 2016;**12**:51-60

[63] Aizawa T, Shiratori T. Plasma 3D-priting of micro-punch array for fine embossing into biomedical titanium works. Advances in Metallurgy Materials Engineering. 2020;**3**(1):95-103

[64] Aizawa T, Yamaguchi T. Plasma nitriding assisted micro-texturing into martensitic stainless steel molds for injection molding. Proceedings of 10th 4M/ICOMM. 2015:454-459

[65] Lambiase F, Ilio AD. Damage analysis in mechanical clinching; experimental and numerical study. Journal of Materials Processing Technology. 2016;**230**:109-120

[66] Aizawa T, Satoh S, Yamaguchi T. Micro-texturing design for joining between polymer components. Proceedings of 9th ICOMM. 2014;**15**:1-8

[67] Balouiri M, Sadiki M, Ibnsouda SK. Methods for in vitro evaluation antimicrobial activity: A review. Journal of Pharmaceutical Analysis. 2016;**6**(29):71-79

[68] Henley T, Thomas K, Bressan M. Microinjection-based system for in vivo implantation of embryonic cardiomyocytes in the avian embryo. Journal of Visualized Experiments. 2019; **144**:e59267

Chapter 18

Fashion: From 3D Printing to Digital Fashion

Maria Victória Rocha

Abstract

Our scope is to emphasize the massive changes brought about by 3D and 4D printing in the fashion world (now connected with augmented reality and virtual reality) by addressing the concept of 3D printing and explaining when and how it started to develop in fashion, becoming the sewing machine of the twenty-first century. It has allowed for great changes in the fashion industry, namely the appearance of more sustainable, and customized apparel and accessories, simpler logistics and less transaction costs. This requires increased creativity, because it demands the cooperation of specialists from many different areas. New and before deemed impossible shapes have arisen. The final idea of prosumer and its meaning is before us. If we add 4D printing, that is, "intelligent" materials, 3D printing enters our day lives, with wearables and smart clothes. All this raises questions regarding copyright and industrial property protection and possible infringement in those areas. Current 4D products also raise questions regarding personal data protection. We will address how to protect intellectual property and personal data in this world of digital fashion, where augmented and virtual reality play a roll of increased importance.

Keywords: fashion law, 3D printing, 4D printing, augmented reality, virtual reality, metaverse; copyright, industrial property, personal data

1. Introduction

Three-dimensional (3D) printing includes several additive manufacture technics, toward a new era of production and consumerism, based on the "make it yourself," thus the expression prosumer (producer and consumer).[1] Three-dimensional printing allows manufacture of apparel, wallets, footwear, and accessories on demand, near the distribution chains, lowering the costs of transport and storage. These fashion items can be almost immediately obtained and customized and achieve creative forms not possible before and in a much more sustainable way because there is almost no waist. Such technology exists for about 40 years but is always evolving and did not enter the fashion industry until 2009, when pioneer Dutch stylist Iris van Herpen started to use it in a regular basis. Before, plastic materials used in 3D were too stiff and inappropriate for fashion [1].

[1] This work is part of the project I + D + i PID2020-112707GB-I00, MODA Y BIENES INMATERIALES, financed by MCIN/AEI/10.13039/501100011033.

What is common in 3D technique is the creation of a computer file with the desired shape (*Computer-Aided Design* file), which passes on to a 3D printer. The most used format of the CAD file is *.stl.* The file can be obtained by several means, namely by creating the object using 3D software, by 3D scanning an existing object, or using 3D sharing websites.

Three-dimensional printing is an additive manufacture technique, usually layer by layer, thus more sustainable with much less waist and costs. Today, almost all materials can be used, even biological ones, and the technique allows to recycle the 3D-printed apparel, footwear, wallets, and accessories, and even to reuse plastics that pollute the seas. 3D allows the user to customize any piece.

Four-dimensional printing (4D) is more recent, and it adds to the 3D technique the use of "intelligent" materials, programmed to adjust to the changing needs of the user, like hydrogel, polymers with memory shape (PMF), and liquid crystal elastomers (LCE). The difference between 3D and 4D has to do with the difference of materials. Three-dimensional (3D) uses traditional materials, while 4D uses pseudo-intelligent materials printed in 3D.

In the last years, 4D technology has been increasingly used in wearables and smart clothes. Examples include apparel and footwear that can adapt to the body of the user and can change with movement, adapt to possible impacts, temperature changes, among others. The Covid-19 pandemic led to very interesting solutions and, even more, accelerated the interest in totally digital fashion items.

Wearables and smart clothes use technology that offers the user new and distinct functionalities, making everyday life more practical [2]. Examples of wearables include: (a) wristbands that monitor exercise, linked to an app, allow, among others, to count the steps, and monitor heart rate, vital parameters, sleep cycles, and calories burned; (b) although more recent, sunglasses, pendants, earrings, and rings, which incorporate technologies are used as physical status trackers, stress managers, and step counters; (c) smart watches that, in addition to showing the hours, provide all the services of smart wristbands and many connect to the user's smartphone, making it possible to make payments, thanks to the embedded chips, see messages, answer calls, and control music through music apps compatible with these devices.

Many smart clothes are created also considering sustainability and compromise with the environment, using sustainable materials, reusing, and recycling the items. The examples go from sportswear to luxury apparels, baby clothes, college uniforms, and more. Examples of smart clothes include: (a) Start up BeClothed created intelligent underwear with a silver compound that eliminates bacteria responsible for bad smell, so the underwear can be used without getting dirty or smelling and does not need to be washed [3]; (b) MOxATech is Portuguese invented mask with an impermeable fabric that integrates waterproof and reusable covers, as well as a coating that inactivates the Covid-19 when it comes into contact with it [4]; (c) Italian Cliu Mask has a Bluetooth system, sensors, and algorithms, connected to an app in the smartphone of the user that gives information about the quality of the air, heart rate, and existence of some corona virus outbreak in the vicinity. With a disinfection system based on ultraviolet light, the mask allows to eliminate any virus it scans in a few minutes. It is a transparent mask because of people with hearing problems [5]; (d) Harvard and MIT also invented masks that, thanks to intelligent sensors, recognize the genetic code of the Covid-19 virus and light up when in contact with particles of the virus. This was a preventive solution for contacts with asymptomatic people. Similar masks are Leaf Mask, Flat Tube Energy, or TrioMed Active Mask ([2], p. 402); and (e) during the pandemic, Anouk Wipprecht created the "Proximity Dress" with many sensors that allow the dress to stretch creating a barrier if someone approaches, thus identifying the infringement of the mandatory safe distance due to Covid-19 .

There are devices with numerous functions, for example, clothing that tells if you have been in the sun too long, the distance you have traveled, speed, calories burned, or even devices that charge the phones or other electronics. Some include geolocation systems, fingerprints, or built-in facial recognition cameras that enable the control and location of their wearer ([2], p. 402).

Garments, footwear, wallets, and accessories with 3D and 4D technology may be protected by copyright and industrial property rights, and especially 4D items may raise problems related to the protection of personal data. In these cases, rules and regulations on personal data protection must be fulfilled with particular care.

2. Changes in fashion industry caused by 3D printing

Technology is evolving ever so rapidly that almost everything may be printed in 3D, in all sorts of materials and at low prices. If 3D and 4D fashion items are cheaper and more practical in daily life, this technology will continue to evolve. Currently, there are many designers involved in 3D and 4D in fashion industry. There are several platforms online where the user may buy 3D and 4D items, mainly in relation to clothes, wallets and purses, footwear, wearables, smart clothes, and accessories. There are several 3D platforms where it is possible to print the 3D CAD designs when the designer does not have a proper 3D printer, because it is too expensive. It is predictable that in the future many will have the economic capacity to buy a good 3D printer and print the CAD file designs from home. The fashion industry has to adapt, namely selling CAD files in a way that will allow users to print authentic item, not counterfeited, instead of buying them in physical stores on online stores. It is desirable that fashion companies will be associated with 3D or 4D printing platforms that may provide legal CAD files, allow customization by each user, and provide 3D or 4D printing services. This is a new paradigm in fashion industry.

2.1 Return to tailor-made

Thanks to 3D printing we may move from a fashion industry almost all based on *prêt-a-porter* to a tailor-made one. Three-dimensional (3D) printers are the sewing machines of the twenty-first century. The items are made according to the body and taste of the user. Prototypes rapidly become the end product. Three-dimensional (3D) fashion modeling is like architecture. One makes 3D scans of the measures and with the software makes the creation, being able to use unbelievable shapes. Cristiano Ronaldo, for example, has 3D-printed Nike soccer boots, customized according to medical advice regarding the weak points of his legs, knees, and feet. Patterns of injuries are detected, and the boots are designed to avoid them. The 3D soccer boots are lighter, totally customized, flexible, and completely adapted to the player's needs and taste ([1], p. 137; [6]). This happens with many important players of various sports.

In the autumn/winter 2018 collection, Demna Gvasalia, creative director of Balenciaga, used 3D scans of the bodies of the models and a CAD program to reach the desired tailoring. Tweeds and velvets were glued to light foam to create almost finished 3D-printed coats ([1], p. 137).

XYZ created the famous *InBloom dress* using an Ultimaker 3D printer, with PLD plastics. It is gorgeous, took 480 h of printing, has 1.7 kg of flexible plastic filaments, and seems a lace dress ([1], p. 137; [7]).

Italian trademark XYZBAG creates and sells 3D-customized bags [8].

The Spanish trademark ZAP&Buj-BUJ STUDIO creates clothes with flexible filaments on an elastic tulle fabric, which allows the adaptation of the piece to the movements of those who wear it [9].

Annie Foo creates high fashion 3D-printed shoes [10].

One of the more recent 3D dresses of Iris Van Herpen is the "Infinity Dress," that changes its appearance with movement [11].

Anouk Wipprecht above-mentioned "Proximity Dress," with many sensors that allow the dress to stretch creating a barrier if someone approaches, thus identifying the infringement of the mandatory safe distance due to Covid-19, is also 3D-printed [12].

2.2 Less logistic and transportation costs

Three-dimensional (3D) technology allows to manufacture less items and to cut on transportation costs. Manufactures and consumers approach, without or, at least, with less intermediaries, mainly transport companies. A fashion company may create the item in a CAD file and print it wherever it wants. If it is a Portuguese company of 3D-printed shoes, for example, with headquarters in Porto, they may be printed here, requiring a small team of experts for quality control. If, for example, this company were to open a branch in South Korea, the file of the shoes may be sent by the internet, namely by e-mail, and only 3D printers, a small team of control experts and distribution at the national scale, besides a place to store printed items, are needed. There is no need to search for the country with the cheapest production costs all over the world. Fashion companies may have reduced distribution chains and less storage needs.

The end user may participate in the process of manufacture, asking for the piece to be customized according to taste and needs. There is no need for large storage of the printed items. One can lower the number of printed pieces according to the requests of the consumers.

The whole manufacturing and distribution processes can be cheaper, sustainable, and democratic. We must not forget that many 3D printers, protected by patents, are falling into public domain, thus cheaper and legally able to be used by anyone, the same is happening with many invented materials.

2.3 Less undifferentiated workforce

As with all technological industries, the new production model needs less workers, especially undifferentiated ones. In the present, few items are done entirely in one single 3D-printed piece. It is still necessary to sew, to connect the 3D-printed parts. Like what happens with the sleeves and buttons of a coat, or the accessories that embellish the dress, for example. Anyway, less undifferentiated workers are needed, and they can work in the country where the items are 3D-printed, namely in the country where the company has its headquarters.

This may imply less work demand in underdeveloped countries, that do not respect human rights of the workers and have cheap labor force, and more demand in more developed countries. Items do not need to be transported to countries like Bangladesh, Pakistan, or Mexico, or even to India or China. In China, technology is abundant, but labor force is cheap, hence the interest in China of many fashion companies. Thus, 3D printing may contribute to the decrease of infringement of worker's rights. Less differentiated workers will still be needed to assemble 3D parts, give the finishing aspects of items, control quality, and work in smaller distribution circuits.

Creating a fashion item will be each time more like creating a work of architecture, or a sculpture, with 2D and 3D phases, and will imply the use of software and large teams of experts. Designers, engineers, software and hardware experts, architects, economists, managers, marketers, big data, and trend experts, among others, can all be, in some way, involved in the fashion industry. We must not forget end users, consumers, that may become co-authors, if the company allows so (prosumers).

Resistance from undifferentiated or less differentiated workers and workers' unions is expected, and that is an important issue to be addressed.

2.4 A more environment friendly fashion industry

Three-dimensional (3D) fashion industry is more environment-friendly, causing much lesser pollution. An additive and non-subtractive technique means much lesser waist.

Plastics, which are everywhere, namely in the oceans, and cause great concern, are reused and recycled into 3D (or 4D) fashion items. These items can also be recycled and reused. Nowadays, 3D fashion also uses biodegradable materials. To mention some of the trademarks that already incorporated 3D fashion and use recycled plastics from the oceans, we may name Nike, Adidas, or New Balance sneakers and sportswear ([1], p. 139; [12]).

2.5 Approaching fashion from the virtual world

If fashion industry changes to be based in 3D and 4D technology, using CAD files, and printing less quantities of goods, many of them "intelligent," as happens with wearables and smart clothes, and if the items are increasingly sold online, namely allowing purchases through download of the files, sharing of the files, or 3D printing online platforms, the fashion industry will progressively dematerialize.

This approach to the virtual world was greatly sped up due to the Covid-19 pandemic, which led to the introduction of more and more augmented and virtual reality (AR and VR) in fashion, besides accelerating the development of wearables and smart clothes. Fashion trademarks are already entering the starting metaverses.

Augmented reality (AR) and virtual reality (VR) are already used in fashion. While virtual reality (VR) is totally immersive, augmented reality (AR) is partially immersive, meaning that the virtual experience overlaps the real word in AR. To have 3D virtual reality experiences, one must use helmets or glasses of VR. AR 3D experiences may be achieved through smartphones, projectors, or other hardware [13].

In fashion, AR may help trademarks to make fashion statements, stand out from the crowd, and connect with clients through new and exciting ways. AR is associated with entertainment, being very gratifying. The technology allows consumers to experience the item before buying it, which can reduce returns dramatically and allows access through social networks (namely Instagram and TikTok). If the experience is good, it has the potential to be shared and sharing can go viral, with new customers wanting to buy the product without the need for marketing or advertising. AR experiences can create fashion trends. There are several fashion companies that are using AR filters to attract new customers and engage existing ones. This was very important during the Covid-19 pandemic and has also to do with the importance of the use of social networks. Trying before buying increases profits by reducing returns, and as said above, this immersive experience can go viral. There is hardly a person that has never used social media filters. Without posting anything, one cannot avoid the temptation to find out how one will look like with blue hair, a certain shape of sunglasses, dress, shoes, and so on. Among fashion trademarks that have these filters and other AR technologies, we find Gucci, Calvin Klein, Louis Vuitton, Zara, Burberrys, and H&M. AR technology increases costumers' confidence and helps to create data for analytics, because it helps to understand reactions of users toward new collections, thus allowing to predict and create trends. AR is good for the environment because it allows creation of digital clothes. Fashion influencers, bloggers, and everybody, can try them on and take pictures without having the physical item. Many AR techniques are being used in fashion. The AR Clothing Try-on is the dream of every passionate shopper, and one can access and experiment on any tablet or phone. For example, the luxury online shop FARFETCH used a particular feature under collaboration with the trademark Off White. The shop allowed costumers to try jackets from the collection using the photo messaging app. The 3D and stimulation technologies made clothes cling to the body on display and move by their actions. Virtual Tailors allows customers to know the exact size of the items based on their body measures. AR-powered virtual tailoring feature is used by Hugo Boss and Superdry. It helped to tailor the items to almost everybody measurements. Nike also combined tailoring and trying-on features, allowing the visitors of the app to measure their feet and find the correct size and virtually try on the item.

There are Colour-Changing apps that allow to change colors and patterns. For example, the American trademark American Apparel offered the costumers the possibility to scan the clothes on display in the shops and change their appearance using the application. AR Virtual Showrooms are a great idea for showing the collections worldwide in a much more sustainable way. Tommy Hilfiger made a digital display of the collection in Amsterdam. The audience could experience clothes by tapping an interactive touchscreen table and seeing every necessary detail on the screen.

Social media AR filters are very important because social media is a marketing star and every company that wants to succeed makes efforts in order to approach the audience in such platforms, for example, by creating custom social media filters. PE Nation, an Australian trademark, lets users try one of the pieces of their collection and imagine themselves being in snowy mountains. Virtual Make Over techniques are also available. Trademarks like Bobbi Brown, L'Oreal, and MAC launched try-on websites allowing customers to try lipstick, blush, or contour sticks before ordering them. AR or even VR is very important for new product launch because they allow for an immersive experience.

AR Mirrors are very interesting in clothing stores. In 2015, start-up Oak Labs presented interactive smart mirrors that allow costumers to see themselves in the reflection. Topshop and UNIQLO use similar technology. In the makeup market, Saks Fifth Avenue's beauty floor and Sephora launched "magic mirrors," allowing visitors to try makeup online. AR also succeeded in saving time when shopping. An In-Store-Navigation feature shows the shop visitors where to find the department they are looking for, or what items have special prices, which is very interesting in big department stores. AR allows better assortment in physical stores. Sometimes, the physical stores do not have all the items of the various trademarks they sell. To show them virtually is more sustainable.

In jewelery, which is expensive and delicate, using AR can be specially interesting because it can be hard to display the items in the physical placements, but costumers can try the items on using AR. Trillion Jewellery uses its app for virtual try-on effortless and fun.

AR expands advertising media options. Marketing is all about emotions, and AR can bring that emotional experience to advertising, and clients want to know more about the company and about it has to offer. Men's clothing Boohooman trademark transformed Black Friday into Hack Friday, creating the digital character and hacker Robin that appeared on all posters with QR codes placed during the campaign. Everybody could scan the code and see the character come to life talking about what the trademark offers. With AR, digital clothing can stand as stand-alone product. Virtual clothes and NFT's are so popular that in 2020 companies could rely on them to be the only selling item, which has also to do with the pandemic and the changes in the way people consume fashion. Studies show that in 2020 costumers started to use fewer products than those they bought, preferring digital items. The Fabricant is an all-digital company that creates virtual clothing since 2018. It became so popular that started to collaborate with trademarks like Puma and Tommy Hilfiger. DressX is also successful. The company uses AR to show that some clothes can exist only in their digital versions, thus contributing also to sustainability. In DressX app, one can find what digital clothes are best using smart AR filters. AR is strictly connected to games (gamification). Gamification in fashion is a big hit with AR technologies. Using AR, the company can create a fun and engaging gaming experience helping costumers to see what it has to offer, many times creating mini games. Victoria's Secret uses the game Pink line. By playing games and unlocking badges, costumers may know more about the company's products and gain exclusive offers [14].

There is an increasing interest of fashion trademarks to dress characters or avatars in games, like what happens with Balenciaga in Fortnite [15]. With games becoming more and more popular, fashion and AR have great potential to join and grow together. Louis Vuitton also created fashion items to the League of Legends, with great success [16]. In many cases, the digital items of the games can also be sold in the physical world and to inspire new collections (like the Balenciaga items). So, AR can come out of the virtual world of games and AR items can be also sold in physical way, namely 3D-printed. We are seeing an interesting and complex mixture between gaming and fashion.

The most recent win of AR happened when the Iridescence virtual dress was sold for $9500 during the Ethereal Summit in New York [14].

Zara also uses AR with great success to bring costumers to stores [15, 17]. AR technology in fashion business is very promising solution if adequately used.

Virtual reality is not so developed in fashion, and elsewhere, because specific hardware is needed, for example, VR helmets or VR glasses. Furthermore, totally immersive experience still has some unattractive reactions, as some individuals experience dizziness or other mild symptoms after experiences in virtual games. Interesting aspects of VR include the possibility of adding gloves with sensors, simulating touch, and creating more interactivity with the virtual items.

During the pandemic, VR started to have an important role in fashion because there could be no presence showrooms and fashion weeks. Virtual runway fashion shows were created to allow the users to virtually attend them, which led to democratization of those fashion shows by trademarks that sent VR helmets or VR glasses of those virtual runways to potential clients. Another important aspect of VR has to do with the buying experience. Companies like Tommy Hilfiger offer their clients the possibility to visit their virtual shop (that is like some of the most emblematic shops of Paris, New York, or London) because in cooperation with technology giants they got the know-how to create the virtual store. The experience allows to hear the music existing in the physical stores and to access parts of the runway, obtaining a 360 view of each design chosen by the client ([2], p. 404).

AR and VR in fashion will extend to metaverses, that is, virtual places where second lives that are starting to appear and attract people that do not want just the real life they live in, and whose avatars must be dressed according to the status they have. Many fashion companies are already investing in metaverse, often described as a network of virtual places linked into a virtual universe, and considered a future version of the Internet. In the metaverse, customizable avatars and dynamic group experiences enable a new era of social interaction. Concerts, parties, happy hours, and religious ceremonies are increasingly taking place virtually, with individuals participating regardless of their geographic location, especially in the aftermath of Covid-19. Metaverse members will engage with and purchase digital and real-world apparel, sporting goods, and other items through virtual shopping malls [18, 19].

3. Fashion and videogames

A videogame is an electronic game that can be seen in a screen. It is a type of audio-visual work and protected as such by copyright. It is a very profitable fast global growing market. The pandemic led to a huge growth of the videogames market. Fashion companies saw the opportunity and joined the videogames world.

Balenciaga was pioneer. Trying to avoid the Covid-19 to affect its fashion market even more, adapted to this new business concept presenting its collection fall/winter 2021 in the videogame "Afterworld. The age of tomorrow."

It is a futurist game, located in 2031, trying to create a future where clothes can transform themselves in order to be reused and last in time. The game is a virtual runway show, where existing and designs of future collections are shown, but the avatar engages in surpassing proofs and in different adventures, like what is common in videogames ([2], pp. 376–377).

Louis Vuitton cooperated with American e-sports *Riot Games* in the creation of the "League of Legends." The cooperation is based in many projects, like the creation of a trunk for the "Summoner's Cup," or the collection *LVxOL,* with a garment style based on the game, which is very appealing to the gamers and to the followers of *Louis Vuitton*. The trademark reaches a public which, otherwise, would not be possible to reach ([2], p. 377).

Celebration 200 years from the birth of the founder of the fashion house, *Louis Vuitton* created a videogame called "Louis the game," that was a continuation of the videogame "Endless Runner" and appeared in 2019 for the Fall and Winter Collection 2019. "Louis the game" is an app of interactive games that pays tribute to the founder of the company through the eyes of *Vivienne*, the mascot of the workshop, that enters a digital world to celebrate the birthday of the founder of the house *Louis Vuitton* ([2], p. 377).

This game uses the *Blockchain* technology, including 30 *Non-Fungible Tokens*, 10 of them created by the American digital artist, graphic designer, and performer Beeper. Nevertheless, the *Non-Fungible Tokens* of this game cannot be sold, and they were created only for the purpose of collection ([2], p. 378).

Moschino decided do dress "the Sims" in 2019. The fashion company included a capsule collection inspired in the famous game. The models entered the virtual world allowing for the characters of the game to be dressed with *Moschino* clothes ([2], p. 378).

In the gaming domain, we must also mention the "Animal Crossing" saga of videogames of Nintendo. Thanks to its popularity many fashion houses decided to use the characters of the games to present their collections in a very creative way. *Valentino* house included its Spring Summer collection 2021 in this videogame allowing the user to immerse in a true photoshoot in which the 3D characters of Nintendo are the protagonists and use items of the collection. The images may then be carried to the Instagram page (@animalcrossingfashionarchive) where, with a certain code, the user may buy the desired item through a web page ([2], pp. 387–388).

4. Fashion, Blockchain, and Non-Fungible Tokens (NFTs)

The *blockchain* technology is being used in fashion in many ways [20–22]:

a. to guarantee the authenticity of the items. One of the main characteristics of the blockchain is immutability; thus, it is possible to make a chain of transactions through the application of blockchain solutions to those fashion items that are object of sale, in order to guarantee their origin, who designed it, the trademark owner, among other things. Therefore, it is simpler to identify and locate counterfeit products or products that present alterations with respect to the product initially marketed by a fashion company. In March 2019, the *LVMH Moët Hennessy-Louis Vuitton (LVMH Group)* released the launch of the project "Aura," developed by Quorum, Microsoft, and Consensys. The project is based in a blockchain solution that allows to verify the authenticity of the products of the fashion company, uniquely identifying its origin and authorship in real time. This procedure favors transparency and confidence of the clients. *Vestiaire Collective* uses the advantages of *blockchain* technology in an *online* platform for the sale of high-end fashion and accessories. The platform has a group of fashion experts that certifies the authenticity of the products. After this process of verification and validation, the article is incorporated with a technological footprint through blockchain technology, accompanying its entire useful life, as a sign of authenticity and quality, against possible counterfeits ([2], p. 388).

b. to guarantee the transparency of the production chain. In the last year, it has become increasingly important for companies to respect ethics, culture, and social aspects. Clients want to be informed about the origin of the products, location of factories, compromises with the environment, and respect of human rights. *Blockchain* technology can help to reassure the transparency, the security, and the legality of the production chain. The technology implies the incorporation in each garment of a QR code (*Quick Response Code*) that contains all information about the entire manufacture process ([2], p. 389).

c. With smart contracts. *Blockchain* technology allows the creation of *smart contracts,* or intelligent contracts, whose main characteristics are its automatic fulfillment and execution. The *smart contracts* run autonomously, via command *if-this-then-that,* when the factual assumption contemplated by the contract is fulfilled. This type of contract is very useful when there is an objective component, such as the activation of the payment because of the receipt of a specific remittance or the payment for the contracting of an insurance or financial product. Intermediaries are eliminated and, therefore, the costs ancillary to the contract are considerably reduced. With the ever-increasing evolution of *Blockchain* technology, *smart contracts* can be utilized and integrated into the domain of *Intellectual Property Rights* to digitize and secure transactions of such intangible assets while ensuring trust, transparency, and safety of both the enforcer and the consumer ([2], p. 390).

The *Non-Fungible Tokens (NFTs)* are digital assets whose value is granted by the user, or whoever is willing to pay at a certain time, since, through this type of tokens, a kind of collectible and unique objects are created. These features are very attractive in fashion and art, where what is exclusive and unique is highly valued, and *NTFs* grant authenticity. Each *NFT* has a single valuation, being part of a certain *blockchain*. Thus, what one acquires buying an *NFT* is, mainly, a license or an asset to sell in the future.

What the sale of NFT and the purchase of haute couture garments through this way entails is that the buyer acquires the use of them, that is, acquires an implicit and nonexclusive license for its use. Copyright is not transferred to the buyer, unless expressly agreed by the seller, as the intellectual property right remains in the possession of the author who created the garments.

This new business model led to the emergence of digitally native fashion trademarks in the market, that is, trademarks that lack physical showcases or recognized names in the conventional fashion industry and sell their products through *NFTs*.

Some examples of these trademarks are: Overpriced, with its sale of a virtual sweatshirt such as TFN by 22,000 euros; RTFKT and its sale of digital sneakers, paying homage to the culture of the game; DressX, a fashion store in which customers buy a garment and send a photo of themselves, receiving in return a digital photo of them wearing the garment they have bought, and its essence lies in the fact that users buy clothes to upload photos on the networks; and Auroboros, as last example, is the first fashion house that merges science and technology with physical haute couture [2].

5. Three-dimensional (3D) fashion and intellectual property

Three-dimensional (3D) fashion must respect copyright, related rights, and industrial property rights, namely patents, utility models, design patents, trademarks, trade dress, logotypes, denominations of origin, geographical indications, rewards, trade secrets, and rules against unfair competition.

Three-dimensional (3D) and 4D files must respect intellectual property rights and whoever 3D prints must have the authorization of the copyright or of the industrial property right (trademark, design, patent, or other right) owner. Because in EU copyright laws register is not mandatory, as a rule, it may be difficult for the 3D printer to know whether the CAD file comes from a legal origin. This is essential to ensure the absence of counterfeit. Industrial property rights, as a rule, must be registered, so it may be easier to know if the item comes from a legal source, at least in the EU [23–26].

The problem is private use. One must deal with the difficulty to grant the enforcement of intellectual property rights when the CAD files and its online sharing and 3D printing are done for strictly private uses. In copyright, the situation is not covered by exceptions or limitations and does not fulfill the three-step test. Regarding industrial property rights, laws must be adapted in order to forbid not only 3D printing with economic or commercial goal but also private uses. A company's profit is affected when the consumer decides to print counterfeited items instead of buying the legal ones. If necessary, industrial property laws must be changed, considering the deliberate 3D printing of counterfeited trademarks illegal, logotypes, or designs, among others, even for private uses [27].

An important way of preventing piracy could be to protect the CAD files that contain a work or other protected industrial property exclusive rights. Protection should also involve the CAD files. The CAD file that has a trademark, for example, will be protected as if it was the trademark itself.

Regarding the price of the 3D printers, that is, the hardware, an amount ought to be for copyrights, related rights, and industrial property rights owners, as it happens with uses for private copying in relation to reproduction right for private use in the EU in copyright and related rights.

Technological measures to protect the CAD files and digital rights management systems would also be important to enforce intellectual property where 3D technology is used, including the possibility of the use of technology of destruction of illegal CAD files.

Blockchain technology grants intellectual property rights owners against non-authorized exploitation by third parties regardless the link in the production chain in which the articles are located. *Smart contracts* can be utilized and integrated into the domain of *Intellectual Property Rights* to digitize and secure transactions of such intangible assets while ensuring trust, transparency, and safety of both the enforcer and the consumer.

Enforcement is essential in the 3D fashion market. Fashion is already very affected by piracy, and hacking is a big issue to address when we are approaching the dematerialized fashion world.

Fashion companies can be affected in their reputation when the pirated items lack quality, and the consumers will associate them to the companies affected by the illegal copies. Additionally, if the CAD files and the 3D-printed items are not approved by a quality department and cause damages to the consumer, the producer will be liable, but we must find an effective way to find who is the actual producer.

6. Personal data protection

Wearables and smart clothes raise questions related to personal data protection [28, 29].

Wearables and smart clothing products come in many shapes and sizes, but all of them have sensors that allow a series of functions. For example, some shirts monitor the heart rate, activity, and breathing, and sleepwear garments can track sleep levels and sync data to the smartphone. Football fans will also have noticed that footballers are increasingly wearing under-shirt garments that resemble a sports bra. It is a form of smart clothing containing a GPS tracking device that records data about the players' movements. This also happens in other sports and activities. Data such as heart rate, daily strides, and calories burned, as well as location and route data are recorded. This is a lot of sensitive information. Manufacturers are bringing more and more new technologies for recording a wide range of body and movement values onto the market. This is also reflected in the sales trend in the wearable sector. The best known and most successful devices in this field are smart watches and bracelets that function as fitness trackers. For the functionality of the wearables and the apps, the operators usually record data as time, location (GPS), heart rate, blood pressure, sleeping behavior, calorie consumption, pedometer, and much more. In addition to smart watches, many fitness trackers also become smart and provide information about incoming calls, upcoming appointments, messages, and notifications from social networks. With all this information, a personal profile of a person can be created. Now rings, earrings, and pendants start to perform the same functions of smart watches and bracelets.

Besides the personal data included in wearables and smart clothes, the way in which the data are sent is very important. These are recorded in the wearables (or smart clothes). The raw data generated by the sensors are then usually transmitted to the own smartphone (or other wearables or smart clothes) via Bluetooth. There the user can then view them in evaluated form. Adding to that, most manufacturers collect and store all captured data in a cloud. For all these reasons, it is fundamental to protect these personal and sensible data.

Despite concerns about data protection, wearables and smart clothes also offer great potential. The technology opened and will open many new areas of application. Especially in the medical field, there are many development opportunities. Sensors can detect diseases at an early stage and support medical care. Each time, more companies are developing clothing equipped with sensors and smart fibers that offer completely new possibilities. Smart garments can detect movements and touches in addition to the recording of classical health data. This offers ideal possibilities for recording and optimizing motion sequences in certain sports, as aforementioned. This means that data protection issues will arise each time more.

The wearable and smart clothes industry is still starting to develop. It has great potential, but with data protection issues will arise. It is fundamental to know how data are obtained, how they are stored and transmitted, and what they are used for. When one buys wearables or smart clothes, it is very important to know the general terms and conditions and data protection declarations. Attention should always be paid to the possibilities for deleting the collected user data to avoid possible future discrimination. Furthermore, there is always the right to request the stored data from the manufacturers.

Legislation is also fundamental. In the EU, there is, among other acts, the important Regulation (EU) 2016/679 of the European Parliament and of the Council. This General Data Protection Regulation *(GDPR)* is a European Union regulation on data protection and privacy in the EU and the European Economic Area (EEA). It is an important component of EU privacy law and of human rights law, in particular Article 8°, n°1 of the Charter of Fundamental Rights of the European Union [30, 31].

7. Conclusions

Three-dimensional (3D) printing became the sewing machine of the twenty-first century, causing great changes in the fashion industry, namely the appearance of more sustainable, and customized apparel and accessories, simpler logistics and less transaction costs, increased creativity, and cooperation of specialists from many different areas. The final idea of prosumer and its meaning is before us. If we add 4D printing, that is, "intelligent" materials, 3D printing enters our day lives, with wearables and smart clothes. If we add the recent developments achieved by augmented reality and virtual reality and metaverses, the fashion world becomes even more virtual and sustainable, and buying is each time more connected to a gaming experience. All this raises numerous questions regarding copyright and industrial property protection and possible infringement in those areas.

Three-dimensional (3D) and 4D files must respect intellectual property rights and whoever 3D prints must have the authorization of the copyright or of the industrial property right (trademark, design, patent, or other right) owner. Because in EU copyright laws register is not mandatory, as a rule, it may be difficult for the 3D printer to know whether the CAD file comes from a legal origin. This is essential to ensure the absence of counterfeit. Industrial property rights, as a rule, must be registered, so it may be easier to know if the item comes from a legal source, at least in the EU.

It is difficult not to infringe intellectual property rights when the CAD files, its online sharing, and the 3D or 4D printing are done for private uses. In copyright, the situation is not covered by exceptions or limitations and does not fulfill the three-step test. Regarding industrial property rights, laws must be adapted to also forbid 3D and 4D for private uses and not only with economic or commercial goals.

An important way of preventing piracy could be to protect the CAD files that contain a work or other protected industrial property exclusive rights. Protection should also involve the CAD files. The CAD file that has a trademark, for example, will be protected as if it was the trademark itself.

Regarding the price of the 3D printers, that is, the hardware, an amount ought to be for copyrights, related rights, and industrial property rights owners, as it happens with uses for private copying in relation to reproduction right for private use in the EU in copyright and related rights.

Technological measures to protect the CAD files and digital rights management systems would also be important to enforce intellectual property where 3D and 4D technology is used, including the possibility of the use of technology of destruction of illegal CAD files.

Enforcement is essential in the 3D fashion market. Fashion is already very affected by piracy, and hacking is a big issue to address when we are approaching the dematerialized fashion world.

Fashion companies can be affected in their reputation when the pirated items lack quality, and the consumers will associate them to the companies affected by the illegal copies. Additionally, if the CAD files and the 3D-printed items are not approved by a quality department and cause damages to the consumer, the producer will be liable, but we must find an effective way to find who is the actual producer.

Wearables and smart clothes raise questions regarding personal data protection. The issues to be raised are multiple because we are faced with sensors or sensor terminals. Data security of the device itself in transmission and storage must be addressed. It is fundamental to know how data are obtained, how they are stored and transmitted, and what they are used for. When one buys wearables or smart clothes, it is very important to know the general terms and conditions and data protection declarations. Deleting the collected user data is fundamental to avoid possible future discrimination. Furthermore, there must always be the enforceable right to request the stored data from the manufacturers.

Legislation is fundamental. In the EU there is, among other acts, the important Regulation (EU) 2016/679 of the European Parliament and of the Council. This General Data Protection Regulation *(GDPR)* is a European Union regulation on data protection and privacy in the EU and the European Economic Area (EEA). It is an important component of EU privacy law and human rights law, in particular Article 8°, n°1 of the Charter of Fundamental Rights of the European Union.

Author details

Maria Victória Rocha
Escola de Direito do Porto (Oporto Law Shool), Universidade Católica Portuguesa (Portuguese Catholic University), Portugal

*Address all correspondence to: virocha@ucp.pt

References

[1] Rocha MV. Moda e Impressão 3D: um novo paradigma? Fashion and 3D Printing a new paradigm? Revista Electórinca de Direito (RED). Outubro. 2018;**3, 17**:112-115. DOI: 10.24840/2182-9845_2018-0003_0006

[2] Prado L, Lahiguera R, Nuevas. Tecnologias y moda (chapter 9). In: Ortega Burgos E, Antón Juárez I, García Perez FJ, editors. Tratado de la Moda (Fashion Law) Vol. I: Propiedad Intelectual, Contratación mercantil, Derechos de las nuevas technologias, penal y aduanas. Pamplona. Aranzadi: Thomson Reuters; 2022. pp. 399-403. ISBN: 978-84-1391-685-9

[3] BeClothed—Premium Technical Apparel [Accessed: [25 November 2022]

[4] Moxadtech a máscara portuguesa que cancela o vírus do COVID-19 How2Go (h2gconsulting.com) [Accessed: 25 November 2022]

[5] CLIU Mask—This is not a Mask [Accessed: 25 November 2022]

[6] 3D Printing we should 'Just Do It'! - Technology and Operations Management (harvard.edu) [Accessed: 25 November 2022]

[7] Printing fashion in 3D: The inBloom Dress – Inside the Collection (maas. museum) [Accessed: 22 November 2022]

[8] Create your 3D bag | XYZ bag [Accessed: 22 November 2022]

[9] Zap&Buj, Making Fashion through Additive Manufacturing—3Dnatives [Accessed: 22 November 2022]

[10] Annie Foo Shoe Design Relies on 3D Printing for Bespoke Styles - 3DPrint. com | The Voice of 3D Printing / Additive Manufacturing [Accessed: 25 November 2022]

[11] Iris Van Herpen Infinity Dress—Procurar Imagens (bing.com) [Accessed: 22 November 2022]

[12] How Trademarks addidas, Nike, Reebok made 3D shoes in their factory | Future 3D printing technology - YouTube [Accessed: 28 November 2022]

[13] Navarro F, Martínez A, Martínez JM, Realidad Virtual y Realidad Aumentada, Desarrollo de Aplicaciones, Ra-Ma, Madrid; 2918

[14] In detail, with images and videos see Augmented Reality (AR) in Fashion - Everything that You Need to Know (wear-studio.com) [Accessed: 29 November 2022]

[15] A very well-known example of AR in fashion is the one of GUCCI, with its virtual sneakers, adding an AR feature to its app to let users 'try on' sneakers. Giving customers a visual representation of how a product will look in real life, the technology can in theory reduce returns and boost customer satisfaction. In recent years of pandemic, Gucci released a pair of virtual sneakers, designed to only be worn and shared online. The 'digital-only' trainers, created in collaboration with AR fashion platform Wanna, could be bought via the Gucci app for $17.99, making it an accessible way for consumers to have the luxury trademark. They could be seen wearing the sneakers, that's what was important. Building on the rise in popularity of both virtual influencers and AR filters on social media, this is one of the first examples of 'digitalfashion'—a concept that some predict will be dominant in

fashion. Gucci releases first Virtual 25 sneaker that can only be worn in AR (dezeen.com) [Accessed: 20 October 2022]. Before that in July 2020, to promote their new sneaker collection, Gucci partnered with Snapchat to launch their first global augmented reality shoe try-on campaign on the platform. They created AR try-on lenses to let Snapchat users see how Gucci sneakers look on their feet. The collaboration was a huge success. Augmented Reality Can Be Real Gucci | by Han Nguyen | Marketing in the Age of Digital | Medium [Accessed: 20 October 2022]

[16] Louis Vuitton is working with video game League of Legends | CNN Business [Accessed: 20 October 2022]. For more examples of fashion trademarks using AR see Augmented Reality In Fashion | Rock Paper Reality [Accessed: 20 October 2022]

[17] Zara's Augmented Reality App Brings Virtual Models to Life in Stores (wbresearch.com) [Accessed: 29 November 2022]

[18] With more detail see Fashion in Metaverse - Fibre2Fashion [Accessed: 30 October 2022]

[19] Peña Ó. Metaversos, la gran evolución imersiva. Madrid: Anaya Multimedia; 2022

[20] See Digital Couture | Auroboros | London [Accessed: 10 January 2023]

[21] Sterner P, T. Blockchain, NFT y Smart Contracts: La guía completa, para principiantes, para comprender las criptomonedas y el concepto de Blockchain. Independently Published; 2021. ISBN-13: 979-8787289367

[22] Charlotte, Fiell P. Blockchain for Business. Boston: Addison Wesley; 2019

[23] Rocha MV. Impressão 3D e Direito de Autor/3D Printing and Copyright, Revista Electrónica de Direito (RED). Junho 2017- 2, pp. 1-28

[24] Herrero C, Teresa M. "El concepto de obra plástica y la impresión en 3D", Propiedad Intelectual en el siglo XXI: nuevos continentes y su incidencia en el derecho de autor, Espín Alba, Isabel (coord.), Colección de Propiedad Intelectual, ROGEL VIDE, C. (Dir.). Reus, ASEDA, Madrid: Fundación AISGE; 2014

[25] Le Goffic C, Vivès-Albertini A. "A impressão 3D e os Direitos de Propriedade Intelectual (1ª Parte)". Propriedades Intelectuais. 2015;**3**:40-55

[26] Le Goffic C, Vivès-Albertini A. A impressão 3D e os direitos de propriedade intelectual (2ª Parte). Propriedades Intelectuais. 2015;**4**:15-25

[27] Alves, Diogo. A impressão 3D e a sua crescente relevância na Propriedade Intelectual (The 3D Printing and the Growing Relevance at Intellectual Property) [master thesis]. 2017. Available from: https://ssrn.com/abstract=3466224; http://dx.doi.org/10.2139/ssrn.3466224

[28] Wearables and Data Protection - What Providers Know About You (ispo.com) [Accessed: 06 October 2022]

[29] Smart Clothing: Should You be Worried About Your Privacy? | Cybernews [Accessed: 06 October 2022]

[30] Voigt P, Bussche A. von Dem, the EU General Data Protection Regulation (GDPR): A Practical Guide. Berlin: Springer; 2017

[31] Kazemi R. General Data Protection Regulation (GDPR). Hamburg: Tredition; 2018

Chapter 19

Additive Manufacturing-Based Supply Chain Configurations

Silvana Gallinaro

Abstract

The topic of the chapter "Additive manufacturing-based supply chain configurations" is about the implementation of additive technologies in conventional supply chains and the possible supply chain configurations that can be generated. As a guideline in the field of supply chain management, this chapter suggests designing the AM-based supply chain configuration according to the supply chain strategy decided by the focal company of the supply chain. Two questions are not fully resolved in doctrine: the first concerns the measurement of the effects of additive manufacturing implementation in conventional companies and supply chains: the second, the relationship between the total average unit cost and the production volume of additive products. In agreement with some scholars, quantitative approach to the analysis of the impact of additive technologies in companies and supply chains is recommended, and the choice of a simulation method for ex-ante assessment of pros and cons of additive technologies over conventional ones is suggested. The goal of this chapter is twofold: to demonstrate that the superiority of additive manufacturing over conventional one cannot be discussed a priori, because it must be proven quantitatively on a case-by-case basis; to support the thesis according to which additive machines can achieve economies of scale.

Keywords: additive manufacturing, supply chain, simulation, economies of scale

1. Introduction

Additive manufacturing (AM) does not take place by removing material from a solid, as in conventional or subtractive production, but by adding material, that is, by overlapping layers of materials (such as ceramics, metals, plastics, etc.) in sequence, starting from a 3D (virtual) design that contains all necessary information to create the product, generated using CAD design system or by scanning already an existing object in 3D (reverse engineering).

For additive machines to understand and process this information, a 3D design file must be converted into an STL file (Standard Triangulation Language). By means of a slicing program, information contained in STL file is converted into G-code (machine language). Each slice of an STL file represents a layer of the product to be made by overlapping and by means of an additive machine, using different methods of material extrusion and layer solidification [1, 2]. Time required by the additive manufacturing process depends on product size; specific additive technology adopted; raw material used and degree of precision required in product realization [2].

IntechOpen

In fact, there are many additive technologies available to companies. Among these, 3D printing, often but erroneously, considered synonymous with additive manufacturing [3].

Additive raw materials can be in form of powders, filaments, liquids, or solids. Additive manufacturing processes also use food composites, such as pasta and chocolate, and metal-ceramics, but also materials that are difficult to process by means of traditional subtractive production processes, such as carbon alloys for high temperatures, as well as, more recently, "living" biological materials, such as cells and biological tissues [4, 5].

Additive machines require little manual labor and few operational skills, but they postulate new work skills (e.g., in design), rich in knowledge, generating new organizational roles and responsibilities. Likewise, they do not require production equipment, thus removing the costs of retooling machines. Additive production is, in fact, also called "fixture-less layered manufacturing" because it occurs simply by sending a digital 3D design file to an AM machine without equipping it with tools and molds.

By conveying the manufacturing process in a virtual environment, additive technologies enable digital transformation of manufacturing.

Implementation of additive technologies in companies and supply chains can produce destructive effects on traditional production paradigms and supply chain structures, foster disintermediation and production decentralization, making supply chains lighter or more concise [4–8]. As it will be better explained later on, effects produced by additive manufacturing implementation in conventional technology supply chains depend on the value chain strategy adopted by the focal company and, therefore, on the logistic tier affected by the chosen additive technology.

In the current stage of the evolutionary process they have reached, additive technologies have not yet completely replaced conventional ones. They often integrate, support, and coexist with conventional technologies in companies and supply chains [9], being used in productions with high product customization (e.g., in the healthcare sector) and/or characterized by low or uncontrollable demand (large spikes and dips in product demand) [10–12].

2. Implementation of additive manufacturing technologies in conventional companies

2.1 Advantages of implementing additive technologies in conventional manufacturing-based companies

As regards the implementation of additive manufacturing technologies, it is necessary to distinguish between two different and distinct levels of analysis: company and supply chain.

Implementation of additive technologies in business processes is the result of a strategic decision of pure or tailored postponement [13] in a context of cost-effectiveness: additive technologies entrust product differentiation to design activities of value creation, optimizing organizational flexibility and, at the same time, making production of units or small series of items requested in a discontinuous and unpredictable way and/or with complex geometries, cheaper than in conventional production [14].

A postponement is an approach to value creation process. It takes place when one or more value creation activities are postponed at the very moment of receivement of customer order: the goal is product customization [15]. The company that postpones value creation activities waits to know what the customer really wants (form), where he wants it (place), and when he wants it (time), before starting the design and/or production and/or delivery of the product, so that this is exactly as requested and/or delivered according to the customer's place and time requests.

In this regard, some scholars describe various types of postponement strategies, each of which is according to the specific activity that is postponed. Then, they talk about form postponement strategy, if manufacturing and/or design is delayed, place and time postponement strategy, if delivery is delayed [16–18].

Order penetration point (OPP)—also referred to as customer order decoupling point (CODP)—is defined as the stage in value chain where customer order enters and product differentiation takes place. The further upstream in value chain the customer's order is placed, the greater the degree of postponement adopted by the company, therefore the greater the differentiation and degree of customization of the product. Pure or tailored product customization is achieved if product design is postponed, allowing the customer to intervene in this phase of value creation.

Postponement strategies make company—but also, as it will be said later on, supply chains—efficient in dealing with demand uncertainty [17–19]: in fact, more precise information about form, time, and place of delivery of product, and also about the amount to be made, can be obtained during postponement period, making company able to satisfy demand in form/time and/or place required, therefore agile and flexible.

Therefore, it can be said that, in contexts of environmental uncertainty, efficient companies (and efficient value chains) are those oriented toward postponement strategies, and that efficiency goes through flexibility.

So, the degree of postponement adopted by a company is a function of the level of external uncertainty to be faced and managed.

In conclusion, postponement is the "keystone" of mass customization and a tool for managing environmental uncertainty [19, 20]. Company that adopts postponement in value creation process, demonstrates that it has strategic capabilities for achieving product differentiation in one of the stages of value creation, downstream or upstream of manufacturing [21].

Implementation of AM technologies in a company allows to entrust product differentiation to value design activities and satisfies organizational flexibility needs of all sorts. These value activities enable the company to pure customization, moving CODP up to the product concept stage.

About that, Olhger [22] and Verboeket and Krikke [23] specify that, when AM replaces conventional production in a value creation process, OPP tends to move upstream, in parallel generating the shift of make-to-stock (MtS) production logic toward make-to-order (MtO) or engineer-to-order (EtO) ones. In other words, additive manufacturing makes internalization of customers in value creation process concrete, which becomes "value co-creation with customer process" as a result [14].

Generally, the full participation of the customer in the value creation process takes place by means of co-design platforms. So, additive technologies reveal the potential for pure customization if co-design platforms are active [14, 20].

Numerous scholars have addressed the issue of the pros and cons of additive manufacturing compared to conventional one. To the aim of implementing an additive manufacturing technology in a company or in a supply chain, and as an indication of a decisional method, pros and cons of a hypothetic additive plant compared to actual conventional one have to be matched with each other in a trade-off evaluation.

The topic of implementation of additive technologies in a company will be dealt with in this paragraph, reserving the following one to that of the implementation of additive technologies in conventional supply chains.

Additive manufacturing allows the company to carry out product differentiation and customization "without extra costs" [24, 25] or makes customization "without penalty" possible [1] because—as it is known—it does not require product-unique equipment and molds, and since the digital redesign of products is not expensive. In additive manufacturing, the degrees of freedom in product creation are exponentially higher than those allowed by conventional production because manufacturing is not limited by the possibilities allowed by subtractive technologies. In other words, additive manufacturing generates products characterized by large freedom and high complexity of geometric features, which are difficult to achieve with conventional technologies.

Ready-made products come out of additive machines, without the need for component assembly, and increasing reliability of products with complex geometries which could be compromised by assembly of components in conventional production. Therefore and ceteris paribus, by avoiding assembly costs, additive manufacturing cuts production costs.

Additive machines can achieve time savings in the creation of pure or tailored customized products: product lead times do not include machine set-up times and component assembly times—despite the generally longer production times of additive machines compared to conventional ones.

Additive machines can also accelerate the time to market of radical and incremental product innovation, thanks to lowering production times of early products, with which to rapidly test and validate product ideas in target markets [20]. Furthermore, additive manufacturing is combined with virtual product prototyping by which to carry out virtual simulation tests, thus eliminating the phase of physical tests which requires time to create prototypes.

Therefore, digitization of manufacturing makes it lean and agile, allowing efficient and fast production on demand of single units or small batches of pure or tailored customized products: that is, additive manufacturing achieves economies of variety.

Additive manufacturing evokes a context of just-in-time production, also generating:

- drastic contraction—up to cut-off—of end-use product and component warehouses and relative costs;

- reduction of warehouses of raw materials and lowering of the relative costs, both for lower consumption of materials per unit of product or component—in fact, these can be made hollow and therefore be lightened—and for less waste of materials compared to subtractive production, thanks to the possibility of reusing waste;

- lowering of manual work in the production processes, thus verifying lower costs of factory staff [2, 21, 22]. Organizational roles, skills, and responsibilities need to be redefined and new knowledge and work skills are required, however more expensive than manual factory ones [8].

In short, additive production verifies a lower dependence on manual work and a greater dependence on skilled work with a high knowledge content, for which it can be said that additive production acts in favor of manufacturing reshoring [26–28].

To all those advantages of additive manufacturing compared to conventional one analyzed so far, we must add the ability of processing materials that cannot be processed with conventional technologies (such as carbon alloys for high temperatures, metal-ceramic, and food composites) or anomalous nature for conventional manufacturing (such as organic cells and tissues); longer life cycle of additive product compared to conventional one, thanks to simpler, cheaper, and faster design upgrades that can slow or revitalize the decline of the product in the market.

Lightened additive components—due to the possibility of being made hollow—reduce fuel consumption of additive products (e.g., in the aerospace and aeronautical industries). In this sense, additive manufacturing has been labeled as functional for environmental sustainability.

From the above, it follows that additive manufacturing can generate greater value for the customer and, therefore, a greater willingness to pay a higher price to obtain highly customized products, whose unit production costs and delivery times are lower than those conventional manufacturing would be able to achieve [29].

2.2 Disadvantages of implementing additive manufacturing in conventional manufacturing-based companies

As anticipated, alongside the advantages (pros) of additive manufacturing over conventional manufacturing, there are numerous disadvantages (cons) that must be measured and compared with the aforementioned advantages in trade-off analysis.

First of all, absence of economies of scale of additive machines. In fact, a widespread thesis among scholars is that additive machines would not allow economies of scale because the average unit additive product cost tends to be invariable as production amounts vary.

In our opinion, the reasoning about economies of scale denied by additive technologies is more complex. Total variable costs of additive manufacturing tend to be higher than total fixed ones, but this does not justify the constant average total unit cost of the additive products as production increases. What is undoubted is that additive technologies make it possible to achieve the minimum efficient scale in correspondence with a very low number of units produced. It is known that the minimum efficient scale is high when the level of fixed capital required for production is high. The higher the minimum efficient scale, the smaller the number of firms on the market. This is valid for conventional technology machines, whose minimum efficient scale is reached in correspondence with a large number of products. All this confirms that additive technologies make the level of capital to be invested to do business low. As a result, additive technologies are able to expand the number of smaller efficient production structures present in a territorial context, because the level of fixed capital required to do business tends to be smaller than that required by conventional production.

Therefore, unlike conventional machines, additive ones are efficient in highly customized unit productions or in low-volume ones, and the convenience of using them in manufacturing processes decreases as demand grows up. Persistent limits of additive machines—such as low throughput times and poor quality standardization of products—are the causes that prevent digital manufacturing from achieving economies of scale. Nothing prevents these limits from being exceeded in the near future.

Another disadvantage of additive manufacturing is the non-standardizability quality of additive products and spare parts in terms of strength, durability, consistency, safety, accuracy, and consequent low certainty of reproducibility of products and spare parts [8], which often generates post-processing costs of products and components (e.g., costs of polishing the surface of products) [30]. Lack of shared quality standards makes quality assurance and product warranty difficult.

Disadvantages of additive manufacturing match with advantages in a trade-off evaluation are also difficulties in finding and the high cost of additive materials;

difficulties of their technical integration [26, 31], as well as high prices of additive machines to be used in production cycles.

In additive manufacturing, the hourly cost of subordinate labor tends to be higher than in the conventional one because the former has a higher skill content [31].

Moreover, there is a non-secondary problem of lack of clarity of intellectual and industrial property of additive product designs, which can represent a bottleneck for the implementation of additive technologies.

At last, the high energy consumption of additive manufacturing and material preparation processes must be considered, as well as the size limits of additive components or end-use products caused by the size limits of AM machines [32].

Table 1 shows the summary table of the pros and cons (advantages and disadvantages) of additive manufacturing compared to conventional one.

	Advantages	**Disadvantages**
Costs	• Minimization of investment in equipment • No assembly costs • Less manual work in production processes • Zero end-use product and component warehouses/just-in-time production of products • Lower consumption of materials per product/waste recovery	• Limited availability and high cost of raw materials • High prices of additive machines • Tendentially higher hourly labor cost • Post-processing costs • High energy consumption • Lower economies of scale
Times	• Shorter lead times and time to market • Longer life cycle of additive product for easy upgrade	• Lower throughput time
Flexibility	• Pure or tailored customization without extra cost • Greater creative freedom and complexity of the geometric characteristics of the products	
Quality	• Higher Reliability of products with complex geometries • Possibility of processing materials that cannot be processed with conventional technologies	• Lower shared quality standards and difficulties in the process of quality assurance and guarantee of additive products • Not always guaranteed availability of raw additive materials • Raw materials available are not always able to generate additive products with qualitative characteristics comparable to those of conventional ones • Non-standardizability quality of additive products and components
Price	• Willingness to pay a premium price	
Other		• Poor clarity of intellectual and industrial property of product designs • Dimensional limits of products and machines

Table 1.
Advantages and disadvantages of implementing AM technologies in conventional companies.

3. Implementation of additive technologies in conventional supply chains: AM-based supply chain configurations

3.1 Centralized additive configurations

If shared across multiple organizations, value creation occurs in a supply chain, which is the value system that integrates intra-organizational value chains.

The implementation of additive manufacturing technologies in conventional technology supply chains generates different AM-based supply chain configurations, depending on the supply chain strategy pursued by the focal company (manufacturer). Therefore, as a guideline in the field of supply chain management, it is suggested to design the AM-based supply chain setup according to the supply chain strategy chosen by the focal company. In other words, the AM-based supply chain setup should be designed to implement the chosen supply chain.

The adoption of additive manufacturing in conventional supply chains, in fact, implements not only and not always product customization strategies, but always occurs at the individual company level, but often cost optimization supply chain strategies.

The supply chain strategy determines the level of the supply chain affected by the implementation of additive technologies: the lower the level of the supply chain affected by the implementation of additive technology, the greater the customization of the product guaranteed to the customer, the faster the delivery time, but the lower the cost savings. Thus, it can be argued that supply chain strategies determine its basic configurations, which are as follows:

- centralized additive configurations, whether additive technologies are implemented in OEM plants for the production of components and products, or in regional distribution centers that produce spare parts on request from local service centers, using the spare part designs provided by the OEM. In this case, the spare parts are delivered to the local service centers before being used for the repair and maintenance of the installed base, that is, the local manufacturing plants [33, 34];

- fully decentralized or distributed additive configurations, whether additive technologies are implemented in production plants close to clients or in service centers close to installed bases;

- hub additive configurations, which are intermediate structures between totally centralized configurations and totally distributed ones, with intermediate locations of additive technologies, that is, between the main plant and local production centers (or service centers, in the case of spare part supply chains) close to customers [6].

AM machines centrally located in a supply chain typically eliminate inventory of components and spare parts that are infrequently or occasionally required by internal and external customers (low-demand products or products with large peaks and drops in demand) and whose response time to demand is not critical [12, 31]. In other words, the central warehouse of components or spare parts of this type is replaced by centralized production capacity and inventory of materials and 3D files for additive manufacturing [28]. The costs of storing raw materials and files for additive manufacturing are lower than those of components with large peaks and dips in demand, so the overall inventory packing and storage costs of centralized AM-based supply chains are likely to be lower than those of conventional supply chains.

	Advantages	Disadvantages
Costs	• Reduced stock-keeping and packaging costs for components, products, and spare parts with fluctuating and unpredictable demand	• Higher centralized fixed costs for investments in additive machines • Higher costs of raw material warehouses • High energy costs
Customization	• Higher degree of customization of the component without extra-costs	
Risks	• Lower risk of stock-out of safety components and spare parts	
Times		• Slower component throughput-times

Table 2.
Centralized additive configurations.

The presence of a centralized additive production capacity reduces the risk of stock-out of safety components and spare parts and increases the possibility of their customization.

On the other hand, costs for fixed investments and the dependence of the supply chain on suppliers of raw materials are growing. The costs of additive machines are high as well as energy consumption, while the impact of additive technologies on the total transport costs of the supply chain must be measured and evaluated according to the specific contextual situation.

Below is a synthesis of the analysis so far (**Table 2**).

3.2 Decentralized additive configurations

Implementation of additive technologies in production facilities close to places where end-use products are used, that is, close to the external or internal customers, draw AM-based decentralized supply chains and support product customization strategies. Bypassing component suppliers and end-use product distributors, AM-based decentralized configurations reduce the number of tiers in the supply chain Therefore, they embody the disintermediation of the supply chains, generating reductions in logistics costs, however in the face of the multiplication of local investments in the supply chain which must therefore be justified by local demands to be met in a specific way and in a short time.

In a nutshell, investments in decentralized additive machines increase the fixed production costs of the supply chain and lower the costs that depend on the logistic levels (e.g., transport costs, warehouse costs).

The lower total stock-keeping costs of AM-based decentralized supply chains, compared to traditional ones, are determined despite the greater presence of inventories of raw materials, often in the form of powders or liquids, instead of inventories of components and sub-assemblies, which are bulkier and more expensive than materials. It follows that whether on the one hand, it is possible to observe a minimum dependence of the decentralized supply chains on the suppliers of components, on the other hand, a greater dependence of the same supply chains on the suppliers of additive raw materials is generated.

Decentralized supply chains based on additive technologies tend to guarantee shorter delivery times of products compared to centralized ones, as a consequence not only of the reduction of logistic tiers, but also of the proximity of AM machines to end users of products or spare parts. Furthermore, it should be considered that, in decentralized AM-based supply, the decrease in transport costs is determined not only by the disintermediation of the supply chains (the decrease in the number of tiers) and by the proximity of the AM machines to the end customers, but also by the object itself of the transport (raw material rather than components and end-use products.)

As a result, the risks of transport damage to components and products are reduced.

Coeteris paribus, by reducing transport and realizing manufacturing in situ, disintermediation of supply chains guarantees lower consumption of fossil fuels and lower carbon dioxide emissions.

Although additive machines consume electricity to a large extent, the total balance in terms of environmental sustainability of AM-based supply chains is proven as tendentially positive by numerous studies—also thanks to the lower weight of vehicles (aircrafts and cars) that have hollow and light components.

Decentralized AM-based supply chains serving the plurality of local markets have also been defined as mini-factory networks [14, 35]. Mini-factories are local production facilities close to the end customers. Additive manufacturing technologies, with which each mini-factory is equipped, reduce customization costs. In addition to production activities, each mini-factory carries out sales and customer assistance, often also digital product design activities. Networks of mini-factories are enterprises of pure and tailored customization [13]. Proximity of the mini-factory to the customer or a local market allows the focal company della supply chain to access customer knowledge which is highly strategic for the purpose of creating customized or highly customer-specific products, but also for innovative processes.

By favoring short delivery times and low costs of production of personalized or unique products, mini-factories embody lean agile factory principles.

As aforementioned, the decision to invest in distributed AM machines—which are expensive at the current stage of development that additive technologies have reached—must be economically justified by the existence of local demands, even if low and fluctuating, which must be satisfied in a personalized way and in a short time, as in the case of spare parts requested in aeronautics or biomedical sectors (heart valves, prostheses, dental implants) [9, 26] or be motivated by demands from places difficult to reach by conventional means of transportation [9, 24, 36]: so, additive manufacturing is certainly the enabling technology of space economy.

The decentralized implementation of digital additive manufacturing technologies in a supply chain implies the replacement (total or partial) of warehouses of components and sub-assembly systems, spare parts, and end-use products, not only with warehouses of raw materials, but also with "virtual warehouses" of digital design files. Raw materials and design files travel faster and more efficiently than physical materials, components and products, along supply chain tiers—the latter reduced in number—and no specific local knowledge is required for on-site production: designs submitted to decentralized production structures through IT infrastructures, incorporate necessary knowledge for product realization. Therefore, as a precondition, implementation of additive technologies in decentralized supply chains requires huge investments in ICT infrastructure, capable of supporting the production and circulation of knowledge incorporated in designs in supply chains themselves. In fact, the lack or weakness of ICT infrastructure of a supply chain is to be considered bottleneck, which hinders the implementation of additive technologies [23].

	Advantages	Disadvantages
Costs	• Lower transport costs • Lower stock-keeping costs • Lower consumption of fossil fuels and lower carbon dioxide emissions = lower environmental impact of manufacturing production	• Multiplication of fixed costs • High costs of additive machines • More investments in IT infrastructures
Customization	• High customization of single units or small batches without extra costs • Access to customer knowledge	
Times	• Shorter delivery times	
Competences	• No specific local competence required for on-site production/ability to produce in hard-to-reach places	
Risks	• Lower dependence on component suppliers • Lower risk of stock-out of components and spare parts • Lower risks of transportation damage to products	• Increased dependence on raw materials for AM suppliers

Table 3.
Decentralized additive configurations.

Additive manufacturing also requires customer digital design knowledge and skills: to be made on demand, that is, according to customer-specific requests, additive products must be designed on the basis of co-design platforms that carry out co-creation of value with customer.

Therefore, a first summary can be drawn from the analysis carried out so far: additive technologies centrally located in a supply chain generally reduce the costs of central component or spare parts warehouses whose response times to demand are not critical, replacing them with fixed investments in machines additive and with warehouses of raw materials and 3D files for additive manufacturing; decentralized additive technologies lower the number of supply chain levels, and therefore all those costs that depend on that number (e.g., transport costs and warehouse costs of the entire supply chain). It can be said that the lower the level of supply chain affected by additive technology implementation, the greater product customization guaranteed to the customer, but the lower cost savings ensured.

Table 3 shows the summary of the pros and cons of decentralized AM-based supply chains versus conventional ones, based on the analysis so far.

3.3 Hub additive configurations

In between the two aforementioned structural configurations are hub configurations.

In logistics, the concept of supply hub is well-established in conventional manufacturing [17]. From the point of view of conventional technology supply chains, the concept of hub refers to supply of components or sub-assemblies to local production plants. Hub functions as a buffer of components for the manufacturing plants in a just-in-time context [6, 17, 37]. In fact, the purpose of supply hub is to satisfy downstream demand in a timely and regular manner, and this occurs both thanks to its proximity to the main plant and by assembling parts purchased from upstream suppliers into sub-assemblies to send downstream [38]. Therefore, the use of a supply hub reduces supply risks and investments in equipment and labor of the entire supply chain compared with structural solutions that involve the direct supply of components from external suppliers. As a result, the complexity of operations at the supply chain level is reduced.

In a nutshell, in multilevel supply chains, a consolidation hub enables smooth and reliable supply of components or sub-assemblies to manufacturing plants and lowering structural costs. All this fosters transition of manufacturing in a just-in-time (JIT) context [37].

Supply hub can belong to one supply chain or serves multiple ones. In the latter case, higher volumes of off-the-shelf components purchased and high production volumes of sub-assemblies supplied to local locations, greatly reduce end-use product unit costs.

In an additive manufacturing context, hub configuration refers to an intermediate supply chain structure between fully centralized configuration and fully distributed one. Hub's AM machines are located in a logistic tier between the main plant and local manufacturing plants. They are oriented toward the realization of various strategic objectives, some of which are pursued by AM-based centralized supply chains, others by decentralized ones.

Therefore hub configuration has some of the main advantages of a centralized manufacturing supply chain configuration (e.g., it requires fewer machines and less manpower to meet total demand than a decentralized supply chain configuration, and thus guarantees better utilization of production capacity), but also some advantages of decentralized configuration compared to the centralized one (lower transport costs and faster, cheaper and smoothly supplies) [2].

4. The decision-making process for the implementation of additive technologies in conventional structures: method guidelines

Distributed or decentralized AM-based supply chains are likely to be more concise than traditional ones—and also than centralized AM-based ones: they cut off component suppliers and distributors, but become more dependent on raw material suppliers. It is known that raw materials for additive manufacturing are expensive, but generate lower inventory costs than the components and end-use products ones. Raw material costs are variable production costs; instead, component, spare part, and product costs include fixed costs. Because decentralized AM-based supply chains have fewer layers than conventional ones, and centralized AM-based ones, the transportation costs of the former are likely to be lower than the latter.

All things being equal, in AM-based supply chains variable production costs are higher than in conventional ones, due to material and labor costs, assessed by Li et al. [31] as higher than those incurred by a supply chain based on subtractive technology. In fact, as anticipated in the previous pages, additive technologies generate a change in the required work roles: less manual work, less knowledge related to machines, but new design and operational roles, generally more expensive than the manual ones of subtractive production.

As regards fixed costs of production, Li et al [31] point out that supply chains based on additive technologies may not be cost-effective compared to those based on conventional technologies: costs of AM machines are still high today and multiply in the case of distributed production.

In our opinion, it should also be considered that additive machines undergo rapid obsolescence due to the acceleration of technological innovation in this field of application, so the incidence of fixed costs on additive product unit cost tends to be more high a fortiori.

Based on what has been described in the previous paragraphs, it can be argued that, although additive manufacturing has many advantages over the conventional one, this is not enough to support the superiority of AM-based companies or supply chains over conventional ones in any business situation and environmental context. In fact, on the basis of the analysis so far, it can be noticed that there are negative quantitative interactions between pros and cons of additive manufacturing, that make AM-based company or supply chain not always feasible or appropriate.

All this for saying that the advantages and disadvantages of implementing additive technologies in conventional companies or supply chains have to be measured ex-ante and assessed in trade-off to decide about additive manufacturing choice.

As guidelines of company or supply chain management, in order to design a decision-making process for the implementation of additive technologies in organizational structures based on conventional technologies, a number of decisions are suggested: firstly, which method of simulation of company or supply chain based on additive manufacturing to adopt, then which specific additive technology to implement. With regards to supply chains, the focal company also has to decide at which level of the supply chain to implement additive manufacturing, that is, which AM-based supply chain configuration to adopt. In the previous pages, we have argued that the logistical level of the conventional supply chain affected by the implementation of additive technologies depends on the supply chain strategy decided by the focal company, in other words, the configuration of the AM-based supply chain that comes about depends on the chosen supply chain strategy.

The decision-making process for the implementation of additive technologies in conventional companies or supply chains always requires the choice of a simulation method on the basis of which to build AM-based company or supply chain models from which to derive cost functions to be compared with those of the status quo. The comparison between the aforementioned cost functions substantiates the ex-ante and trade-off assessment of the advantages and disadvantages of the additive option. Furthermore, the decision-making process described up to now must be repeated for all the different and possible additive technologies that can be adopted, because the effects that each of them can produce on conventional companies or supply chains can be so different as to imply different decisions.

In another study of ours [21], we specified that only a small number of scholars deal with the evaluation of the pros and cons of the implementation of AM from a quantitative point of view, thus filling a gap present in the studies on additive technologies. Among them are Li et al. [31], who suggest that the choice of one supply chain configuration can only be taken by resorting to mathematical simulation models of supply chains from which to derive functions of quantitative variables to be compared, mainly cost functions.

5. Conclusions

5.1 Recent advances

- Centralized implementation of additive technologies in a conventional supply chain lightens and simplifies its structure without changing the number of logistic tiers (component supplier-manufacturer-distributor-customer), because it reduces the intervention of suppliers of components and spare parts required fluctuating and in small quantities by the manufacturer. The result is reduced inventory costs, as centralized additive machines reduce the risk of stock-out of discontinuously requested, low-quantity spare parts and components. In a centralized AM-based supply chain, inventory costs are therefore likely to be lower than in a conventional one, although they are partly replaced by AM machine costs. Product customization is allowed not only by the assembly of modules but also by the potential of additive machines. The production context is that of modularity. Balance between the advantages and disadvantages of a centralized additive choice has to be measured to decide on its implementation.

- Distributed or decentralized AM-based supply chains are likely to be more concise than traditional ones and centralized AM-based ones: they cut off component suppliers and distributors, but become more dependent on raw material suppliers. It is known that raw materials for additive manufacturing are expensive, but generate lower inventory costs than the components and end-use products ones. Raw material costs are variable production costs; instead, component and product costs include fixed costs. The customization of the product is made possible by the large geometric freedoms with which the product can be made, allowed by the additive machine near the customer. So the production context is that of pure customization. Since decentralized AM-based supply chains have fewer levels than conventional ones, as well as centralized AM-based supply chains, any cost dependent on those levels (i.e., transportation cost) is likely to be lower in the former than in the latter. Balance between the advantages and disadvantages of a decentralized additive choice has to be measured to decide on its implementation as well.

- No AM-based supply chain configuration is better than the other because the advantages and disadvantages of each, inferred from simulation models, must be measured quantitatively and weighed in trade-off assessments. Thus, in certain organizational and environmental contexts, conventional supply chains can be better than additive configurations.

Referring to Verboeket and Krikke [23], we specify that the replacement of the central warehouse with centralized additive machines can represent a first step in the process of implementing additive technologies in supply chains and that, over time, a configuration centralized AM can evolve into a distributed one.

5.2 Future directions

The development of additive manufacturing technologies is moving toward the maturity stage, therefore, a rapid improvement in the technical characteristics of additive machines on which economies of scale depend, such as throughput time and standardization of production, is likely.

As a result, an increasingly vast adoption of additive machines in companies and supply chains is foreseeable, accompanied by a progressive decrease in prices.

Nowadays the consumer increasingly wants to participate in the business process of value creation, postulating personalization, short delivery times, and, at the same time, convenient prices. Additive technologies allow for the most suitable product solution for this market context.

The near future is a complete outsourcing of the production process to the consumer. So, distributed manufacturing can be considered an intermediate stage in the evolutionary process of manufacturing toward pure customized production made directly by the consumer, using design files downloaded from open design platforms, and by home printers or nearby 3D printing shops [39].

Author details

Silvana Gallinaro
Department of Economics and Statistics, University of Turin, Turin, Italy

*Address all correspondence to: silvana.gallinaro@unito.it

References

[1] Gibson I, Rosen DW, Stucker B, editors. Additive Manufacturing Technologies. Business opportunities and future directions. New York: Springer; 2010;**20**:437-446

[2] Alammar A, Kois JC, Revilla-León M, Att W. Additive manufacturing technologies: Current status and future perspectives, March 2022. Journal of Prosthodontics. 2022;**31**:4-12

[3] Lipson H, Kurman M. Fabricated: The New World of 3D Printing. Haboken: Wiley; 2013:1-320

[4] Robinson D, Lagnau A, Boon W. Innovation pathways in additive manufacturing: Methods for tracing emerging and branching paths from rapid prototyping to alternative applications. Technological Forecasting and Social Change. 2019;**146**:733-750

[5] Javaid M, Haleem A. Additive manufacturing applications in medical cases: A literature-based review. Alexandria Journal of Medicine. 2018;**54**(4):411-422

[6] Khajavi SH, Partanen J, Holmstrom J. Additive manufacturing in the spare parts supply chain. Computers in Industry. 2014;**65**(1):50-63

[7] Holmström J, Partanen J. Digital manufacturing-driven transformations of service supply chains for complex products. Supply Chain Management. 2014;**19**(4):421-430

[8] Cohen D, Sargeant M, Somers K. 3-D printing takes shape. The McKinsey Quarterly. 2014;(January)

[9] Holmström J, Holweg M, Khajavi SH, Partanen J. The direct digital manufacturing (r)evolution: Definition of a research agenda. Operations Management. 2016;**9**(1-2):1-10

[10] Sirichakwal I, Conner B. Implications of additive manufacturing for spare parts inventory. 3D Printing and Additive Manufacturing. 2016;**3**(1):56-63

[11] Ford S, Despeisse M. Additive manufacturing and sustainability: An exploratory study of the advantages and challenges. Journal of Cleaner Production. 2016;**137**:1573-1587

[12] Liu P, Huang SH, Mokasdarb A, Zhoub H, Houc L. The impact of additive manufacturing in the aircraft spare parts supply chain: Supply chain operation reference (Scor) model based analysis. Production Planning and Control. 2014;**25**(13-14):1169-1181

[13] Mintzberg H. Generic strategies: Toward a comprehensive framework. Advances in Strategic Management. 1988;**5**:1-67

[14] Gallinaro S. Dai modelli lineari di business alla piattaforma di progettazione e manifattura. Gli effetti delle tecnologie additive sulla logica di creazione del valore delle imprese manifatturiere. ImpresaProgetto – Electronic Journal of Management. 2019;**2**:1-41

[15] Van Hoek RI. The rediscovery of postponement: A literature review and directions for research. Journal of Operations Management. 2001;**19**(2):161-184

[16] Zinn W, Bowersox DJ. Planning physical distribution with the principle of postponement. Journal of Business Logistics. 1988;**9**(2):117-136

[17] Lee HL. Postponement for Mass Customization: Satisfying Customer

Demands for Tailor-Made Products. In: Gattorna J, editor. Strategic Supply Chain Alignment. Aldershot: Gower; 1998:77-91

[18] Waller MA, Dabholker PA, Gentry JJ. Postponement, production customization, and market-oriented supply chain management. Journal of Business Logistics. 2000;**21**(2):133-160

[19] Yang B, Burns ND, Backhouse CJ. Postponement: A review and an integrated framework. International Journal of Operations & Production Management. 2004;**24**(5):468-487

[20] Gallinaro S. Produzione. Torino: Giappichelli; 2015

[21] Gallinaro S. Catene di fornitura basate sulla produzione additiva. ImpresaProgetto – Electronic Journal of Management. 2021;**1**:1-28

[22] Olhager J. Strategic positioning of the order penetration point. International Journal of Production Economics. 2003;**85**(3):319-329

[23] Verboeket V, Krikke H. The disruptive impact of additive manufacturing on supply chains: A literature study, conceptual framework and research agenda. Computers in Industry. 2019;**111**:91-107

[24] Holmström J, Liotta G, Chaudhuri A. Sustainability outcomes through direct digital manufacturing-based operational practices: A design theory approach. Journal of Cleaner Production. 2017;**167**:951-961

[25] Weller C, Kleer R, Piller FT. Economic implications of 3D printing: Market structure models in light of additive manufacturing revisited. International Journal of Production Economics. 2015;**164**:43-56

[26] Berman B. 3-D printing: The New Industrial Revolution. Business Horizons. 2012;**55**(2):155-162

[27] Fratocchi L, Ancarani A, Barbieri P, Di Mauro C, Nassimbeni G, Sartor M, et al. Motivations of manufacturing reshoring: An interpretative framework. International Journal of Physical Distribution and Logistics Management. 2016;**46**(2):98-127

[28] Fratocchi L. Is 3D printing an enabling technology for manufacturing reshoring? In: Vecchi A, editor. Reshoring of Manufacturing: Drivers, Opportunities, and Challenges. Berlino: Springer; 2017:99-124

[29] Franke N, Piller FT. Value creation by toolkits for user innovation and design. The case of the watch market. The Journal of Product Innovation Management. 2004;**21**(6):401-415

[30] Chekurov S, Metsä-Kortelainenb S, Salmia M, Rodac I, Jussila A. The perceived value of additively manufactured digital spare parts in industry: An empirical investigation. International Journal of Production Economics. 2018;**205**:87-97

[31] Li Y, Jia G, Cheng Y, Hud Y. Additive manufacturing technology in spare parts supply chain: A comparative study. International Journal of Production Research. 2017;**55**(5):1498-1515

[32] Gebhardt A. Rapid Prototyping. Cincinnati: Hanser Gardner Publications Inc.; 2003

[33] Emelogu A, Marufuzzaman M, Thompson SM, Shamsaei N, Bian L. Additive manufacturing of biomedical implants: A feasibility assessment via supply-chain cost analysis. Additive Manufacturing. 2016;**11**:97-113

[34] Holmström J, Partanen J, Tuomi J, Walter M. Rapid manufacturing in the spare parts supply chain: Alternative approaches to capacity deployment. Journal of Manufacturing Technology Management. 2010;**21**(6):687-697

[35] Reichwald R, Piller FT, Jager S, Zanner S. Economic evaluation of miniplants for mass customization. In: Piller FT, editor. The Customer Centric Enterprise. Advances in Mass Customization and Personalization. Berlin: Springer; 2003:51-69

[36] Ryan MJ, Eyers DR, Potter AT, Purvis L, Gosling J. 3D printing the future: Scenarios for supply chains reviewed. International Journal of Physical Distribution and Logistics Management. 2017;**47**(10):992-1014

[37] Naylor JB, Naim MM, Berry D. Leagility: Integrating the lean and agile manufacturing paradigm in the total supply chain. International Journal of Production Economics. 1999;**62**(1-2):107-118

[38] Creazza A, Dallari F, Melacini M. Evaluating logistics network configurations for a global supply chain. Supply Chain Management: An International Journal. 2010:**15**(2):154-164

[39] Rayna T, Striukova L. From rapid prototyping to home fabrication: How 3D printing is changing business model innovation. Technological Forecasting and Social Change. 2016;**102**:214-224